Migration Strategies of Birds of Prey in Western Palearctic

Editors

Michele Panuccio
Mediterranean Raptor Migration Network, Italy
Department of Earth and Environmental Sciences,
University of Pavia, Pavia, Italy

Ugo Mellone
Mediterranean Raptor Migration Network, Italy
Vertebrates Zoology Research Group, University of Alicante, Alicante, Spain

Nicolantonio Agostini
Mediterranean Raptor Migration Network, Italy

CRC Press
Taylor & Francis Group
Boca Raton London New York

CRC Press is an imprint of the
Taylor & Francis Group, an **informa** business
A SCIENCE PUBLISHERS BOOK

Cover photo: European honey buzzard (*Pernis apivorus*). © Michele Panuccio

Book: Panuccio M., Mellone U. and Agostini A., 2021. Migration Strategies of Birds of Prey in Western Palearctic. CRC Press, Boca Raton, Florida.

Chapter: Onrubia A. and B. Martin, 2021. Black kite *Milvus migrans*. Pp. 165-179 in: Panuccio M., Mellone U. and Agostini A. [eds.]. Migration Strategies of Birds of Prey in Western Palearctic. CRC Press, Boca Raton, Florida.

First edition published 2021
by CRC Press
6000 Broken Sound Parkway NW, Suite 300, Boca Raton, FL 33487-2742

and by CRC Press
2 Park Square, Milton Park, Abingdon, Oxon, OX14 4RN

ISBN: 978-1-138-49572-2 (hbk)
ISBN: 978-0-367-76543-9 (pbk)
ISBN: 978-1-351-02362-7 (ebk)

Typeset in Times New Roman
by Radiant Productions

Foreword

Sadly, I never met Michele Panuccio. But I was well aware of the significant contribution he was making to our knowledge of raptor migration.

Michele died in June 2019. He was just 42. Already he had spear-headed many important migration studies, encouraged a generation of young migration researchers in his native Italy and beyond, and seen his dream realised—establishing MEDRAPTORS, the Mediterranean Raptor Migration Network, which is now such an important organisation.

At the time of his death he was pursuing another ground-breaking project that was eventually to become this book.

In inviting me to write this Foreword, Ugo Mellone, wrote:

> "*Losing Michele destroyed his friends and colleagues, but we are determined to do our best to keep alive the scientific and human legacy of Michele. His family asked me and my colleague Nicolantonio Agostini to complete his book.*"

This they have done and I'm sure Michele would be proud of the result—a compilation that brings together our knowledge of raptor migration in Europe in chapters written by some of Europe's acknowledged experts. I need say no more, for you, the reader, will quickly become aware of the book's importance.

As I pored over the draft chapters it brought home to me just how dramatically the study of raptor migration has advanced in the last 50 years. Back in the mid-1960s, when I started marvelling at the migration over Turkey's Bosphorus, raptor watching and counting was a rare event, unheard of in most countries. Moreover, we were only just starting to discover how to identify these tantalising 'dots' in the sky. In those days, and the following decade, ensuring that birds of prey were being identified correctly was a primary objective. Birds that we now know as Lesser Spotted and Steppe Eagles were being documented as Greater Spotted, whilst Steppe Buzzards were almost unheard of on migration, being logged as Long-legged Buzzards and the separation of 'ring-tail' harriers in flight was rarely attempted.

Now things have moved on apace with advances in identification, aging and sexing, widespread systematic counting at bottleneck sites, radar tracking and, of course, modern technology. Our knowledge would be so much the poorer without the ability to follow the migration and fortunes of these magnificent birds remotely by satellite telemetry.

And so I dedicate this Foreword to Michele. What a fitting tribute it is that his colleagues have honoured him by completing the work he started. True friendship and a lasting legacy.

Richard Porter
Norfolk, England, September 2020

Preface

This book is the result of a search for answers. Throughout several years observing migratory raptors crossing mountains, seas and deserts we generated countless questions. Some of them have been answered, while several are still to be solved. In any case, the knowledge about these fascinating journeys improved dramatically and the chapters written for this book by some of the most prominent researchers in the field are a proof of it.

Studies on raptor migration in the Western Palearctic experienced a significant increase in the last decades. In particular, besides the traditional method of visual observations carried out at watchsites, the migration strategies of several species have been investigated in detail thanks to the use of modern technologies, such as tracking by radar and satellite telemetry. The results revealed species-specific behaviours, especially among long distance migrants, shaped mostly by the interaction of morphological characteristics with ecological barriers (such as the Mediterranean basin and the Sahara desert). In this scenario a wide array of behavioural responses evolved, concerning travel distances, geometry of the routes, social learning, effect of weather conditions and optimization according to time or energy minimization.

After a general introduction where we made a synthesis about different aspects of raptor migration, this book summarizes the available information published in the literature until 2020, through a review for each species of migratory raptor of Western Palearctic, and also for some species that breed in Eastern Asia. Species-specific chapters have been written by several researchers, which reviewed the knowledge made available by papers published on peer-reviewed journals, as well as on the so-called "grey literature". The aim has been to obtain the most comprehensive and up-to-date synthesis on this topic, as well as a source of references for those who want to go even more in depth.

The target audience of this book is very wide, spanning from movement ecologists, to professional ornithologists and amateur birdwatchers, who want to approach the study of migration strategies of these charismatic birds. We hope this book will become a tool to transmit passion and knowledge to new generations of ornithologists, and to support the conservation of Palearctic raptors and of the regions they use throughout their lives.

Unfortunately, our friend Michele Panuccio cannot see this book, but we believe he would have been more than satisfied, and already looking forward going back in the field to fill new knowledge gaps, or simply to enjoy raptors approaching the coast after the crossing of the Mediterranean Sea.

Ugo Mellone
Nicolantonio Agostini

Contents

Editors' Biographies

Michele Panuccio graduated from the University of Camerino, Italy in Natural Sciences. He obtained his Ph.D. in Experimental Ecology at the University of Pavia. He was a co-founder member of the Mediterranean Raptor Migration Network (MEDRAPTORS), a non-profit association working to improve research and protection of birds of prey in Western Palearctic. He carried out research on migratory strategies of diurnal birds of prey from 1998 using different methodologies: direct visual observations, ringing recoveries, radar and GPS tracking. He was also involved in research on breeding and wintering ecology. In later years he focused on research both as post-doctoral researcher at the University of Pavia and as research coordinator of MEDRAPTORS. While working on this book and conducting a research on waterbird movements in ISPRA (Italian Institute for Environmental Protection and Research), he passed away in 2019, after having battled with ill health for a long time.

Ugo Mellone graduated from the University of Pavia, Italy in Natural Sciences and is co-founding member of MEDRAPTORS. He carried out research on raptor migration from 2003. In 2013, he got his Ph.D. in Biology from the University of Alicante, Spain, with a thesis focused on movement ecology on migrating raptors studied by means of satellite telemetry. He developed research stays also in the Centre for Animal Movement Research (Lund, Sweden), Ecotono (Bariloche, Argentina), and Estación Biológica de Doñana (Seville, Spain). He currently combines the development of MEDRAPTORS project with his job as a freelance photo journalist. Aimed at raising awareness on conservation issues and endangered species, he produces stories that are disseminated through magazines, books and exhibitions.

Nicolantonio Agostini graduated from the University of Parma, Italy in Natural Sciences. He obtained his Ph.D. in Experimental Ecology at the University of Pavia. Between 1986 and 1989 he was member of the Scientific Coordination of the Raptor Rescue Centre in Parma, managed by LIPU-Birdlife Italy. In this period he studied predatory and social behaviour of both captive wild diurnal and nocturnal birds of prey. Co-founder member of MEDRAPTORS, since 1989 he has been studying migration of diurnal birds of prey by direct visual observation at several watchsites in Italy, Tunisia, Malta, Greece, Turkey and Georgia. In recent years he has been involved in studies carried out by radar and GPS tracking such as in ecology research of birds of prey at wintering sites. Between 1984 and 1996 he was active in protectionist actions against poaching of diurnal birds of prey at the Strait of Messina together with LIPU-Birdlife Italy.

Editors' note

MEDRAPTORS (Mediterranean Raptor Migration Network) is a network of ornithologists involved in the research and protection of migrating birds of prey, especially in the Mediterranean basin. It has been founded by the Editors in 2004. More information is available at www.raptormigration.org.

The maps of this book have been prepared by Martina Scacco (Max Planck Institute of Animal Behavior) and Ugo Mellone with input from the authors of the chapters, using BirdLife data as a basis to illustrate species ranges (which have been modified). [BirdLife International and Handbook of the Birds of the World (2018) Bird species distribution maps of the world. Version 2018.1. Available at http://datazone.birdlife.org/species/requestdis.]

1

Raptors as Models to Study Animal Migrations

Michele Panuccio,[1,2,†] *Nicolantonio Agostini*[1,*] and *Ugo Mellone*[1,3]

A Brief History of Raptor Migration Studies in the Old World

First for their spectacular concentrations en route and, later, for their size (enabling to track them through remote tracking devices) diurnal raptors have been always among the main models to study animal migrations. Here, we briefly trace the research made since the first half of the 20th century on migratory raptors moving in the Afro-Palearctic system, through the different techniques that have been used: visual counts, trapping and ringing, radar tracking and satellite telemetry. The count of visible migrants at traditional bottlenecks, but also along minor migration corridors, has been the earlier approach to this research field, being the easiest possible method, just with the aid of binoculars and telescopes (Rudebeck 1950). Since the 1930s large numbers of birdwatchers have been able to identify migrating raptors correctly thanks to the first popular field guide published by Roger T. Peterson (1934), increasing the reliability of the surveys, often involving trained volunteers (Bildstein 1998). Moreover, since the 1970s, the knowledge improvement on flight identification of Western Palearctic raptors (Porter et al. 1974), allowed to study this phenomenon by direct visual observation, also focusing on differential migration in relation to age and sex classes (Kjellén 1992). As a result, besides the traditional bottlenecks where tens of thousands of raptors concentrate annually (Porter and Willis 1968, Bernis et al. 1980), the discovering of tens of minor "hawkwatches", mostly from the 1960s onwards, has better depicted the paths of long-distance migrants (Zalles and Bildstein 2000). Within the Afro-Palearctic system, surveys revealed the passage of more than 100,000 raptors at seven bottlenecks, one located between Spain and Morocco (Strait of Gibraltar), two in Turkey (Bosphorus and Borçka), one in Georgia (Batumi), two in Israel (Eilat and Northern Valleys) and one between Yemen and Djibouti (Bab-el-Mandeb Strait; Bildstein 2018; Fig. 1). In the Mediterranean region, surveys made at the Strait of Messina and over the island of Malta began in the 1980s, starting from conservation efforts against poachers, and the same occurred at Batumi (Georgia) since the beginning of this century (Bildstein 2006). In addition, simultaneous counts made at several watchsites since 1990s, mostly focused on European honey buzzards (*Pernis apivorus*), short-toed snake eagles (*Circaetus gallicus*) and western marsh harriers (*Circus aeruginosus*), well depicted their flyways highlighting the importance of social learning, and the existence of age-related migration patterns

[1] MEDRAPTORS (Mediterranean Raptor Migration Network), Rome, Italy.
[2] Department of Earth and Environmental Sciences, University of Pavia, Pavia, Italy.
[3] Vertebrates Zoology Research Group, University of Alicante, Alicante, Spain.
[†] Deceased
[*] Corresponding author: nicolantonioagostini@gmail.com

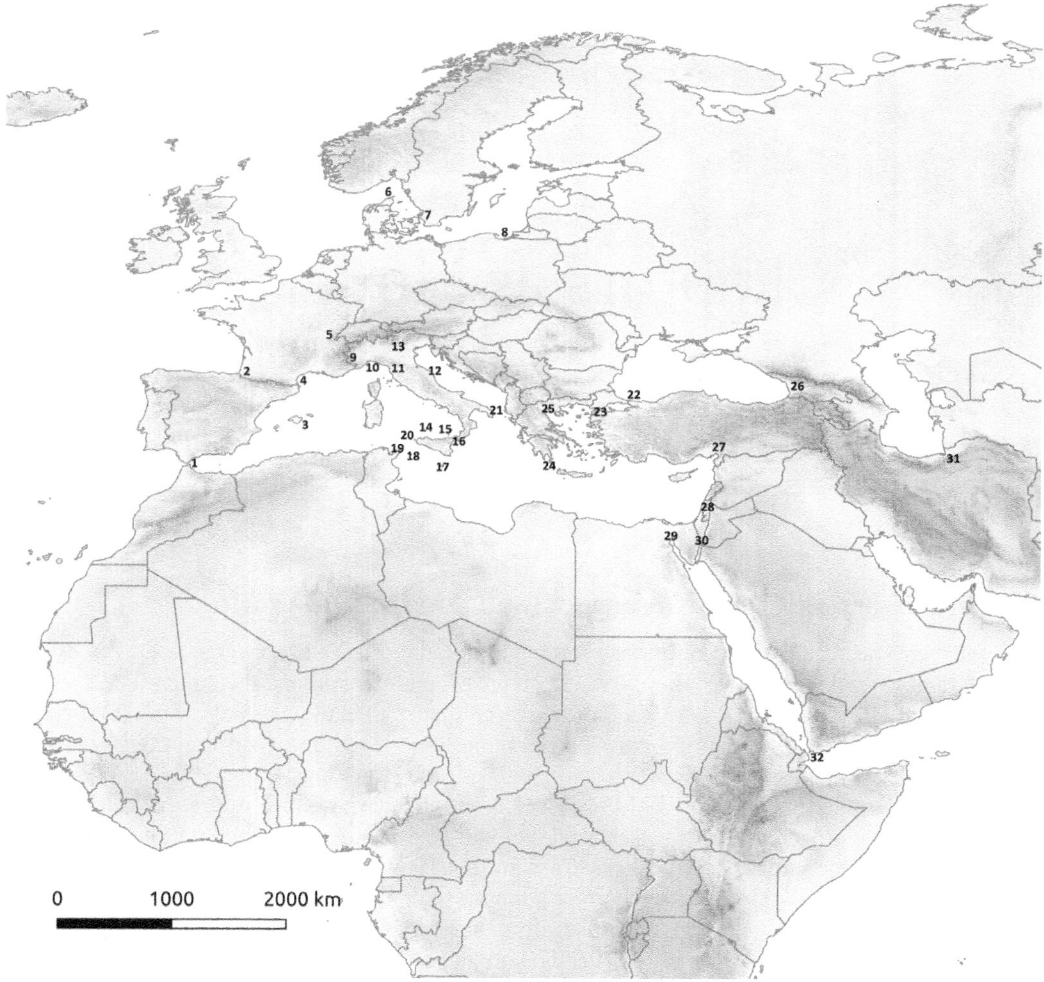

Figure 1. The main raptor migration watchsites cited throughout the book. (1) Strait of Gibraltar, (2) Organbidexka, (3) Mallorca, (4) Leucate, (5) Fort l'Ecluse, (6) Skagen, (7) Falsterbo, (8) Vistula Spit, (9) Stura Valley, (10) Arenzano, (11) Capriglia/Alpi Apuane, (12) Conero, (13) Ponti sul Mincio, (14) Ustica, (15) Panarea/Aeolian islands, (16) Strait of Messina, (17) Malta, (18) Pantelleria, (19) Cap Bon, (20) Marettimo, (21) Capo d'Otranto, (22) Bosphorus, (23) Dardanelles, (24) Antikythira, (25) Mount Olympus, (26) Batumi, (27) Iskenderun/Sarimazi, (28) Northern Valleys, (29) Suez, (30) Eilat, (31) Southeast Caspian, (32) Bab-el-Mandeb.

(Agostini 2004, Chapters 4, 7, 15). Along the Western flyway, long term autumn counts made at Falsterbo (1942–1997), in Southern Sweden, and at the Strait of Gibraltar (1999–2014) allowed to study population trends and interspecific differences in response to climate change (Kjellén and Roos 2000, Panuccio et al. 2016). Finally, thanks to migration counts made at several watchsites in Spain (Strait of Gibraltar), Italy (Arenzano, Strait of Messina and several circum-Sicilian islands), Greece (Mount Olympus, island of Antikythira), Turkey (Bosphorus, Dardanelles) and Georgia (Batumi), it was possible to investigate the influence of weather on the migration magnitude such as on the visible migration at regional and local scale (Fig. 1; Miller et al. 2015, Agostini et al. 2016b, Panuccio et al. 2017, Vansteelant et al. 2014). Trapping and ringing has been used in bird migration studies since the end of the nineteenth century. Even if systematic ringing effort has been mainly directed especially to passerines (but see Saurola et al. 2013), ringing recoveries helped to get a first idea of migratory connections also in several raptor species. The first Atlas summarizing the fieldwork carried out in the Western Palearctic was published by Schütz and Weigold in 1931. Since the beginning of this century, several national ringing atlases have been published (Fransson and Pettersson 2001, Wernham et al. 2002, Bakken et al. 2003, Bønløkke et al. 2006, Spina and Volponi 2008, Cepák et al. 2008, Csörgö et al. 2009, Saurola et al. 2013) providing a general picture of the spatio-temporal migration patterns of European raptors. Interestingly, one of the pioneering experimental studies on the orientation and navigation abilities of birds was carried out using a raptor as model species in 1930s by R. Drost (1938). The German ornithologist trapped and ringed 209 Eurasian sparrowhawks (*Accipiter nisus*) on an island (Helgoland) off the coast of Germany during the autumn migration. The birds were then transported about 600 km east to Western Poland and finally released. A total of 36 individuals were later recovered. Juveniles tended to move along an inherited NE-SW axis while adults flew more towards W as if they were trying to locate the axis along which they had migrated in previous years, showing true navigational abilities. More recently, ringing recoveries analysis has been used to study age-dependent migration behaviour also in the European honey buzzards (Schmid 2000) and the differential wintering area selection in the western marsh harrier (Panuccio et al. 2013d). Finally, a study made on four species belonging to the populations breeding in Sweden compared ringing recoveries with satellite tracking, showing a very similar spatio-temporal distribution, confirming the reliability of this method in studying raptor migration despite the obvious bias caused by the different distribution of human population and therefore recapture probability (Strandberg et al. 2009d).

Thanks to radar devices, aided by visual observation to identify the species, it is possible to determine accurately the flight behaviour of migrants such as their altitude, the ground speed, the air speed and the climbing rate in thermal circling (Spaar 1997). The first study on raptor migration by radar was made in 1970s by E. Houghton at the Strait of Gibraltar (Kerlinger 1989). Subsequently, radar studies were made in Switzerland (Schmid et al. 1986), Sweden (Kjellén et al. 2001) and Israel (Leshem 1987), the last one using an airport surveillance radar. Between 1990s and the beginning of this century, researchers of the Swiss Ornithological Institute carried out several studies in Israel (Negev Highlands and Arava Valley), Spain (southern coast near Malaga) and at the Strait of Messina by using a radar derived from military equipment allowing to follow a single bird up to about 8 km away (Spaar 1997, Meyer et al. 2000, Mateos-Rodríguez and Liechti 2012). Some of these studies focused on the water-crossing behaviour and on the flight performances in relation to wind and thermal conditions. More recently, studies were made using a marine surveillance radar (Capotosti et al. 2019) inland of the Strait of Messina, mostly focusing on flight altitudes and migration paths, in relation to topography, flock size and weather conditions (Panuccio et al. 2016c, 2018a, 2019b, Agostini et al. 2017b, Becciu et al. 2018).

Since the beginning of the first half of the 1990s, the use of satellite telemetry revolutionized the study of raptor migration, providing detailed information that before was unthinkable. These first studies were carried out on large and medium-sized species using locations obtained from solar-powered Platform Transmitter Terminals (PTTs) and provided by the Argos Doppler-based system. In those pioneering studies, very few individuals were tracked and the spatial and temporal resolution was quite low, even if they allowed depicting for the first time the migratory routes of several species and getting some basic travel parameters (Meyburg and Meyburg 1999, Kjellén et al. 2001). In the last decade there has been a proliferation of studies involving several species, sometimes analyzing tens of tracks belonging to different individuals and even repeated journeys from the same ones, improving the spatial and temporal resolution thanks to the GPS technology (Mellone et al. 2012a). Nowadays, the availability of different kind of instruments is much higher: every year different companies produce devices that are lighter (therefore allowing the tracking of smaller species) and more precise, including not only a geolocation system, but also other sensors, e.g., accelerometers. The third dimension is now included in several research thanks to the precise altimeters. Data can be stored and sent to researchers worldwide also through GSM network, or at short distances at a base station, using different systems (Klaassen et al. 2017, Sará et al. 2019). The costs are lower than before and therefore hundreds of individuals from almost all migratory raptor species of the old world have been tracked. A great part of these data is uploaded on facilities like MoveBank (Lopez-Lopez 2016), enabling easier collaboration among researchers. Moreover, the availability of large datasets of meteorological and other environmental data, together with the improvement of statistical techniques, is making it possible to disentangle with great precision which factors affect the migratory behaviour of raptors at a very fine scale. Despite this impressive development of remote tracking techniques, we believe that field observations at watchsites remain an invaluable tool to monitor long term population trends (Vansteelant et al. 2020b), response to climate change (Panuccio et al. 2016b) and for general education/conservation purposes (Bildstein 1998), and should be continued at least in the key sites of each flyway, following standardized protocols (Panuccio et al. 2018a).

Migration Patterns in Space and Time

The way in which bird populations migrate and spend different life stages in different geographical areas have been classified according to different categories. First of all, some populations are just partially migratory, that means that only a portion of the individuals, generally those breeding at the northernmost latitudes, migrate, mainly on short and medium distances. Some examples among those shown in this book are the Eurasian sparrowhawk (Chapter 14), the common buzzard (*Buteo b. buteo*, Chapter 23) and the common kestrel (*Falco tinnunculus*, Chapter 25). Instead, the western marsh harrier (Chapter 15) is a partial migrant that shows an important proportion of its European population migrating long distances. In this way, these birds spend the winter beyond resident populations, or overlapping with the most southern ones, shaping therefore a leap-frog migration pattern. In Europe, species which are considered as fully migratory are generally also long-distance migrants—that means they spend the winter south of the Sahara. However, these definitions are not unambiguous: a merlin (*Falco columbarius*) migrating between Northern Europe and Spain covers longer distances than a short-toed snake eagle breeding in Spain and wintering in the Sahel! Long-distance migrating individuals of the osprey (*Pandion haliaetus*, Chapter 2), which are the majority in the Europe, show a textbook example of leap-frog migration, "jumping" over the resident, or partially migratory, Mediterranean population. Some raptor species included in this book, have more puzzling patterns, in which it is sometimes difficult to distinguish migration from dispersal. For example, the Eurasian griffon vulture (*Gyps fulvus*, Chapter 6), the eastern imperial eagle (*Aquila heliaca*, Chapter 12) and the saker falcon (*Falco cherrug*, Chapter 32) show long-distance movements mostly in juveniles, and only in a portion of their populations, with the same individuals travelling thousands of km in their first years of life becoming resident ones as adult. Given that these movements show a strong directionality, repeatability, and connect the breeding range with a non-breeding one, they can be considered as another variant of partial migration. Finally, all Western Palearctic migratory raptors have their wintering grounds located closer to the equator than the breeding ones, apart the recently discovered case of long-legged buzzards (*Buteo rufinus*, Chapter 22) from Israel, migrating northward after the end of their breeding season.

Morphology, Flight Styles and Detours

Among migrating raptors breeding in the Palearctic, several species travel thousands of kilometers twice a year between breeding and wintering areas with considerable energetic costs and high mortality risks. Under the pressure of natural selection, birds have developed a range of morphological and behavioural adaptations to cope with their journeys in order to overcome barriers and adverse weather conditions. However, since migration time covers only a fraction of the life of a bird, morphological characteristics have been shaped by different ecological pressures also affecting non-migratory foraging flight. As a result, smaller and lighter species with higher aspect ratio wings (relatively long and pointed; e.g., falcons) largely use the flapping flight, migrating on a broad front and undertaking long water-crossings, sometimes also during the night (Bildstein 2006); in this scenario, indeed the longer distances covered over water have been recorded in species such as the Eleonora's (*Falco eleonorae*) and the Amur falcon (*Falco amurensis*), capable of crossings lasting 1,500–2,500 km (see Chapters 27, 28). As a matter of fact, by reducing the induced drag, slender and pointed wings contain the energetic cost of flapping flight which, conversely, increases rapidly with the increase of the body mass and in shorter-winged (low aspect ratio) species. For this reason, heavier species with relatively broad and rounded wings (e.g., buzzards, vultures and eagles) are obliged to use the low-cost soaring-gliding flight (Kerlinger 1989, Bildstein 2006, Pennycuick 2008, Agostini et al. 2015a, Nourani and Yamaguchi 2017). In so doing, they use circle and straight-line (slope) soaring by exploiting thermal currents and deflection updrafts, which mostly form over land and especially along mountain chains, a sort of highway for soarers when aligned along the principal axis of migration. In addition, migrants are able to exploit lee waves, particular deflection updrafts which form on the downwind side of a mountain chain, which extend for several kilometers and at high altitudes also over a water surface (Evans and Lathbury 1973). The soaring-gliding flight is optimal for heavier species also because the climbing rate in thermal circling is not affected by body mass, while they glide faster than smaller ones, reaching a cross-country speed comparable with the highest speed of flapping migrants (Hedenström 1993, Spaar 1997). However, because of their size, larger species show a greater circling radius and rarely soar in weaker thermals having smaller diameters in the early morning and late afternoon. As mentioned above, vertical currents form mostly over land and, even if recent research reveal striking exceptions (Duriez et al. 2018), thermal updrafts are very weak or absent over water, especially in temperate zones (Bildstein 2006). For this reason, water surfaces act as ecological barriers for typical soarers which perform long and sometimes circuitous detours to avoid the crossing. Moreover, water surfaces do not offer birds the possibility to land or feed. As a result, in the Afro-Palearctic migration system, the largest concentrations of soaring migrants occur at geographical bottlenecks mostly located along the western and eastern shorelines of the Mediterranean Sea, where the crossings are shorter than in the middle of the basin, on the Sicily Channel. However, some medium-sized raptors showing intermediate morphology between obliged soarers and typical flappers, such as honey buzzards, black kites (*Milvus migrans*), harriers (*Circus* spp.) and ospreys, are capable of long non-stop flapping flights over water showing more or less broad fronts of migration and a more flexible flight strategy (Mellone 2020). In particular, western marsh harriers often migrate almost matching the beeline between breeding and wintering grounds (chapter 15). Conversely, the short-toed snake eagle, an obliged soarer, performs one of the most detoured migrations to avoid a long water-crossing through the Mediterranean, with birds flying up to the 150% of the beeline between their breeding and wintering grounds during both spring and autumn (Chapter 7).

Time and Energy Minimization

Time, energy and safety are the main currencies according to which a migratory journey is arranged (Alerstam and Lindström 1990): if a bird wants to save time, it must spend more energy but, on the other hand, saving energy and minimizing risks entails a higher consumption of time. Within the context of this book, a classic example of risk minimization is the avoidance of sea-crossings that lead some raptors to perform long, and time consuming, detours (e.g., the short-toed snake eagle, Chapter 7). Generally, during the pre-breeding migration, birds try to minimize time, in order to reach the nesting grounds as soon as possible and occupy the nest territories: this statement is based on the fact that, generally, spring migration speeds are higher than autumn ones (Nilsson et al. 2013). Among long-distance migrating European raptors, too, this is a widely represented pattern, independently from the fact that wind assistance may be more favourable during autumn (Mellone et al. 2012a) or during spring (Vansteelant et al. 2015). Exceptions to this rule are the booted eagle (*Hieraaetus pennatus*, Chapter 10) and the Montagu's harrier (*Circus pygargus*, Chapter 18), in which the seasonal difference seems to be related to longer stopover during spring, at least in the latter species (Klaassen et al. 2017). However, speed may not be the only useful parameter to judge which criteria a bird is following: Sergio et al. (2014), showed that early departure is the key to occupy the best breeding territories and that late departing individuals could not reach earlier ones even when showing significantly higher travel speeds. Finally, the seasonal differences in the water-crossing tendency (which is ultimately related to risk avoidance) shape detours and loop migration patterns, sometimes with conflicting evidence: for example, in agreement with the widely accepted theory of time minimization being more common during the pre-breeding journey, European honey buzzards breeding in Hungary show more direct routes, involving longer sea crossings, during spring. However, the opposite pattern occurs in Finnish individuals, mainly crossing the Eastern Mediterranean during autumn, and detouring around it during spring (Chapter 4). This suggests that the relationships between costs and benefits of early arrival vary according to the population, and that, together with spatial and seasonal variation of wind conditions, they may promote very different strategies even within the same species.

Flocking Behaviour, Orientation, Navigation and Social Learning

Flocking (join together in groups) can be observed in several species of migrating birds, as a behaviour underlying a social function (Kerlinger 1989, Bildstein 2006, Newton 2008). Among migrating raptors, the flocking tendency varies in accordance with their behavioural ecology and their dependence on soaring-gliding flight. In species often breeding in colonies, such as the black kite, birds tend to together leave the breeding grounds migrating in mixed-age flocks (Panuccio et al. 2014, 2018b). The timing of migration of flocking species is more restricted than that of non-flocking migrants. As regards the relation with the flight style, it is well-known that soaring raptors are more efficient in thermal localization when they travel in flocks rather than alone. Birds flying in flocks are more likely to encounter random thermal currents than an individual flying alone, since larger flocks sample wider areas (Kerlinger 1989). In addition, large flocks engaged in thermals can be located from the distance, and anyone who has spent some time watching migrating raptors will have had the opportunity to observe individuals diverting their course in order to join a flock circling in a thermal. Consequently, as mentioned above, flocking is more evident in species strictly using soaring-gliding flight, which sometimes form huge flocks during migration, even if they do not breed in colonies. Super-flocks containing hundreds of lesser spotted eagles and black kites (sometimes up to one thousand), and flocks of thousands of both steppe and honey buzzards, sometimes passing in uninterrupted streams, are regularly observed at traditional bottlenecks (Bildstein 2006). However, super-flocking can have some disadvantage, especially in enormous mixed-species groups. As a matter of fact, collisions among individuals engaged in a thermal or in inter-thermal flights sometimes occur due to their differences in flight mechanics, some species being less agile than others during both soaring and gliding flight (Bildstein 2006). The species which have a more flexible flight behaviour during migration (e.g., harriers), tend to migrate solitary or in small loose flocks, rarely exceeding a few dozens of individuals (Kerlinger 1989). Interestingly, flocking also affects the water-crossing behaviour of two species which, despite using soaring gliding flight over land, are capable of long water-crossings, the European honey buzzard and the black kite. When facing the open sea, these raptors often hesitate to undertake the crossing, especially in suboptimal weather conditions (e.g., moderate-strong lateral winds), stopping migrating even for several hours or days and building large flocks (Agostini et al. 1994, Agostini and Duchi 1994). When it occurs, the first individuals which finally undertake the crossing are followed by the others, and therefore the flock size might affect their decision (crossing or not) because the larger the flock, the greater the chance that some individual will begin the crossing dragging the others. Moreover, during the crossing, flocking can improve their flight efficiency because a larger flock size will lead to a lower flapping rate (Bildstein 2006, Careau et al. 2006).

As shown by the "displacement experiment" carried out on the Eurasian sparrowhawk, during their first migration juveniles largely rely on innate compass orientation to move along a north south axis, while adults display true navigation abilities, modifying their direction of migration and changing their behaviour in relation to their previous experience. However, in species migrating in mixed-age groups, flocking allows the information transmission between experienced and inexperienced individuals. In particular, by following the elders, juveniles of soaring species can learn the better pathways to avoid long water crossings and find suitable wintering grounds. The benefit of social learning is particularly evident in the case of juvenile short-toed snake eagles belonging to populations breeding in Italy and Greece, since, in loose flocks, they follow elders along extremely detoured pathways avoiding the long water-crossings through the Mediterranean Sea (Chapter 7). A high mortality risk during sea-crossings is a strong selective pressure, which may promote the overlap among age classes and, thus, social learning: such process has been highlighted in juvenile Egyptian vultures (*Neophron percnopterus*) born in the Balkans (Chapter 3). Similar processes occur also in the lesser spotted eagle (*Clanga pomarina*, Chapter 8). On the other hand, in a species with a more flexible flight style as the European honey buzzard, the overlap among age classes is an exception rather than a rule, and juveniles generally successfully accomplish their first migration, including also long sea-crossings, without following adults (Chapter 4).

Differential Migration by Age and Sex

Among migratory birds, both age and sex dependent migration behaviour has been observed (Kerlinger 1989, Cristol et al. 1999). In particular, during spring adults are more motivated and tend to migrate earlier and faster than immatures as a result of a time minimization strategy, to reach as early as possible their breeding grounds. Moreover, in species showing a delayed maturation, many, if not nearly all, immatures mainly remain in their wintering quarters (e.g., European honey buzzard, Egyptian vulture), probably to avoid the flight effort that migration entails, and the intraspecific competition with adults in the breeding grounds. During autumn migration, in short distance and partial migrants, which finish moulting at the breeding sites (e.g., the sparrowhawk), juveniles migrate earlier than adults. Conversely, among long-distance migrants, which suspend moult during migration (e.g., the osprey and the European honey buzzard; Chapters 2 and 4), adults leave before juveniles (Kjellén 1992). In addition, in some species, adults tend to migrate shorter distances than juveniles and, among adults, males longer distances than females, or vice versa (Bildstein 2006). To explain such a complex picture, hypothesis related to earlier arrival at breeding grounds (adults, generally males, travelling shorter distances than juveniles and adult females), behavioural dominance and tolerance to cold climates (larger females vs smaller males) have been suggested. In particular, as regards behavioural dominance and climate hypothesis, it has been suggested that juveniles outcompete with adults, and smaller males with larger females, in closer or preferred wintering areas (Bildstein 2006). In this scenario, the smaller males would be less able to survive during the cold season in wintering areas located at higher latitudes (e.g., the western marsh harrier, Chapter 15).

Fasting vs Foraging

There are two distinct feeding strategies among migrating raptors: some species regularly hunt during migration either in concrete stopover areas or through a fly-and-forage strategy, while others store fat before migrating, hunting episodically and opportunistically en route, but regularly drinking at stopover sites both in Africa and Middle East (Dupuy 1969, Clark and Gorney 1987, Yosef 1996, Panuccio et al. 2006). In general, soaring, flocking species migrating long-distances belong to the latter group, and to complete their migration they require fat deposition amounting to 15–25% of lean body mass (Gessaman 1979, Candler and Kennedy 1995, Smith et al. 1986). Conversely, raptors migrating alone, on shorter-distances and those mostly using flapping flight en route (e.g., falcons, Ospreys, harriers and sparrowhawks), accumulate a smaller amount of fuel than long-distance soarers, and regularly feed en route (Blanco 1994, 1997, Yosef 1996, Alerstam and Hedenström 1998, Mellone et al. 2013b, Klaassen et al. 2017, Vincze et al. 2019). As mentioned above, flapping flight has higher energetic costs than soaring-gliding flight, so that the same amount of fat can sustain a soaring raptor up to four times longer than a flapping one (Smith et al. 1986). For this reason, raptors using the most expensive flight style (e.g., harriers), need to refuel and often migrate alone or in small loose flocks, probably to lower the competition with conspecifics when feeding at stopover sites. Moreover, species regularly hunting birds on migration, as the peregrine falcon (*Falco peregrinus*) and the Eurasian sparrowhawks, apparently synchronize their passage with that of their prey, interrupting the journey at stopover sites also for several days (Bildstein 2006).

Influence of Weather Conditions

Several radar studies showed that 80–90% of migrating raptors fly within 1000 m of the ground, being strongly affected by weather conditions both at local and regional scale (Kerlinger 1989, Spaar 1997, Shamoun-Baranes et al. 2017). Thermals used by soaring species form on sunny days with high atmospheric pressure, being stronger during midday and early afternoon when migrants show higher migration speeds. As regards the influence of horizontal winds, the direction and speed of a flying bird is the sum of the wind vector and the bird's own flight vector relative to the surrounding air (Alerstam 1990). It is easily understandable that head winds makes a flight more difficult since birds move in a gaseous medium that is itself in motion against their direction of travel, while crosswinds can shift migrants from their path. Indeed, wind has been highlighted as the most important single factor increasing the daily flight distance of raptors crossing the Sahara desert (Mellone et al. 2012a). In addition, long-distance migrants in the Afro-Palearctic migration system face different prevailing winds during the journey, such as westerly winds prevailing between 30° and 60° of latitude north and south, and easterly winds prevailing between 5° and 30° of latitude north and south. As a result, migrants often modulate their flight behaviour to exploit as much as possible a tail wind support especially when travelling through ecological barriers to reduce the energetic cost (e.g., deserts, large water bodies). In so doing several long-distance migrants perform loop migrations, changing their behaviour in relation to crosswinds showing drift, compensation and overcompensation during different legs of their seasonal journeys. Such pattern is evident in all trans-Saharan raptors crossing the Mediterranean at Gibraltar; in particular, some raptors cross the desert with a stronger tail wind support during autumn while others show the opposite, but all migrate closer to the Atlantic coast during spring (see Vansteelant et al. 2017b for a detailed discussion). Unlike adults, juveniles flying alone or in flocks made entirely of juveniles during their first migration, are wind drifted in strong crosswinds probably also because they have not a fixed goal to reach (e.g., ospreys and Europeann honey buzzards, see Chapters 2 and 4). The influence of weather conditions is even more evident when approaching a water-crossing, given the impossibility to land in case of adverse conditions, making individuals much more selective. In particular, birds can choose among different paths over water in response to different wind directions, undertaking longer crossings when they can exploit a tail wind support and using small islands as stepping stones especially when facing suboptimal weather conditions (e.g., European honey buzzard; see Chapter 4). Finally, as regards species that usually fly through wide water bodies, they choose to cross in strong crosswinds if they provides a tail support (e.g., western marsh harriers; see Chapter 15) and select their paths over water to avoid the passage through low pressure cells (e.g., the Eleonora's falcon, Chapter 28).

Loop Migration

When birds follow different flyways in autumn and spring, at least during a part of the journey, they are showing a loop migration strategy. These patterns have been suspected for decades on the basis of visual counts, but recently, thanks to satellite telemetry, some of them have been confirmed, and new ones have been discovered. Loop migrations may be shaped by wind conditions, or by seasonal differences in the distribution of trophic resources. Even when wind conditions do not vary significantly among seasons, they interact in different ways according to birds' heading. For example, it seems that all raptors following the western flyway follow different routes during the crossing of the Sahara Desert (where prevailing winds blow from the north-east), following an eastern corridor during autumn, and a more western one during spring, shaping therefore a narrow clockwise loop (see especially Chapters 4, 10, 15, 20, 24). Larger scale loop take place in trans-equatorial migrants:

for example, the red-footed falcon (*Falco vespertinus*, Chapter 26) shows a clockwise one, while the Eleonora's falcon an anti-clockwise one, which seems to be determined by trophic resources, rather than wind conditions (Chapter 28). Prevailing easterly winds at Bab-el-Mandeb Strait force some species that crossed it during autumn, as the steppe buzzard (*Buteo b. vulpinus*, Chapter 23) and the steppe eagle (*Aquila nipalensis*, Chapter 11), to come back to Asia through the Suez isthmus, forming therefore an evident loop. Another stunning case in which wind conditions play an important role is the one of crested honey buzzards (*Pernis ptilorhynchus*, Chapter 5) breeding in Japan, which avoid the longer sea crossing during spring, when the wind support would be much weaker than in autumn.

Migratory Connectivity

Few studies investigated the degree of migratory connectivity and the existence of migratory divides within Western Palearctic raptor species. Connectivity is strong when different breeding populations follow different migratory routes, leading to different wintering areas. It seems that for soaring species entering Africa mainly from two access points (Spain and the Middle East), as the short-toed snake eagle (Chapter 7) and the Egyptian vulture (Chapter 3), a migratory divide exists, and wintering populations are segregated in two major clusters, one in the Western Sahel, and another one in more eastern countries. Two comprehensive analyses have been carried out for two broad front migrants: the Montagu's harrier (*Circus pygargus*, Chapter 18) and the lesser kestrel (*Falco naumanni*, Chapter 25). In the first case, connectivity was more diffuse (Trierweiler et al. 2014), while in the second one rather stronger values have been found (Sará et al. 2019). These differences occurred probably because in the lesser kestrel the distance between the breeding and wintering grounds is shorter (leaving lower possibilities of shifting the route) and the longitudinal spread of the sampled populations wider than in the Montagu's harrier. When the wintering area is very restricted, as in the Eleonora's falcon (Chapter 28), connectivity is practically absent. Further research on this topic should be carried out on long distance migrants breeding and wintering across wide areas, for example the steppe eagle (Chapter 11).

Individual Flexibility

In the last two decades, telemetry studies allowed to know the whereabouts of single raptors even for several years in a row, opening thus the possibility to study consistency among and within individuals of a given species. Recent studies found contrasting evidence: for example, western marsh harriers and Egyptian vultures exhibit a high individual repeatability in departure dates, thus suggesting that timing of migration is likely to be endogenously determined, while routes, which are highly flexible, are more influenced by weather conditions. On the other hand, ospreys show the opposite pattern, with routes being more repeatable than timing, probably because of their fidelity to intermediate individual stopover sites (López-López et al. 2014, Vardanis et al. 2016). Therefore, in the first case, inter-year route variation within the same individual may be much stronger than inter-individual variation within the same year, especially when different individuals are travelling in the same region at the same time: in this scenario, an interesting example is provided by Eleonora's falcons crossing the Indian Ocean (Chapter 28).

Recent Changes and Plasticity at a Population Scale

Due to shifts in climate, sea level and in the distribution of major biomes, migration systems have been showing great changes over a long time scale (Sutherland 1998). Currently, the ongoing climate change is significantly altering migration patterns, even during a relatively short time span. Among Western Palearctic raptors, those species that are adjusting their phenology, advancing the autumn migration dates, are also experiencing a positive population trend (Panuccio et al. 2016b). Other species, especially short-distance migrants as the common buzzard, are reducing their migration distance, eventually switching even to sedentariness (Martin et al. 2014). Recently, it has been possible to witness the formation of species-specific migratory flyways that did not exist until a few years ago, promoted by the establishment of new breeding or wintering grounds. Good examples of this are the significant numbers of booted eagles from the Spanish population, which winter in Southern Italy, and reach these areas migrating along the Mediterranean coastline (Chapter 10). Another example is provided by the pallid harrier (*Circus macrourus*), which recently established a breeding population in Finland, and reaches Africa through the Iberian Peninsula, where it was just a vagrant species until a few years ago (Chapter 17).

Mortality

Migration is among the most risky stages of the life cycle of birds. Satellite telemetry allowed, in recent years, to detect where and when a death occurs, at an individual level. A comprehensive study involving only adult individuals of three raptor species highlights that spring migration has the highest death rates, especially during the Sahara crossing (Klaassen et al. 2014). However, when taking into account also individuals with different ages, not unexpectedly, mortality is higher in first year birds during their first migration (Sergio et al. 2019). Death during migration may occur due to natural (adverse weather conditions, extreme effort; Zu-Aretz and Leshem 1983, Strandberg et al. 2009e, Oppel et al. 2014) or man-induced causes, like poaching (Panuccio 2005) or collisions with wind turbines (Martin et al. 2018; see below).

Conservation of Migratory Raptors

The conservation of migratory birds, in general, implies serious challenges. They spend different parts of their life cycle in different places located thousands of kilometres apart each other, without taking care of political boundaries, which makes the coordination among different administrations and conservation agencies even more difficult. They are exposed to a high variety of risks, and conservation efforts aimed at protecting them, for example, during the breeding season, can be jeopardized by threats occurring during the wintering one, and vice-versa. Carry-over effects, e.g. events occurring in a given place during a given season, but having an impact also on processes occurring elsewhere in another season, can trigger serious demographic consequences. It has been clearly shown that long-distance migrants are declining at a faster rate than other species (Sanderson et al. 2006). Daily mortality rate is higher during migration than during breeding or wintering, but these stationary periods are longer and therefore show higher absolute values (Klaassen et al. 2014). During winter, electrocution in concrete areas may have serious detrimental consequences, as seen for Egyptian vultures wintering in Sudan (Angelov et al. 2013b). Many raptor species included in this book concentrate in narrow hot-spots in order to avoid the crossing of ecological barriers. A very restricted area may be crossed in a few days by a significant proportion of the global population of a given species (Vansteelant et al. 2020b). European honey buzzards in Italy and Eastern Europe increased significantly their populations as a consequence of actions against poaching undertook at the Strait of Messina (Agostini et al. 2007). However, this is still a serious issue, threatening raptors especially along the Eastern flyway. Infrastructures as wind-farms are another source of mortality at bottle necks, apparently more deleterious for local populations, rather than for those passing through during migrations (Martin et al. 2018). A recent research suggests that wind turbines may have a more subtle impact, too, as they reduce the availability of soaring areas (Marques et al. 2020), with unfavourable consequences on optimal flight behaviour. On the other hand, raptors concentrations at bottle necks are also an opportunity, as they allow to study their behaviour and to get robust population estimates in a very efficient way. They also offer unparalleled educational and recreational possibilities, being a spectacular sight that impacts everybody, from laymen to serious biologists (Bildstein 2006).

2

Osprey *Pandion haliaetus*

Flavio Monti

ABSTRACT

The osprey (*Pandion haliaetus*) is a cosmopolitan raptor species exhibiting a leap-frog migration pattern, with both long- and short-distance migratory as well as resident populations. It is a versatile flyer representing an exception among raptor species since it is capable of flying over several hundred kilometres of open water. The species generally migrates on a broad front, performing a fly-and-forage strategy combined with stopover periods. In the Western Palearctic, its breeding range stretches from Northern Europe to Northern African coasts and oceanic islands on the west, and to Red Sea and Persian Gulf on the east. Three main migration routes, partially overlapping across populations, have been revealed by long-term ringing and satellite tracking studies. The first leads from Western Europe to Western sub-Saharan Africa passing through the Iberian Peninsula; the second from Western and Central Europe to Western and Central sub-Saharan Africa passing through Italian coasts and/or Balkan Peninsula; the third from Central and Eastern Europe to Western, Central and Eastern sub-Saharan Africa through Middle East. Wintering grounds of European migratory populations spread along a wide belt from west to east in sub-Saharan Africa and can partially overlap among populations. Ospreys breeding at lower latitudes, such as in the Mediterranean basin, are sedentary or perform short-distance inter-breeding movements, mostly wintering at temperate latitudes along the North African coasts or in Southern Spain and Italy. Ospreys from Cape Verde, Canary Islands, Red Sea and Persian Gulf represent no-migratory populations, breeding at lower latitudes for the Western Palearctic.

Introduction

The osprey (*Pandion haliaetus* Linnaeus, 1758) is a cosmopolitan medium-sized raptor species that has evolved specialized morphological characteristics and exhibits unique behaviour to assist in catching prey, consisting of live fish only (Cramp and Simmons 1980, Poole 1989). Adaptations to its specialized manner of feeding are particularly manifest in the structure of the tarsus that is provided, as well as the lower surface of toes, of sharp spicules essential to catch and manipulate slippery fishes. In addition, the outer toe is reversible and permits to the bird to grip the prey with two toes forward and two toes back (Cramp and Simmons 1980, Poole 1989). Despite its high specialization as a piscivore, it is an opportunistic forager that can feed in both freshwater and marine environments

University of Siena, Department of Physical Sciences, Earth and Environment, Via Mattioli, 4, 53100 Siena – Italy.
Email: flaviomonti00@gmail.com

(Poole 1989). Due to its particular evolutionary history and morphological characteristics, the species is considered sufficiently distinct from other raptor species (family *Accipitridae*) to merit a monotypic family (*Pandionidae*). A recent study relying on an extensive dataset of both ancient and fresh samples collected from many populations around the world, revealed four main genetic evolutionary clades (on the basis of mtDNA) representing quasi non-overlapping geographical regions (Americas, Indo-Australasia, Europe-Africa and Asia) and shed light on how such a specialized raptor was able to colonize most of the globe (Monti et al. 2015). Similarly, on the basis of morphometric and plumage characteristics the osprey has partitioned into 4 subspecies: *P. h. haliaetus* (Linnaeus 1758) in the Palearctic from Europe, northwest Africa, and Asia north of the Himalayas, *P. h. carolinensis* (Gmelin 1788) in North America, *P. h. ridgwayi* (Maynard 1887) in Caribbean Islands, and *P. h. cristatus* (Vieillot 1816) in the Indo-Pacific and Oceania. Holarctic ospreys from high latitudes in America, Europe and Asia are long-distance migratory, whereas those breeding at lower latitudes carry out short-distance migrations or can be sedentary (Poole 2019). Generally, breeders at warmer, drier subtropical latitudes are paler than breeders at cooler and higher latitudes and tend to have smaller body sizes (Prevost 1983). However, such differences are not always straightforward, with some morphological characteristics overlapping across populations and among individuals (Strandberg 2013). The species has relatively long and narrow wings (wingspan of 145–170 cm) and an average body mass of 1.5–1.8 kg (Cramp and Simmons 1980). Females are slighter bigger than males, even if biometric measures can overlap between sexes (Poole 1989). Even though is currently globally listed as "Least Concern" according to IUCN criterions, it is considered a priority species for conservation across its distributional range and it has been included in the Annex I of the "Birds Directive" of the European Union 70/409/EU as well as in many other international conventions.

In the Western Palearctic, its breeding range stretches from Northern Europe to Northern African coasts and oceanic islands on the west, and to Red Sea and Persian Gulf on the east. In Northern Europe, it is a relatively common species building tree-nests in forest habitats, and is generally associated with freshwater lakes and rivers for feeding during spring (but using mostly saltwater habitats on the wintering grounds; Poole 1989, Wahl and Barbraud 2014). At lower latitudes, as in the Mediterranean, the osprey breeds mostly on rocky pinnacles within a fragmented coastal habitat, being closely linked to marine environments for fishing (Cramp and Simmons 1980, Monti et al. 2018d). During the 19th and in the early 20th century, some populations declined because of heavy persecution. Nowadays, the number of osprey nesting pairs in Europe, Northern Africa, and the Middle East ranges between 9500 and 11500 (Schmidt-Rothmund et al. 2014). The bulk of the European population breeds in Scandinavia and in some other countries of central Europe (Saurola 2005, Bai and Schmidt 2011). In the Mediterranean region, few breeding nuclei remain and the species is still threatened by local extinction and affected by multiple factors limiting the numbers increasing again (e.g., Monti et al. 2018d). In this region, although direct management actions allowed a partial recovery, the osprey is still an uncommon breeding species with a vulnerable conservation status (Thibault et al. 2001): less than 80–100 breeding pairs distributed between Corsica, the Balearics, Morocco and Algeria (Monti et al. 2018d). In the last decades, thanks to reintroduction programs, the species has returned to breed in mainland Spain (Muriel et al. 2010), Portugal (CIBIO 2011) and Italy (Monti et al. 2014). More recently, other reintroduction projects have been launched in Northern Spain (Urdaibai Biosphere Reserve—Basque Country in 2013; Galarza and Zuberogoitia 2012), in Switzerland (in 2013; Strahm and Landenbergue 2013), in Southwestern France (Aquitaine region in 2018; Région Aquitaine 2018—unpublished technical report) and in Eastern Spain (Valencia region in 2019; Rafel Triay pers. comm.). However, beside numerous threats on the breeding grounds (e.g., Monti et al. 2018b), it is likely that other threats during the non-breeding season, including migration, may hinder the recovery of these populations, even after 30 years of legal protection (Monti et al. 2018d). Finally, few pairs breed in Canary Islands (Triay and Siverio 2008, Rodriguez et al. 2013) and Cape Verde (Siverio et al. 2014), whereas larger resident populations exist in the Red Sea (estimated population size as per number of active pairs: 600–800; Poole 2019) and Persian Gulf (100–200 active pairs: Poole 2019). In general, the breeding season begins between February and April (according to the latitude of breeding areas): on average three eggs are laid, which hatch in May. Chicks fledge at the end of June or early July and migrate between August and October (Poole 1989).

Knowledge Status

From the last decades of the past century, the osprey has been one of the most studied raptor species concerning migration strategies. First, through visual observations, radar measurements and ringing recoveries, and subsequently it became one of first species followed with satellite telemetry (Österlöf 1977, Kjellén 1992, Kjellén et al., 2001).

This species is known to migrate on a broad front and through three major migratory flyways/routes in the Western Palearctic (Fig. 1): (i) a western route passing through the Iberian Peninsula until sub-Saharan wintering grounds in Western Africa, followed by Western and Central European ospreys (e.g., Alerstam et al. 2006, Bai and Schmidt 2012); (ii) a central route passing through Italy and/or Balkan Peninsula until sub-Saharan wintering grounds both in Western and Central Africa, followed by Central European ospreys (e.g., Bai and Schmidt 2012); and (iii) an eastern route passing through

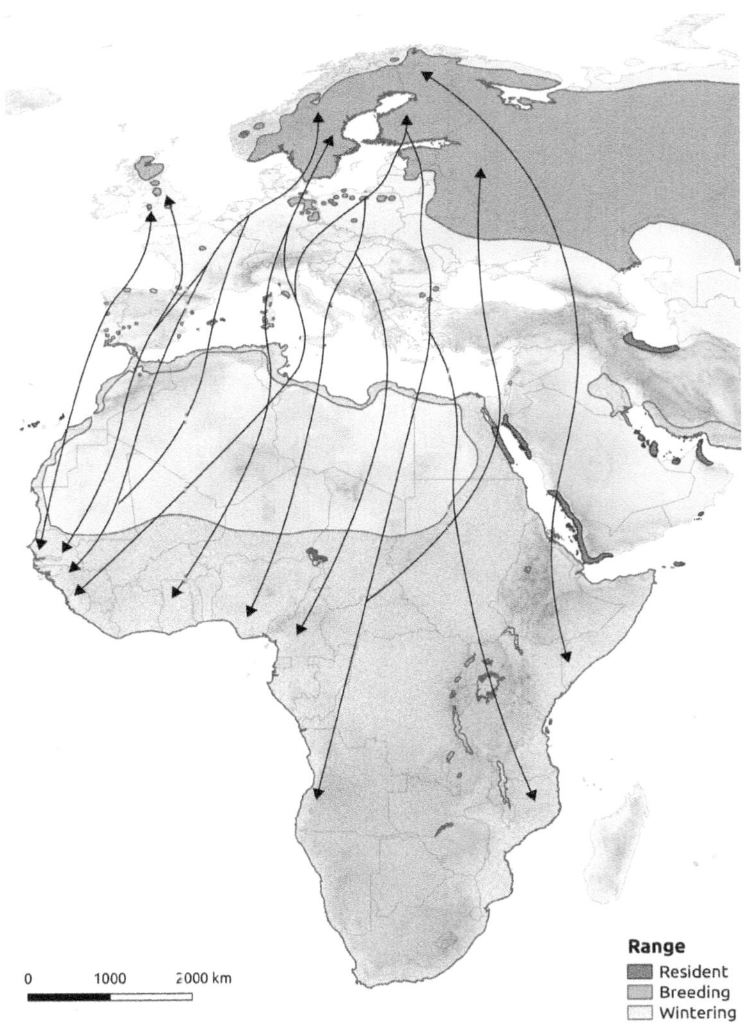

Figure 1. Schematic representation of the routes followed by long-distance migrating populations of ospreys breeding in Europe.

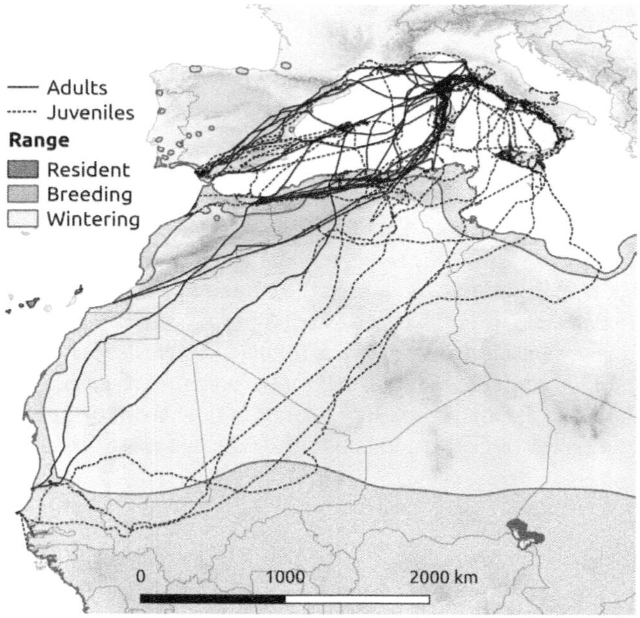

Figure 2. Post-breeding migration routes followed by GPS-tracked ospreys belonging to Mediterranean populations.

Middle East and Red Sea until sub-Saharan wintering grounds in Central and Eastern Africa, followed by Central and Eastern European ospreys (e.g., Newton 2010b, Väli and Sellis 2018, Babushkin et al. 2019). These main migratory routes as well as wintering quarters can partially overlap among populations (Zwarts et al. 2009).

Beyond these major migratory routes, there are populations breeding at lower latitudes, such as those of the Mediterranean basin, that are sedentary or perform short-distance inter-breeding movements, mostly wintering at temperate latitudes along the North African coasts or in Southern Spain and Italy (Monti et al. 2018d; Fig. 2). Breeding ospreys in Cape Verde, Canary Islands, Red Sea and Persian Gulf are known to be mostly resident, staying at breeding sites all year round (Rodriguez et al. 2013, Siverio et al. 2014, Poole 2019); inter-breeding movements are likely to occur, but future studies should focus on gathering more information on the spatial ecology of these populations that are to date still poorly known.

As a general picture, ospreys breeding at different latitudes in the Western Palearctic use different migration strategies, which probably evolved as a response to different ecological conditions encountered at breeding sites. Palearctic ospreys display a leap-frog migration system, with populations breeding at higher latitudes over-passing the latitudes where southern populations live. This implies that northern birds have to cross the Sahara Desert, a major ecological barrier that the Mediterranean ospreys (at least the majority of them) do not have to confront with (Monti et al. 2018c).

Given its morphology (i.e., slender wing-shape with a high aspect ratio between 8 and 9 similar to that of a seagull; Kerlinger 1989, Agostini et al. 2015a), the osprey is a versatile flyer representing an exception among large raptor species since it is capable of flying over several hundred kilometers of open water (Horton et al. 2014, Duriez et al. 2018). Being less dependent on thermal conditions than typical soaring species, the osprey regularly exhibits more travelling hours (Mellone et al. 2012a), sometimes extending them also through the night (Galarza et al. 2017). Thus, it is not obliged to concentrate at main migratory bottlenecks such as the Strait of Gibraltar and the Strait of Messina (Cramp and Simmons 1980), though it is regularly observed at these spots and on several Mediterranean islands during migratory passages (Zalles and Bildstein 2000, Panuccio et al. 2005a). In the Western Palearctic, Falsterbo (Sweden), is the watchsite with the highest counts (Kjellén 1992), with an average of 289 individuals in the 2010–2018 (Kjellén 2018). However, it has been calculated that the observed individuals represent just the 2% of the source breeding population (Kjellén 1997). Ospreys are not gregarious migrants: observations at migratory watching-points exclusively involve solitary individuals. A large variation in routes between individuals and between siblings from the same broods is known as well as a high degree of fidelity to individual-specific stopovers and wintering areas (Vardanis et al. 2016). The former is explained by the presence of high-quality foraging sites widespread throughout Europe, which could promote broad-front migration, whereas the latter by the fact that hunting efficiency may improve with increasing local knowledge and experience about the area, and thus, would explain high fidelity to specific stopover sites or wintering grounds year after year (Vardanis et al. 2016, Poole 2019). A high inter-sexual variation in timing of migration also exists for long-distance migratory populations: generally, females are the first leaving breeding grounds, followed by males and juveniles. This is probably due to the fact that in long-distance migratory populations, the female can start moulting some feathers during the incubation being ready to migrate before males that, on the contrary, have to keep feathers intact in order to maintain peak fishing efficiency for the entire duration of chick rearing period. Accordingly, males postpone the moult and put back the onset of the autumn migration (Hake et al. 2001, Kjellén et al. 2001). Finally, although the osprey is normally restricted to diurnal migration, there are several confirmed records of nocturnal flights over both land (Alerstam et al. 2006, Galarza et al. 2017) and sea (Monti et al. 2018d, Duriez et al. 2018). The longest non-stop protracted flight over sea was of 1,558 km in 43.9 h carried out by an American osprey crossing the Western Atlantic Ocean during autumn migration (Horton et al. 2014).

Migration Patterns

Western flyway

From west to east, this migratory route principally interests ospreys from UK, continental France, Germany, Norway, Sweden and, to a lesser extent, some Finnish ospreys. Thanks to extensive banding programs and long term GPS tracking, ospreys' movements along this migratory flyway is well known and a lot of information both for autumn and spring journeys is available (Fig. 1). The first satellite tracking study on Palearctic ospreys was carried out in Scandinavia, with 18 ospreys tagged in Sweden between 1995–1998 (Hake et al. 2001, Kjellén et al. 2001). These and subsequent studies have revealed that Northern and Central European ospreys undertake long-distance migratory journeys and can cover ca. 6,000 km towards sub-Saharan wintering grounds (Hake et al. 2001, Kjellén et al. 2001, Alerstam et al. 2006, Strandberg and Alerstam 2007, Klaassen et al. 2008, Bai and Schmidt 2012, Østnes et al. 2019). There are records of few Northern and Central European ospreys wintering in Southern Spain and Italy, respectively, especially males from UK and Germany (MacKrill 2017, Bai and Schmidt 2012). Such numbers seem to have increased more recently, likely in relation to climate warming effects and other land use changes, although they represent only about 3% of the total breeding population estimated for Western Europe (Martín et al. 2019, Birdlife International 2019b). The majority of European ospreys in fact keep moving crossing the Mediterranean Sea and then the Sahara Desert until sub-Saharan Africa for wintering (Poole 2019).

The autumn migration commonly starts in August and can last until October–November. Mean departures dates can span over a large temporal period of c. 50 days (Alerstam et al. 2006). Migratory paths are mainly conducted over land and the entire journey can last around 2 months (61.0 ± 17.8 [SD] days, N = 24; Monti et al. 2018c, Klaassen et al. 2008). The duration of migration is slightly longer in juveniles, which tend to wander more than adults do, often because they are diverted by winds (Klaassen et al. 2011). While crossing Europe, these birds make between 2–5 stops at stopover sites, mainly located in Central Europe (e.g., France and Germany as first stopovers) and Southern Mediterranean region (in Southern Spain or North Africa as successive stops, in preparation of the crossing of the Sahara Desert). Time spent at stopover sites can account for the 50% of the autumn migration duration (31.5 ± 13.3 days), so that only half of the migration time is spent in active flight

(Monti et al. 2018c). Also in this case, juveniles seem to stop for more days than adults. In order to enter in Africa, some ospreys fly through the Strait of Gibraltar (Onrubia 2016), while the majority cross at other places along the coasts, flying over the open sea. For some ospreys, like those from the UK, the autumn migration entails a first sea-crossing at higher latitudes, in the Atlantic Ocean over the Bay of Biscay before they run down into the Iberian Peninsula (a flying bout of ca. 500–800 km; MacKrill 2017, Poole 2019).

These ospreys reach wintering grounds located in sub-Saharan Africa, between October and November. Wintering grounds of ospreys migrating through the western flyway end up in Senegal, The Gambia, Ivory Coast and/or Ghana. There, adults stay for about five-six months until March–May (range: 04/03–10/05; e.g., for Swedish ospreys; Alerstam et al. 2006) when they start the backward migration towards breeding grounds. On the other hand, juveniles tend to skip the first spring migration, staying at wintering grounds for c. 18 months before returning to their natal area for the first time, once the sexual maturity is reached (Poole 1989). Spring migration is not just the simple reversal of autumn migration: spring journeys are shorter, lasting about 21–33 days (Alerstam et al. 2006), and include less stopover days (mean 4 days, range 1–7 days) than the autumn journeys (mean 14 days, range 0–28 days). In spring, the timing of migration is considerably less variable than in autumn. Departures in spring of the same individuals are restricted to a short period (range: 19 March–12 April; Alerstam et al. 2006).

Central flyway

This migratory route is mostly followed by ospreys from Finland, German, Poland and, at lesser extent, by Swedish and Norwegian ones. Big numbers of these ospreys cross central Europe in a north-east/south-west direction, flying over the Mediterranean region via the Balkan countries and Italy, then crossing the Sahara Desert to reach wintering grounds located in sub-Saharan Africa, along the western coasts of Senegal or in the Gulf of Guinea (e.g., Ivory Coast, Ghana and Cameroon; Fig. 1). In 2001, a satellite project started in Finland for gathering information about the migration ecology of ospreys from Lapland (i.e., the northernmost region in the world hosting breeding ospreys) and, following a latitudinal gradient within the country, from the southern regions of Satakunta, Kainuu and Häme. Looking at the GPS tracks of 25 ospreys from these regions (Saurola 2014, https://www.luomus.fi/en/finnish-satellite-ospreys), Finland seems to be at the middle of a migratory divide: 12% (3 ind.) of these ospreys took a western flyway, a 44% (11 ind.) a central one and the remaining 44% (11 ind.) an eastern flyway (details shown below). Also in this case, timing of autumn migration shows large variation among individuals and age classes. Those following the central flyway took similar routes in both autumn and spring, except for few of them. Similar results in individual variation of migratory routes were found in a more recent study conducted on juveniles from a re-established breeding area in central Norway (Østnes et al. 2019). In contrast, a differential timing of migration in both spring and autumn was demonstrated for German ospreys, using a long-term ring-recovery database (Bai and Schmidt 2012). In autumn, adult females leave the breeding area earlier than adult males, while the migration of first-calendar year ospreys takes place over a wide time-span. In this study, the fact that no osprey was recovered in the ringing area in February, and that the 3cy males already arrive at their summer locations in March, indicates that the migration of 3cy male ospreys is concentrated in late February and early March (Bai and Schmidt 2012). Adult (third year and older) males return earlier to the breeding area in spring, followed by adult females and then two year-old birds. This is also because 3cy males stay closer to the ringing area than females and younger birds for winter (Bai and Schmidt 2012). The sexual difference in the timing of spring migration and in the choice of wintering grounds in ospreys is consistent with the arrival-time hypothesis, with males selected for prior access to breeding resources (Bai and Schmidt 2012, Poole 2019). Within this large ring-recovery database, only two autumn recoveries had southeasterly directions and were both dead 1cy birds, which probably reflect an unfavourable orientation for central European ospreys.

Eastern flyway

Little is known about osprey's migration strategies across the Eastern European-East African flyway. Very few Russian and other Eastern European ospreys have been banded or tagged, but first interesting data are sorting out. As described above, some Finnish ospreys are known to migrate across this flyway as well as few from Norway and Sweden, despite conspecifics from the same country can select both a western or central flyway independently. The individuals that decide to migrate across this route mostly cross the Black Sea (or circumnavigate it) to enter in Turkey over Anatolia, or either in Greece through the Bosphorus (Fig. 1). Recently, few adult birds were equipped with GPS-based satellite-transmitters at their breeding sites in Estonia (Northeastern Europe) and tracked to their wintering grounds in Africa and back (Väli and Sellis 2016). Departure times, migration routes, as well as wintering and stopover sites varied remarkably between individuals. As found for other flyways, autumn migrations were of longer duration and had more stopovers than those in spring. Time spent at stopovers ranged between 2–0 days; stops were made mostly in Europe and less in the Middle East (Turkey) and Northeastern Africa (Egypt). The Nile Delta was the only stopover site in Africa used for consecutive years by two birds. The majority of Estonian ospreys follow the Eastern European flyway, flying directly to the south and winter in Eastern, Central or South-Western Africa.

Between 2015 and 2017, four individuals from north-west Russia were GPS-tagged making available detailed autumn migratory patterns for this population for the first time (Babushkin et al. 2019). Similarly to other northern populations, the mean start date of autumn migration occurred in August. The mean distance of the migratory journey was 6,258 km and, also in this case, the wintering locations were spread over a broad belt: one wintered in Angola, two in Sudan and Egypt and one in Israel. These ospreys migrated over central Europe (two of them stopping already in Russia for c. 2 weeks), crossed the Black Sea and then the Mediterranean via Suez and Cyprus, stopping again in Turkey or Israel before entering in Africa.

One of the longest migratory journey of a Western Palearctic osprey known so far occurred along this migratory flyway (Saurola 2014, https://www.luomus.fi/en/finnish-satellite-ospreys): a male osprey breeding in Lapland in Northern Finland and having wintered on the coast of Mozambique. According to satellite data, this bird travelled 10,392 km in autumn and very nearly the same distance (10,420 km) in spring on its way back to Lapland. In spring, starting on 28 March, it migrated for 50 days progressing at an average of 208 km per day including stopovers. On its way north, after having stopped over in Northern Egypt, beside the Suez Canal, it crossed the Mediterranean *via* Cyprus and continued north into Turkey, around the northern edge of Istanbul and then west along the Black Sea coast crossing into Bulgaria, progressing north through central Europe, arriving back in Lapland on the 16th of May. Among Swedish ospreys, a minor proportion is known to have a southeastern orientation (Hake et al. 2001, Thorup et al. 2003): a Swedish female osprey tagged in 1995 and migrating east to Russia until Kazakistan took an extremely eastward migratory route. From here the bird abruptly turned south-west across Georgia, Iraq and Saudi Arabia, over the Red Sea and into Eritrea, then still well south until Mozambique as final winter destination (Hake et al. 2001).

Mediterranean basin

Sporadic ring recoveries and anecdotal observations collected in the past, suggested variations in migration and wintering strategies for the Mediterranean osprey populations (Thibault et al. 1996, 2001). Some ospreys have been observed to stay in their breeding territories during winter and others have been located several hundreds of km from their nest-sites (Thibault et al. 1996). Such fragmentary and contrasting data precluded a complete and comprehensive understanding of their year-round ecology for a long time. The first detailed analysis of the migratory and wintering ecology of these osprey populations started in 2012: between 2012 and 2019, more than 60 individuals (both adults and juveniles) have been tagged throughout Corsica, Balearic Islands and Italy (Monti et al. 2018d; Fig. 2). This study revealed that the Mediterranean ospreys are partial migrants having distinctive migratory and wintering strategies compared to northern conspecifics, with main differences residing in route choice, orientation and timing. Mediterranean ospreys are short distance migrants (73% of the individuals) wintering at temperate latitudes along Mediterranean shores or resident (27%) staying at breeding grounds all year round. The main wintering grounds were located in Southern Spain (Andalusia), Sardinia, Sicily and northern coasts of Morocco and Algeria. Pre- and post-migratory journeys to secondary sites were recorded during both autumn and spring migration (Monti et al. 2018d). Before autumn migration, 90% of migratory adults performed pre-migratory trips to a secondary site, on average 103.0 ± 65.4 (SD) km from their nest, located at the seashore or inland near a lake or river. Most birds spent on average 23.2 ± 22.6 days (range: 5–60 days) at these secondary site. For migratory individuals, an extremely short duration of migration (5.2 ± 2.6 days) was recorded (Monti et al. 2018d). Stopovers were very rare and performed by only few individuals. Autumn migration was concentrated in time, with mean departure date from breeding sites around the 12th of August (range for departures: adults = 24 June to 9 November; juveniles = 29 July to 28 August) and mean arrival dates to wintering sites occurring around the 22th of August (range for arrivals: adults = 25 June to 17 November; juveniles = 3 August to 2 September). Thus, in the Mediterranean populations, juveniles do not depart significantly after the adults as commonly happens for Northern European osprey populations. This is especially evident also because many Mediterranean adult ospreys are resident and remain at breeding sites throughout the year.

Mediterranean Ospreys were able to perform long non-stop flights over the open sea, sometimes conducted overnight (Duriez et al. 2018). During winter, they showed high plasticity in habitat selection, using marine bays, coastal lagoons/marshland and inland freshwater sites along the coasts of different countries of the Mediterranean basin. Movements and home range areas were restricted during the wintering season. The short duration of trips and high levels of variability in migratory routes and wintering grounds, revealed high behavioural plasticity among individuals, probably promoted by the relatively low seasonal variability in ecological conditions throughout the year in the Mediterranean region, and weak competition for non-breeding sites. Since distances to reach wintering sites were relatively short and little time was required, individuals could choose to invest part of their energy to cross the sea when there were favourable winds, otherwise selecting a safer but longer route over land (Monti et al. 2018c). For a short migration, they may concentrate their efforts on flying at sea for a limited time, even if it is energetically more demanding. These results showed that the migratory behaviour of Mediterranean ospreys was highly flexible and could be adapted to local circumstances. Only a few cases of trans-Saharan migration (following a central flyway) were recorded, indicating that at least some individuals from the Mediterranean region do winter in sub-Saharan Africa, as their conspecifics from Northern Europe commonly do. However, the great majority of Mediterranean ospreys wintered within the Mediterranean basin, at temperate latitudes. In spring, adult ospreys left their wintering grounds in mid-February (mean departure date = 21 February ± 16.4 [SD] days; range 6 February to 25 March) and arrived at breeding sites 5.8 ± 4.3 days later (mean arrival date 27 February; range 6 February to 4 April). Only one bird used a stopover site. Two of six migratory adults from Corsica performed a loop-migration, taking different routes in autumn (crossing the Mediterranean Sea directly from Corsica to reach Spain) and spring (crossing Spain over land until the Pyrenees, and then reaching Corsica from the continental French shores). The detour in spring, probably due to bad weather conditions across the Mediterranean Sea, added 93 km and a 2-day delay for one bird, and 299 km and a 5-day delay for the second one. As for autumn migration, some Corsican adults were observed making repeated post-migratory round trips (ca. 100 km from their nest, following the seashore or going inland near a lake or a river) performed after spring migration and before the onset of the breeding season (Monti et al. 2018d). Juveniles tend to behave differently: those that survived, spent their second calendar year entirely at their wintering area and came back to their natal area in their third calendar year, after 18 months (except for two cases of an early return at 9 and 11 months, respectively; Monti et al. 2018d).

Resident populations

At the very low latitudes for the Western Palearctic, some other breeding osprey populations exist in Cape Verde and Canary Islands on the west (Atlantic Ocean) as well as on the islands and coasts of Red Sea and Persian Gulf on the east (Fisher et al. 2001). These southern populations, mainly confined to islands, are sedentary and usually breed on coastal cliffs, or occasionally inland with nests that can be built directly on the ground (Poole 1989). In Canary islands, adults have been observed staying at breeding sites all year round (Siverio et al. 2018), with local movements up to 40 km away from their territories, as per field observations. However, extra inter-insular movements are very likely to occur, especially for juveniles in dispersal. For example, a juvenile from Tenerife (Canary Islands) was observed in Dakhla, Morocco (Manuel Siverio pers. comm.). In Cape Verde and the Middle East, specific tracking studies have not been conducted yet. Future research aiming at obtaining detailed information on the spatial ecology of these populations is necessary.

Behavioural and Ecological Remarks

In contrast with other raptor species, the osprey shows low repeatability in timing and duration and a high repeatability in routes, with a high degree of fidelity to individual-specific stopovers and wintering areas (Vardanis et al. 2016). However, at inter-individual level, there are significant differences in migratory routes (longitude) during both autumn and spring migration (Vardanis et al. 2016).

At the Western Palearctic scale, a leap-frog migration system exists with populations breeding at higher latitudes over-passing the latitudes where southern populations breed and, to a certain degree, also winter. Ospreys breeding at different latitudes use different migration and stopover strategies. Northern European populations clearly differ from Mediterranean osprey populations with respect to both temporal (total migration speed and use of stopover) and spatial (distance, routes and main direction) components of migration strategy. In a comparative satellite tracking study on both Scandinavian (LDM) and Mediterranean birds (SDM), it was found that the total speed of migration of northern ospreys was 2.4 times slower compared to southern ones, though northern migratory ospreys travel distances five times longer than their Mediterranean counterparts (Monti et al. 2018c). The time constraint related to a longer migratory journey was not the main factor driving the total speed of migration. The main difference in total speed of migration was due to a greater use of stopovers (in both number and duration of stopovers).

Long distance migratory populations show regional variations in time budgets on migration performances. In Europe, where suitable feeding habitat is abundant, ospreys behave like "fly-and-forage" migrants, trading-off travel time with feeding opportunities (Strandberg and Alerstam 2007). On the contrary, in the Sahara Desert where feeding habitat is virtually lacking, ospreys behave like 'sprinting migrants', using all potential hours to travel to ensure a rapid crossing of this ecological barrier (Klaassen et al. 2008, Mellone et al. 2012a). The average cross-country speed of migrating ospreys significantly differs in relation to the overflight habitat. Although the average cross-country speed is 30.7 ± 11.7 [SD] km/h and daily distances travelled are in the order of 200 km/day both for LDM (202.4 ± 41.1 km d–1) and SDM (211.4 ± 87.7 km d–1), cross-country speed is on average 10 km/h higher during sea-crossing than over land (sea = 38.2 ± 11.3 km/h vs land = 29.8 ± 11.5 km/h; Monti et al. 2018c), as calculated within the Mediterranean region. Anyway, there is significant variation in migration, and both nutrition and body condition may play a role in explaining these differences. For example, one satellite-tagged male flew from Sweden to Spain in a single day, a distance of 746 km, while another one made a trip from Southern Sweden to Guinea Bissau, West Africa, in just 14 days, averaging 431 km per day (Hake et al. 2001).

Short distance migratory populations, such as those of the Mediterranean, can cover very long distances over open sea, as already observed in other osprey populations worldwide (DeCandido et al. 2006b, Horton et al. 2014). The protracted flights over sea recorded for Mediterranean ospreys suggest that these birds build fat stores before departing. This hypothesis is supported by the observation that most Mediterranean individuals performed pre-migratory movements to a secondary feeding site, several dozen of km from their nests (Monti et al. 2018a). The function of such movements might be related to the necessity of building fat stores before crossing the barrier and/or to wait for favourable weather conditions. During sea crossings, ospreys mostly adopt an active flapping flight, as suggested by higher airspeeds over sea than over land, and partially benefit from the use of tailwinds. However, a more recent study has shown that ospreys are able to find and use thermal uplift over the open sea. Five Italian juvenile ospreys equipped with GPS-Accelerometer-Magnetometer loggers and tracked while crossing the Mediterranean Sea, were able to find and use thermal uplift on average 7.5 times per 100 km, and could reach altitudes of 900 m above the sea surface. Their climb rate was 1.6 times slower than over land, and birds kept flapping most of the time while circling in the thermals, indicating that convections cells were weaker than over land (Duriez et al. 2018). The frequency of thermal soaring was correlated with the difference between the sea surface and air temperature, indicating that atmospheric convection occurred when surface waters were warmer by at least 3°C than the overlaying air. These observations help explain the transoceanic cosmopolitan distribution of osprey, and question the widely held assumption that water bodies represent strict barriers for large raptors (Duriez et al. 2018).

As a general pattern, sex and age variations in migratory performance are also evident: (a) even though males and females travel at about the same speed, males tend to arrive few days ahead at nest, suggesting they leave wintering grounds earlier, to return to their nest well in time (Poole 2019). A faster return to breeding grounds increases the chances to have access at best nest sites, and contributes to an early start of the breeding season and its success. Also without taking into account the sex of the individuals, it has been shown that Swedish adult ospreys migrate more quickly over the Sahara desert during spring rather than during autumn (daily distance: 309 vs 275 km), even if winds are less favourable, suggesting an extra effort leading to a time minimization strategy (Mellone et al. 2012a)

These differences, however, are less manifest in the Mediterranean populations (Monti et al. 2018d). (b) adults tend to travel faster and show straighter migratory paths than juveniles, which conversely display a greater variance in the major axis of migration (Alerstam et al. 2006, Monti et al. 2018c). This suggests that adults can rely on their experience from previous years to reach a precise goal, while strong divergence of juveniles suggests strong influence of sidewinds on migratory route of naïve juveniles (Thorup et al. 2003). It is not unusual for young to end up well west of their migratory path. As a reference example, in 2008 a satellite-tagged juvenile from Scotland was diverted by strong winds while crossing the English Channel, being drifted over ocean waters in the Atlantic Ocean where, after approximately two weeks of no-stop flight, it got exhausted and drowned off the Azores islands (RSPB 2008).

This explains the high juvenile mortality recorded for the species, being roughly 50% during the first year (Poole 2019). A significant portion of overall mortality seem to occur in the first year, especially during autumn migration and/or wintering, for both LDM and SDM populations. Data show that one in three juvenile Swedish ospreys die crossing the Sahara in their first migration south (Strandberg et al. 2009e, Klaassen et al. 2014). Spatio-temporal patterns in mortality are also evident for these long distance migrants, with spring mortality occurring mainly in Africa in association with the crossing of the Sahara desert, while most mortality during autumn took place in Europe (Klaassen et al. 2014). For Mediterranean ospreys, mortality was usually directly or indirectly associated with human activities: illegal shooting, electrocution at a power lines, collision with wind turbines or trapped and drowned in a net in a fishpond (Monti et al. 2018d). Such events have an important impact on the population dynamics and their persistence over-time.

Beyond differences in migratory strategies, some genetic differences between populations exist at the Palearctic scale. In a recent study, aiming at quantifying the connectivity between osprey populations by means of polymorphic microsatellites specifically developed for the species, a high degree of genetic divergence has been estimated between populations of the Western Palearctic, with a special interest for the Mediterranean area (Monti et al. 2018a). Two sub-entities geographically featured were found in the Palearctic, with a first sub-group from north-east Europe (EURASIAN group) and a second sub-group in the Mediterranean area (MEDIT group). Although the two sub-groups were partially interconnected, the estimated gene flow appeared very limited and asymmetric (less than 4% and mainly directed from Mediterranean to Eurasia). Factors influencing such genetic diversity and structuration were probably the result of migratory and dispersal strategies on the one hand, and of the strong philopatric behaviour, on the other hand. In the same study, the role of sex in shaping genetic structure was also evaluated. Despite a strong philopatric behaviour, the osprey shows a certain degree of dispersal that allows populations to maintain genetic variability and admixture. As for other raptor species, dispersal in ospreys is sex-biased in favour of females that tend to cover greater distances than males, which are more philopatric (Martell et al. 2002, Monti et al. 2014). A Bayesian analyses performed with Structure indicated that more females (about 11%) than males (2%) have been identified as highly introgressed, and thus can be considered as migrants (Monti et al. 2018c). These results suggest that females contribute much to maintain a certain degree of admixture between populations. These findings are in line with ringing data and GPS tracking, showing females dispersing more than males.

3

Egyptian vulture *Neophron percnopterus*

Steffen Oppel,[1,*] *Evan R. Buechley,*[2,3] *Pascual López-López,*[4] *Louis Phipps,*[5] *Volen Arkumarev,*[6] *Anastasios Bounas,*[7,8] *Freddie Williams,*[1] *Vladimir Dobrev,*[6] *Dobromir Dobrev,*[6] *Stoycho Stoychev,*[6] *Elzbieta Kret,*[9] *Anna Cenerini,*[10] *Guido Ceccolini,*[10] *Victoria Saravia*[7] and *Stoyan C. Nikolov*[6]

ABSTRACT

The Egyptian vulture (*Neophron percnopterus*) is the only regular long-distance migrant amongst the Palaearctic vulture species. The species has a very broad distribution, with resident populations in Africa, Arabia, Southern Asia, and some islands, and migratory populations in Europe, the Middle East and Central Asia. Egyptian vultures generally migrate singly or in small groups and over a broad front, except where detours around major water bodies become necessary. The European population is divided into two major flyway populations and a remnant population in Italy that differ in the complexity and duration of their respective migrations. The western population, breeding in Southern France and the Iberian Peninsula, has a fairly direct migratory route across the Strait of Gibraltar to wintering areas in West Africa, a journey that usually takes 2–3 weeks and covers ~ 3,500 km. The eastern population, breeding in the Balkans and Turkey, migrates around the eastern coast of the Mediterranean towards the Red Sea, which is either bypassed near Suez in the north, or via the Bab-

[1] RSPB Centre for Conservation Science, Royal Society for the Protection of Birds, David Attenborough Building, Cambridge, CB2 3QZ, United Kingdom.

[2] HawkWatch International, 2240 South 900 East, Salt Lake City, UT 84106, USA.

[3] Department of Biology, University of Utah, 257 S. 1400 E., Salt Lake City, UT 84112, USA.

[4] Movement Ecology Lab. Cavanilles Institute of Biodiversity and Evolutionary Biology, University of Valencia, C/ Catedrático José Beltrán 2, E-46980, Paterna, Valencia, Spain.

[5] Vulture Conservation Foundation, Wuhrstrasse 12. CH-8003 Zurich, Switzerland.

[6] Bulgarian Society for Protection of Birds/BirdLife Bulgaria, Yavorov complex, bl. 71, vh. 4, PO box 50, 1111 Sofia, Bulgaria.

[7] Hellenic Ornithological Society, Themistokleous 80, GR-10681 Athens, Greece.

[8] Department of Biological Applications and Technologies, University of Ioannina, 45110, Ioannina, Greece.

[9] WWF Greece, 21 Lambessi, Gr 117 43 Athens, Greece.

[10] Association CERM Endangered Raptors Centre, Via Santa Cristina, 6, 58055 Rocchette di Fazio, Italy.

* Corresponding author: steffen.oppel@rspb.org.uk

el-Mandeb Strait, between Yemen and Djibouti, in the south. This migration route is more convoluted, usually taking 3–4 weeks and covering a distance of more than 5,000 km. This chapter presents an overview of the current knowledge, the timing of migration at major migration bottlenecks, and highlights that very little information exists about the migration strategies and routes of Egyptian vultures breeding in Central Asia.

Introduction

The Egyptian vulture (*Neophron percnopterus* Linnaeus, 1758) is a long-lived opportunistic scavenger and the only regular long-distance migrant amongst the Palaearctic vulture species. The species has a very broad distribution range, from the Iberian Peninsula in the Western Palaearctic through the Italian and Balkan peninsulas and Central Asia to Northern India in the Indo-Malayan region, and resident populations in many parts of Africa, the Middle East, and Southern Asia, as well as on Atlantic and some Mediterranean islands. The species has already disappeared from northern (i.e., Ukraine) and southern (i.e., South Africa) range limits, and many populations are declining worldwide (Ogada et al. 2015, Velevski et al. 2015, Botha et al. 2017).

Egyptian vultures typically breed in warm and arid areas, where they find food resources such as livestock or wild mammal carcasses, dead birds or reptiles, and a broad range of other animals, with remnants of > 100 taxa having been recorded during the breeding season (Hidalgo et al. 2005, Margalida et al. 2012, Sanchis Serra et al. 2014, Dobrev et al. 2016, Di Vittorio et al. 2017). While Egyptian vultures are very opportunistic in their diet choice, food availability decreases during the typically cold winter in the northern parts of the Egyptian vulture's range, i.e., most of Europe, Turkey, the Caucasus region and Central Asia. The populations breeding in those areas therefore migrate to arid or semi-arid regions in sub-Saharan Africa and Southern Asia (Newton 2010a, Phipps et al. 2019).

Egyptian vultures are soaring migrants that rely primarily on thermal and orographic updrafts to gain altitude, and then glide to cover distance for large portions of their migration. This form of movement is extremely energy efficient and allows large birds like vultures to travel much longer distances than birds of a similar size that use flapping flight (Alerstam et al. 2003). However, by relying on thermal updraft the birds generally cannot travel at night, and avoid crossing large bodies of water where thermals are limited (López-López et al. 2014, Agostini et al. 2015a, Meyburg et al. 2017a). Egyptian vultures therefore migrate diurnally with intermittent nocturnal roosts on trees, cliffs, power pylons, or other structures that offer safety from terrestrial predators. Their migration routes cover a broad front in continental areas, but converge at narrow land passages where larger water bodies such as the Mediterranean, Black, Red, or Caspian Seas form major barriers to direct travel (Buechley et al. 2018b). Birds originating from similar breeding areas can winter > 3,000 km apart in Africa, and birds from different breeding areas can winter in the same areas, leading to very diffuse migratory connectivity at the scale of sub-populations, but strong separation between sub-populations (Webster et al. 2002, Finch et al. 2017, Phipps et al. 2019).

In certain concentration areas in Africa, such as in Ethiopia (Arkumarev et al. 2014, Buechley et al. 2018a), or along flyway bottlenecks in the Middle East (Buechley et al. 2018b), the northern migrants may outnumber the resident Egyptian vulture populations. The high relative abundance of migratory birds in some parts of Africa is accentuated by the fact that young and immature birds may spend up to four years roaming around wintering areas in Africa before returning to breeding areas (Yosef and Alon 1997, Oppel et al. 2015).

The Egyptian vulture is a long-lived species with delayed maturity and a life span of about 20–30 years (Mebs and Schmid 2006). Birds start prospecting around breeding areas at the age of about three years, but may not acquire a territory and start breeding until they are five to seven years old (Grande et al. 2009, Lieury et al. 2015, Sanz-Aguilar et al. 2017). Despite many years of absence, and journeys that can cross up to three continents (Buechley et al. 2018b), most young Egyptian vultures ultimately return to breed in the same region where they were raised (Grande et al. 2009, Lieury et al. 2015). So far, only one long-distance natal dispersal event (550 km) has been documented for this species (Elorriaga et al. 2009), but ongoing satellite tracking studies of juvenile birds will likely provide new information on this poorly known aspect of the species' biology (Phipps et al. 2019).

The ecology of the species has mostly been studied in migratory populations in Europe, with a few exceptions from resident populations in the Balearic, Canary and Arabian islands, and the South-Western Iberian peninsula (e.g., Donázar et al. 2002, Agudo et al. 2011, Porter and Suleiman 2012, Angelov et al. 2013c, Sanz-Aguilar et al. 2015a, McGrady et al. 2018). The Egyptian vulture breeds once a year and typically raises one brood of 1–2 chicks, the only European vulture species that can raise two fledglings. The chicks are fed in the nest for 68–80 days before they fledge, and are then supported by their parents for another 2–4 weeks in the breeding territory (Donázar and Ceballos 1989). However, once migration starts, usually in mid to late August or early September, the parents migrate without their offspring and young Egyptian vultures therefore need to navigate independent of their parents on their first migration. As soaring birds, Egyptian vultures have difficulties crossing the Mediterranean Sea, with most experienced birds opting to navigate around either the eastern or western coast. However, inexperienced juveniles often attempt to cross the Mediterranean, resulting in many young birds dying *en route*, which is a typical pattern for many species and symptomatic for the cost of migration (Klaassen et al. 2014, Oppel et al. 2015, Sanz-Aguilar et al. 2015a).

Partly because of mortality during migration, many Egyptian vulture populations are decreasing and the species is globally threatened with extinction. Population declines are however not restricted to migratory populations, but also affect resident populations in Africa (Thiollay 2006b, Virani et al. 2011, Ogada et al. 2015, Di Vittorio et al. 2018) and India (Cuthbert et al. 2006, Galligan et al. 2014). In Europe, large populations on the Iberian Peninsula appear to be relatively stable or locally increasing (Mateo-Tomás et al. 2010, Zuberogoitia et al. 2014, Lieury et al. 2015), while the small remnant populations in Italy (Liberatori and Penteriani 2001, Sarà and Di Vittorio 2003), the Balkans (Velevski et al. 2015), and in Southern Spain (Carrete et al. 2009, Sanz-Aguilar et al. 2015b) are still declining. For other large populations, for example in Turkey (Balaban and Yamac 2018, Katzenberger et al. 2019), or in Central Asia (Abuladze and Shergalin 1998), only very limited data exist on population size and no information on trends (Abuladze 2013, Karyakin et al. 2018a).

Egyptian vulture population declines are caused by a range of different threats, including most prominently poisoning, followed by direct persecution, electrocution and other accidental mortality caused by human infrastructure, and loss of habitat and food resources through changes in farming systems (Cuthbert et al. 2006, Mateo-Tomás and Olea 2010, Angelov et al. 2013b, Ogada et al. 2015, Sanz-Aguilar et al. 2015b, Buechley and Şekercioğlu 2016, Saravia et al. 2016, Oppel et al. 2017). As a consequence of population declines, traditional migration systems where juveniles reduce the difficulty of navigation by following experienced birds may no longer function as well as when populations were larger, although this effect has yet to be demonstrated. Young birds of several raptor species follow experienced migrants to navigate along their first migratory journey (Mellone et al. 2011a, Panuccio et al. 2012, Mueller et al. 2013, Agostini et al. 2017c, Meyburg et al. 2017a). However, juvenile birds may struggle to find and associate with experienced migrants if the number of experienced migrants has become too small. Under such circumstances juvenile birds may have to rely mostly on innate compass navigation systems to reach their wintering grounds. Inexperienced juvenile migrants in populations that require a fairly convoluted circumnavigation of water bodies, such as the Western Balkan population that migrates around the Eastern Mediterranean, may therefore face much more profound challenges than inexperienced birds on other flyways with fewer or easier water barriers (Oppel et al. 2015, Meyburg et al. 2017a, Phipps et al. 2019). In this chapter we summarise the migration routes and seasons of Western Palaearctic Egyptian vulture populations to provide the background for future studies to support the conservation of this charismatic species.

Knowledge Status

Knowledge about the migration of Egyptian vultures is deeply rooted in traditional folklore and wisdom of ancient human populations. For example, in Albania and parts of Greece the species is commonly referred to as the 'Cuckoo's Horse', because the return of the Eurasian cuckoo, *Cuculus canorus*, in spring typically coincided with the return of the much larger and more powerful Egyptian vulture, which were traditionally believed to have carried the much smaller cuckoo across the sea and the desert (Lukopulos 1940, Stara et al. 2016).

In recent times, more detailed knowledge about the migration of Egyptian vultures has been collected through regular observations at migration observation points (Porter and Willis 1968, Finlayson 1992, Leshem and Yom-Tov 1996, Shirihai et al. 2000, Martín et al. 2016), and through satellite tracking data (e.g., Meyburg et al. 2004, García-Ripollés et al. 2010, López-López et al. 2014, Oppel et al. 2015, Buechley et al. 2018b, Karyakin et al. 2018, Phipps et al. 2019). Both approaches offer a number of advantages and disadvantages, but together provide complementary information on the phenology and abundance of Egyptian vultures as well as their migration routes and speed from departure to arrival at wintering or breeding grounds.

Data collected at fixed observation points—usually at migration bottlenecks or on islands—provide a time series of the number of birds passing over a narrowly defined observation point. These observations are generally not targeted at a single species and record all migratory birds. Given that vultures comprise generally < 1% of the total number of soaring migrants (Finlayson 1992), these observations may overlook rarer species (such as Egyptian vultures) in large flocks of more numerous species (e.g., white stork *Ciconia ciconia*, European honey buzzard *Pernis apivorus*, black kite *Milvus migrans* or lesser spotted eagle *Clanga pomarina*). However, the main benefit of observation point time series is that they provide data that allows us to delineate the phenology of the species at a given geographic location by recording the first and the last migrants and the days with peak migration intensity. In addition, the data can also be used to compare the importance of flyways in terms of population size, and assess population trends over time (Martín et al. 2016).

Migration observation points where data on Egyptian vulture migration have been collected consistently over at least one migration period (Table 1) are located in Gibraltar (Spain) (Finlayson 1992, Bensusan et al. 2007, Martín et al. 2016, Panuccio et al. 2016b), Sicily (Corso 2001a, Agostini et al. 2000, 2004c, Agostini 2005), Antikythira (Greece) (Lucia et al. 2011), the Bulgarian Black Sea coast (Michev et al. 2011), the Bosphorus (Collmait and Croxall 1967, Porter and Willis 1968, Porter and Beaman 1985, Fülöp et al. 2014), the Caucasus (Verhelst et al. 2011, Heiss 2013), Southern Turkey (Sutherland and Brooks 1981, Can 2001, Oppel et al. 2014), Israel (Safriel 1968, Frumkin et al. 1995, Leshem and Yom-Tov 1996, Shirihai et al. 2000, Alon et al. 2004), Egypt (Hilgerloh 2009, Hilgerloh et al. 2011a, Megalli and Hilgerloh 2013), and Djibouti (Welch and Welch 1988, 1998). Some of these observatories have accumulated long time series over many years or decades, whereas others have only provided data for one season. However, from the existing records the phenology of Egyptian vultures can be summarised quite reliably for both the Western and Eastern Mediterranean Flyway (Table 1). By contrast, very little information exists on the migration between central Asian breeding grounds and wintering areas on the Arabian Peninsula, Africa, or Southern Asia, with relatively few Egyptian vultures migrating through the Georgian Caucasus region (Verhelst et al. 2011, Abuladze 2013, Karyakin et al. 2018). The Bab el-Mandeb Strait is a bottleneck for birds moving between the Arabian Peninsula and Africa (Welch and Welch 1988) that funnels an as yet unknown portion of migrants originating from Eastern European, Middle-Eastern and Central Asian breeding grounds (Angelov et al. 2013a, Rayaleh et al. 2013, Buechley et al. 2018b, Karyakin et al. 2018). However, very little information exists on the migration routes and phenology of birds breeding in Kazakhstan, Uzbekistan, Turkmenistan, Kyrgyzstan and Tajikistan (Abuladze and Shergalin 1998, Meyburg et al. 2003). Regular monitoring of soaring migrants across the Bab-el-Mandeb Strait would be a high priority to monitor population trends of Egyptian vultures across a large, but poorly known, breeding region.

Table 1. Overview of the timing of spring and autumn migration of Egyptian vultures from ground observations at selected migration bottlenecks along the Western and Eastern Mediterranean flyways. Note that the date 'range' is dependent on observation effort and may not encompass the entire temporal window of migration. For more detailed data visit www.trektellen.nl.

Country	Location	Spring migration		Autumn migration		Source
		peak	range	peak	range	
Western Mediterranean						
Italy	Sicily	27-Apr	26-Mar–28-May	02-Sep	24-Aug–12-Sep	(Corso 2001a, Agostini et al. 2004c, Agostini 2005)
Spain	Gibraltar	08-Mar	20-Feb–09-May	8-Sep	25-Jul–15-Oct	(Finlayson 1992, de la Cruz et al. 2011b, Onrubia 2015, Martín et al. 2016, Panuccio et al. 2016b)
Eastern Mediterranean						
Georgia	Batumi	*no data*		13-Sep	30-Aug–26-Sept	(Verhelst et al. 2011, Abuladze 2013)
Greece	Porto Lagos	26-Mar	15-Mar–31-May	*no data*		(de Nobel et al. 1990)
Greece	Antikythira	no migration		05-Sep	21-Aug–20-Sep	(Lucia et al. 2011)
Turkey	Bosphorus/Dardanelles	22-Apr	23-Mar–23-May	10-Sep	22-Aug–6-Oct	(Nisbet and Smout 1957, Porter and Willis 1968, Üner et al. 2010, Fülöp et al. 2014, Panuccio et al. 2017)
Turkey	Gulf of Iskenderun	*no data*		15-Sep	20-Aug–6-Oct	(Cameron et al. 1967, Sutherland and Brooks 1981, Can 2001, Oppel et al. 2014)
Israel	Northern Valleys	01-Apr	20-Feb–31-May	20-Sep	24-Aug–8-Oct	(Frumkin et al. 1995, Leshem and Yom-Tov 1996, Shirihai et al. 2000, Alon et al. 2004)
Israel	Eilat	02-Apr	2-Feb–31-May	12-Oct	25-Aug–30-Nov	(Shirihai and Christie 1992, Shirihai et al. 2000)
Egypt	Gulf of Suez/Zait Bay	30-Mar	1-Mar–2-May	02-Oct	4-Sep–31-Oct	(Bijlsma 1983, Hilgerloh et al. 2009, Hilgerloh et al. 2011a, Megalli and Hilgerloh 2013)
Yemen	Tihama Foothills	*no data*		07-Nov	28-Oct–18-Nov	(Porter and Christensen 1987)
Djibouti	Bab-el-Mandeb	March		18-Oct	26-Sep–9-Nov	(Welch and Welch 1988, 1998)

By contrast, technological advances in the last 30 years have enabled tracking studies to provide highly accurate information about the route and speed of migration of individuals (Meyburg et al. 2004a). Migratory Egyptian vultures have been tracked predominantly from Spain (García-Ripollés et al. 2010, López-López et al. 2014), Bulgaria and Greece (Oppel et al. 2015, Buechley et al. 2018b), Turkey, Armenia, and Djibouti (Buechley et al. 2018a, Buechley et al. 2018b), with few deployments in Albania, Ethiopia, North Macedonia, Israel, Iran, Jordan, France, Portugal, Italy, and Russia until 2018 (Oppel et al. 2015, Karyakin et al. 2018, Phipps et al. 2019). Given the generally small sample sizes, most studies are not fully representative of the variation in timing and route choice of the entire population. Nonetheless, tracking data provide highly detailed information on the entire annual cycle of individuals and are therefore an excellent addition to the more traditional time series data at fixed observation points (Table 2).

Post-breeding Migration

The autumn migration of Egyptian vultures from European breeding grounds starts in mid to late August (García-Ripollés et al. 2010, López-López et al. 2014), with juvenile birds on average migrating slightly later than adults (Meyburg et al. 2004a, Oppel et al. 2015, Buechley et al. 2018b), although breeding adults are usually the last birds to leave their breeding territories.

On their way south to Sub-Saharan Africa, the European breeding birds face the obstacle of the Mediterranean Sea, which they generally avoid by using the narrowest water crossings at the Strait of Gibraltar, Sicily, or the Bosphorus and Dardanelles (Fig. 1). The migratory route is most straightforward for birds breeding on the Iberian Peninsula and Southern France (Phipps et al. 2019). The majority of these birds depart from their breeding grounds in late August to early September in a southerly direction, cross the Strait of Gibraltar in mid September (average counts during 2009–2012: 2,482 individuals; Onrubia 2016) and arrive within two weeks at their wintering grounds in Southern Mauritania, Mali, Senegal, Niger or Burkina Faso (Meyburg et al. 2004a, García-Ripollés et al. 2010, López-López et al. 2014). Egyptian vultures appear to be able to cross the Atlas mountains in Morocco without major difficulties and then head straight across the western regions of the Sahara desert to their wintering destination (García-Ripollés et al. 2010, López-López et al. 2014). This journey is usually around 3,000–3,500 km long, but can be covered in almost a week less than the corresponding spring migration (Mellone et al. 2012a, López-López et al. 2014) (Table 2). However, accounting for the unfavourable winds across the Sahara in spring (Vansteelant et al. 2017c), birds may actually travel with greater urgency in spring than in autumn (Mellone et al. 2012a, López-López et al. 2014, Phipps et al. 2019). Captive-bred juveniles released in Italy migrated either via Cap Bon in Tunisia or via Libya (Fig. 1) to wintering areas in Mauritania, Mali and Niger.

By contrast, the migration route for birds breeding in Eastern Europe is more complex because they first have to negotiate the Mediterranean and then the Red Sea, which renders their journey more hazardous and longer in both time and distance (Fig. 1, Table 2). Adult birds tracked with satellite transmitters from the Balkan Peninsula, Turkey, and Armenia travelled on average 23 days over a distance of 4800 km (Oppel et al. 2015, Buechley et al. 2018b), with some important differences compared to their spring migration (Phipps et al. 2019) (Table 2). For Balkan birds, the most critical part of the journey is the Aegean part of the Mediterranean Sea, which requires birds from Western Greece, Albania, and North Macedonia to depart into a counter-intuitive easterly or even north-easterly direction. Only very few Egyptian vultures migrate through the Peloponnese and the Greek islands, although single, mostly juvenile, birds are observed every year at Antikythira (Lucia et al. 2011) and Crete. Tracked juveniles that used this route originated more frequently from western breeding locations in the Balkans, and suffered high mortality when crossing the 300 km of open water between Crete and Africa (Oppel et al. 2015). Once the birds have entered the Asian continent via either the Bosphorus or the Dardanelles, they travel through Turkey on a broad front into a south-easterly direction until they hit the coast of the Mediterranean (Fig. 1). Birds then follow the coast around an important bottleneck at the Gulf of Iskenderun (> 800 individuals counted in 2018

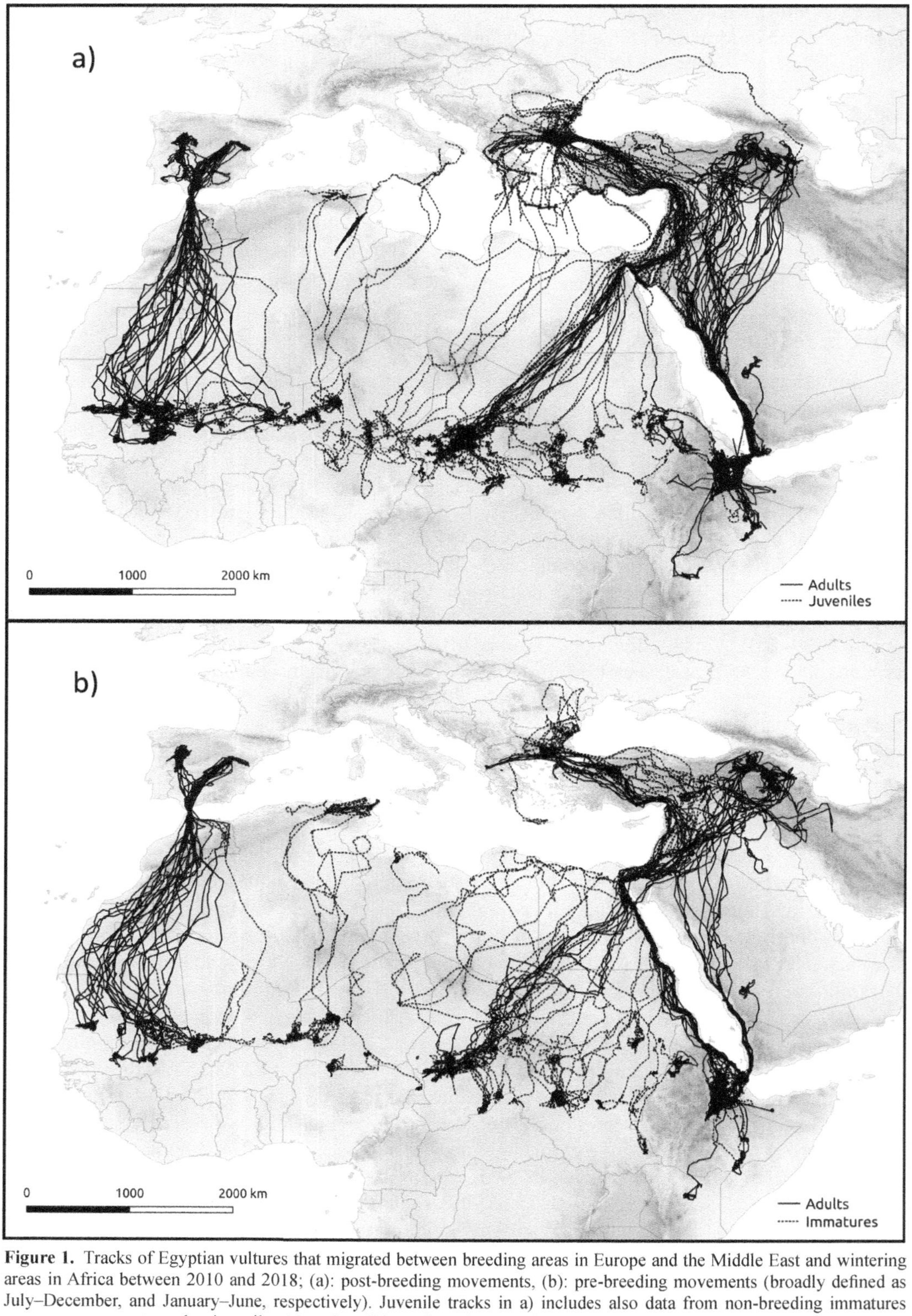

Figure 1. Tracks of Egyptian vultures that migrated between breeding areas in Europe and the Middle East and wintering areas in Africa between 2010 and 2018; (a): post-breeding movements, (b): pre-breeding movements (broadly defined as July–December, and January–June, respectively). Juvenile tracks in a) includes also data from non-breeding immatures tagged as juveniles. Note that all juveniles migrating from Italy were of captive origin.

Table 2. Overview of spring and autumn migration distances and speeds of adult Egyptian vultures along the Western and Eastern Mediterranean flyways based on satellite tracking data. The Eastern Mediterranean data are from birds equipped in Bulgaria, Greece, Turkey and Armenia (Buechley et al. 2018b), the Western Mediterranean data are from Spain (López-López et al. 2014).

Migration parameter	Spring migration		Autumn migration	
	mean	range	mean	range
Western Mediterranean (n = 6)				
Travel distance (km)	3,580	3,119–4,400	3,256	2,803–3,958
Travel duration (days)	19	13–26	13	9–23
Speed (km/day)	190	131–275	256	184–352
Straightness	0.8	0.70–0.89	0.85	0.71–0.95
Eastern Mediterranean (n = 12)				
Travel distance (km)	5,304	3,797–7,395	4,843	3,302–6,409
Travel duration (days)	25	15–47	23	12–50
Speed (km/day)	223	157–350	231	115–361
Straightness	0.67	0.42–0.85	0.7	0.45–0.91

and 2019; Oppel et al. 2014, Buechley et al. 2018b), from where birds cross the Amanos mountain range in a few different locations (Cameron et al. 1967, Sutherland and Brooks 1981). The further migration south through the Middle East and the Arabian Peninsula occurs again on a broad front, where birds from Eastern Europe are joined by birds from Eastern Turkey, the Caucasus region, and presumably also from Central Asia (Karyakin et al. 2018). Unlike during spring migration, when most birds follow the western coast of the Red Sea through Egypt, during autumn most birds follow the eastern coast of the Red Sea in Southern Saudi Arabia to Yemen (Buechley et al. 2018b), from where they cross via the Bab-el-Mandeb Strait into Africa (554 counted in 1987; Welch and Welch 1988). Some birds, especially those that winter farther west in Africa (Niger, Nigeria, Chad or Sudan), follow a similar route as in spring through Northern Egypt (Hilgerloh 2009, Hilgerloh et al. 2011a) before crossing the Sahara in a broad front along tracks that lead directly to their wintering areas. Overall, the migration route of the eastern birds is substantially less straight (ratio of travel distance over linear distance = 0.5–0.7) compared to birds tracked from the Iberian Peninsula (straightness = 0.7–0.9) (Phipps et al. 2019).

Compared to spring migration, the timing of autumn migration is slightly more concentrated with a distinct peak, which nonetheless spans a time window of 2–4 weeks at most ground-based observatories along major flyways. Along the Western Mediterranean flyway, migration begins in July and peaks in late August to early September in Gibraltar, where up to 3,800 Egyptian vultures can be observed in a given year (Finlayson 1992, de la Cruz et al. 2011a, Onrubia 2016, Martín et al. 2016). The individuals migrating via the Mid-Mediterranean flyway (74 in 2003) are also observed between the end of August and early September (Agostini et al. 2000, 2004c). Along the eastern flyway, Egyptian vultures start passing the Turkish Straits (Porter and Willis 1968, Fülöp et al. 2014), Greek Islands (Lucia et al. 2011), and the Caucasus (Verhelst et al. 2011) from late August, with a peak in early September. The birds migrate 100–300 km per day (Table 2), so that the peak of migration gradually shifts towards mid to late September in Southern Turkey (Cameron et al. 1967, Sutherland and Brooks 1981, Can 2001, Oppel et al. 2014) and Northern Israel (Frumkin et al. 1995, Leshem and Yom-Tov 1996a, Alon et al. 2004). In Southern Israel and Egypt, the majority of birds pass in late September and early October (Table 1), but at Eilat the temporal window appears to be extended by birds joining from Central Asian or other Middle Eastern breeding areas (Shirihai and Christie 1992, Shirihai et al. 2000).

Egyptian vultures arrive in their African wintering grounds between mid-September to mid-October, with a few immature or juvenile birds arriving until late November (Phipps et al. 2019).

Especially young and immature birds may perform substantial movements while in Africa (Meyburg et al. 2004a, Buechley et al. 2018a), and may not settle at a single destination. Similar to other species, immature Egyptian vultures may use the winter for a period of vagrancy during which they learn about multiple sites which they may rely on as key wintering sites in future years (Oppel et al. 2009, Trierweiler et al. 2013).

Pre-breeding Migration

Spring migration of adult Egyptian vultures from their African wintering grounds starts from late January to late February in West Africa (García-Ripollés et al. 2010, López-López et al. 2014), but in early to mid-March in East Africa. This difference in the onset of migration is surprising because the birds wintering in East Africa have a much longer spring journey back to their breeding grounds and might be expected to depart earlier. Birds breeding on the Iberian Peninsula arrive at their breeding grounds in late February and early March, following a journey of ~ 3,500 km along a similar direct route as their autumn migration (García-Ripollés et al. 2010, López-López et al. 2014). Meanwhile, birds breeding in the Balkans and the Middle East arrive at their breeding grounds in late March to mid April after travelling 3,800–7,400 km from Central and East Africa, along routes that in some cases differ substantially from autumn migrations (Phipps et al. 2019) (Tables 1,2).

For birds wintering in West Africa the only major water barrier on their way to breeding grounds on the Iberian Peninsula or Southern France is the Strait of Gibraltar, which is a relatively narrow (14.4 km at the narrowest point) barrier and does not require a major detour along the migration route north. The diffuse spread of migration paths by different individuals across the Sahara in West Africa generally converges towards a narrow corridor south of the Atlas mountains in Southern Morocco (García-Ripollés et al. 2010, López-López et al. 2014). Spring migration tracks are usually slightly further west than autumn migration tracks of the same birds (López-López et al. 2014) as a result of seasonally varying winds over the Sahara (Klaassen et al. 2010, Mellone et al. 2012a, Vidal-Mateo et al. 2016, Vansteelant et al. 2017c). In addition, some individuals may migrate around the Atlas Mountains to the north-east, and then follow the coastline of the Alboran Sea westward to cross at the Strait of Gibraltar. Once birds have reached the Iberian Peninsula around Tarifa or Gibraltar, further progress north is in a broad front to the ultimate destination, as there are no other major barriers to migration. Birds breeding in Italy likely depart from Cap Bon in Tunisia to cross 150 km of sea to reach Sicily.

Conversely, the birds wintering in East Africa face major barriers at the Red Sea, the Mediterranean Sea, and the Caspian and Black Seas farther north, depending on the breeding grounds to which they migrate. Egyptian vultures using the Eastern Mediterranean—Red Sea flyway winter as far west as Southern Niger and Northern Nigeria (Fig. 1; Oppel et al. 2015, Buechley et al. 2018b), and begin their spring migration generally by flying north-east. Some individuals wintering in Djibouti, Eritrea, Ethiopia, and Kenya cross the Red Sea at the Bab-el-Mandeb Straits to reach the Arabian Peninsula (Welch and Welch 1998, Rayaleh et al. 2013, Buechley et al. 2018b), where further progress north initially follows the Red Sea coastal plain, before eventually dispersing into a broad front across the Arabian desert, Jordan, Iraq, and Syria. Other birds wintering in East Africa fly north-east until they reach the Red Sea, and then follow the western coastline of the Red Sea northwards until they reach Suez in Egypt (Hilgerloh et al. 2009, Megalli and Hilgerloh 2013, Buechley et al. 2018b). Once these birds have passed the barrier of the Red Sea, the next barrier is the Mediterranean Sea, which requires a slight detour eastward. In North-Eastern Egypt, Jordan and Southern Israel, the flyways begin to separate depending on whether birds head to breeding areas in Central Asia, the Middle East, or Western Turkey and the Balkans. Egyptian vultures from the Balkans, and presumably also from western Turkish breeding grounds, generally follow the eastern coastline of the Mediterranean north through Jordan, Israel, Lebanon and Syria to the Gulf of Iskenderun, from where they turn north-westward towards their breeding grounds. Although Istanbul is generally recognised as a very important migration bottleneck for large soaring birds (Collmait and Croxall 1967, Porter and Beaman 1985), most Egyptian vultures using this flyway reach the European continent via the Dardanelles at the south-western margin of the Marmara Sea. Egyptian vultures breeding in Western Greece or Albania generally migrate around the northern coast of the Aegean Sea (de Nobel et al. 1990) and have among the longest and most convoluted migratory journey of Egyptian vultures tracked so far (Phipps et al. 2019).

Birds breeding in the Middle East and in the Caucasus region take a more direct route from Northern Egypt, migrating in a direct line towards their breeding grounds (Buechley et al. 2018b). No Egyptian vultures have been tracked from Central Asian breeding grounds, but their migration might be similar to Steppe Eagles (*Aquila nipalensis*) and lead through Iraq and Iran to navigate around the southern coast of the Caspian Sea (Meyburg et al. 2003). Some birds use the Strait of Hormuz to cross the Persian Gulf from Arabia into Iran (McGrady pers. comm.).

The general timing of spring movements is also apparent from the observation point stations from which spring migration data are available. In East Africa, the first northward spring migrants pass through Egypt from late February until early May (Hilgerloh et al. 2009), with a peak in the second half of March. In the spring of 2012 a total of 1,128 Egyptian vultures were recorded in March and April at the northern tip of the Gulf of Suez (Megalli and Hilgerloh 2013). Most of these birds then migrate through the Judean desert or past the well-known bottleneck at Eilat in Southern Israel (Frumkin et al. 1995), where three migration waves have been identified in mid-March, early April, and early May (Shirihai et al. 2000). However, most adults pass before mid-April, and the late 'migrants' observed in April in Egypt and May in Israel may be immature birds that either only migrate part of the way, or even roam around Africa and the Middle East (Yosef and Alon 1997, Shirihai et al. 2000, Buechley et al. 2018b). However, there may be large numbers of Egyptian vultures migrating via the Bab-el-Mandeb Strait and across a broad front towards the Middle East and Central Asia that are not recorded along the Red Sea or in Israel. For example, in 1998, > 700 Egyptian vultures were observed during just three days in both Djibouti (5–7 March) and Yemen (22–25 March), and 1,167 Egyptian vultures were observed over 9 days (2–10 March) in 2013 (Welch and Welch 1998, Rayaleh et al. 2013). The timing of these large movements is too late for birds breeding in Eastern Europe, which frequently arrive on breeding grounds in late March.

Relatively few systematic observations are available from migration observation points north of Israel (Shirihai et al. 2000). At the Turkish straits (Bosphorus and Dardanelles) there are now too few Egyptian vultures to determine a migration peak between late March and late May (Collmait and Croxall 1967, Üner et al. 2010), but in the late 1980s the peak migration at the northern coast of the Aegean Sea was around 26 March (de Nobel 1990). These dates match with the typical arrival of Egyptian vultures in breeding territories across the Balkans during the last week of March or early April (Oppel et al. 2017, Buechley et al. 2018b). While none of the migration bottlenecks along the Eastern Mediterranean flyway have recorded large concentrations of Egyptian vultures in spring, the number of birds using this flyway might be similar to the number using the Western Mediterranean flyway (Finlayson 1992).

In the Western Mediterranean, spring migration observations exist only from Gibraltar, with some records from Sicily. At Gibraltar, the spring migration starts in February and peaks in March and early April, with a continuous passage through May and some late migrants occurring as late as July (Finlayson 1992, Garcia and Bensusan 2006, Bensusan et al. 2007, Onrubia 2016). Here, during 2009–2012, an average of 1,282 migrating individuals has been estimated during spring (Onrubia 2015). The spring migration at Gibraltar is very protracted and migrants can also be observed outside the main migratory periods, but from mid-April onwards an increasing proportion of migrants are immature birds or non-breeders (Finlayson 1992), which is similar to the pattern observed at Suez and Eilat on the Eastern Mediterranean flyway (Shirihai et al. 2000, Hilgerloh et al. 2009). Owing to the relatively long crossing between Cap Bon and Sicily, and the very small Italian population of Egyptian vultures, there are only very few birds that migrate via Sicily (Corso 2001a, Agostini 2005).

Behavioural and Ecological Remarks

During the pre-migratory period, both juveniles and adult birds often gather at communal congregation sites. Although most data regarding Egyptian vulture communal roosts come from populations in Western Europe (Ceballos and Donázar 1990, Margalida and Boudet 2003), pre-migratory movements likely also occur elsewhere along the species' range, but are very poorly known. For example, an adult bird breeding in Northern Greece was observed at a feeding station in Kresna Gorge, Bulgaria, in mid-August 2014, indicating that it moved approximately 100 km north from its breeding territory prior to southbound migration (Bounas et al. 2015).

Like many other soaring migratory raptors, the Egyptian vulture exhibits a rapid diurnal migration without extended stopover periods, but requires safe roosting sites every night. The migration is usually solitary or in small groups and follows the main migration routes of many soaring migrants around the Mediterranean and Red Seas, likely because Egyptian vultures have a very similar flight performance as other soaring migrants (Spaar 1997). As in other species, migratory performance likely develops gradually over time with juvenile and immature birds often taking suboptimal or more circuitous routes than adults (López-López et al. 2014, Sergio et al. 2014, Vansteelant et al. 2017a). There appears to be great phenotypic plasticity both at the individual level in different years (López-López et al. 2014), and in the degree and extent to which different populations migrate (Phipps et al. 2019). This flexibility is also reflected in the variability among immature birds to return to breeding grounds, and may also allow the rapid adoption of new wintering areas once they become climatically suitable: during the last decade, Egyptian vultures have been observed in Southern Spain, Southern Turkey, Crete and Sicily during the winter (Di Vittorio et al. 2016a). Because Egyptian vultures can overwinter on other Mediterranean islands (Sanz-Aguilar et al. 2015a), survival of these wintering individuals may ultimately lead to a regular wintering or even resident population on those islands (Kretzmann et al. 2003, Agudo et al. 2011), or facilitate short-stopping of formerly long-distance migrants in Southern Spain. Such behavioural flexibility in migration may also affect the temporal adjustment of migration timing in response to climate change, but changes in the timing of migration will be difficult to assess unless more long-term monitoring data become available (Panuccio et al. 2016).

Juvenile Egyptian vultures remain in wintering regions for at least 1.5 years and generally do not attempt spring migration in the year after their first arrival in Africa (Oppel et al. 2015, Buechley et al. 2018b). However, for some birds the first spring migration and subsequent roaming started in May of the bird's second year but the first spring migration was highly heterogenous and differed markedly among individuals (Bougain 2016). Some birds remained largely stationary for > 3 years in Africa, while other individuals migrated only half-way within Africa, thus crossing the Sahara twice within a month, and others migrated farther north but did not reach their natal breeding sites (Phipps et al. 2019).

For adult breeding birds, the spring migration typically ends when they arrive at their breeding grounds, but for immature birds without a fixed breeding area and breeding responsibilities, the spring movements can continue throughout the summer and transition into autumn movements. For example, one Egyptian vulture marked as a fledgling in Bulgaria in 2012 returned to Turkey for the first time in 2014 at the age of two years and performed an extensive journey across 12 countries that are known to host breeding Egyptian vultures or used to have breeding populations in the past. The immature bird first travelled north-westward through Turkey, upon reaching the barrier of the Aegean Sea it turned back and travelled east along the coast of the Black Sea to central Turkey where it headed south and visited several known breeding areas in Turkey (Balaban and Yamac 2018, Katzenberger et al. 2019). After three weeks in Turkey, the bird continued to Greece and Bulgaria in June and flew north to Romania and Moldova before returning to Bulgaria. While temporarily residing in Bulgaria throughout July, the bird undertook several forays to Serbia and North Macedonia, eventually returning to its place of origin in North-Western Bulgaria in late July. In the beginning of August, the bird flew north-east through Romania, traversed Southern Ukraine in seven days, continued southwards through Russia for about 10 days, until crossing the Caucasus on 18 August into Azerbaijan. After a few brief stops in Iran, the bird arrived at a roost site 50 km south of Kirkuk (Iraq) on 29 August, where it remained for two weeks before continuing its autumn movements. Overall, the bird travelled 20,481 km in 4.5 months and returned via Saudi Arabia to its wintering areas in Yemen, Djibouti and Ethiopia at the end of September. Another bird marked as a juvenile in Bulgaria in 2012 explored and extensively used wintering areas in Nigeria, Niger, Northern Cameroon, Chad, Sudan, South Sudan, and Ethiopia over six years before returning to occupy a territory in Greece. This individual wintered in all parts of the Balkan's population wintering area at least once during its immature life stage and therefore demonstrates the diffuse migratory connectivity at the sub-population level in this species (Phipps et al. 2019).

Documenting these movements for more individuals between their first migration and recruitment to a breeding population will provide useful information in future years about the patterns and intensity of natal philopatry (Weatherhead and Forbes 1994). From the limited data available so far, immature birds appear to prospect around many potential breeding areas that can be > 1000 km away from their natal site of origin. While dispersal events of up to 550 km have been recorded using ringed birds (Elorriaga et al. 2009), satellite telemetry may reveal even longer natal dispersal events among adjacent populations in the Balkans, Middle East and Central Asia in the future. Such dispersal events may provide sufficient gene flow to limit the genetic isolation of the different populations. Species like the Egyptian vulture with large breeding ranges must maintain connectivity between geographically disparate populations to prevent genetic or behavioural divergence and subsequent speciation (Kretzmann et al. 2003, Méndez et al. 2015). Currently, the extent of genetic similarity between populations that use different migration flyways is unknown. For many large raptor species, dispersal generally occurs during the immature life stages (González et al. 2006, Cadahía et al. 2010, Newton 2010a). The exploratory migrations of young and immature Egyptian vultures prior to breeding, combined with the diffuse migratory connectivity within sub-populations (Phipps et al. 2019), are the most likely mechanisms that allow the species to maintain gene flow throughout European and Asian breeding ranges. Whether this dispersal of individuals will also be sufficient to rescue certain populations that are close to extinction (e.g., Di Vittorio et al. 2016b) will require more information about the rate of natal philopatry and abmigration (Thomson 1931). Interestingly, the French Egyptian vulture population has recently recovered as a consequence of immigration (Lieury et al. 2015), and if source populations produce sufficient birds that survive the first migrations and return to breed, then the recovery of some populations may be facilitated by the large distances that Egyptian vultures can disperse.

Acknowledgements

We thank our volunteers and collaborators including Kuzey Doğa Society (Turkey), Iğdır Directorate of Nature Conservation and National Parks (Turkey), American University of Armenia, Ethiopia Wildlife Conservation Authority, Ethiopia Wildlife and Natural History Society, International Avian Research, and Djibouti Nature, and our colleagues who assisted with Egyptian vulture trapping, including Emrah Çoban, Lale Aktay, Kayahan Ağırkaya, Berkan Demir, Mete Türkoğlu (Turkey); Karen Aghababyan, Anush Khachatrian, Garo Kurginyan (Armenia); Sisay Seyfu, Alazar Daka Ruffo, Yilma Dellelegn Abebe, Girma Ayalew (Ethiopia); Mike McGrady and Houssein Rayaleh (Djibouti); Svetoslav Spasov, Ivaylo Angelov, Saniye Mumun, Tsvetomira Angelovan, Ewan and Jenny Weston, Emil Yordanov (Bulgaria); Vasilis Sideris, Giannis Chondros, Christos Lambris, Antonis Vroikos, Lavrentis Sidiropoulos, Dimitris Vavylis, Angelos Evangelidis (Greece), Mirjan Topi (Albania), Metodija Velevski and Zlatko Angeleski (North Macedonia), and L. Bolonio, J. De Lucas, V. García, R. Ibáñez, M. Nieto, Á. Vela, C. García-Ripollés and V. Urios (Spain). This work was logistically and financially supported by the LIFE+ projects "The Return of the Neophron" (LIFE10 NAT/BG/000152), "Egyptian vulture New LIFE" (LIFE16 NAT/BG/000874), and "Conservation of Egyptian vulture and Bonelli's Eagle in the Douro Valley" (LIFE14 NAT/PT/000855) funded by the European Union and co-funded by the AG Leventis Foundation and MAVA Foundation, Sociedade Portuguesa para o Estudo da Aves (SPEA); Associação Transumância e Natureza (ATN); Associação de Conservação da Natureza e do Património Rural (Palombar); Guarda Nacional Republicana, Portugal (GNR); Fundación Patrimonio Natural de Castilla y León (FPN); EDP Distribuição – Energia SA (EDP-D); Instituto da Conservação da Natureza e Florestas, Portugal (ICNF); Junta de Castilla y León, Spain (JCyL); the vulture Conservation Foundation; the US National Science Foundation, the Christensen Fund, SALORO S.L.U. (Javier Garcia Fernandez, Jose Jambas e Isidoro Carbonell Alanis), National Geographic Society, the Whitley Fund for Nature, Faruk Yalçın Zoo and KuzeyDoğa's donors (in particular Bilge Bahar, Devrim Celal, Seha İşmen, Lin Lougheed, Burak Över and Batubay Özkan). We are grateful to Turkey's Ministry of Forestry and Water Affairs General Directorate of Nature Conservation and National Parks and NorthStar Science and Technology for donating three transmitters each.

4

European honey buzzard *Pernis apivorus*

Wouter Vansteelant[1,2,*] and *Nicolantonio Agostini*[3]

ABSTRACT

The European honey buzzard (*Pernis apivorus*) is a medium-sized raptor, summer visitor to Europe and Asia, until Kazakhstan, with wintering grounds located in Sub-Saharan Africa. Having an intermediate morphology between obligate-soaring species and those largely using flapping flight, honey buzzards migrate mostly using thermals and deflection updrafts over land but are also capable to cross large water bodies using long non-stop flapping flights. As a result, this is the commonest species observed at nearly all Western Palearctic watchsites, both at the major bottlenecks, where obligate-soaring migrants converge (e.g. Strait of Gibraltar, Batumi, Eilat), and along minor corridors involving several Mediterranean islands. Several study techniques, ranging from visual and radar observations to ringing and biologging, have been used to study flight parameters and migration pathways sometimes at a fine scale, revealing age-dependent migration behaviours such as differences in pathways used among seasons and shaped by wind regimes, geography and related to time and/ or risk minimization strategies. As regards the age-dependent migration behaviour, during their first migration juveniles migrate about two weeks later than adults and generally cannot learn the shortest overland route to cross the Mediterranean Sea by social learning. As a result, they migrate on a broad front, largely drifted by prevailing winds and undertaking the water crossings as soon as they hit the Mediterranean coast. Satellite telemetry devices deployed on adult individuals belonging to multiple populations have shown an extreme flexible migration behaviour between and within populations, with some using the same pathways during both spring and autumn and others performing large scale loop migrations. They exploit the tail support of prevailing winds but sometimes choose to fly in suboptimal weather conditions to follow more direct flyways between wintering and breeding grounds. Future research should determine how maturing honey buzzards learn individual and population-specific loop migration patterns, and quantify migratory connectivity between breeding sites, flyways and non-breeding areas.

[1] Dept. of Wetland Ecology, Doñana Biological Station (EBD-CSIC), 41092 Sevilla, Spain.
[2] Institute for Biodiversity and Ecosystem Dynamics (IBED), University of Amsterdam, 1098XH Amsterdam, the Netherlands.
[3] MEDRAPTORS (Mediterranean Raptor Migration Network), Rome, Italy.
* Corresponding author: vansteelant.wouter@gmail.com

Introduction

The European honey buzzard (*Pernis apivorus* Linnaeus, 1758; hereafter: honey buzzard) is a widespread summer visitor to Europe, breeding as far north as Central Scandinavia and Finland, to as far south as Portugal and Greece, and from the British Isles in the west to Kazakhstan in the east. Further east, they are replaced by a morphologically distinct sister species, the crested honey buzzard (*Pernis ptilorhynchus*, Chapter 5), with a poorly known hybridization zone occurring in Central Asia (Ferguson-Lees and Christie 2001, Faveyts et al. 2011). Honey buzzards spend the winter in sub-Saharan Africa. Weighing on average 790 grams (male: 440–943 g; female: 450 g–1.05 kg) and with a wing span of 1.3 meters (Clark 1999, Ferguson-Lees and Christie 2001), the honey buzzard is a little lighter while also being slightly larger than the common buzzard (*Buteo buteo*), to which it also bears a strong resemblance in plumage. Honey buzzards, however, bear little relation to true buzzards (*Buteo*), and diverged from an old branch of mostly tropical raptor species including kites and bazas about 6.7 Million years ago (Jetz et al. 2012). In fact, it is thought that the plumage of juvenile honey buzzards is a form of Batesian mimicry, mimicking the appearance of more powerful common buzzards in their breeding range, or perhaps even small African eagles in their non-breeding range, and thus reducing predation risk (Ferguson-Lees and Christie 2001).

Honey buzzards stand out among similarly sized European raptors due to their slow reproduction (max. 1 clutch of 2 eggs per year), their remarkable feeding strategies, and, as will be discussed at length in this chapter, their complex migration strategies. Like all representatives of the *Pernis* genus (3 other species including multiple resident subspecies occur in Southeast Asia), honey buzzards are specialized at plundering larvae from nests of paper wasps as well as hornets, bumblebees and bees. Scale-like feathers at the base of the bill, an exceptionally narrow nostril and short, flat nails to facilitate digging, are just some of their protective adaptations to this remarkable feeding strategy (Mebs and Schmidt 2006). Honey buzzards do supplement this diet with small vertebrate prey such as frogs and nestlings of pigeons and thrushes. But while their preferred Hymenopteran preys are mostly active during mid-summer, honey buzzards are one of the last long-distance migrants to arrive to Europe in spring (May–June). After a remarkably short breeding season, adult honey buzzards are also among the first long-distance migrants to leave Europe in autumn (starting mid-August). As such, adult honey buzzards spend more than half the year in the forests and tree savannahs of sub-Saharan Africa, with some individuals going as far as Southern Africa, in particular females and immatures (Howes et al. 2019). And while honey buzzards are notoriously hard to study on the breeding grounds, we have even less information about their lives in the African non-breeding range. Juvenile honey buzzards migrate on average two weeks later than adults (Kjellén 1992, Agostini and Logozzo 1995, Schmid 2000, Vansteelant et al. 2020b) and thus have to find the way to Africa by themselves, after which they will typically reside in the tropical forests and woodland savannahs of Africa for at least 18 months before they first return to Europe (Panuccio et al. 2006, Corso 2012, Panuccio and Agostini 2012, Strandberg et al. 2012, Wright et al. 2019).

Knowledge Status

The honey buzzard is well known for its spectacular and highly synchronized passage through geographical bottlenecks such as sea-straits and peninsula's throughout Europe and the Middle-East (Zalles and Bildstein 2000, Verhelst et al. 2011). In fact, the honey buzzard is among the most abundant soaring birds at nearly all major migration watch-sites in Europe and the Middle-East. This typically involves some 1000′s of individuals per autumn or spring at watchsites in the Central and Eastern Mediterranean, to 10,000′s of individuals passing through the Strait of Gibraltar in the west, and several 100,000′s of individuals moving along the eastern shores of the Black Sea and Mediterranean Sea.

Honey Buzzards have historically been hunted at various migration bottlenecks to be used as food, trophies, or simply for recreational purposes. In Falsterbo (Sweden) Honey buzzards were caught at night on their roosts sites and salted until as late as the 19th Century (Karlsson and Bentz 2004). At the Strait of Messina, until the beginning of the 1990's, each spring thousands of honey buzzards fell victim to illegal shooting, mostly along its continental slope, where at least 2,000 illegal hunting posts were used by poachers. Since then, thanks to the effort of international volunteers and wardens, shooting has decreased and today few cases are still recorded (Giordano et al. 1998). This valuable conservationist effort was probably the reason of the notable increase of both migrant and breeding honey buzzards reported in the Central-Eastern Mediterranean region between 1990's and early 2000's (BirdLife International 2004, Agostini et al. 2007). However, substantial numbers of honey buzzards are still illegally shot on migration nowadays, most notably in places like Georgia and Lebanon, where community-based conservation efforts are being developed to solve this problem (Sándor et al. 2017, 2018). That said, and although it is hard to obtain good population estimates for this secretive woodland species, the honey buzzard population seems stable throughout most of its range (Status: Least Concern; online datazone BirdLife 2020). Indeed, despite the ongoing illegal hunting pressure, the autumn records of mature individuals at Batumi Raptor Count (Georgia) alone account for at least 127% of the global population estimate of this species (280,000–420,000 breeding individuals) and show no clear downward trends since between 2011–2018 (Vansteelant et al. 2020b). As such, the global breeding population, and the eastern populations in particular, must be substantially larger than currently estimated, even if we assume many of the individuals observed on migration are non-breeders (Keller et al. 2020).

Due to their aggregation at geographical bottlenecks honey buzzards are often lumped in with eagles and *Buteo* species as 'obligate soaring migrants'. However, thanks to their size and morphology, intermediate between true buzzards and harriers, they are in fact quite versatile flyers capable of extended periods of flapping flight (Agostini et al. 2015a), and capable of exploiting weaker thermals than other soaring migrants with higher wing loads (Bruderer et al. 1994, Spaar 1997, Horvitz et al. 2014, Nourani et al. 2020). As such they will start travelling shortly after sunrise, when large eagles, storks and pelicans are typically still waiting for strong thermals to form (Spaar 1997, Vansteelant et al. 2014, 2015). Contrary to the latter species, honey buzzards will also exploit the weaker updrafts at the bottom and near the top of thermals (Shamoun-Baranes et al. 2003) and they are much less reluctant to cross open water than other thermal-soaring raptors, even under suboptimal weather conditions using small islands as stepping stones (Agostini et al. 2016b).

Besides visual observations at geographical bottlenecks, the route choice of juvenile and adult honey buzzards has been studied through ringing recoveries (Schmid 2000). However, there are only very few (~dozens) of recoveries from the African wintering grounds. The most extensive and detailed tracking studies to date concern adult honey buzzards breeding in the Netherlands (N = 13) that use the West African-Eurasian flyway, crossing the Mediterranean at the Strait of Gibraltar during both autumn and spring migration (Vansteelant et al. 2015, 2017b, 2017c). In addition, small numbers of adult honey buzzards (N < 10) have been tracked by other researchers in the Netherlands (Willem Bouten, Jan van Diermen, Rob Bijlsma, Willem van Maanen and Stef van Rijn, unpublished data), the United Kingdom (Roy Dennis, personal website), Belgium (Stef van Rijn, unpublished data), Spain (Bermejo and De La Puente 2016), Sweden (Hake et al. 2003), Germany (Meyburg et al. 2010), Austria (Anita Gamauf, unpublished data), Hungary (Agostini et al. 2019) and Finland (Nourani et al. 2020, Patrik Byholm, unpublished data). Juveniles have been tracked from breeding areas in the United Kingdom (N = 11, Roy Dennis personal website), Sweden (N = 3, Hake et al 2003, Thorup et al. 2003) and Finland (N = 34, Vansteelant et al. 2017a, Patrik Byholm unpublished data). Finally, radar studies were carried out in the Negev Highlands and in the Arava Valley (Israel; Bruderer et al. 1994, Spaar 1997, Bruderer and Boldt 2001) and at the Strait of Messina (Southern Italy; Mateos-Rodríguez and Liechti 2012, Panuccio et al. 2016c, 2018, Becciu et al. 2018) during both spring and autumn, and along the southern coast of Spain, 140 km north-east of Gibraltar, in autumn (Meyer et al. 2000).

Post-breeding Migration

In Northern Europe a significant number of honey buzzards originating from Scandinavia converge through the Falsterbo peninsula (Southern Sweden), where on average 4,434 birds were counted in the autumns of 2010–2019 (Falsterbo Fågelstation 2020). Count data from migration watch-sites across Western Europe suggests the bulk of these birds then travel along a relatively narrow migration corridor over Western Germany into France, only deviating further westward over the Netherlands and Belgium when confronted with adverse weather along their habitual route (Desmet and Vansteelant 2014, see also www.trektellen.nl). The bulk of the birds originating from Scandinavia, the Low Countries, France and Germany moves into the Western African-European flyway, travelling into the Iberian Peninsula through both sides of the Pyrenees (Fig. 1). Here, substantial aggregations of honey buzzards are monitored every year at Organbidexka, Lindux (Western Pyrenees) and Eyne (eastern Pyrenees), with an average count of 10,812, 7,940 and 8,630 birds in the period 1981–2008 (max. annual count 21,504 in 1991), 1987–1994 (max. annual count 10,794 in 1991) and 1987–1991

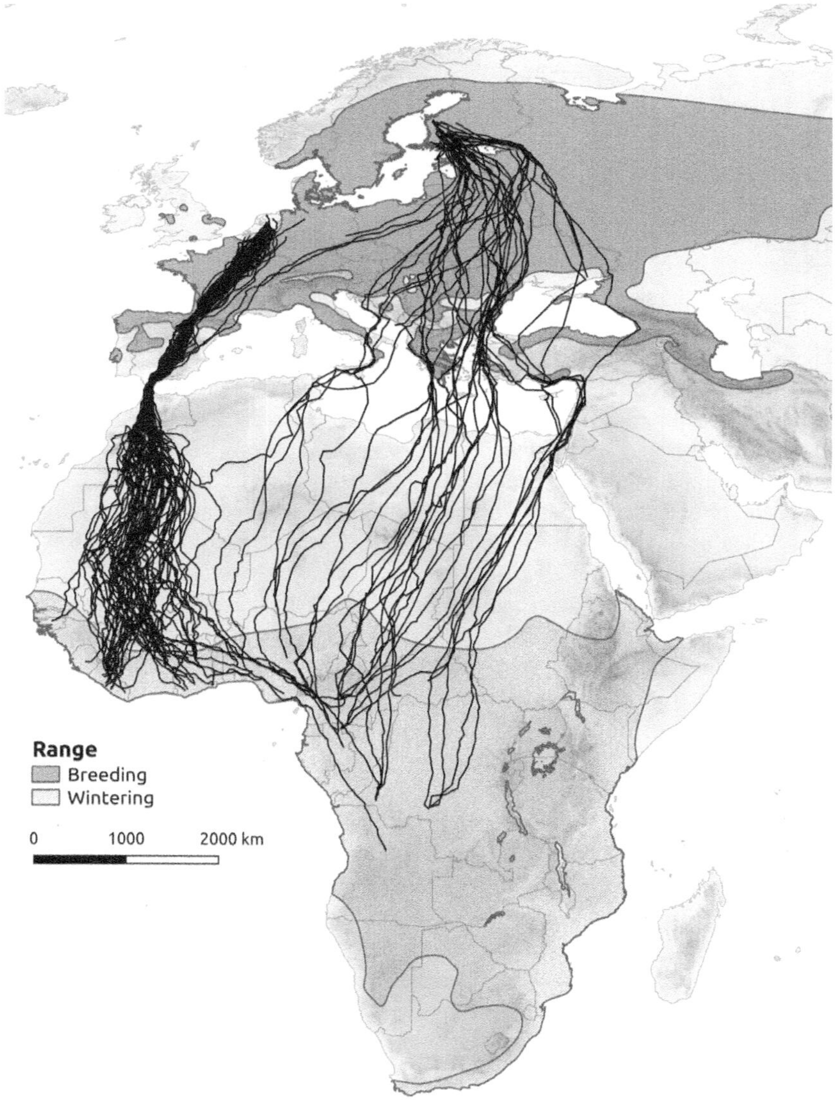

Figure 1. Autumn migration tracks of adult honey buzzards breeding in Finland and the Netherlands (courtesy Patrik Byholm, Jan van Diermen, Willem Bouten and Wouter Vansteelant).

(max. annual count 14,800 in 1990), respectively (Zalles and Bildstein 2000, Filippi-Codaccioni et al. 2010). The highest concentration along the west African-European flyway occurs at the Strait of Gibraltar (Finlayson 1992, Zalles and Bildstein 2000). On average 86,700 birds were reported at the site in 1970's (max. annual count 116,000 in 1972), 57,259 each autumn season in the period 2009–2012 from two watch points (on average 48,854 from a single point) (Zalles and Bildstein 2000, Miller et al. 2015). Although honey buzzards significantly advanced their median autumn passage date through the Western Pyrenees by ~ 3.5 days between 1981–2008, the autumn passage at the Strait of Gibraltar appears to have remained stable between 1999–2014 (Panuccio et al. 2016b).

Further east, the passage in the Central Mediterranean region is concentrated through the Sicily Channel (150 km wide), between Western Sicily and the Cap Bon Peninsula (Tunisia), with substantial numbers concentrating over the islands of Marettimo and Pantelleria (Agostini et al. 2005b). Most of honey buzzards following this migration flyway are expected to reach the Italian Peninsula from the Balkans, passing through the Adriatic Sea using some Croatian islands as stepping stones (Schneider-Jacoby 2001). En route to Western Sicily they follow the peninsula passing through the Marcellinara Isthmus (Panuccio et al. 2010) and concentrate at the Strait of Messina, between Southern continental Italy and Eastern Sicily, where on average 17,232 birds were counted in the autumns of 2011–2015 (max. annual count 26,155 in 2013; Agostini et al. 2015b, Pelle et al. 2014, Brichetti and Fracasso 2018). In Northern Italy, observations made at several watchsites revealed the passage of thousands of birds breeding in Central-Eastern Europe en route to the Strait of Gibraltar (Fig. 2). In this area, the highest concentration has been reported along the moraine hills south of the Lake Garda (Lombardy), where on average 10,500 birds were counted each autumn in the period 2005–2015 (maximum annual count 22,621 in 2007). Conversely, through the Western Alps the passage occurs on a broader front (max. annual count 11,670 in 2019 and 6,441 in 2016 in the Stura Valley and Po Valley, respectively; Brichetti and Fracasso 2018, Panuccio et al. 2018a, Scalco et al. 2019).

Compared with the migratory flow reported at the Strait of Gibraltar, relatively few honey buzzards leave Europe via the Eastern Mediterranean corridor into Turkey, with surprisingly low counts reported from Burgas Bay (Bulgaria) and at the Bosphorus (Turkey). At Bourgas Bay, on average 6,716 birds were counted in the period 1979–2003 (max. annual count 23,759 in 1996; Michev et al. 2011). At the Bosphorus, nearly all honey buzzards move through the southern side of the Channel (Çamlicas area), although substantial numbers are expected to cross the Marmara Sea

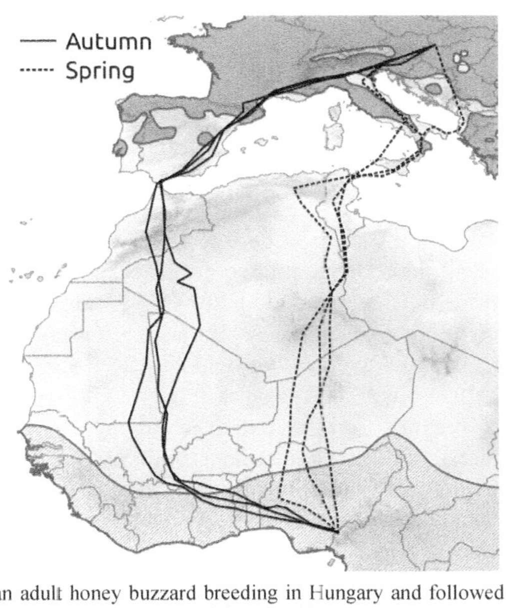

Figure 2. Migratory routes of an adult honey buzzard breeding in Hungary and followed for three years (modified from Agostini et al. 2019).

accepting a significant wind drift by dominant northerly winds (Panuccio et al. 2017). Observations made in 1960's and early 1970's reported on average 9,000 birds each autumn season (maximum annual count 25,745 in 1971; Porter and Willis 1968, Bijlsma 1987, Zalles and Bildstein 2000), while recent surveys made in 2005 and 2006, reported a total of 5,527 and 5,515 individuals, respectively (Panuccio et al. 2017). The main reason for the limited convergence of honey buzzards through the Bosphorus seems to be that many individuals originating from Northern and Eastern Europe cross the eastern Mediterranean directly via Southern Greece (Fig. 1). Surveys made in 2007 and 2008 from the island of Antikythira, located between the Peloponnese and Crete, revealed the passage of more than 1,000 adults each season en route to North Africa, probably via Crete (Agostini et al. 2012). However, this number represents only a part of the Greek population, accounting 1,000–2,000 pairs (BirdLife International 2004), in agreement with the conclusion of a broad front of migration of this species along the Central-Eastern Mediterranean flyway, using an island-hopping strategy, including also adults from Northern Europe (Fig. 1; Nourani et al. 2020). Further east, also the island of Cyprus seems to be interested by a significant passage of adult birds in this period. Observation made in 2005 at Cape Greco, the tip of a peninsula located on the southeastern coast of the island, reported 3,302 honey buzzards peaking early September (2,250 on 4 September; Roth and Corso 2007). Although these birds were not aged, the migration period clearly indicates that the vast majority were adults which, once they reached the Mediterranean coast, chose to move directly between Southern Turkey and the Middle-East coast (Agostini and Panuccio 2008).

Extremely large concentrations of honey buzzards occur even further east, and this is during both spring and autumn. In fact, the most concentrated autumn passage of the species takes place through the so-called Batumi bottleneck, a corridor of roughly 15 km wide, situated between the Eastern Black Sea coast and the Lesser Caucasus range in the Republic of Georgia (Andrews et al. 1977, Verhelst et al. 2011). At Batumi, on average 530,568 individuals were counted in the period 2011–2018 (Vansteelant et al. 2020b). These birds ultimately converge with conspecifics from Eastern Europe at the Belen Pass in Southern Turkey, at the easternmost point of the Mediterranean Sea (Gulf of Iskenderun; maximum annual count 15,970 in 1976; Oppel et al. 2014). From there all birds head southward across Israel, where the migratory flow is most concentrated at Kfar Qassem (max. annual count 476,565 in 1993) and Northern Valleys (max. annual count 522,555 in 1997; Shirihai et al. 2000), although much less concentrated than in Southwestern Georgia and Northeastern Turkey. The bulk of these birds eventually reach Africa via the Isthmus of Suez (Bijlsma 1983). Consistent with recent studies at the western end of the range (Panuccio et al. 2016b), the migration period of honey buzzards has not significantly advanced with climate change in the eastern flyway between 2011–2018 (Vansteelant et al. 2020b).

One of the most striking aspects of the honey buzzard migration is the particularly large difference in autumn migration timing and route choice between adults and juveniles. We have long known from observations at geographical bottlenecks that adult honey buzzards leave Europe in a remarkably synchronized manner in late August-early September, while juveniles leave on average two weeks later (Kjellén 1992, Agostini and Logozzo 1995, Trabalon and García 2015, Vansteelant et al. 2020b). Moreover, at all the aforementioned bottlenecks, juveniles account for only a small fraction of migrants (Schmid 2000, Agostini et al. 2004b, Vansteelant et al. 2020b). At Batumi, for example, the core migration period of elder honey buzzards, representing 75% of the annual passage, lasts less than a week, occurring on average between Aug 28th and Sept 5th, with peak migration days in excess of 100,000–200,000 birds. By contrast, only 30,000 of the 500,000+ honey buzzards that are recorded there each year are juveniles, and nearly all these juveniles are seen after more than 95% of elder birds have already passed (Vansteelant et al. 2020b). This stands in stark contrast to all other medium-sized and large soaring raptors. Indeed, in most migrant raptors the core migration periods of adults as well as juveniles lasts at least two weeks, showing a much greater overlap in migration windows between age groups, giving juveniles an opportunity to follow elder conspecifics and to learn overland detours at an earlier age than honey buzzards do. Migration counts in the Central Mediterranean region also provide a clear example of the remarkably large difference in adult

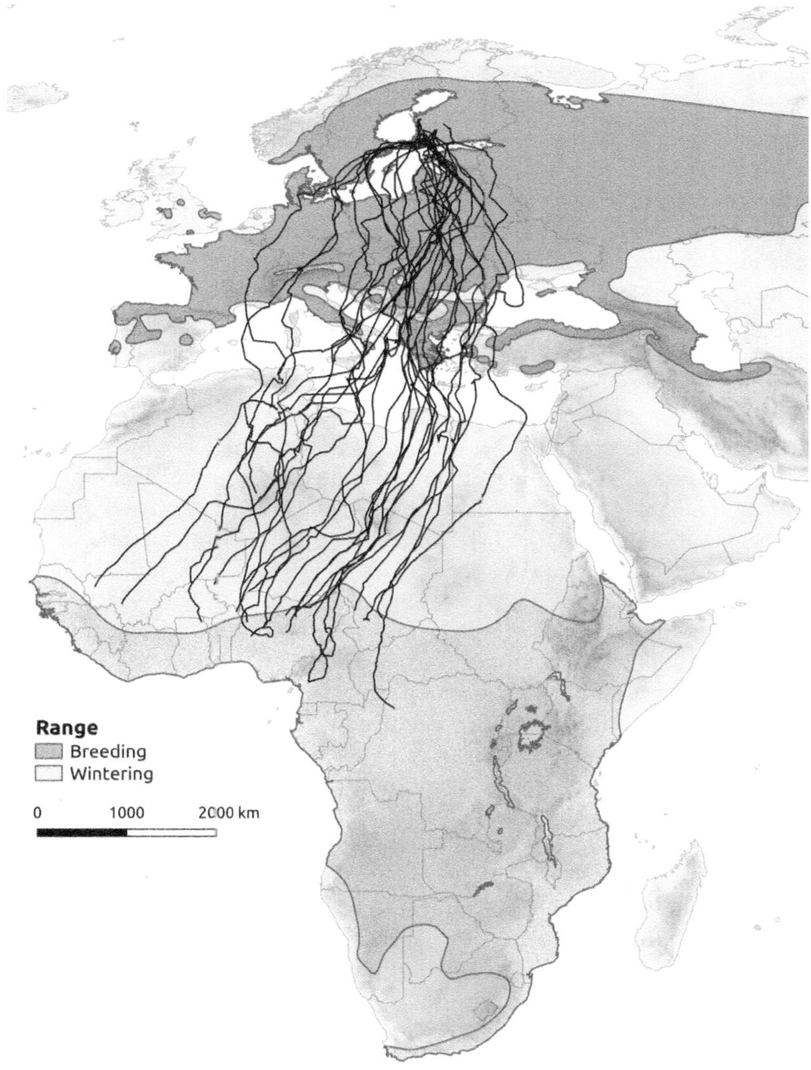

Figure 3. Autumn migration tracks of juvenile honey buzzards tagged in Finland (courtesy Patrik Byholm).

and juvenile migration strategies of honey buzzards. For example, adults that travel through Italy purposefully turn west when they reach the Strait of Messina, thus heading towards the shorter sea-crossing between Western Sicily and Tunisia, whereas juveniles orient along an inherited direction of migration moving along a southwestward axis, thus undertaking the sea-crossing as soon as they reach the coastline (Fig. 3), being commonly observed virtually on any Mediterranean island during mid-September (for a review see Agostini 2004). In so doing, juvenile honey buzzards behave like typical broad-front migrants such as the western marsh harrier (*Circus aeruginosus*; Panuccio and Agostini 2010a). Juveniles passing through Southern continental Italy undertake the longer sea-crossing between Southern Sicily and Libya (about 450 km) concentrating over the island of Malta. Only few juveniles migrate early, dragged in flocks of adults, allowing them to learn the shortest sea-crossing (via the Sicily Channel) through social learning.

The notion that young honey buzzards travel later and over a much broader front on their first migration than adults has also been confirmed by ring recoveries (Schmid 2000) and by all satellite tracking studies to date (Hake et al. 2003, Vansteelant et al. 2017a). In the early 2000's, for example, a

Swedish satellite-tracking study of six adult and three juvenile honey buzzards showed that juveniles were very sensitive to wind drift, whereas elder breeding birds purposefully compensated for side winds to circumnavigate the Pyrenees and the Mediterranean Sea *en route* to their individually fixed wintering destinations (Hake et al 2003, Thorup et al. 2003). A more recent and larger Finnish tracking study also corroborated that juveniles do not have a predetermined destination. In fact, these juveniles already diverged over a front more than 1,000 km wide by the time they reached Central Europe, because juveniles that left Finland in westerly winds crossed the Gulf of Finland into Estonia or Russia, whereas individuals leaving in easterly winds were more likely to depart along a westward course into Scandinavia or across the Baltic Sea. By the time they reached tropical Africa the wind drift accumulated by juveniles caused them to disperse over a longitudinal front more than 3,400 km wide, even though they had all hatched in Southern Finland (Vansteelant et al. 2017a). The sensitivity of juvenile honey buzzards to wind drift is also evidenced by some remarkable influxes of the species in the United Kingdom. For example, in 2000 an estimated 1,879 honey buzzards reached Eastern and Southern England between 20th September and 1st October due to the prevalence of unusually strong easterlies during the juvenile migration period (Fraser and Rogers 2002, Brown and Grice 2005).

A particularly surprising finding from the Finnish tracking study is that very few juveniles die during their first autumn migration. Indeed, out of 34 juveniles tracked on autumn migration, only 7 died before reaching sub-Saharan Africa (Vansteelant et al. 2017a), which is relatively few compared to juveniles of larger raptor species (Oppel et al. 2015, Meyburg et al. 2017a), but comparable to juvenile survival rates of black kites *Milvus migrans* (Sergio et al. 2019). Moreover, such as for Finnish adults, none of the juvenile mortality events were due to birds drowning in the sea, despite the fact that juveniles, unlike adults, initiate the sea-crossing as soon as they hit the Mediterranean coast (Fig. 3). The main reason for this is undoubtedly that the prevailing autumn winds over the Mediterranean generally provide good tailwind support during sea-crossings for both age groups (Agostini et al. 2012). Intriguingly, however, a recent study also found that during the juvenile migration period the sea surface in the Mediterranean is often warmer than the air above it, which can lead to the formation of (weak) sea-thermals (especially over the Eastern Mediterranean) whereas the opposite occurs during most of the adult migration period (Nourani et al. 2020). What's more, one juvenile that was tracked at high resolution effectively engaged in thermal-soaring for more than 100 km over the Mediterranean Sea (like ospreys; Duriez et al. 2018), while such behaviour has not yet been recorded for adults, despite the fact that more adults have been tracked at high-resolution over the Eastern Mediterranean Sea (Patrik Byholm, unpublished data). Of course, we cannot exclude the possibility that (late) adults would also use sea thermals if they had the opportunity to do so. That said, the ability of honey buzzards to exploit weak thermals at sea may be an overlooked factor contributing to the ability of this species to migrate independently from elders on its first migration, as opposed to large eagles and vultures that would not be able to exploit such weak thermals due to their high wing loading (in addition to the fact those species are also unable to sustain long periods of flapping flight). The question how young honey buzzards eventually learn to compensate for wind drift, and how they find the strategic overland detours that are used by most adults, however, has not been solved yet. That is because until recently, all satellite-tracked juvenile honey buzzards died or stopped transmitting data before they first returned to Europe (Strandberg et al. 2012). The fact that immature honey buzzards typically stay in Africa during their 2nd calendar year, and the fact they attain mature plumage in their 3rd calendar year, makes it extra challenging to track migratory development in this species.

Tracking studies have also revealed a striking difference in the extent to which adult honey buzzards from various populations rely on overland autumn migration routes. More specifically, all adults tracked from Spain, UK, Belgium, the Netherlands and Sweden, as well as most adults tracked from Germany and an adult male from Hungary, crossed the Mediterranean via Gibraltar in autumn (e.g., Vansteelant et al. 2015, 2017a, Agostini et al. 2019; Fig. 1 and 2). Further east, however, a majority of experienced adults in the Finnish breeding population was found to readily engage in long sea-crossings, leaving Europe mostly via Greece on a broad front, probably taking advantage of tailwinds, which occur frequently during their post-breeding migration (Agostini et al. 2012). In so doing, birds crossing the Aegean Sea used several islands as stepping stones, but only one passed over the Antikythira watchsite (Nourani et al. 2020). In fact, it seems that the passage of adult honey buzzards over the Eastern Mediterranean could be underestimated due to the fact birds can use several islands as stepping-stones to cross the Aegean Sea, broadening their migration front, and being thus unlikely to be observed from a single watchsite like Antikythira. The most striking finding regarding the migration behaviour of adult honey buzzards from Finland, however, may be that the autumn route choice of these birds is surprisingly flexible. For example, while many birds cross the Eastern Mediterranean via Southern Greece on most of their autumn migrations, several of these birds switched to other routes via the Strait of Messina, or via the Bosphorus and the Eastern Mediterranean coast across the Isthmus of Suez into Africa in some years, while one individual consistently used the overland detour via Gibraltar on two consecutive journeys (Nourani et al. 2020). At the same time, some of these birds switch between completely distinct autumn flyways between years, likely depending on weather conditions, but probably also depending on the collective experience of other individuals with which they flock along the way. This surprising flexibility in route choice may compromise the ability to infer behavioural strategies from geographical and seasonal patterns in migratory aggregations, as well as to infer population trends from watchsites, especially if such flexibility is more common in North and East European populations of the species. However, honey buzzards originating from the northernmost limits of the breeding range in Finland may be more likely to diverge into distinct flyways than individuals breeding further south, simply because they need to cover more distance before reaching the Mediterranean. As such, it seems reasonable to assume that route choice will be more consistent between years within individuals and populations breeding in Southern and Eastern Europe. But this still needs to be confirmed by tracking large samples of individuals there (Agostini et al. 2019).

In terms of migratory performance, adult honey buzzards tend to reach their highest travel speeds while travelling in good soaring conditions and northeasterly trade-winds across the Sahara in autumn, reaching speeds in excess of 250 km day^{-1} (Vansteelant et al. 2015). Only after crossing the desert adults start compensating for sidewinds to navigate back to their individually fixed wintering areas (Vansteelant et al. 2017c). Over the Sahel and the tropical forest, they typically achieve travel speeds of 170–180 km day^{-1}. Overall, adults breeding in the Netherlands typically complete the migration in approx. 25 days (Vansteelant et al. 2015), which is quite a fast migration for a bird that needs to stop every night to roost. They achieve this fast migration in large part by not making long stop-overs along the way and thanks to largely supportive wind conditions en route (Vansteelant et al. 2017b). In fact, tracked adults have rarely been observed to linger for more than one day on autumn migration. Honey buzzards are known to store large fat reserves prior to (autumn) migration, only occasionally feeding en route (Panuccio et al. 2006). The most common field observations of stop-over behaviour in honey buzzards concern individuals that briefly interrupt travel at freshwater ponds just prior to or after the desert crossing (Yosef 1996). Adults tracked from Finland behave in a very similar way, so that they complete their slightly longer autumn migration with just a few more travel days. Regardless of whether they make long sea-crossings during these fasting flights, adult honey buzzards typically show very high survival rates during their non-stop autumn migration, with a few recorded mortality events being related to persecution (i.e., shooting) and other human activities. This makes sense given that adult honey buzzards typically experience good soaring conditions and supportive winds during most of their autumn migration. Juvenile honey buzzards complete their first migrations at a comparable pace as adults, despite their lack of experience, their use of different routes, and the weaker overland soaring conditions during their late migration period (Vansteelant et al. 2017a). In contrast to adults, however, they probably achieve their high performance primarily by gaining more tailwind assistance, since they drift with prevailing winds for most of their migration.

Once they reached their non-breeding range some adult honey buzzards will stay at the same site until the start of spring migration, using relatively small home ranges of just a few km^2. However, there are large individual differences in the mobility of honey buzzards within sub-Saharan Africa, with some individuals exploiting up to three distinct wintering sites each year (Vansteelant, unpublished data). We still do not fully understand how environmental factors and habitat quality determine these individual differences in intra-African mobility. However, there seems to be an important innate component driving this behaviour, as females tend to move further south than males during the non-breeding season, leading to a very strong over-representation of females among honey buzzards in Southern Africa (Howes et al. 2019). It seems likely that males stay further north because then they need to cover less distance than females in spring, allowing them to reach their breeding site a few days before the females, even if both sexes leave their respective non-breeding sites at the same time and achieve similar migration speed (Vansteelant et al. 2015). Juvenile honey buzzards ultimately settle their primary non-breeding site where-ever they happen to end up after their first migration (Vansteelant et al. 2017a). Before their first return to Europe, however, some immatures are more mobile than others, and in general immature birds are more explorative than adults during the non-breeding season (Strandberg et al. 2012) and more likely to venture as far as Southern Africa (Howes et al. 2019). This may be due to adults outcompeting juveniles at high-quality non-breeding sites, although any interpretation of intra-African movements remains highly speculative. Either way, it is likely during these first months in Africa that immatures settle individual territories to which they will continue to return for the rest of their lives. Half of the juveniles that were tracked from Finland to Africa died during this settlement period, more specifically during the first 6 months in Africa (Patrik Byholm, unpublished data). As such, claiming high-quality non-breeding sites seems to be a major bottleneck in the life-cycle of the honey buzzard, more so than the first autumn migration.

Pre-breeding Migration

The spring migration of adult honey buzzards is relatively well synchronized, with most adult breeders departing from sub-Saharan Africa around mid-late April. The principal passage at Gibraltar and other bottlenecks occurs in the first week of May, and the main arrival to the breeding grounds taking place in mid-late May. The central variation in schedules seems to be latitudinal, with individuals breeding in Finland departing and arriving around one week later than adults breeding in the Netherlands (P. Byholm unpublished data). As regards migration counts, at the Strait of Gibraltar much lower numbers have been reported in spring than in autumn, with 17,042 and 15,086 individuals observed during spring 2008 and 2009 from six watch points used at the same time (De La Cruz et al. 2011a). In Southern France, on average 10,000 individuals were recorded at Leucate, along the southeastern coast, in the period 1982–1991 (max. annual count 20,000 in 1991; Zalles and Bildstein 2000). The migratory flow of birds using the western flyway and passing through Northern Italy is much less consistent in spring, with on average 2,941 birds observed during the peak passage (9–18 May) in the period 2004–2012 (maximum annual count 6,072 in 2005) along the Ligurian Apennines (Northwestern Italy; Agostini et al. 2006, Brichetti and Fracasso 2018). In the Central Mediterranean, the spring migratory flow is more consistent than the autumn passage. Each spring 30,000–40,000 individuals cross the Sicily Channel, although they show a relatively broader front of migration than in autumn, with thousands bypassing the Strait of Messina and instead undertaking long flights across the Tyrrhenian Sea when they can exploit favourable winds (Agostini et al. 2016b). More specifically, on average 24,100 birds were counted at the Strait of Messina in springs 1996–2016 (max. annual count 38,929 in 2015; Corso 2005, Brichetti and Fracasso 2018), while in the period 2006–2013 on average 4,597 honey buzzards were detected bypassing the Strait and concentrating over the islands of Ustica (max. one day count 2,165 on 2 May 2010) and Panarea (max. one day count 3,454 on 6 May 2011), which probably account for a fraction of birds crossing the Tyrrhenian Sea each spring (Agostini et al. 2016b). Individuals that do not breed in Italy reach their territories in Eastern Europe crossing the Adriatic Sea or, to a lesser extent, through Northern Italy (Fig. 2; Agostini et al. 2019, Premuda et al. 2004). Such as during autumn migration, the highest number of honey buzzards was reported along the eastern flyway, but in this season the front is clearly most concentrated in Eilat (Israel), where a world-record honey buzzard migration count of 851,598 was conducted in 1985

(Shirihai et al. 2000). It is still unclear what routes this vast number of birds uses as they progress further north. On the western side of the Black Sea, at the Bosphorus, the most comprehensive spring count to date was conducted in 2006, revealing a passage of about 10,000 birds, mostly concentrated along the northern side of the Channel (Üner et al. 2010). This is of the same order of magnitude as the autumn passage at this site. However, at Batumi, along the Eastern Black Sea coast, only 89,366 honey buzzards were recorded during the only full spring season count to date at this location, conducted in 2019 (Tal and Jansen 2019). That said, Abuladze (2013) reports a formidable spring passage of honey buzzards through Georgia also in spring, but exact numbers are lacking.

Migration surveys clearly show seasonal differences in the extent to which honey buzzards converge at geographical bottlenecks. Most notably, the passage of honey buzzards through the Central Mediterranean is greater during spring while the opposite occurs at the Strait of Gibraltar and in Northern Italy. This indicates that a substantial number of honey buzzards switch between different flyways depending on the season, undertaking longer sea crossings and showing a broader front of migration during spring movements. As such, it has been suggested that thousands of individuals, probably belonging to populations breeding in some countries of Central-Eastern Europe, may migrate via the Strait of Gibraltar in autumn, but then minimize the duration of their return migration to Europe along a more direct route through the Central Mediterranean flyway in spring (Agostini and Panuccio 2005, Bensusan et al. 2007). Such a strategy has recently been found for an adult male that was satellite-tracked across three consecutive migration cycles between its breeding area in Hungary and non-breeding area in Cameroon (Fig. 2). This bird performed the predicted migration strategy during each migration cycle, passing through Northern Italy and crossing the Mediterranean at the Strait of Gibraltar each autumn (covering on average 7,046 km in 43 days), but choosing the Central Mediterranean flyway each spring (covering on average 5,598 km in 23 days) (Agostini et al. 2019). During spring this bird actively followed the direct route back to its breeding site, even if that required additional energy expenditure through compensation for side winds in Africa. Once he had reached the Mediterranean coast at the Cap Bon Peninsula, he chose between alternative routes over water, heading northeastward through the northern side of the Sicily Channel and through the Tyrrhenian Sea in southerly winds, while conversely travelling through the Strait of Messina when facing northwesterly ones. The total migration distance in autumn was on average 49% longer than the straight line (loxodrome) between breeding and wintering location, while the spring migration was on average only 19% longer. Although future research should verify how widespread this loop migration pattern is in this species, the thousands of adult honey buzzards seen passing through Northern Italy each autumn using the same pathway of the Hungarian individual, agree with the idea that at least substantial numbers of birds breeding north of the Balkan Peninsula perform this strategy. Conversely, it is likely that birds passing during both spring and autumn between Italy and Tunisia mostly consists of individuals breeding in Central Italy and in countries of ex-Yugoslavia, for whom the cost of detouring via Gibraltar or the Bosporus is probably considerably higher than for birds breeding in Western or Eastern Europe.

Similar to the Hungarian bird, some individuals that were satellite-tracked from Germany travelled via Gibraltar in autumn but returned via the Central Mediterranean flyway. Moreover, one male that was tracked over three consecutive spring migrations returned via Gibraltar only on its first journey, while choosing a slightly more direct route across the Western Mediterranean and the Balearic Islands on its two following migrations (Meyburg et al. 2010). Such findings corroborate the notion that at least some honey buzzards belonging to specific populations prefer to minimize travel duration by choosing more direct routes in spring. However, for German birds the seasonal difference in the extent of detours is not nearly as great as for the Hungarian bird, and it was also reported that the majority of German birds actually do return via the Strait of Gibraltar in spring (Meyburg et al. 2010), which also applies to all adults that have ever been tracked from the Netherlands until now, even those that winter far east in Central Africa and that could take a much shorter route back home by travelling via the Central Mediterranean flyway or across the Western Mediterranean Sea (Vansteelant et al. 2015). Dutch birds that winter in Central Africa do sometimes take a seemingly risky route straight across the

Central Sahara in order to reach Gibraltar, but in most cases they already initiate their western detour prior to the Sahara crossing (Vansteelant et al. 2017c). In fact, these birds typically overcompensate for sidewinds at the start of their spring migration, moving westward until they get an opportunity to cross the desert in supportive tailwinds, which occur more frequently in the westernmost regions of the Sahara where anticyclones roll in from across the ocean (Vansteelant et al. 2017c). This is best understood as an energy-minimizing strategy, with the additional benefit that birds can better avoid the risk of encountering a dust-storm over the interior of the Sahara (Vansteelant et al. 2017c). And as such it is still unclear how common it is for honey buzzards to follow a time-minimizing strategy by returning along more direct routes in spring, or whether this perhaps only applies to certain populations such as those in the Balkans.

The picture is further complicated by the extremely variable and flexible strategies of adult honey buzzards breeding in Finland. The majority of these birds actually make much longer detours in spring than they do in autumn, circumventing the Sahara and the Mediterranean Sea, rather than attempting a direct crossing of the Eastern Mediterranean Sea (Fig. 4). More precisely, six of nine satellite-tracked Finnish adults crossed the Mediterranean directly in autumn while detoured via Suez and the Eastern Mediterranean flyway in spring (Byholm unpublished data, Vansteelant et al in prep). The role of wind conditions in shaping this eastern detour has not yet been formally assessed, but this strong seasonal difference in the sea-crossing propensity shown by the Finnish adult honey buzzards can probably be explained in large part by the fact that wind regime over the Central-Eastern Mediterranean is dominated by northerlies in both seasons (see Agostini et al. 2012). Such winds are stronger in autumn (Meltemi wind), offering tailwind support for honey buzzards to make long sea-crossings between Greece and North Africa in this season (Panuccio et al. 2016a). In spring, although much weaker, these northerlies result in headwinds for birds attempting a direct crossing of the Mediterranean, which clearly increases the risk off fatigue and drowning. Perhaps this sea-crossing (comparable, in terms of length, to the Tyrrhenian Sea one) is performed during spring just by individuals breeding in Greece, for which making a detour would be suboptimal in terms of both time and energy (Agostini et al. 2012). This pattern confirms again that costs and benefits related to the use of different flyways may be different across different populations, therefore is not surprising that the results shown by individuals tagged in different regions may seem conflictive.

Behavioural and Ecological Remarks

Social learning

Although the role of social information remains to be proven unequivocally, most evidence to date suggests that honey buzzards may learn strategic overland detours synchronizing their seasonal departures with more experienced individuals, or meeting them along the route. For example, juveniles that pass earlier through Southern Italy in flocks of adults may learn the shortest water-crossing already on their first migration (Agostini et al. 1999, 2004b, Agostini 2004). However, in general the majority of juveniles only start migrating when most adults have left Europe and nearly all juveniles that have ever been tracked failed to learn strategic detours on their first migration (Agostini and Logozzo 1995, Schmid 2000, Hake et al 2003, Vansteelant et al. 2017a, 2020). As such, social learning during the first migration is definitely not the primary mechanism for route learning in this species (as opposed to larger obligate soaring species, such as the short-toed snake eagle; Chapter 7). This may seem surprising since one would expect such a slowly reproducing species to experience strong selection for strategies that enhance juvenile survival. However, compared to similarly slowly reproducing species, juvenile honey buzzards are much better equipped to engage in prolonged periods of flapping flight over open water, and juveniles rarely drown in the Mediterranean despite engaging in seemingly risky and long sea-crossings (Vansteelant et al. 2017a). Moreover, juvenile honey buzzards have been recorded to engage in thermal-soaring flight over the Mediterranean for distances up to 100 km (Nourani et al. 2020, Byholm unpublished tracking data). Consequently, it

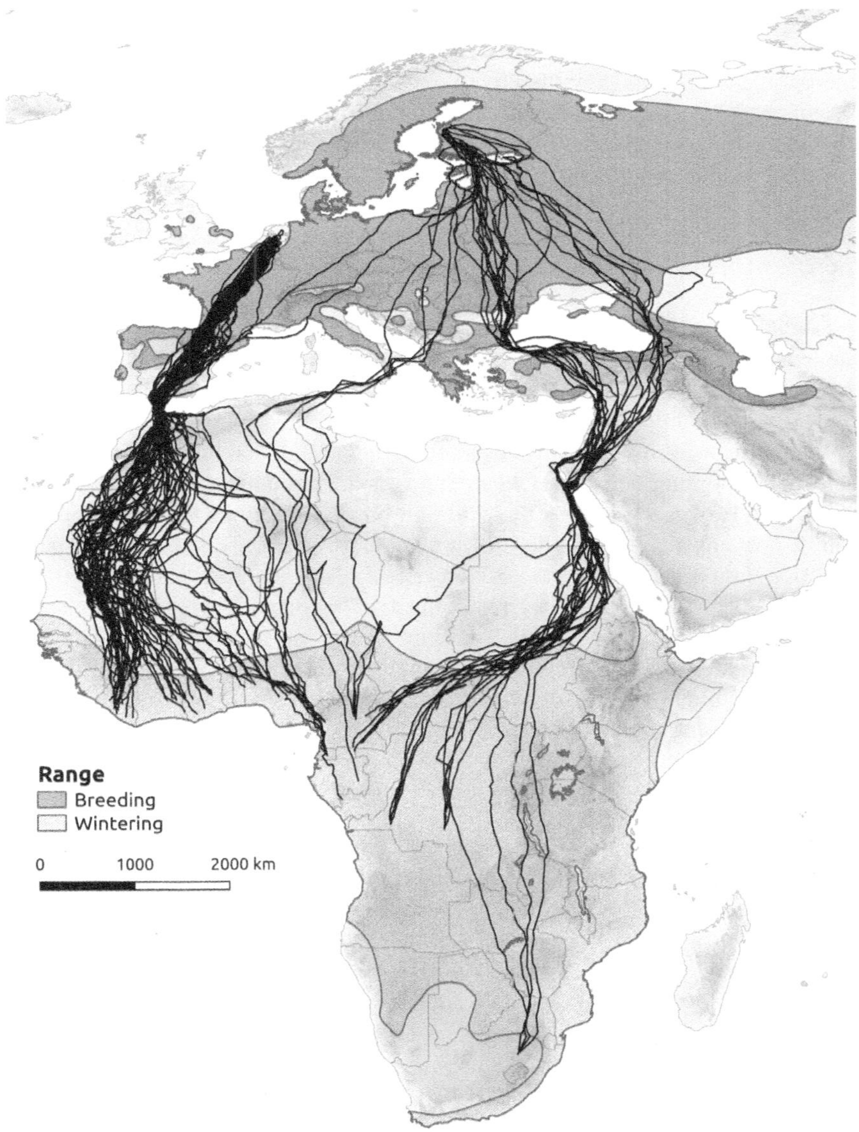

Figure 4. Spring migration tracks of adult honey buzzards breeding in Finland and the Netherlands (courtesy Patrik Byholm, Jan van Diermen, Willem Bouten and Wouter Vansteelant).

appears that synchronization of adult and juvenile migration would not yield a much higher juvenile survival. Finally, during their first migration, juveniles may discover more direct paths between breeding and wintering areas which they can use later in their life as alternative flyways to avoid long overland detours (Agostini and Panuccio 2005).

Fine-scale soaring-gliding behaviour

Honey buzzards are capable of sustaining soaring flight, and achieving high soaring performance under a much broader range of weather conditions than obligate soaring migrants (Spaar 2007, Shamoun-Baranes et al. 2003, Horvitz et al. 2014). In particular, in his study made by radar in Southern Israel, Spaar (1997) provided detailed information concerning flight parameters in thermal circling and inter-thermal gliding flight. Honey buzzards often do glide more slowly between thermals than they should to maximize their cross-country ground speed (which is what they are expected to do if they aim to minimize the total duration of their migration; Horvitz et al. 2014). However, by gliding below time-optimal airspeeds they effectively extend their gliding range after leaving a thermal, thus increasing the probability of encountering another strong thermal before running out of altitude, and thus reducing the need to supplement soaring-gliding with flapping flight. Honey buzzards do glide closer to their time-optimal airspeed than larger, heavier migrants under comparable weather conditions (Horvitz et al. 2014), and this makes sense given that the latter species stand to lose much more energy if they have to supplement soaring with flapping flight (Agostini et al. 2015a). In line with predictions of optimal migration flight theory, radar studies at the Strait of Messina have shown that honey buzzards also decrease their gliding airspeed between thermals under increasing tailwind support, presumably lowering the energetic cost of flight (Becciu et al. 2018) as such a strategy allows the birds to extend their gliding range after leaving a thermal whilst simultaneously achieving high ground speeds (cf. Vansteelant et al. 2017b). Using high-resolution GPS-tracking, researchers have now also been able to study how honey buzzards modulate their fine-scale soaring strategies along the entire West African-Eurasian flyway. This further confirmed that honey buzzards are most likely to glide at near time-optimal airspeeds when thermals are strong and abundant, for example when crossing the desert. However, they readily reduce gliding airspeed with tailwind support, as observed at the Strait of Messina, and they will glide at much slower airspeeds relative to the time-optimal strategy in those parts of their migration cycle where soaring conditions are poor, allowing them to extend their gliding range, and thus reducing the need for flapping flight (cf. Vansteelant et al. 2017b). Meyer et al. (2000) estimated that in this species the energy consumption of the flapping flight at an average airspeed of 11 m/s is six times higher than that required for the soaring-gliding flight over land, with birds assuming to maximize cross-country speed in soaring-gliding flight. Consequently, in terms of energy expenditure, a water-crossing 100 km long is convenient when avoiding detours exceeding 600 km overland, assuming no wind.

Flight altitude and wind selectivity

Birds can maximize tailwind support by adjusting their flight altitude to the altitude with most wind support or by adjusting their route choice to prevailing wind conditions. Only a few studies have assessed wind selectivity in terms of altitude adjustments in case of the honey buzzard, but in a recent study using optical range-finders, researchers revealed that birds passing through the Tyrrhenian Sea and using the island of Ustica as a stepping stone are more prone to leave that island when flying at higher altitudes, in larger flocks and during weaker winds. Probably, during such wind conditions, thermals are stronger over the island which are more attractive for honey buzzards as they can gain altitude soaring in larger flocks (Panuccio et al. 2019c). By departing at higher altitude, the birds would also be able to cover a larger section of the sea-crossing in gliding rather than flapping flight. However, radar studies conducted during spring migration at the Strait of Messina have also shown that honey buzzards tend to fly below the optimal flight altitude that would maximize tailwind assistance while crossing the sea (Mateos-Rodriguez and Liechti 2012). The reason for this is probably that honey buzzards would have to invest too much energy to stay at, or climb to, the high altitudes required to maximize wind support whilst over the sea. Once these birds reach land, however, they do modulate the response to wind drift depending on time of day and convenience in relation to exploitation of updrafts (Panuccio et al. 2016c).

At the Cap Bon peninsula (Tunisia), during spring, honey buzzards facing the Channel do not necessarily undertake the water-crossing as soon as they reach the coastline during strong crosswinds, but wait for better wind conditions (e.g., weaker winds) gathering in large flocks before leaving the mainland (Agostini et al. 1994). In particular, during moderate (4–5 m s^{-1}) northwesterly winds, many use the island of Pantelleria (approximately 70 km ESE of the Cap Bon Peninsula) as a stepping stone (max. one day count 5,748 on 2 May 2012), concentrating the passage through the southern side of the Channel (Agostini et al. 2016b).

5

Crested honey buzzard *Pernis ptilorhynchus*

Elham Nourani

ABSTRACT

The crested honey buzzard (*Pernis ptilorhynchus*) is an Eastern Palearctic species with a wide distribution range throughout the Asian continent. The migratory sub-species within *P. ptilorhynchus, P. p. orientalis*, is mostly known from East and South Asia, but is increasingly recorded in Central Asia and the Middle East. The migration flyway in East Asia is relatively well documented through observational and remote-tracking studies of the Japanese population. Birds breeding in Southern Siberia, Southern and North-Eastern China, Japan, and Korea winter in India and Southeast Asia. A smaller number of birds follow a western migration flyway through Central Asia. These birds probably breed in the species' westernmost Siberian breeding grounds and migrate through Central Asia to winter in the Middle East. The species has also been recorded in the African-Eurasian flyway, where it is probably caught in flocks of migrating European honey buzzard *P. apivorus* and migrates with them to the African wintering grounds. These vagrant birds are sometimes thought to be hybrids, but no DNA analysis has confirmed this. This chapter presents a summary of the current knowledge of the migratory behaviour of the migratory sub-species of the crested honey buzzard in East Asia, with an overview of available information in its western distribution range and within the Western Palearctic. The information presented here is gathered from resources available in English, and is therefore not a complete synthesis of all the existing knowledge about the migration behaviour of the crested honey buzzard.

Introduction

The crested honey buzzard (*Pernis ptilorhynchus* Temminck, 1821) is an Eastern Palearctic species, closely related to the European honey buzzard *P. apivorus*. There are six sub-species within the *P. ptilorhynchus*, but only the northern subspecies, *P. p. orientalis* (Taczanowski 1891) is migratory (Faveyts et al. 2011). Some sources consider *P. p. orientalis* a separate species and refer to it as *P. orientalis* (Brazil and Nurney 2009). The species has a broad distribution range. In East Asia, it breeds in Southern Siberia to Sakhalin, southern to North-Eastern China, Japan, and Korea and winters in the Indian sub-continent and Southeast Asia, particularly in Indonesia and the Philippines. The

Max Planck Institute of Animal Behavior, Am Obstberg 1, 78315 Radolfzell, Germany.
Email: mahle68@gmail.com

distribution and migratory behaviour in the western limits of the crested honey buzzard's distribution range is not well-known. A small number regularly migrate through Central Asia and winter in India and the Middle East.

The crested honey buzzard is morphologically similar to its western counterpart, the European honey buzzard. It generally has a heavier, bulkier body that makes it stand out in a flock of European honey buzzards. It also has a shorter tail and broader wings, resembling an eagle in flight (Faveyts et al. 2011, Zannetos et al. 2018), with six fingered primaries compared to the five in the European honey buzzard. Unlike the European honey buzzard, the crested honey buzzard lacks a dark carpal patch. Despite these pronounced differences, distinguishing the two species or their potential hybrids is not always straight-forward. Juvenile crested honey buzzards have narrow wings and can be mistaken for European honey buzzards. Moreover, high individual plumage variation (DeCandido et al. 2015) makes identifying crested honey buzzards difficult (Faveyts et al. 2011, Zannetos et al. 2018).

The diet of the crested honey buzzard is also similar to the European honey buzzard. Their specialist diet consists mostly of bees and wasps, particularly their larvae, and even pieces of honey comb and honey. Crested honey buzzards also feed on other insects and occasionally vertebrates such as lizards, frogs, small mammals, and birds (Orta et al. 2019). The timing of breeding in this species is related to the availability of bees and wasps. This specialized diet is also reflected in the spatial consistency in migration routes that allows the birds to take advantage of the known resources on stopover sites (Sugasawa and Higuchi 2019).

From the available estimates, population size of the crested honey buzzard appears to be very large and population trends stable (BirdLife International 2016a). Although threats to this species have not been fully documented, it is known to be vulnerable to wind energy development (BirdLife International 2016a). However, a recent GPS-tracking study in Japan showed that the energetic cost of detouring wind turbines along migration routes does not largely increase flight costs of the individual birds (Kikuchi et al. 2019). The species is evaluated as Least Concern by IUCN (BirdLife International 2016a), mainly due to its large population size and wide distribution range.

This chapter provides an overview of the current knowledge of migration ecology and behaviour of the crested honey buzzard available in English. The bulk of the information presented here is from the East Asian flyway, particularly the satellite-tracked Japanese population. A small section has been devoted to summarizing the little knowledge that exists about the migratory behaviour of this species in the western limits of its range.

Knowledge Status

The Asian-Australasian migratory flyways are relatively little studied compared to the Eurasian-African and the American migration systems. Consequently, not all aspects of migratory behaviour of the crested honey buzzard are known. Basic information on migration timing, intensity, and ecology has come from observations at migration bottlenecks in Southeast Asia, particularly Thailand, Malaysia, and Indonesia, where crested honey buzzards are observed in flocks of 100s strong (Puan et al. 2014). Yet, migration counts at these watchsites do not have a long history and have not been carried out consistently (see Table 1 for a summary of phenology and counts at key watchsites). Despite the large distribution range of the crested honey buzzard, only the population breeding in Japan has been studied using remote-tracking technology.

The first remote-tracking study of the crested honey buzzard started in 2003 (Higuchi et al. 2005). Three birds (one juvenile, one adult male, and one adult female) were captured in breeding grounds in Japan and were equipped with Argos platform transmitter terminals (PTT). This study provided the first complete picture of the migratory routes, timing, and behaviour of this species. The dataset then grew to include over 30 individuals (Fig. 1) and has provided much opportunities to investigate the unique loop migration and sea-crossing behaviour shown by individuals from the Japanese population (Yamaguchi et al. 2008, Yamaguchi et al. 2012), the effects of weather and climate change on the

Table 1. Overview of the timing of spring and autumn migration of crested honey buzzards from ground observations at selected migration bottlenecks along the East Asian flyway. Note that the date 'range' is dependent on observation effort and may not encompass the entire temporal window of migration.

Country	Location	Migration peak	Range of count days	Average number	Source
Spring					
Malaysia	Tanjung Tuan 2°24.4'N, 101°55.3'E	14–15 Mar	Mid-Feb–Mid-Apr	47,563	(Puan et al. 2014)
Thailand	Promsri Hill, Chumphon 10°30'N 99°04'E	25 Mar	Late Feb–early Apr	5,501	(DeCandido and Nualsri 2009)
Autumn					
Thailand	Khao Dinsor, Chumphon 10°28.40'N 99°13.26'E	2-8 Oct	Late Aug–Mid-Oct	35,000	(DeCandido et al. 2015)
Malaysia	Selangor plains	NA	Oct–Nov	121,000	(Wells 1999) cited in (DeCandido et al. 2015)
Indonesia	Dieng Mountains, Central Java	25 Oct	28 Sep–17 Nov	2,500 (extrapolated)	(Nijman 2004)
Indonesia	Tinjalas, East Bali 08°24'S 115°41'E	NA	3 Oct–25 Nov	1,608	(Germi 2005)

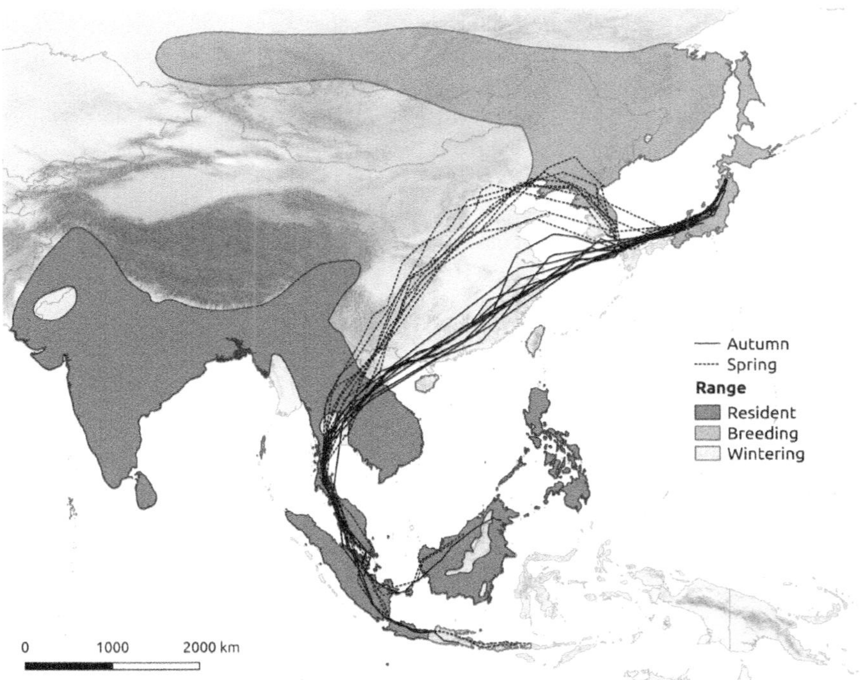

Figure 1. Routes followed by 10 satellite tracked crested honey buzzards breeding in Japan.

species along its migration route (Nourani et al. 2016, Nourani et al. 2017), site fidelity (Shiu et al. 2006), and consistencies in temporal and spatial migration patterns (Sugasawa and Higuchi 2019). A recent study GPS-tracked eight birds in Japan in June 2017 (Kikuchi et al. 2019) to assess the potential cumulative barrier effect of wind farms on energy expenditure during migration.

Post-breeding Migration

Based on satellite-tracking data, the autumn migration of the crested honey buzzard from Japanese breeding grounds starts early September and all individuals leave by the end of the month. They take 51–79 days to complete migration, covering 8,515–12,502 km (Higuchi 2010). Birds congregate in large numbers at the south-western most part of Japan, the Goto Islands (Personal observations), before starting their ca. 680 km journey over the East China Sea. They then follow a route inland through China, Vietnam, Laos, and Thailand until they reach the Malay Peninsula. The migration route from Japan to the southern part of the Malay Peninsula is very consistent among individuals. Afterwards, the birds follow different routes to winter in the Malay Peninsula, the Philippines, or Borneo (Yamaguchi et al. 2008). Overall, females tend to arrive at the wintering sites and depart from them earlier than males. There is however a large individual variation in arrival and departure dates, as well as the duration of stay at wintering sites (Yamaguchi et al. 2008). The onset of autumn migration is highly repeatable, though slightly less so than in spring (Sugasawa and Higuchi 2019).

Observations at watchsites in Southeast Asia suggest a bimodal distribution in the number of counts over time in autumn. This can be due to the age-related differences in the timing of migration, with an early peak of adults and a later peak for juveniles (Nijman 2004, DeCandido et al. 2015). This pattern is similar to the age-specific migration strategy of the European honey buzzard (Agostini 2004, Trabalon and García 2015). Another similarity with the European honey buzzard is that juvenile birds spend their first spring in wintering grounds and return to breeding grounds after 14 months (DeCandido et al. 2015).

As in other large raptors, crested honey buzzards are soaring birds that rely on thermal uplift and wind support to save energy while migrating. Yet, they perform a 680-km sea-crossing in autumn that is one of the longest recorded among *Accipitriformes* (Agostini et al. 2015a). Agostini and Mellone (2007) speculated that such long sea-crossing is the result of a time minimization strategy, given that during winter the birds have to compete for suitable territories with the conspecific resident populations. In addition, this seasonal sea-crossing behaviour is facilitated by favourable wind conditions over the East China Sea in autumn (Yamaguchi et al. 2012, Nourani et al. 2016), as well as the potential for uplift over the water surface (Yamaguchi et al. 2012). This behaviour is shown in adults as well as in the one juvenile that has been tracked (Higuchi et al. 2005). Nourani et al. (2017) expect that climate change can impair the suitability of atmospheric conditions over the East China Sea for autumn migration of crested honey buzzards as early as the mid-21st Century.

Pre-breeding Migration

Satellite-tracking data suggest that crested honey buzzards depart from their wintering areas in Southeast Asia from mid-February to mid-March. The birds take 66–85 days to complete their 8,690–12,350 km journey back to breeding grounds in Japan (Higuchi 2010). On spring migration, the birds fly back from scattered wintering grounds in Southeast Asia to the Malay Peninsula and go through Thailand, Laos, and Vietnam to arrive at inland China. In China, they follow a detoured route compared to autumn (Fig. 1). That is, they avoid flying over the East China Sea. Instead, they proceed northward within China to arrive at the northern part of the Korean Peninsula, where they then fly south over the Korean Peninsula and cross the Korean/Tsushima Strait to reach Japan. All arrival dates in breeding areas are concentrated in late May, with females arriving earlier than males (Yamaguchi et al. 2008). Spring migration in crested honey buzzards is highly repeatable temporally, slightly more so than in autumn migration (Sugasawa and Higuchi 2019).

It is not clear why the species performs a loop migration. The detoured migration route in spring can be partly explained by the lack of supportive wind over the East China Sea in this season that hamper the birds' ability to sustain a 680-km long flight over the sea (Yamaguchi et al. 2012, Nourani et al. 2016). Stationary weather fronts over the sea in spring cause rain and storm (Yamaguchi et al. 2008), making flight over the sea more costly than taking a longer route over land. Nonetheless, a small number of individuals have been observed to cross the sea and arrive in south-west Japan in late March, about 45 days earlier than the tracked birds (Yamaguchi et al. 2008). It is not known whether atmospheric conditions over the sea earlier in spring promote this sea-crossing.

The birds have significant stopovers in spring that last 1–5 weeks, mostly in Southeast Asia, before travelling to China (Yamaguchi et al. 2008). Stopover locations and timing are very repeatable, which allows the birds to take advantage of predictable resources, a behaviour expected from such a specialist species (Sugasawa and Higuchi 2019).

Movements in Central Asia, the Middle East, and the Western Palearctic

Hundreds of crested honey buzzards are observed each year flying through Central Asia (Schweizer and Mitropolskiy 2008, Faveyts et al. 2011). The origin of these birds may be the western limits of the species' distribution range in Siberia. This flyway remains poorly studied. Most information comes from the Chokpak Pass in Kazakhstan, a major bottleneck for migration of crested honey buzzards in Central Asia, where hundreds of birds are observed every September (Forsman 1994). In spring, tens of birds are observed occasionally in Kazakhstan and other Central Asian countries (Faveyts et al. 2011). Crested honey buzzards winter in small numbers (10s–few 100s) in the Middle East, with records from the Arabian Peninsula (Babbington and Campbell 2016) and Iran (Khaleghizadeh et al. 2011).

Crested honey buzzards have been sporadically recorded in Europe, e.g., in Greece, Cyprus, and Southern Italy (Orta et al. 2019). They are also increasingly recorded in migration watchsites in the Western Palearctic (Faveyts et al. 2011), including Eastern Turkey, Eilat in Israel, and Batumi in Georgia (www.batumiraptorcount.org). These records, as well as some records in Africa (Gabon: Clark and Christy 2006; Egypt: Faveyts et al. 2011; Sudan and Kenya: Kennedy and Marsh 2016), indicate that some birds may join flocks of European honey buzzards and migrate with them to Africa. Alternative explanations point to the possibility that these vagrant individuals are hybrids between the two species, as their breeding ranges overlap in Siberia, though this has not been confirmed by DNA studies.

Records of the crested honey buzzard in the western limits of its range are increasing. This could be the result of improved identification knowledge of birdwatchers and bird counters (Faveyts et al. 2011), but could also indicate a westward expansion of the species, potentially due to the increase in suitable anthropogenic habitat, such as irrigated tree plantations in Middle Eastern countries (Babbington and Campbell 2016).

Behavioural and Ecological Remarks

Crested honey buzzards on migration in East Asia are likely to benefit from thermal activity when crossing migration bottlenecks. Peak passage through these bottlenecks is during mid-day, when thermal activity is most favorable (DeCandido et al. 2004a, Puan et al. 2014). In fact, strong thermals sometimes allow the birds to fly at very high altitudes, beyond the detection range of counters, thus resulting in underestimations of bird numbers (Germi 2005). Thermals and orographic updrafts also enable the birds to combine soaring and flapping flight while flying over bottlenecks (Puan et al. 2014). In Tanjung Tuan, Malaysia, fewer crested honey buzzards are counted under strong winds. Puan et al. (2014) suggest that this could be due to the disturbance of thermals by strong winds, which makes sea-crossing difficult for the birds. In contrast, supportive winds have been shown to facilitate Crested honey buzzards' sea-crossing over these bottlenecks (DeCandido et al. 2004a, DeCandido et al. 2004b, DeCandido et al. 2006a, Puan et al. 2014). To date, remote-tracking studies have only looked at the effects of atmospheric conditions on the sea-crossing behaviour of the crested honey buzzard over the East China Sea (Yamaguchi et al. 2012, Nourani et al. 2016). Future studies could uncover this species' use of uplifts and wind when sea-crossing further south along its migration route.

Feeding ecology of the crested honey buzzard on autumn migration is another behaviour that requires further investigation. Observations at watchsites in Thailand suggest that the species adopts a fly-and-forage strategy in autumn, as many migrating crested honey buzzards are observed with a full crop or carrying pieces of honey comb (DeCandido et al. 2015). Satellite-tracking data suggest that the birds rarely stopover during autumn migration (Yamaguchi et al. 2008), although they sometimes slow down while continuing in the direction of migration. It is not clear from satellite-tracking data if this is due to adverse weather conditions or foraging activity (Yamaguchi et al. 2008).

6

Eurasian griffon vulture *Gyps fulvus*

Alejandro Onrubia

ABSTRACT

The Eurasian griffon vulture (*Gyps fulvus*) is a long-lived species characterized by delayed maturity, long-lifespan, long breeding performance and low reproductive rates. Most of the studies have focused on breeding ecology of the species, but less is known about its movements. It is a large raptor with relatively high wing loading that is favourable for obtaining high inter-thermal gliding speed, but limits the ability to use weak thermals and greatly increases the energetic cost of flapping flight compared with smaller soaring species. The griffon vulture is a sedentary or partially migratory species, with a complex movement pattern, not entirely well known, which includes also movements assimilable to a true migration because of their directionality, regularity in spatio-temporal patterns and its population scale. Such migratory movements connect the breeding grounds in Europe with sub-Saharan Africa, and occur mainly during the first years of life, becoming mostly sedentary after first breeding, though some adults eventually engage in long-range forays. The movements of juveniles and immatures (and some adults) resemble the spatial pattern of other large migratory birds of prey flyways (Western Mediterranean, Eastern Mediterranean and Middle East, Caucasian and Central Asian flyways), being even more constrained by sea-crossings and therefore the distribution of land masses.

Introduction

The Eurasian griffon vulture (*Gyps fulvus* Linnaeus, 1758), hereafter simply "griffon vulture", is a large bird of prey (93–110 cm), with a wingspan of 230–270 cm and a weight ranging from 5.5 to 10.7 kg. It is a robust bird with a white head, neck and tail, robust body and brown plumage, and black flight feathers. Juveniles are similar to adults, but with darker tones, brown collareta and black bill color. It is very similar to other vulture species (gen. *Gyps*), slightly smaller and brownish than *G. himalayensis* and bigger and paler than *G. bengalensis*, *G. indicus*, *G. rueppellii* and *G. africanus* (Orta et al. 2020). Two subspecies are recognized: *G. f. fulvus* (Hablizl 1783) and *G. f. fulvescens* (Hume 1869). For flying, the griffon vulture is characterized by a relatively high wing loading that is favourable for obtaining high inter-thermal gliding speed, but limits the ability to use weak thermals and greatly increases the energetic cost of flapping flight compared with smaller soaring–gliding species (Ruxton and Houston 2004, Duriez et al. 2014, Agostini et al. 2015a).

As a breeder, the griffon vulture is distributed throughout the mountainous massifs of the circum-Mediterranean arch (North West Africa, Iberian Peninsula, Southern France, islands of Majorca,

Migres Foundation. Tarifa-11380. Spain.
Email: aonrubia@fundacionmigres.org

Sardinia, Sicily, Crete and Cyprus), Balkans, Turkey, Caucasus, Middle East, Arabia, Zagros Mountains and Elburz, Kan plateau and Pamir plateau to the southeast of Kazakhstan. The subspecies *G. f. fulvescens* occupies the mountains of the Hindukush and its surroundings (Afghanistan, Pakistan and northwest India). During the winter, a part of the population (usually juveniles and immature) moves to the south and can also be found in the Sahel belt, the entire Arabian Peninsula, Iran, Pakistan and the foothills of the Himalayas, from Northwestern India (Punjab, Rajasthan, Himachal Pradesh, Uttar Pradesh, Bihar) to Bhutan (Orta et al. 2020).

It usually occupies extensive open areas and a wide variety of habitats, including mountain ranges, plateaus, steppes and semideserts with the presence of large herbivores, occupying abrupt rocky areas such as crags, cliffs and canyons for nesting or roosting. It nests in a wide range of altitudes, from cliffs at sea level (for example, Spain, Sardinia, Croatia) to heights above 2,750 m in the Caucasus (Orta et al. 2020). Being a large and heavy bird makes its movements soaring and gliding, it needs strong updrafts associated with thermal or orography for its flights, which is why it is usually associated with dry and sunny areas.

Their diet consists almost exclusively of carrion and feeds mainly on the muscles and viscera of the carcasses of medium and large-sized mammals, usually ungulates. Originally it depended on wild preys (mostly goats, deer, gazelles, antelopes, and sometimes lagomorphs, carnivores, and even cetaceans), but since the Neolithic revolution, domestic ungulates (sheep, goats, cows, horses, pigs) have become important, and are now dominant in its diet which has become highly dependent on their availability (Cramp and Simmons 1980, Donazar 1993, Orta et al. 2020). This circumstance has recently caused certain problems due to the limitations imposed by sanitary regulations in the European Union, which has motivated the search for alternative food sources, such as rabbits affected by diseases, or even garbage (Garrido et al. 2002, Donazar et al. 2010).

It is a gregarious species and different individuals cooperate in searching for food, dispersing great distances from the breeding colonies or roosts, exploring wide areas meticulously. Usually the griffon vulture breeds in colonies, although isolated pairs are not uncommon. It nests in crags or cliffs, preferring a protected ledge or small cave, where it installs a simple branch platform. Mating season takes place in winter and the laying dates are framed between December and March (Cramp and Simmons 1980, Donazar 1993, Orta et al. 2020). After breeding, most of adults extend their domains, making extensive daily movements but stay close to the breeding areas which are usually linked to the colonies or communal roosts. Seasonal or nomadic movements also occur and are frequent and are in relation to climate and movements of livestock herds (Bernis 1983, Donazar 1993). In autumn, juveniles start a period of dispersal that can lead them to cover long distances, sometimes far away from their usual range, for example, wandering in Central and Northern Europe or on the islands of the Mediterranean (Terrasse 2006, Danko et al. 2013, de Juana and Garcia 2015). However, part of these juveniles perform what has been assimilated to a true migration, and not to dispersal, due to different characteristics: (1) the magnitude (hundreds or thousands of individuals involved), (2) the entity of the distances travelled, (3) the spatio-temporal regularity, and (4) the directionality, connecting the breeding range (Europe) with an area only used during a non-breeding stage (Sahel) (Bernis 1983).

The griffon vulture is a very long-lived species and there are known records of individuals living up to 37 years old in captivity (Orta et al. 2020). Sexual maturity is reached around 4–5 years of age, although the first reproduction usually occurs later. Approximately 70% of pairs lay eggs and breeding success is usually 30–35% with a productivity of 0.3–0.7 chicks fledged per clutch (Donazar 1993, Orta et al. 2020).

The griffon vulture has a world population that has been estimated at over 640,000 mature individuals (BirdLife International 2019c). Although it is a species favored by the increase in the domestic livestock herd and the extension of the grasslands, it suffered a sharp decline throughout the 20th century in Europe, North Africa and the Middle East, becoming extinct or significantly rare in large areas (Morocco, Algeria, Balkans, Eastern Europe), mainly due to direct persecution and the effect of the use of poison to control predators, and in some areas was linked to the reduction of food availability due to changes in livestock and carcass management or affected by different sanitary measures (Thiollay 2006, Azmanis et al. 2009, Donazar and Margalida 2009, Donazar et al. 2010, Arrondo et al . 2018, Orta et al. 2020, Ogada et al. 2011). More recently, the use of veterinary drugs in livestock (Non-steroidal anti-inflammatory drugs—NSAIDs, for example flunixin or diclofenac) has been identified as a serious threat to this species (Blanco et al. 2009, Zorrilla et al. 2015, Ławicki and van den Berg 2016), being responsible for severe declines of the Asian vultures (Ogada et al. 2011). It is also very vulnerable to the impact of wind farms (De Lucas et al. 2011, Strix 2012) and to electrocution and collision in power lines (Global Raptors Information Network 2019). However, it is probably stable in most of Asia and has undergone a recent spectacular increase in some populations in Western Europe, with increased protection measures and sometimes aided by reintroduction or population reinforcement projects (Italy, Austria, Bulgaria, Sardinia) (Terrasse et al. 2004, Genero 2009, Andevski 2013, Orta et al. 2020). Currently the most important populations are in the Iberian Peninsula (more than 30,900 couples in Spain; Del Moral and Molina 2018), Southern France (2,000 pp), Balkans (> 380 pp), Greece (> 350 pp), Mediterranean islands (30 pp Sardinia, > 90 pp in Sicily), Turkey (> 500 pp), Russia (> 300 pp), Israel (> 70 pp) and Azerbaijan (> 100 pp) (BirdLife International 2019c). No information is available on the populations of the Middle East and Asia Minor, but they could be numerous. It is not currently considered a threatened species and has been listed as Least Concern (BirdLife International 2019c).

Knowledge Status

Knowledge about the complex movements of griffon vultures has been gained by direct observation, often under the umbrella of migration-specific monitoring programmes at observatories located in bottlenecks, corridors, or concentration points (Bijlsma 1987, Zalles and Bildstein 2000), either through marking with rings and special marks and, more recently, from marking vultures with radiotransmitters or with satellite-transmitters (Alarcón et al. 2018). Furthermore, the griffon vulture is a target species for the study of flight dynamics in soaring birds, using modern GPS devices with accelerometers and other sensors (Nathan et al. 2012, Treep et al. 2016).

The first evidences of the long-distance movements of griffon vultures derive from direct observation and from records of changes in their distribution in various areas (Bernis 1983, Dendaletche 1988). The appearances of vultures in territories far from their breeding areas, e.g., in Central and Northern Europe, evidenced certain erratic movements, while other periodic records revealed summer nomadism associated with the seasonal movement of livestock in mountain and foothill areas (transhumances) involving long displacements related to the climate and the movement of livestock herds (Bernis 1983, Mendelsshon and Leshem 1983, Donazar 1993). Raptor migration surveys carried out at different watchsites in the Mediterranean basin revealed regular movements of griffon vultures, especially through the Strait of Gibraltar and in the Suez Canal (Egypt).

Nowadays, the annual crossing of several thousands vultures (mainly juveniles and immatures) through the Strait of Gibraltar between Spain and Morocco has been estimated, in what is currently the most important migration place in the world for griffon vultures (Griesinger 1996, Onrubia 2016, Ramirez et al. 2019). On the other hand, reflecting their absence in Northern Europe, vultures are not recorded at the Falsterbo and Baltic observatories (Kjellen and Roos 2000). The low numbers recorded at the observatories of the Central Mediterranean (Strait of Messina, Sicily Channel, Cap Bon, Malta; Corso 2001a, Sammut et al. 2013) shows the limitation that long-distance flight over water (> 25 km) means for the species, and the barrier effect of the central Mediterranean sector (Agostini et al. 2015a).

In the eastern route, there are some griffon vultures that pass through the western shores of the Black Sea, in Dobrogea region (Romania) and further south in Bourgas Bay (Bulgaria), which sum a few dozens in the Bosphorus in Turkey and lower numbers in Dardanelles (Michev et al. 2011, Fülöp et al. 2014, Fülöp et al. 2018, Panuccio et al. 2017, Uysal and Tosunoğlu 2018). Similarly, a few dozens are recorded crossing the Caucasus mountains, in Batumi (Georgia) and Armenia (Verhelst et al. 2011). Dozens of griffon vultures are also recorded passing through the eastern coasts of the

Table 1. Overview of the timing of spring and autumn migration of griffon vultures from ground observations at selected migration bottlenecks along the Western and Eastern Mediterranean flyways. Note that the date 'range' and 'magnitude' are dependent on observation effort and may not encompass the entire temporal window of migration.

Country	Location	Spring migration				Autumn migration				Source
		Magnitude	Range	Peak	Range	Magnitude	Range	Peak	Range	
Western Flyway										
Spain	Gibraltar	8,700	6,000–9,000	8-May	07 Apr–30 May	9,800	8,000–12,000	30-Oct	28 Jul–18 Nov	Onrubia 2015, El Khamlichi 2017, Ramirez et al. 2019
Eastern Flyway										
Romania	Dobrogea					< 5				Fulop et al. 2018
Bulgaria	Bourgas Bay					< 10		24-Sep	24 Aug–21 Oct	Michev et al. 2011
Georgia	Batumi					< 10				Verhelst et al. 2011, Werham et al. 2019
Turkey	Bosphorus	< 10	5–20	5-Apr	27 Mar–23 Apr	55	5–170	15-Oct	5 Sep–25 Oct	Collmait and Croxall 1967, Porter and Willis 1968, Shirihai et al. 2000, Üner et al. 2010, Fülöp et al. 2014, Panuccio et al. 2017
Turkey	Dardanelles	< 5				< 5				Panuccio et al. 2017, Uysal and Tosunoğlu 2018
Turkey	Northern Pontics					30				Shirihai et al. 2000
Turkey	Gulf of Iskenderun/ Belen Pass					< 10	5–125			Sutherland and Brooks 1981, Shirihai et al. 2000, Oppel et al. 2014
Lebanon	Harissa					50				Shirihai et al. 2000
Israel	Kfar Qassem/ Northern Valleys					45	10–76			Shirihai et al. 2000, Alon et al. 2004
Israel	Eilat	10	2–22		21 Feb–20 May	17	2–45	30-Sep	20 Sep–5 Nov	Shirihai and Crhistie 1992, Shirihai et al. 2000
Egypt	Gulf of Suez/Zait Bay	140					0–600	25-Oct	23 Sep–5 Nov	Bijlsma 1983, Shirihai et al. 2000, Megalli and Hilgerloh 2013
Djibouti / Yemen	Bab Al Mandeb					< 5	1–15	30-Oct	25 Oct–10 Nov	Porter and Christensen 1987, Shirihai et al. 2000, Welch G. and Welch 1988, 1998
Asian Flyway										
Nepal	Thoolakharka						11–85		10–30 Nov	Subedi and De Candido 2013, 2014, Subedi and Gurung 2018, Gurung et al. 2019

Mediterranean in the Middle East, crossing the Gulf of Iskenderun (Belen Pass, Turkey), the coasts of Syria, Lebanon (Harissa), Kfar Qassem (Israel), Rif corridor (Israel, Jordan) or crossing the Sinai Peninsula with records of several dozens in Eilat (Israel) or a few hundreds across the Gulf of Suez or Zait Bay (Egypt) (Shirihai et al. 2000, Zalles and Bildstein 2000, Megalli and Hilgerloh 2013, Kirwan et al. 2015). Observations and censuses of griffon vultures played an important role, even now, for describing the patterns of their movements, but they exhibit some important limitations. On the one hand, they do not allow determining clearly the type of movement—feeding, foraging, nomadism, dispersal or true migration—of birds observed. On the other hand, they do not say much about the origins and destinations of the vultures in motion, for which the individual tagging of vultures with different techniques has played a crucial role, either with metal rings, special markings or remote monitoring.

The first ringing campaigns of griffon vultures were carried out in Spain in the 1960s and 1970s, and already with the first regular ringings in the 1960s, a clear trend was found for young vultures born in northern areas of the Iberian Peninsula to move in their first autumn to the southwest, some coming to cross the Strait of Gibraltar (Elosegui and Elosegui 1977). Subsequently, Bernis (1983) exposes the information on several more recoveries and confirms the same trend. Likewise, Alonso (1984) cites how two ringed chicks in Cádiz (Southern Spain) were recovered in their first winter in Senegal, 2,740 and 2,860 km away from their birthplace. Later works by Griesinger (1996 and 1998) marking 31 young vultures in the north of the peninsula with colored rings and wingtags showed that the majority of these young birds were re-sighted in Southern Spain, where a large part of them cross to the African continent through the Strait of Gibraltar. From the 1990s to the present, the marking of griffon vultures with metal and plastic rings and wingtags has spread throughout Europe, providing novel data about vulture movements (see for example, Spina and Volponi 2008, Kralj et al. 2013, Zuberogoitia et al. 2013).

SEO/BirdLife (2012a) ringing database for Spain includes 805 vulture recoveries in the period 1960–2010, showing a complex movement pattern (Fig. 1). Marked vultures in Spain have given recoveries throughout the Iberian Peninsula, North Africa (Tunisia, Algeria, Morocco), Sahel (Mauritania, Mali, Senegal), but also in France (Pyrenees, Massif Central, Alps), Corsica, Netherlands,

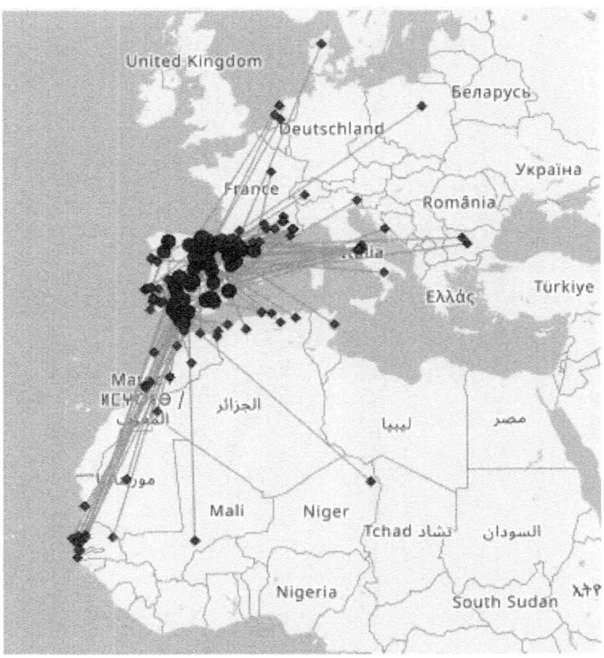

Figure 1. Ringings (dots) and recoveries (squares) of griffon vultures born in Spain (SEO/BirdLife 2012a).

Poland, Italy, Croatia and Germany (SEO/BirdLife 2012a, Bairlein et al. 2014, De Juana and Garcia 2015). Ringed vultures in France have been recovered in the Iberian Peninsula (Portugal, Spain), the United Kingdom, Belgium, the Netherlands, Germany, Switzerland, Italy and Austria (Dendaletche 1988). In Italy, the ringing of the Sardinian population (the only one of natural origin although with successive reinforcements) show local movements within the island (Spina and Volponi 2008). Ringed or recovered vultures in continental Italy (reintroduction projects in Abruzzo) show movements between the Apennines and the Alps (both western and eastern), with connections with the Eastern Adriatic (Croatia, Montenegro) (Spina and Volponi 2008, Genero 2009). Regarding the population of griffon vultures in Croatia (100–150 pp), it had intense ringing or special markings programs that, between 1990 and 2009, produced more than 1,400 records of tagged griffon vultures (Kralj et al. 2013). Of these, 90% of young birds marked with rings or wing-marks performed extensive journeys, which produced records in autumn from Slovenia, Spain, France, Poland, Bulgaria, Greece, Romania, Turkey, Ukraine, Israel and Chad; winter records from Serbia, Italy, Spain, France, Greece, Turkey and Israel; and records in spring and summer from Slovenia, Italy, France, Poland, Austria, Hungary, Bulgaria, Romania, Russia and Greece (Kralj et al. 2013).

Furthermore traditional ringing schemes, radio-tracking studies have shown that the species ranges extensively, particularly juveniles (e.g., Bahat et al. 1993, Griesinger 1996 and 1998, Gil et al. 2009). The study of Griesinger (1996 and 1998), who investigated the migration of griffon vultures from the north of the Iberian Peninsula by marking 16 juveniles with radio transmitters between 1990 and 1993, was pioneering. The results showed that more than 90% of the juveniles left their natal colonies at the end of October, and two or three weeks later, a large part of them reached the area of the Strait of Gibraltar in the south of the Iberian Peninsula, travelling average distances of 30 to 50 km/day. However, the limitations of this radiotelemetry technique did not allow knowing the complete route in detail.

Through satellite telemetry and GPS data-loggers different movement patterns of griffon vultures have been described, showing the complexity of this behaviour, consisting in a continuum between foraging movements and likely true migration (Fig. 2). During foraging movements normally griffon vultures do not move beyond 200 km from their home range center (Xirouchakis and Andreou 2009,

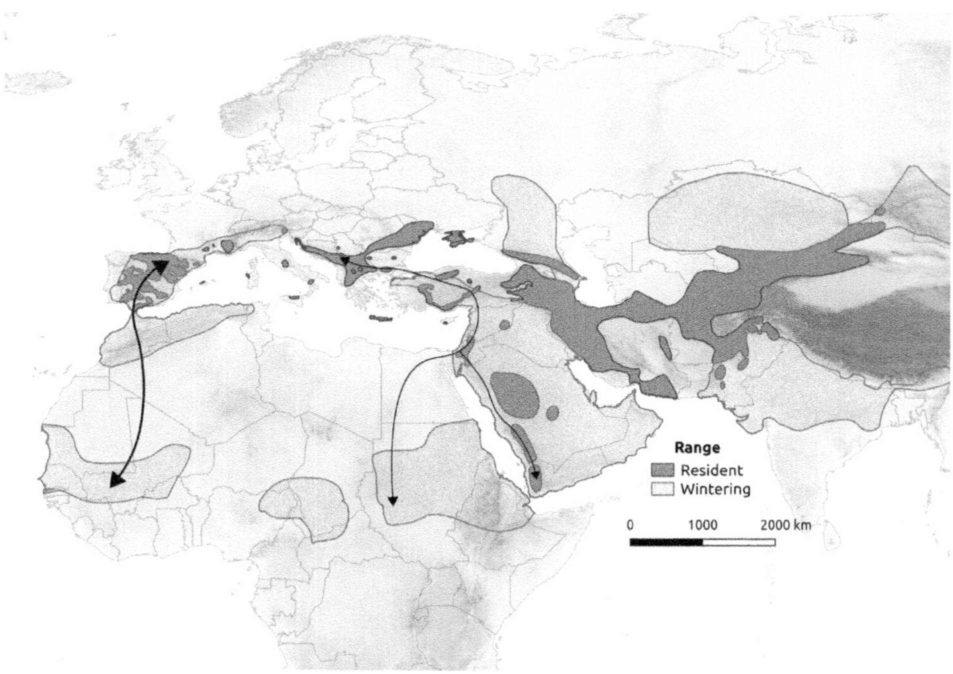

Figure 2. Schematic representation of the known routes followed by griffon vultures during long-distance displacements.

Garcia-Ripolles et al. 2011, Nathan et al. 2012, Monsarrat et al. 2013, Harel et al. 2016a, Peshev et al. 2018). However, griffon vultures can also perform long-distance foraging movements, probably related to non-breeding individuals, which may be either dispersing juveniles, or adults that failed the breeding season (Bahat 2001, McGrady and Gavashelishvili 2006, Spiegel et al. 2013, Harel et al. 2016a, Garcia-Ripolles et al. 2011). Nathan et al. (2012) describe the movement during autumn of four vultures tagged in Israel (less than 2.5% of the sample) flying almost directly south-southeast for 2–10 days, 400–1,600 km from their usual activity center, foraging for 11–68 days, before returning home in a journey lasting 12–13 days. Finally, some *migration movements* have been recorded using this GPS technology. Muñoz et al. (2016) show the details of a complete migration cycle of an adult griffon vulture from Spain to Senegal, while Arkumarev et al. (2019) described the autumn migration of a juvenile from Bulgaria to South Sudan.

According to this information griffon vultures can be considered as partial migrants during the first years of life, becoming mostly sedentary after first breeding, though some adults eventually engage in long-range forays, even migration cycles, presumably reflecting unsuccessful breeding attempts or visits to foraging hot-spots.

After leaving the nest, the young vultures can disperse over a very wide territory, with variable directions. A portion of these juvenile birds, which can number several thousands individuals, carry out movements that can be assimilated to a true migration, due to their directionality, magnitude, regularity in space-time patterns, and the fact that they imply a "stationary" period within a non-breeding area (Western Sahel) (Bernis 1983). Griesinger (1998) estimates that 90% of the juvenile population in the north of the Iberian Peninsula leaves their breeding colonies towards the end of October and moves south, where a large part of them cross to the African continent, to spend the winter months scattered in the Maghreb and in the westernmost portion of the Sahel (Senegal, Mauritania, Mali, Niger; Bernis 1980 and 1983, Terrasse and Thauront 1988, Thévenot et al. 2003, Orta et al. 2020). It is estimated that 30% of the Iberian juvenile population could cross the Strait of Gibraltar each autumn (Griesinger 1996). Also, Kralj et al. (2013) estimate that 90% of young vultures in Croatia leave the breeding grounds of the Balkans and move in different directions, either towards Italy, France and Spain, or to Greece and the Middle East. Moreover, vultures tagged in Israel, either resident or migrating birds, showed a wide array of destinations, ranging from Turkey to Yemen, or even Chad (Bahat 1997 and 2001, Bahat et al. 2001, Nathan et al. 2012, Kirwan et al. 2015). Juveniles tagged in the Upper Caucasus moved to Iran, and some individuals even reach the central zone of Saudi Arabia (McGrady and Gavashelishvili 2006).

The return of juvenile vultures to the north usually occurs in the following spring, in their second calendar year, although a part of them may remain for a variable time (2–4 years) in wintering areas or in intermediate territories (Donazar 1993, McGrady and Gavashelishvili 2006, Zuberogoitia et al. 2013, Ramirez et al. 2019). Although the general patterns of the complex movements of griffon vultures are known, there are still gaps, especially concerning smaller populations (Alps, Balkans, Caucasus) moving mainly through the Eastern flyway, where low numbers are registered. Likewise, migration timing of vultures is different to other soaring birds of prey, being usually very late in both autumn and spring, and out of most of the observatory monitoring protocols. And all this aggravated with the difficulty of evaluating in the observatories what corresponds to foraging movements of local birds and true migratory displacements of foreign birds.

Post-breeding Migration

The migratory routes of griffon vultures follow the general patterns of other soaring raptors, distinguishing a western route and several eastern routes, and the absence of a route in the Central Mediterranean (Fig. 2). In general, griffon vultures tend to fly over land masses in order to make use of thermal currents. The Western flyway gathers mainly young vultures from the Iberian and Southern French populations, which cross the Iberian Peninsula in a broad front, and concentrate in Southern Spain. Part of them crosses the Strait of Gibraltar to North Africa. An unknown portion of these vultures cross the Sahara desert until reaching the steppes and savannas of Western Sahel. Such movements involve also French birds and even Italian and Balkan ones (Terrasse 2006, www.migraction.net).

Within the Iberian population, juvenile griffon vultures mainly start their southward movements since the end of July. The transit of the Iberian Peninsula extends until December. A significant part of this flow is funneled through the Strait of Gibraltar to cross into Africa (Fig. 2). In the 1970s, it was estimated that 1,000 to 2,000 griffon vultures crossed the Gibraltar Strait in the autumn (Bernis 1980, 1983), although today the figure ranges between 8,000 and 12,000 individuals (Onrubia 2016, El Khamlichi 2017, Ramirez et al. 2019), in what is the largest concentration of griffon vultures worldwide.

Daily monitoring in the Strait of Gibraltar between 2008–2019 ("Migres programme") indicates that autumn migration extends for about 5 months (146 days), from July 11 (first contact crossing) to December 3 (last contact), although 80% of the griffon vultures groups cross in 29 days, between October 17 and November 14, with the peak of migration around October 30 (Table 2, Fig. 3). The maximum number of vultures crossing in one day was recorded at 28-October-2015, when 3.200 vultures crossed the Strait after five days of bad weather. Almost 90% of passage concentrates on 10:00 to 15:00, with maximums at 11:00–14:00, when the stronger thermals occur (Fig. 4). When passing through the Strait of Gibraltar, the griffon vultures appear travelling in groups of variable size, from a few individuals to several hundreds. During post-nuptial migration, the average size of the groups is 33.1 ± 5.96 (SE) individuals (range of 1 to 746 birds), and 78% of the flocks are made up of less than 20 vultures. In general, the size of the flock during post-breeding migration is markedly greater than during the pre-breeding one (Fig. 5). Within a sample of 2,610 aged birds, more than 90% were not adults (76% juveniles, 17% immatures). Regarding the magnitude of the passage, the available data in the period 1970–2019 indicate a clear increase in the passage of griffon vultures through the Strait of Gibraltar in recent years definitely related to the increase of the source population.

Bildstein et al. (2009) concluded that the Strait is a significant barrier to the vultures, and the width of the Strait at its narrowest point (14 km) is close to the maximum that Griffon Vultures are capable of crossing regularly. Usually, large numbers may build up in the hinterland at peak migration periods while the birds wait for suitable crossing conditions and follow the coast, west or east, perhaps searching for other place for crossing. Griffon vultures are very common in Southern Portugal in October and November, and hundreds of vultures sometimes wander as far as the Sagres Peninsula, in the extreme south-west, some of which may eventually return eastwards and cross the Strait later (Palma and Beja 1994, Tomé et al. 1998, De Juana and Garcia 2015).

Table 2. Timing of Eurasian griffon vultures passage through the Strait of Gibraltar (according to Migres Program database).

	Spring migration	Autumn migration
Years	2009–2012	2008/2015–2018
Passage duration (nº days)	126	146
Duration 95%	64	55
Duration 90%	55	41
Duration 80%	47	29
First contact	5-Feb	11-Jul
1%	19-Mar	29-Jul
5%	6-Apr	14-Oct
10%	13-Apr	17-Oct
50%	8-May	4-Nov
Máximum	25-Apr	28-Oct
90%	28-May	14-Nov
95%	30-May	23-Nov
99%	31-May	2-Dec
Last contact	10-Jun	3-Dec

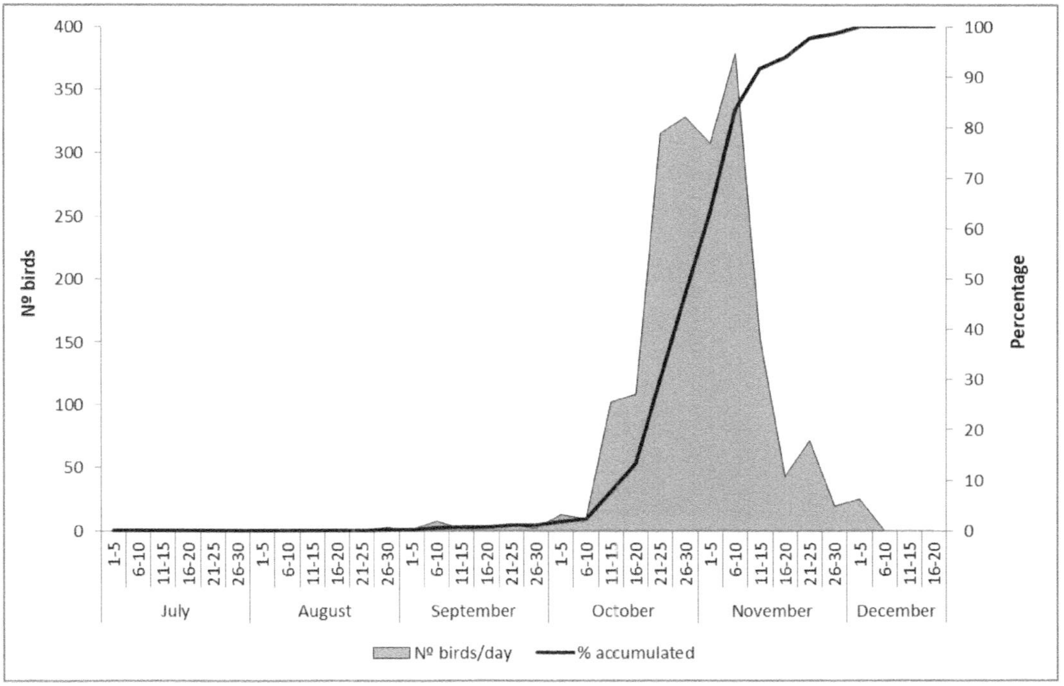

Figure 3. Timing of griffon vulture during autumn migration through the Strait of Gibraltar (average daily passage, period 2008–2019).

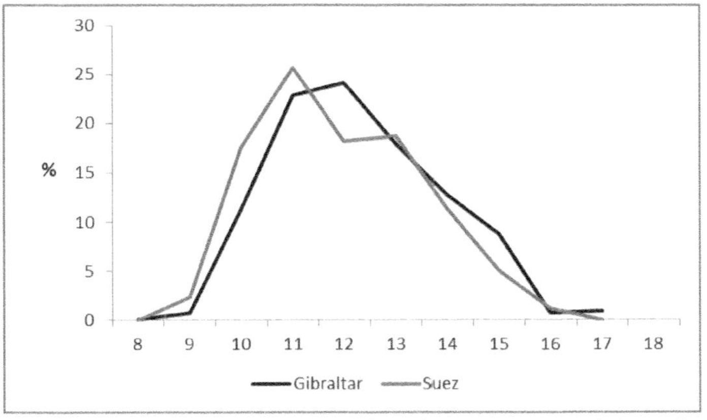

Figure 4. Time schedule in percentage of birds per hour of griffon vulture passage through the Strait of Gibraltar -Spain- (n = 7,494 vultures) and Suez –Egypt- (n = 1,284) (Onrubia 2016, Bijlsma 1983).

In the Central Mediterranean flyway there are no records of regular crossing of griffon vultures in the Sicily Channel, between Italy and Tunisia (> 150 km), and also from the islands where they breed (Mallorca, Sardinia, Crete, Cyprus), which shows the limitations of the species for crossing long distances over the sea (Corso 2001a, De Juana and García 2015). Vultures from the colonies of Italy or the Alps disperse towards the south of France and Iberia or towards the Balkans (Spina and Volponi 2008).

A tracked adult vulture from Spain needed 13 days for the complete autumn journey from Iberian Peninsula to Senegal flying a daily average of 206 km/day (maximum distance covered in one day—323 km), spending 4 days for crossing the north-african mountains in Morocco and 8 days for the

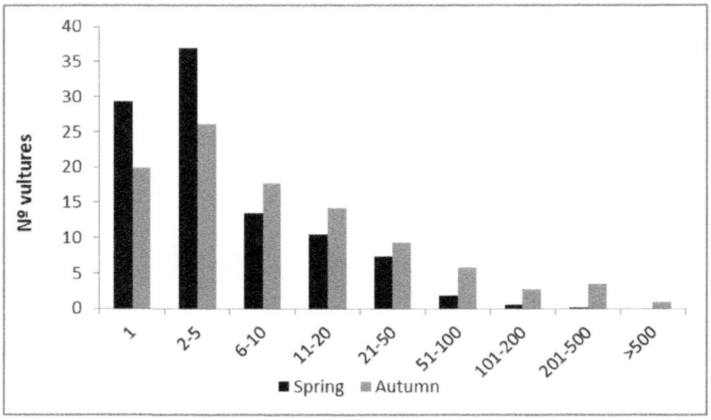

Figure 5. Gregarism of griffon vultures passage through the Strait of Gibraltar (Size of groups, for n = 7,494 vultures (autumn) and n = 10,395 (spring), period 2008–20019).

Sahara crossing. The bird crossed the Strait at mid of November and reached the wintering areas at the end of November (Muñoz et al. 2016). The average flight speed varied between 37 and 45 km/h, with a maximum of 65.5 km/h (Muñoz et al. 2016).

There are three clear wintering areas of the Spanish and Central European populations: (1) the Sahel in Senegal, Mauritania, Mali, Burkina Faso, and Niger; (2) the north of Morocco; and (3) the Iberian Peninsula, especially Southern Spain.

As regards the eastern flyway, which involves mainly individuals from the Balkans and Middle East, it is more discrete, summing probably no more than 1,300 birds (Shirihai et al. 2000). Less than 10 vultures are recorded between late August till end of October in observatories located at both sides of Black Sea in Romania, Bulgaria or Batumi (Michev et al. 2011, Verhelst et al. 2011, Fülop et al. 2018, Werham et al. 2019; Table 1), summing up no more than 200 vultures in Bosphorus and Dardanelles in Turkey, with the peak at mid-October (Collmait and Croxall 1967, Porter and Willis 1968, Shirihai et al. 2000, Üner et al. 2010, Fülöp et al. 2014, Panuccio et al. 2017, Uysal and Tosunoğlu 2018; Table 1). Low numbers (< 350 birds) are also recorded in the Gulf of Iskenderun (Turkey), and Israel (Shirihai et al. 2000). In the Middle East the vulture passage concentrates from the beginning of September to early November, and crosses into Africa, through Suez, where up to 1,200 vultures have been recorded (Megalli and Hilgerloh 2013). The majority of the birds consisted of juveniles and immatures (97%, n = 816), usually forming groups freely mixed with the stream of large eagles (Bijlsma 1983). A juvenile tagged in Bulgaria with a GPS transmitter started the journey in mid September and arrived in the wintering grounds in South Sudan six weeks later, travelling almost 6,000 km, the longest distance travelled in a single day was 346 km on 28 October, over the Sahara Desert (Arkumarev et al. 2019). The populations of Caucasus are migratory and mostly head south, following the western coasts of the Caspian Sea until reaching the Elburz (Lashgarak) mountains, or crossing the lower Caucasus, Armenian Highlands, and the mountains of Kurdistan until reaching the great mountain range of the Zagros in Iran, or even the central plateaus of Saudi Arabia (McGrady and Gavashelishvili 2006).

Pre-breeding Movements

In their spring migration, the griffon vultures probably follow the same routes they had used previously, but in the case of the Sahara crossing there is a tendency that migration occurs closer to the coast given the prevailing winds at this time (Muñoz et al. 2016), as happens with almost all trans-Saharan migratory raptors. For birds wintering in West Africa, spring migration of griffon vultures starts in March, arriving to the Iberian Peninsula in late May and June, while birds wintering in East Africa

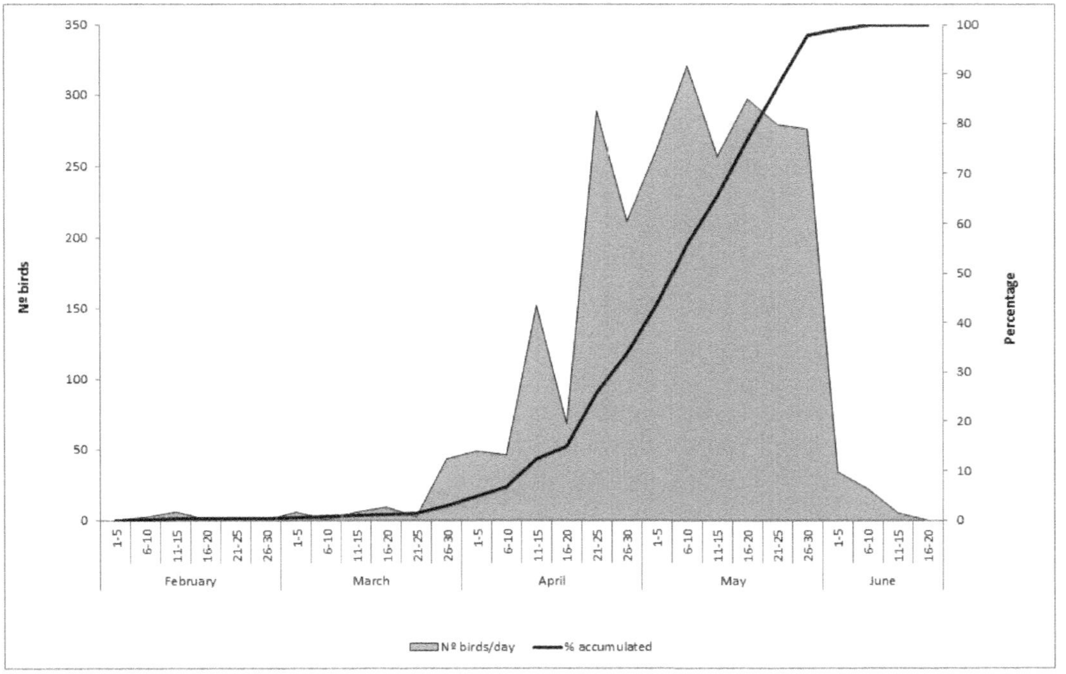

Figure 6. Timing of griffon vulture during spring passage through the Strait of Gibraltar (average daily passage, period 2009–2012—Migres Program database).

and Middle East reach Eastern Europe and Caucasus between late February and late May (Table 1; Fig. 2). There are not many details about this spring migration, but a tracked adult vulture from Senegal showed a very similar pattern of migration comparing to autumn performance: it took 14 days to complete the spring journey from Senegal to Iberian Peninsula flying a daily average of 216 km/day (maximum distance covered in one day—446 km), spending 8 days for Sahara crossing and 6 days for passing the north-african mountains in Morocco, finally crossing the Strait of Gibraltar at early April (Muñoz et al. 2016).

Data obtained in the daily monitoring program in 2009–2012 (Migres Program), showed that spring migration in the Strait of Gibraltar extends for about 4 months (126 days), from February 5 (first contact crossing) to June 10 (last contact), although 80% of the griffon vultures groups cross in 47 days, between April 13 and May 28, with the peak of migration around May 8 (Table 2, Fig. 6). The maximum number of vultures crossing in one day in spring was recorded on 25 April 2012, when 756 vultures crossed the Strait (Migres Programme database). When passing through the Strait of Gibraltar, the griffon vultures fly in groups of variable size, from a few individuals to several hundreds vultures. During spring migration, the average size of these groups is 8.5 ± 0.5 (SE) individuals (range of 1 to 260 birds), and 90% of the flocks are made up of less than 20 vultures (Fig. 6). Within a sample of 280 aged birds, more than 94% were not adults. In the last decade, the number of griffon vultures recorded crossing during spring migration ranged between 6,000 and 9,800 individuals, with an average of 8,700 individuals in passage (Onrubia 2016). A more recent study carried out during spring 2017 by Ramirez et al. (2019), reported 6,191 griffon vultures crossing and 800 individuals aged through photographs, of which 67.6% were second calendar-year birds and 32.4% were third calendar-year birds or older.

Vultures wintering in East Africa enter Asia through Israel, where some dozens are recorded every year from late January to May, but mainly end February to mid March (Bijlsma 1983, Shirihai et al. 2000, Megalli and Hilgerloh 2013). Birds wintering in the Arabian Peninsula progress north following the Red Sea coastal plain, before eventually dispersing into a broad front across the Arabian desert,

Jordan, Iraq, and Syria (Fig. 2). The flyways begin to separate depending on whether birds head to different breeding areas located in Central Asia, the Middle East, or Western Turkey and the Balkans. Most vultures using this flyway reach the European continent via the Bosphorus or Dardanelles, where less than 20 birds are recorded between mid March to end of April (Collmait and Croxall 1967, Porter and Willis 1968, Üner et al. 2010, Fülöp et al. 2014, Uysal and Tosunoğlu 2018).

Interestingly, one immature vulture tracked from Central Saudi Arabia, started migrating at mid-January and reached the breeding area in the Caucasus at mid February, 2,000 km and 27 days later (McGrady and Gavashelishvili 2006).

Behavioural and Ecological Remarks

The griffon vulture is a very large raptor (6–11 kg) with relatively high wing loading that is favourable for obtaining high inter-thermal gliding speed, but limits the ability to use weak thermals and greatly increases the energetic cost of flapping flight compared with smaller soaring-gliding species (Duriez et al. 2014, Agostini et al. 2015a). Vultures are large obligate scavengers that rely on thermal and orographic uplift for their soaring flight, where they can ascend at high speed (1.2–2.4 m/s), reaching variable heights above the ground, usually between 340–460 meters, occasionally above 1,800 m and exceptionally up to 8,000 m, then gliding at moderate speed (16.2–16.7 m/s) with high glide rates (8–10 m horizontally by 1 m vertical) (Harel et al. 2016 a,b, Treep et al. 2016, Péron et al. 2017). The result is a low-energy flight that leads them to make long daily movements with little energy consumption (Duriez et al. 2014). Energy-efficient soaring flight appears to be a key innovation in raptors in permitting large numbers of birds of prey to complete long-distance migratory movements with little or no refueling en route (Kerlinger 1989, Bildstein 2006).

Usually griffon vultures are especially active at midday, when the thermal currents are more intense, making daily movements of 90–160 km (Harel et al. 2016a). During migration, the temporal pattern obtained by direct observations and marked birds shows periods of intense activity of about 5 hours (typically from 10:00 a.m. to 3:00 p.m.), with cross-country speeds of about 40 km/h, allowing them average displacements of about 200 km/day and maximum of 450 km/day (Griesinger 1996, Muñoz et al. 2016). As the soaring-flight type consumes little energy, these birds do not need to accumulate many energy reserves in the form of fat stores like other migrants and they do not make frequent or long-term stopovers, although they can make stops to feed, if they find suitable places, well in form of carrion, rubbish dumps or garbage (Garrido et al. 2002, Azmani et al. 2009, Donazar and Margalida 2009).

The dependence of these large raptors on the existence of updrafts, poses serious limitations for performing long movements over large bodies of water where there are no thermals. Most long-distance migratory birds of prey, particularly obligate soaring migrants, appear reluctant to fly even short (i.e., < 25 km) distances over water, and instead migrate mainly, if not entirely, over land (Kerlinger 1989, Bildstein 2006). Only three griffon vultures crossed the 30-km Bab-el-Mandeb at the southern end of the Red Sea during 2 years of autumn counts there (Welch and Welch 1988), suggesting that the limit for regular water crossings may be close to being reached at the 14-km Strait of Gibraltar.

At Gibraltar, the species migrates later in the autumn than do most raptors and some individuals overwinter in Southern Spain rather than continuing on into Morocco and beyond (Griesinger 1996, Garrido and Sarasa 1998, Garrido et al. 2002). Numbers of griffon vultures on the Spanish side of the Strait increase through autumn as individuals wait to cross until conditions that favour soaring flight occur (Bernis 1983, Griesinger 1996, Finlayson 1992). Most crossing attempts are made at the narrowest stretch of the Strait (Bernis 1980, Finlayson 1992). Even so, many individuals attempting to cross turn back upon reaching the coastline (Bernis 1983). Water-crossing attempts were restricted to times between 11:00 and 14:00 h on days with specific winds conditions, attempting to cross in large flocks and rarely attempting to do so alone (Bernis 1983, Griesinger 1996, Bildstein et al. 2009, Ramirez et al. 2019). With optimal conditions, vultures needed 18–25 minutes to cover the 15 km across the Straits, with speeds of 36–50 km/h, starting from 500–1,000 m.a.s.l. and arriving on the Maroccan coast at an altitude of 20–100 m.a.s.l., soaring and climbing rapidly as soon as they reach the coastline (Griesinger 1996, Onrubia *pers.obs*). Griffon Vultures usually hesitated before water-crossing and exhibited a strong tendency to remain in flocks as they approached the Strait. Indeed, flocking behaviour appeared to increase the likelihood of water crossing attempts significantly, especially if some adult are involved (Griesinger 1996, Bildstein et al. 2009). Flocking behaviour may enhance the likelihood of water-crossing behaviour in obligate soaring migrants at least in part because increased flock size increases the probability that the group will contain one or more individuals with a pronounced tendency to continue migrating over water (Agostini and Duchi 1994, Agostini et al. 1994). Many vultures soared during crossing attempts, but most engaged in considerable flapping flight when attempting to cross, not flapping continuously, but intermittently in brief bouts of flapping interspersed with periods of gliding or soaring flight (Bernis 1980 and 1983, Griensinger 1996, Bildstein et al. 2009). However, many vultures aborted attempted crossings and returned to coast in intermittent flapping and gliding flight, and there were numerous reports of vultures falling into the Strait and drowning while attempting to cross, many times mobbed by seagulls, as well as reports of returning vultures collapsing on the beaches and shores (Bernis 1980 and 1983, Finlayson 1992, Griesinger 1996, Bildstein et al. 2009). The magnitude of birds that fall into the sea is unknown, but preliminary data indicates that they could account for 0.5–1% of vultures crossing (Migres Foundation, unpublished data). The 14-km-wide sea channel significantly limits the migration of the species into Africa, with many birds attempting repeated passage for weeks before crossing, and others not crossing at all and overwintering in Southern Spain. These observations indicate that passage of griffon vultures at the Strait of Gibraltar is limited by the species' over-water flapping-flight abilities, and even relatively short sea crossings represent significant obstacles to migrating vultures (Griesinger 1996, Bildstein et al. 2009).

It is common in large migratory birds of prey that both juveniles and adults migrate annually, although juveniles may remain in wintering areas or intermediate areas for a variable time in subsequent spring migration, usually until sexual maturity is reached (Bildstein 2006, Newton 2016). However, in the case of the griffon vulture, adults behave mainly as residents. Only some of them can make long-distance movements, following the same patterns of juveniles, although it is not clear if they are specimens with previous experience. In any case, the fact of having a prolonged reproductive process (5–7 months) together with the fact that the breeding season begins in the middle of winter (December–March), limit the possibilities of seasonal migration by reproductive adults. On the other hand, juvenile griffon vultures probably remain in wintering regions or intermediate areas for a variable period, ranging from 0.5–3 years, and some do not attempt spring migration in the year after their first arrival in Africa (SEO/BirdLife 2012a). For many birds the first spring migration and subsequent roaming starts in February–April of the bird's second or third year (Ramirez et al. 2019). Some birds remained stationary for > 3 years in Africa, while some birds migrated only half-way within Africa, but without reaching their natal breeding sites.

The migrations of juvenile griffon are the most likely mechanisms that allow the species to maintain gene flow throughout different breeding populations, given the mixing occurring during the non-breeding period and the fact that these individuals are responsible to colonize new areas (Donazar 1993).

7

Short-toed snake eagle *Circaetus gallicus*

Nicolantonio Agostini

ABSTRACT

The short-toed snake eagle (*Circaetus gallicus*) is distributed in the Palearctic and Indomalayan region. Palearctic populations are migratory, overwintering mainly in the Sahel region south of Sahara. Because of its size and morphology, this species mostly uses soaring-gliding flight over land during migration, avoiding long sea-crossings. As a result, large concentrations of migrants are observed at straits and along coastal and isthmus areas, with the highest counts reported at the Strait of Gibraltar during both spring and autumn. Migrating individuals belonging to populations of the Iberian Peninsula, France and Italy converge at this site. There is no evidence of a migratory flow across the Sicily Channel. Those breeding in Eastern Europe concentrate at the Bosphorus, part of those of Central-Western Asia at Batumi (Georgia). In the Middle East and North Africa, thousands are observed around the Gulf of Iskenderun, in Israel, in the area of Suez and along the western coast of the homonymous Gulf. Populations breeding in Italy and Greece use extremely detoured routes concentrating at the Strait of Gibraltar and at the Bosphorus, respectively, during both spring and autumn. Large numbers of juveniles learn these migration pathways following elders encountered en route. Recent studies made by satellite telemetry and involving mostly juveniles and immatures well described migration pathways of this species.

Introduction

The short-toed snake eagle (*Circaetus gallicus* Gmelin, 1788) is a medium-sized raptor, showing a little reversed sexual dimorphism with an average body mass of 1.7 kg (male: 1.2–2.0 kg; female: 1.3–2.3 kg) and a wing span of about 1.8 m. They are distributed in the Palearctic and Indomalayan region (Bruderer and Boldt 2001, Ferguson-Lees and Christie 2001). It is a highly specialized species, preying predominantly on snakes, but occasionally hunting rodents, lizards and tortoises (Bakaloudis and Vlachos 2011). Individuals of Palearctic populations are long-distance migrants, breeding in Northwest Africa, Southern Europe, Eastern Europe and Central-Western Asia, and overwintering in the Sahel and savannah region of sub-Saharan Africa, north of equator. Few birds spend the winter in Southern Europe. For these populations the wintering range is less than one-third the size of the

Mediterranean Raptor Migration Network, Rome, Italy.
Email: nicolantonioagostini@gmail.com

breeding range (Newton 1995). The other five member of the genus *Circaetus*, all distributed in Afrotropical, are sedentary with some population being nomadic (Ferguson-Lees and Christie 2001).

The European population has been estimated in 17,600–20,900 pairs, the whole population size in at least 100,000 pairs. The population trend is stable and, having an extremely large range, its status has been classified as Least Concern (Birdlife International 2017a, 2020). However, a linear increase reported in counts made at different bottleneck sites (Vansteelant et al. 2020b) suggests that this species is becoming more abundant and is probably expanding its distribution as a result of a warming in the cooler northern ranges, temperature being a key factor affecting both distribution and abundance of snakes (Martín et al. 2016). The breeding season consists in one clutch with a breeding success of 0.3–0.9 chicks fledged per territorial pair (Ferguson-Lees and Christie 2001, Bakaloudis et al. 2005). Whereas the female takes care of the young, the male provides most of the food, with the daily prey delivery rate synchronised with the diurnal activity pattern of reptiles (Bakaloudis and Vlachos 2011). This species exhibits a delayed maturation, normally breeding for the first time at the fourth calendar year or even later (Campora and Cattaneo 2005, Mellone pers. comm.). The breeding habitat is a mosaic of shrub-lands, where they can easily capture their prey, and forests, which provide suitable nesting sites (Moreno-Rueda and Pizarro 2007, Cecere et al. 2018). Human pressure has a negative effect on the occurrence of this species, especially because of urbanization close to forests and reduction of suitable foraging habitats (Bakaloudis 2009, López Iborra et al. 2011). Adults and juveniles differ in plumage and it is possible to age migrants when passing close and in good light (Forsman 1999, 2016). Because of its size and morphology, short-toed snake eagles mostly use soaring-gliding flight over land during migration, displaying an innate aversion to fly over water even when approaching relatively short stretches of sea (Agostini et al. 2016b). In doing so, they limit the energy consumption of the flapping flight and also the mortality risk associated with long sea-crossings (Kerlinger 1989, Bildstein 2006, Agostini et al. 2015a). In particular, using the procedure described by Pennycuick (2008), it has been calculated that for this eagle the energy consumption of the flapping flight is 8.7 times higher than that required for the soaring-gliding flight, assuming no wind (Panuccio et al. 2012).

Knowledge Status

Studies on the short-toed snake eagle migration have been made by direct visual observation, satellite and radar tracking. As the species avoid large sea-crossings (Fig. 1), substantial counts (> 1,000 individuals) were reported at several watchsites: for example, in Spain (Strait of Gibraltar; autumn: Martin et al. 2016; spring: De la Cruz et al. 2011a), Italy (Arenzano, Capriglia; Agostini et al. 2017c, Premuda et al. 2015), Northeastern Turkey (Bosphorus; autumn: Fülöp et al. 2014, Panuccio et al. 2017; spring: Üner et al. 2010), Southern Turkey (Gulf of Iskenderun; autumn: Oppel et al. 2014), Georgia (Batumi; autumn: Agostini et al. 2017c, Vansteelant et al. 2020b), Israel (autumn: Kfar Qassem, Northern Valleys; spring: Bet Shemesh, Beer Sheva; Shirihai et al. 2000, Alon et al. 2004), Egypt (autumn: Suez, Bijlsma 1983, 1987; spring: Ayn Sokhna, Megalli and Hilgerloh 2013) and Yemen (Bab-el-Mandeb Straits; autumn: Welch and Welch 1988). Moreover, a substantial passage has been recorded in the Stura Valley (Maritime Alps, North-Western Italy; Baghino et al. 2012) and at Burgas Bay (western coast of the Black Sea, Bulgaria; Michev et al. 2011) during post-reproductive movements, at Mount Olympus (North-Eastern Greece; Panuccio et al. 2012) and at the Eastern Pontics (Southern Turkey; Shirihai et al. 2000) during both spring and autumn. As regards the radar study, it was made in Southern Israel during three migration seasons (autumn 1991, 1992, and spring 1992) at two migration sites, in the Negev Highlands near Sede Boqer and in the Arava Valley near Hazeva (Spaar 1997). In addition, the flight altitude of migrants was recorded at the Bosphorus during autumn 2010 using an optical range finder (Panuccio et al. 2017). Finally, several short-toed snake eagles, have been tagged with satellite transmitters. Those using the Western Flyway (coming mainly from Italy and Spain) spent the winter in Western Africa, from Senegal to Niger (Meyburg et

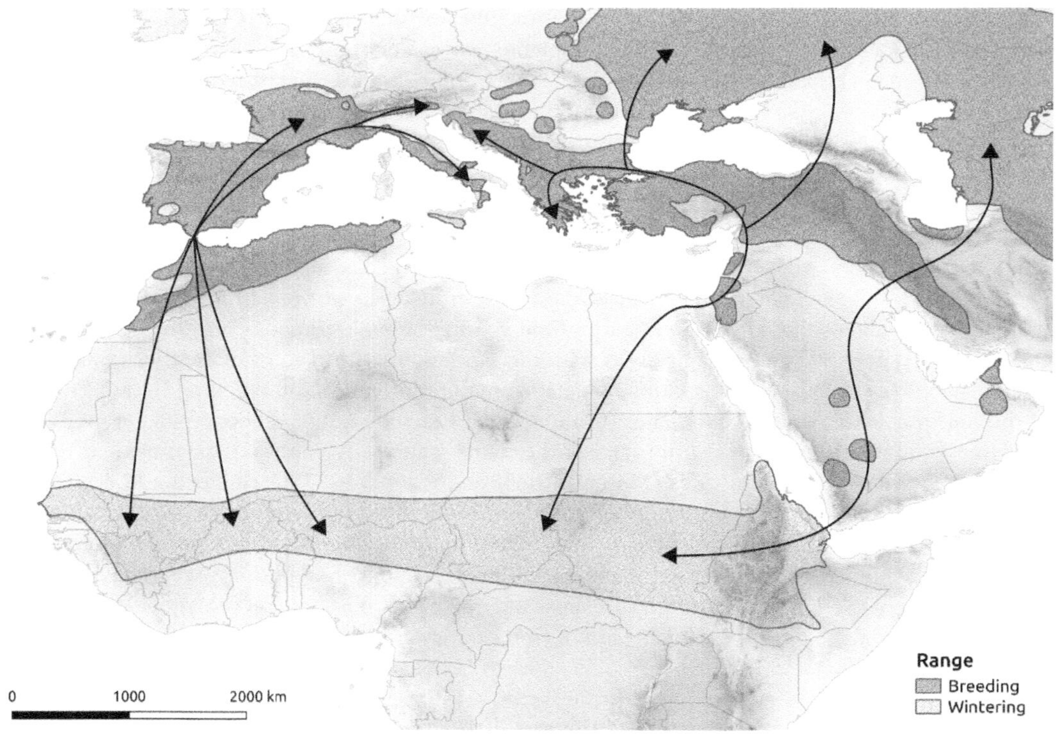

Figure 1. Schematic representations of the main routes followed by adults short-toed snake eagles, and by the majority of juveniles.

al. 1998a, Mellone et al. 2016), while eagles tagged in Croatia and Israel settled in Chad and Sudan (Fig. 1; McClain 2015, Kapelj et al. 2017) .

Post-breeding Migration

Short-toed snake eagles migrate solitary or in small flocks, mostly containing 2–3 individuals (Agostini et al. 2002a, 2002b, 2017c, Panuccio et al. 2012), concentrating at the sea-crossings at the Strait of Gibraltar (14 km wide), at the Bosphorus (0.8–3.3 km wide) and at the Bab-el-Mandeb Straits (approximately 20 km wide). Along both western and eastern flyways, the peak occurs between the second half of September and early October (Ferguson-Lees and Christie 2001). Observations made at Batumi in Georgia, Mount Olympus in Greece (eastern flyway) and Arenzano in North-Western Italy (western flyway) showed a large overlap in the migration periods of adults and juveniles. In particular, at Batumi and Arenzano, the proportion of juveniles increases at a similar daily rate throughout the migration season suggesting that they tend to leave the breeding grounds just few days later than their parents (Agostini et al. 2017c). However, at least a fraction of juveniles migrates much later in the season (Agostini et al. 2004a, 2009, Mellone et al. 2016). Populations of the Iberian Peninsula, France and Italy use the western flyway concentrating at the Strait of Gibraltar (Fig. 1). While birds originating from Iberia and France can easily locate the narrowest sea-crossing heading south-west until they have no other option than to cross the Strait of Gibraltar, those originating from Central and Southern Italy have to undertake an extremely detoured migration, flying hundreds of kilometers heading NW along the peninsula, and then moving westward and southwestward passing through Northwestern Italy and Southern France (Agostini et al. 2002a, Mellone et al. 2011a). In doing so they avoid the long oversea flight from Sicily to North Africa (at least 150 km) as a result of a conservative strategy (Agostini et al. 2002b, Mellone et al. 2011a). A long water crossing greatly increases mortality risks especially among large soaring birds (Zu-Aretz and Leshem 1983, Oppel et al. 2015). Large numbers of juveniles learn this flyway by following elders encountered en route (Agostini et al. 2004a, 2017, Mellone et al. 2011a, 2016, Panuccio et al. 2012). This peculiar migration probably retraces the colonization process of the Italian Peninsula (from north to south) which perhaps is still ongoing (Agostini and Mellone 2008). In Italy the density of this species increases with latitude despite the existence of large areas of suitable habitat in Southern Italy (Panuccio et al. 2015). As regards migration counts along this detoured pathway, observations made at Capriglia (Apuane Alps; Central Italy) and Arenzano (Ligurian Apennines; North-Western Italy) between 2008 and 2019 reported on average 1,899 (max. 3,345 in 2017) and 1,388 (max. 1,919 in 2016) short-toed snake eagles each autumn season, respectively, showing an increasing trend (Baghino 2016, Premuda 2017b, Premuda et al. 2015, http://www.areeprotettealpimarittime.it/ente-di-gestione-aree-protette-alpi-marittime/pubblicazioni/infomigrans). In North-Western Italy, at least part of birds moving along this circuitous route converge with those breeding in the Alps concentrating in the Stura Valley (Maritime Alps), where the passage of 2,558 short-toed snake eagles has been reported in autumn 2019 (Scalco et al. 2019). At Capriglia, among aged birds, about 20% were juveniles, often seen migrating in mixed age flocks (Agostini et al. 2002b, 2004, Baghino et al. 2012, Panuccio et al. 2012). However, as mentioned above, a fraction of juveniles migrates much later in the season. As a result, at least some of those, probably belonging to peripheral populations and/or migrating much later than adults, are not able to learn this flyway probably because they have limited opportunities for social learning compared to those hatched elsewhere and/or migrating earlier. Since juveniles do not follow their parents, social learning is a casual event based on random circumstances (Mellone et al. 2016). Moving without following their elders, these birds head south, influenced by an inherited axis of migration, and once crossing the Strait of Messina, between the "toe" of the peninsula and Eastern Sicily, follow the southern coast of this island finally reaching Marettimo, a small mountainous island off of Western Sicily located about 130 km NE of the Cap Bon Peninsula (Tunisia, Fig. 2). There, each October, flocks of juveniles are seen moving along its mountainous slopes hesitating in front of the open sea, sometimes hunting and flying back to the mainland (Agostini et al. 2004a, 2009, Panuccio et al. 2011, Agostini and Panuccio pers. obs., Mellone et al. 2016; Fig. 2). Conversely, the passage over Malta is negligible and occurring only in some years, probably because this island is not visible from Sicily as is Marettimo (Beaman and Galea 1974, Sultana and Gauci 1982, Agostini et al. 2002a, Coleiro 1999). The bimodal orientation pattern of juveniles has been recently well described through satellite telemetry. In particular, seven juveniles were tagged in their natal sites in Southern Italy; among them, five used the detoured pathway crossing the sea at the Strait of Gibraltar, two others, despite showing periods of northward movements, finally headed south and spent the winter in Sicily, about 3,000 km north of birds using the detoured route (Mellone et al. 2016; Fig. 2). In agreement with previous field observations (Agostini et al. 2009) these two birds reached the island of Marettimo two and three times, respectively, travelling back and forth between Sicily and the small island for about one month before stopping migration and wintering in Sicily (Fig. 2). One of these birds learned the detoured flyway the following autumn. Details of the migration pathways of two birds using the detoured route (Mellone et al. 2011a) seem to confirm that the safety of the flight over land could have been the main selective force that led to the evolution of this detoured route (Agostini et al. 2002b, Panuccio et al. 2015).

As mentioned above, the largest concentration of short-toed snake eagles along the western flyway occurs at the Strait of Gibraltar. There, during surveys made between 1999 and 2013, a mean of 12,792 individuals was counted each autumn, showing an increasing trend such as in North-Western Italy (Martin et al. 2016). Despite the short water-crossing (14 km wide), birds reaching the coastline are often seen hesitating, sometimes for several days, especially in strong easterly winds (Finlayson 1992). Very few birds stop migration and winter in Spain (Martinez and Sanchez-Zapata 1999, SEO/BirdLife 2012b). Like the birds of peninsular Italy, short-toed snake eagles of the Greek population use a detoured flyway to avoid the long sea crossing of the Central-Eastern Mediterranean, converging on the Bosphorus together with other populations of Eastern Europe. During observations

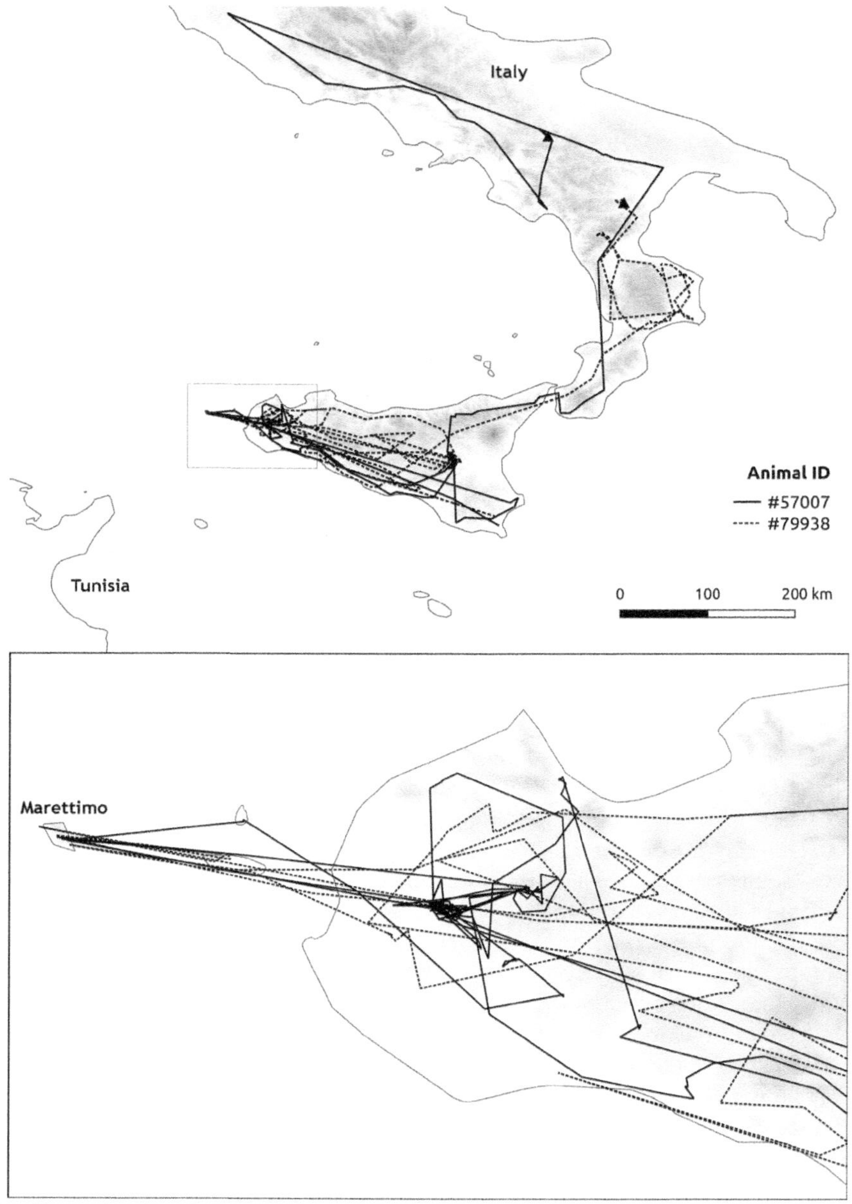

Figure 2. Satellite tracks of two juvenile short-toed snake eagles born in Italy during their first migration (modified from Mellone et al. 2016), showing a southbound direction. After several crossing attempts via the island of Marettimo (see inset), they spent the winter in Sicily. Triangles indicate nest locations.

made in autumn 2009 along the eastern slope of the Mount Olympus (NE Greece), a total of 376 short-toed snake eagles were seen heading NE; 23 mixed age flocks were reported and 37% of aged birds were juveniles (Panuccio et al. 2012). Some juveniles belonging to peripheral populations and/or migrating much later than adults do not learn the detoured flyway and head south passing over Antikythira, a small island located 33 km NW of Crete (Lucia et al. 2011, Panuccio et al. 2013b). As juveniles wintering in Sicily, these birds are expected to avoid the long sea crossing towards Africa, and may spend the winter in Southern Greece, perhaps in Crete (Londei 1991). Besides the Greek population, other ones converge at the Bosphorus, from Eastern Europe. There, during a survey made between 22 September and 10 October 2008, a total of 4,242 short-toed snake eagles were counted from four observation posts located along its Asian side (Fülöp et al. 2014). Moreover, in Southern Turkey, 6,257 short-toed snake eagles were counted in the area of the Gulf of Iskenderun from 30 August to 26 September 2013 (Oppel et al. 2014). A recent study made by satellite telemetry well described the eastern pathway of a juvenile short-toed snake eagle born in Croatia (Kapelj et al. 2017) and overwintering in Southern Sudan. This bird covered a distance of 6,054 kilometers in 25 days, showing the highest total migration speed (km/day) among nine satellite-tagged short-toed snake eagles for which details are available in literature. It passed over the Bosphorus, Iskenderun, Syria, Jordan and Suez. When the tagged Croatian short-toed snake eagle reached the southern tip of the Sinai Peninsula, perhaps having lost his elder, it flew northward along the western coast for almost 300 km and reached Africa through the Isthmus of Suez avoiding the crossing of the Gulf of Suez, that is about 26 km wide at the narrowest point. This movement, aimed at avoiding the water crossing, has been recorded in other species migrating in the area and resembles the behaviour of juveniles short-toed snake eagles in Sicily (Meyburg et al. 2002, Agostini et al. 2009, Mellone et al. 2016). Besides birds coming from the Bosphorus, the area of Suez is also crossed by birds from Central-Western Asia, some of which are observed at Batumi. In Suez the passage of 12,136 short-toed snake eagles was reported in autumn 1984 (Bijlsma 1987, Newton 2008). On the other hand, in Batumi, along the eastern coast of the Black Sea, during surveys made between 2011–2018, an average of 1,427 short-toed snake eagles per season were counted and, among aged birds, 16.5% were juveniles (Vansteelant et al. 2020b). At Kfar Qassem and Northern Valley (Israel), other watch-points along the eastern flyway, surveys were made in the period 1982–1987 and 1990–1999, respectively, with a maximum annual count of 8,045 birds in 1986 (Shirihai et al. 2000). Finally, 1,202 birds, probably belonging to the most eastern populations of the species, were seen heading towards the Bab-el-Mandeb Straits (Yemen) during autumn 1987 (Welch and Welch 1988). Migration durations of satellite tracked birds may vary a lot according to the distance between the breeding area and the wintering one: it ranged from less than two weeks, for Spanish birds (Pavón et al. 2010), to one month and a half, for Italian detouring juveniles (Mellone et al. 2011a). An Italian individual tracked also during adulthood improved the performance significantly, reaching Africa in a much shorter period than during its first migrations (Mellone unpub. data).

Pre-breeding Migration

Short-toed snake eagles leave their wintering grounds in February, reaching the breeding areas in late March. Spring migration pathways along the western flyway largely retrace those of autumn movements. As regards immature birds, among eight tracked by satellite (one from France and seven from Spain), one (from France) remained in the wintering area (in Mali), six moved northward but stopped migration before reaching the breeding area (mostly in Morocco), whereas one completed the spring migration but remained for a shorter period than breeders (Pavòn et al. 2010, Jiguet et al. 2011, Mellone et al. 2011b, Yáñez et al. 2014). At the Strait of Gibraltar, the peak of migrating adults occurs during the first half of March. In agreement with satellite tracking data, immatures (second and third calendar year birds) are much fewer, and delay their spring movements, being still observed in June (Finlayson 1992). Such as during autumn, at this watch-point surveys were made along its European side. In particular, during observations made from 1st February to 10th June 2008 and 2009 using six watch points, from Bolonia to Punta Carnero, a total of 9,907 and 10,406 short-toed snake eagles has been counted, respectively (De la Cruz et al. 2011a). As expected, birds breeding in Central and Southern Italy use the same detoured flyway of the autumn, following the Mediterranean coast. These birds, concentrate at Arenzano and Capriglia where, between 2008 and 2019, on average 1,962 (max. 2,600 in 2013) and 1,853 (max. 2,910 in 2018) birds were counted per season, respectively, peaking during the third week of March and showing an increasing trend (Premuda et al. 2015, http://www.areeprotettealpimarittime.it/ente-di-gestione-aree-protette-alpi-marittime/pubblicazioni/infomigrans). Along this flyway the passage of tens of immatures has been recorded in May (Premuda et al. 2010).

On the other hand, few surveys have been made along the eastern flyway in this period. Along the western coast of the Gulf of Suez, at Ayn Sokhna, 5,301 individuals were counted from 1st March to 2nd May 2012 peaking in mid March (Megalli and Hilgerloh 2013), while in the area of Suez 3,063 individuals were counted in 1982 (Bijlsma 1983, Newton 2008). A total of 226 individuals were reported crossing the Bab-el-Mandeb Strait from 28 Fabruary to 14 March 2013 (Rayaleh et al. 2013). In Israel, short-toed snake eagles move on a broader and more eastern front than in autumn, with the highest counts reported at Bet Shemesh (2,071 in 1982) and Beer Sheva (1,759 in 1985), while fewer are observed at Eilat (max. 345 in 1982; Shirihai et al. 2000). At the Bosphorus, during a survey made from 18 March to 31 May 2006 by a single observation post located at the European side (Keskin Viraij), 473 short-toed snake eagles were reported (Üner et al. 2010). Finally, along the detoured route used by the Greek population, 602 individuals were seen heading SSW along the eastern slope of Mount Olympus between 6–26 March 2010 (Panuccio et al. 2012).

Behavioural and Ecological Remarks

The detoured routes performed by short-toed snake eagles breeding in Italy and Greece provide some of the best examples at a global scale of how water barriers and social learning can shape the evolution of a migration pathway in soaring birds. Several studies investigated the influence of other external factors on the migration strategies of this species. Satellite studies showed that the daily distance covered greatly varies during migration. In particular, that covered by the adult tracked from France to Niger and by the juvenile tracked from Croatia to Sudan, varied from 17 to 467 km and from 5 to 598 km, respectively (Meyburg et al. 1998a, Kapelj et al. 2017). Since this species rarely makes stopover (Pavòn et al. 2010, Yosef 1996), the variation in daily distances covered is probably affected mainly by atmospheric conditions and ecological barriers. This is evident at the Strait of Gibraltar where, as mentioned above, short-toed snake eagles facing the water surface sometimes interrupt migrating for several days especially under strong easterly winds (Finlayson 1992). Unfortunately, as data are available for just an adult short-toed snake eagle (Meyburg 1998), further studies are needed to investigate other aspects of migration, such as eventual seasonal speed differences due to the reproductive motivation. Juveniles and immatures tracked over the Sahara were positively affected by tail winds, but not by day length (Mellone et al. 2012a, Vidal Mateo et al. 2016). However, as expected for a soaring bird, such studies showed that short-toed snake eagles take advantage of the stronger thermals that occur during the central hours of the day and over the Sahara. In particular, peaks in hourly speed and altitude occur just after midday and over the desert, where birds sometimes approach 2,000 m above the ground and may reach hourly speeds of about 80 km/h, which lead to daily flight distances averaging 230 km (Pavòn et al. 2010, Mellone et al. 2012a, Mellone pers. comm.). The radar study made in Southern Israel reported other flight characteristics, as a climbing rate in thermal circling of 1.85 ± 0.68 (mean $\pm$ SD) m/s, a ground speed in inter-thermal gliding of 51.3 ± 11.8 km/hr and a cross-country speed relative to the ground of 39.5 ± 9.6 km/hr (Spaar 1997). Because of the daily variation in the altitude and speed, short-toed snake eagles may pass undetected by direct visual observation at midday and early afternoon, affecting migration counts especially at watchsites located along a mountain chain, as suggested by observations made at Arenzano (Ligurian Apennines) and at Mount Olympus (Panuccio et al. 2013a). In particular, the number of eagles observed decreased drastically when temperatures were higher than 24°C, which is common in September but not in March, at the Mediterranean latitudes. As regards the effect of crosswinds, at a local scale, the lack of difference in number of short-toed snake eagles observed at Mount Olympus in spring during westerly crosswinds compared to other wind directions suggests that these birds may be able to compensate for drift effect toward the Aegean Sea. On the other hand, wind strength and lateral northerly wind negatively affected the number of migrants observed at Arenzano during spring movements. Perhaps, in this period, strong lateral winds pushed migrants toward the coastline, slowed their travel speed and/or influenced them to fly at very low altitudes (below the observation post located along the mountain chain) to avoid being blown out over the

Ligurian Sea, thus passing undetected (Panuccio et al. 2013a). During a study made at the Bosphorus the flight altitudes of short-toed snake eagles have been recorded using an optical range finder, and the results showed that they flew at lower altitudes during strong tail winds. However, birds migrating at this watchsite later in the season, when the proportion of juveniles without elders increases, did not compensate the drift effect of lateral winds perhaps because inexperienced individuals cannot take advantage of the positional information that allows adults to detect, and correct for, lateral displacement (Panuccio et al. 2017).

8

Lesser spotted eagle *Clanga pomarina*

Bernd-U. Meyburg

ABSTRACT

The lesser spotted eagle (*Clanga pomarina*, until recently *Aquila pomarina*), is a medium-sized eagle with a relatively small distribution in the west of the Palaearctic, spanning from Central Europe through Eastern and South-Eastern Europe and Turkey to the Caspian Basin in the north of Iran.

The majority of the European population moves across the Bosphorus, then crosses Turkey in a south-easterly direction to continue on through Syria, Lebanon, Israel and Egypt close to the Mediterranean Sea. Africa is reached through a very narrow flyway at Suez. Nearly the entire world population of this species passes over Israel and the other countries on the Eastern Mediterranean coast twice a year.

From Suez, the migration route continues in a narrow front over the mountains about half way between the Nile and the Red Sea, crossing the Sahara and finally arriving in the south of Sudan, where some of the birds, especially the younger ones, sometimes perform a stopover. The following part of the journey also takes place in a narrow front along the Rift Valley, at least for all the satellite-tracked adult birds, through Uganda, further parallel to Tanzania's western border, east past Lake Victoria and along the eastern shore of Lake Tanganyika, until its southern tip, where the routes start to split towards different wintering areas. Some younger eagles winter north of the equator.

The spring migration largely retraces the autumn route, except that adults get close to the Red Sea in Eritrea and then parallel to the coast to Suez, creating a distinct loop.

A considerable part of the European population also migrates along the east side of the Black Sea, concentrating in Batumi and then crossing Turkey to join the birds on the main flyway.

Few individuals, mainly juveniles that do not learn the optimal route from adults, also migrate, not always successfully, through the Mediterranean Sea (Greece, Italy, Spain), to almost unknown wintering areas in Africa.

The total world population of the LSE is estimated at 40,000–60,000 mature individuals. There are mixed pairs of *C. clanga* x *C. pomarina* in apparently increasing number.

BirdLife Germany (NABU), PO Box 33 04 51, 14174 Berlin, Germany.
Email: bumeyburg@aol.com

Introduction

The lesser spotted eagle (*Clanga pomarina* Brehm, 1831), hereafter LSE, was included in the genus *Aquila* until a few years ago. Now it forms its own genus together with the greater spotted eagle (*C. clanga*) and the Indian LSE (*C. hastata*), formerly a subspecies of *C. pomarina. C. pomarina* and *C. clanga* often hybridize in Europe, probably more so in recent decades. The LSE is a long-lived (up to at least 30 years), medium-sized eagle and a long-distance migrant wintering in sub-Saharan Africa. It is the eagle species in Europe that regularly migrates furthest south in Africa.

Meyburg (1996) for the first time estimated the total world population at about 20,000 breeding pairs, thus clarifying the hitherto unexplained difference between high migration numbers in Israel (e.g., Christensen and Sorensen 1989) and the low estimated population in the individual countries. Such contrasting numbers were probably related to breeding of this species in dense woodlands, making the detection of nests not easy. Autumn censuses in Israel 1982–1987 had yielded about 100,000 individuals, a number which was severely questioned by experts in the individual countries of occurrence, but which was later confirmed in terms of magnitude. The Meyburg's estimation was obtained by comparing results of roadside counts in various countries with relatively reliable populations in those countries and extrapolating the populations of all countries of occurrence by means of the countries' surface area. This estimation has been recently adopted by BirdLife International (2020) which classifies the status of this species as "least concern" estimating the world population at 40,000–60,000 mature individuals. The European population has been estimated in 16,400–22,100 pairs (BirdLife International 2017a). This is a rare case where a raptor species described in the literature as scarce and even endangered has turned out to have a rather important world population. No count of a bird species from a migration survey ever aroused such fierce resistance from some established specialists working in the breeding areas. The western distribution limit of the species nowadays runs through north-east Germany from the Baltic Sea near Rostock to the north of Berlin. Formerly the species bred much further west into France. In the south of Germany, the species was still breeding until the first world-war, in Bavaria.

East of Germany, the breeding area extends in a narrow front along the Baltic Sea through Pomerania in Poland to the west of Gdansk. Further east in Masuria and to the south, a densely populated but narrow distribution area exists along the eastern border of Poland up to Slovakia. The range of the LSE in Russia presented in handbooks and websites is based on obsolete data. It is obvious that the eastern border of the range is located further east than currently documented. A considerable eastward expansion of the range since the early 1990s is taking place, reaching the 47th meridian, 600Km further east than Moscow. In the eastern part of the range, within the limits of Central Russia (Upper Volga area), the LSE is a rare nesting bird species (Mischenko and Melnikov 2019, Melnikov and Mischenko 2019). Further to the south, the LSE occurs as a breeding bird in Central and Eastern Slovakia, parts of Hungary, the Balkans, Greece, Turkey and the Southern Caspian lowlands of Iran, as well as in relatively high density in the Caucasus (Meyburg and Meyburg 1988, Melnikov and Mischenko 2019). There is a regular isolated breeding occurrence in France near the Swiss border (Michelat 2005, 2019) with individual breeding attempts in north-east Spain (Bosch and Meyburg 2012) and perhaps in Algeria (Isenmann and Moali 2005).

The breeding season lasts from April to September and consists in one clutch, nearly always with two chicks hatched, but with only one surviving, the second one normally being victim of cainism (Meyburg 1974, 2001, 2002). The nest is built almost always on trees and preferably close to the forest edge. This species is normally observed hunting in open areas on relatively small prey (e.g., *Microtus* sp.; Palasthy and Meyburg 1973, Zub et al. 2010). Its frequent hunting inside the forest is very difficult to observe and only by chance. Both female and young are provisioned by the male during much of nestling period. In the wintering grounds the diet includes termites and nestling queleas (Kemp 2000, Meyburg et al. 2010).

Because of its size and morphology, the LSE is a typical soarer species exploiting thermal currents and deflection updrafts over land during migration, thus avoiding water crossings, which require the energy demanding flapping flight (Agostini et al. 2015a). In this chapter I summarise the seasonal migration routes of LSEs to provide the background for future studies to support the conservation of this long-distance migrant species. I also discuss questions of orientation and social learning, partly in relation to insight gained from translocation experiments.

Knowledge Status

Traditionally, for more than half a century, bird migration research, and thus also LSE migration research, has been predominantly based on ringing data. LSEs have been ringed in various countries for many decades, but the recovery rate is only about 2% (Danko et al. 1996, Meyburg et al. 2005). Of a total of 1,649 ringed LSEs in Europe, only 41 were recovered by 1996 (Danko et al. 1996). The same recovery rate has been found for 633 individuals ringed in Germany (Bairlein et al. 2014). Due to these few recoveries, LSE migration routes were only partially and very roughly known. In particular, there were hardly any recoveries from Africa.

For many decades, insights have been gained also through observations at raptor watchsites, the so-called bottlenecks. These observations are mainly used to obtain data on population size and migration phenology, etc. (Table 1; Sierdsema and van Loon 2008), which need objectives and field protocols for sharing objectives (Panuccio et al. 2018a). There is a large variety of raptor bottle necks. Only the most important ones can be discussed here.

For a long time, LSEs were first and foremost observed at the concentration points on the main migration route at the Bosphorus, in Israel, but also in Burgas (Bulgaria) and near Suez (Egypt; Table 1). The most important migration route runs along the west coast of the Black Sea. In Bulgaria observations have been carried out in spring (Shurulinkov et al. 2011), most often however in autumn near Burgas (Roberts 1979, Michev et al. 2011, Iankov et al. 2019).

The next most important watchsites are at the Bosphorus near Istanbul where observations were started as early as the 1860 (Alléon and Vian 1869, 1870). Other early papers were published by Steinfatt (1932) and by Mauve (1938). The population decline is impressively demonstrated by the strong differences in the number of passing migrants compared to today (e.g., Van den Bossche and Lens 1994, Porter and Willis 1968, Üner et al. 2010). At the Dardanelles much less individuals are observed (Table 1).

In Southern Turkey most of the birds cross the Taurus mountains near the Göksu river valley, in spring as well as in autumn, which indicates a good watchsite. From there they follow the coast close to the mountains. Fewer birds migrate more directly from the Bophorus to Iskenderun (Meyburg and Meyburg 2009b). Near Samandag, the southernmost city of Turkey, however, located near the Mediterranean Sea, there is a hitherto undescribed bottleneck of bird of prey migration. Unfortunately, this area is also a most important hotspot of illegal hunting (Meyburg 2005). In a very narrow front, the LSEs then move through Syria, Lebanon, Israel and Egypt, in order to reach Suez (Meyburg et al. 1995, Meyburg unp. data, Shirihai et al. 2000), where they enter the African continent. Recently, the migration has been very well studied in the Lebanon, mainly by telemetry but also field observations (Meyburg et al. 1995, Meyburg unp. data). Several watch points are found in Israel (e.g., Alon et al. 2004), where for decades the most intensive observations have been and still are made, especially in autumn (see also Table 1 for some of the references), and finally to a much lesser extent near Suez in Egypt (e.g., Simmons 1951, Bijlsma 1983, Bruun 1985), where all LSEs concentrate and virtually the whole world population migrate past (Meyburg unp. data).

Other migration corridors have been discovered, e.g., the second most important one on the eastern edge of the Black Sea (Batumi area) where LSEs are also observed in considerable numbers (e.g., Verhelst et al. 2011, Vansteelant et al. 2014, Wehrmann et al. 2019). LSEs have been recorded in negligible numbers passing through the Central Mediterranean region (Table 1; Giordano 1991, Corso 2001a, Agostini 2005, Beaman and Galea 1974, Sammut and Bonavia 2004, Sammut et al. 2013).

Formerly no LSEs were observed at Gibraltar (Evans and Lathbury 1973, Bernis 1975, Finlayson 1992). In more recent time LSEs have been counted there mainly in autumn (Onrubia et al. 2011a,

Table 1. Bottleneck migration sites for the Lesser Spotted Eagle.

Bottleneck/Country	Period	Spring migration	Autumn migration	Peak/range	Source
Western Mediterranean flyway					
Gibraltar/Spain	1998–2009	-	1–8	5–24.IX/6. VIII-12.X	Onrubia et al. 2011a
Central Mediterranean flyway					
Strait of Messina /Italy	1984–1990	0–5	-	-	Agostini 2005
Strait of Messina/Italy	1996–2000	0–5	-	-	Agostini 2005; Corso 2001a
Strait of Messina/Italy	2016–2019	0–1	0–2	-	www.straitobservatory.com
Marettimo/Italy	2003		3		Agostini et al. 2004c
Malta	1969–1973	0	0–2	20.IX–23.X	Beaman and Galea 1974
Malta	1998–2002		0–5		Sammut and Bonavia 2004
Malta	2005–2012		0–12		Sammut et al. 2013
Greece					
Antikythira/Greece	2007–2009; 17.VIII–30. IX.2017	0	0–11		Lucia et al. 2011, Evangelidis 2017
Mount Olympus/Greece	2009		13		Panuccio et al. 2013b
Western Black Sea flyway					
Burgas/Bulgaria	1979–2003		1,687–25,786	23.IX – 2.X/26. VIII-21.X	Michev et al. 2011
Burgas/Bulgaria	13–30.IX.2014		51,850		Iankov et al. 2019
Burgas/Bulgaria	2019		41,967	20.IX/6.VIII-4.X	trektellen.org
Eastern Black Sea flyway					
Batumi/Georgia	2011–2018		avg. 7,715		Wehrmann et al. 2019
Batumi/Georgia	2019–2020	1087–1961			https://trektellen.nl/
Southwest coast of Caspian Sea					
Shirvan NP/Azerbaijan	2007	-	daily up to 1400 Steppe eagles	-	Heiss and Gauger 2011
Southeast corner of Caspian Sea					
SE corner of Caspian Sea/Iran	8–12.X.2017	-	56	-	Panuccio et al. 2018b
Bosphorus					
Bosphorus/Turkey	31.VIII-25.X.1966	-	4,309	23.IX/31. VIII-9.X	Porter and Willis 1968
Bosphorus/Turkey	22.VIII-1.X.2010	-	19,325	-	Panuccio et al. 2017
Bosphorus/Turkey	22.IX-10.X.2008	-	58,327	-	Fülöp et al. 2014
Bosphorus/Turkey	18.III-31.V.2006	15,232	-	End of March	Üner et al. 2010

Table 1 contd. ...

...Table 1 contd.

Bottleneck/Country	Period	Spring migration	Autumn migration	Peak/range	Source
			Dardanelles		
Dardanelles/Turkey	22.VIII-1.X.2010		20		Panuccio et al. 2017
Dardanelles/Turkey	12.III-11.V; 11.VIII-18.X	2,797	809		Uysal and Tosunoğlu 2018
Dadia forest/Greece	2003–2005	4–15			Schindler et al. 2015
			Belen pass		
Belen pass/Turkey	2.VIII-23.IX.1976	-	3,875 (66.5% *C. clanga/ pomarina*)	12–23 (study end).IX	Sutherland and Brooks 1981
Gulf of Iskenderun/ Turkey	30.VIII-26.IX.2013	-	22,197	25.IX/ 3–26 (study end).IX	Oppel et al. 2014
			Israel		
Northern Valleys/Israel	1982–2012		54,358–141, 868	21.IX-8.X/24.VIII-15.X	Krumenacker 2013
Eilat/Israel	1977–1988	21–74	-	26.III-5.IV	Shirihai and Christie 1992
			Suez		
Suez/Egypt	23.II-15.V.1982	12,489	-	28.II-4.III/28.II-18.IV	Wimpfheimer et al. 1983
Suez/Egypt	4.IX-5.XI.1981	-	22,000	29.IX-9.X/ 10.IX-31.X	Bijlsma 1983
Ayn Sokhna/Egypt	1.III-2.V.2012	6,536	-	21–25.III/11.III-2.V	Megalli & Hilgerloh 2013
			Bab el Mandeb, Djibouti		
Bab el Mandeb/Djibouti	3.X-9.XI.1987	-	< 31	15.X-7.XI	Welch and Welch 1988
Bab el Mandeb/Djibouti	2–10.III. 2013	3	-	2.III	Rayaleh et al. 2013

De la Cruz 2011b), with very few or no individuals in spring (e.g., Bensusan et al. 2007, Garcia et al. 2006).

A massive breakthrough in migration research was achieved with the beginning of satellite telemetry (overview on technique: Meyburg and Fuller 2007, Meyburg and Meyburg 2013). LSEs have been detected by telemetry in a number of countries from which there have been little or no observations (Figs. 1 and 2). Satellite telemetry began with the tracking of the first young bird in 1992 within the German satellite tracking project of the World Working Group on Birds of Prey (WWGBP; Meyburg et al. 1993, Meyburg and Meyburg 2009b, Meyburg et al. 2017a, Meyburg unp. data). As early as 1994/1995, the first complete annual route and wintering of an adult bird was documented (Meyburg et al. 1995). Further adult and juvenile birds could be followed over large parts of the migration route during the same time. With rapid technical improvements, individual LSEs could finally be tracked over many years (up to 12) in a row and with the help of GPS localization very accurately (Meyburg et al. 2004, 2007, Meyburg unp. data). The latest tags provide GPS fixes up to once per minute, including flight altitude, direction, speed and accelerometry data (Meyburg unp. data). From 1992–2019 a total of 54 adult and 57 juvenile LSEs were fitted with satellite and GSM transmitters by the author, primarily in Germany, but also in Poland, Slovakia, Latvia and Namibia as part of the German Raptor tracking project of the World Working Group on Birds of Prey (WWGBP), but in larger numbers also by others from Estonia and Romania (Milvus group), and a few eagles in

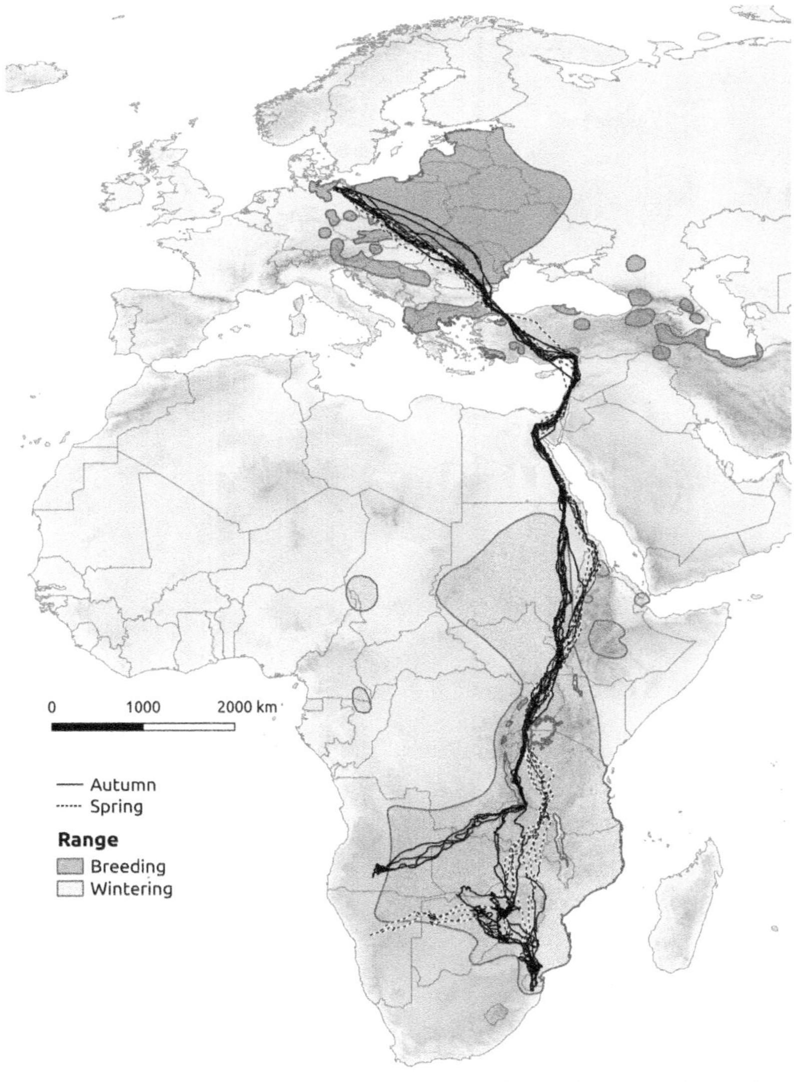

Figure 1. Migration tracks belonging to two adult lesser spotted eagles followed during multiple years.

Poland, Slovakia and the Czech Republic. Only a very small part of this data has been published so far; essential findings promise to come to light with their publication. Most of the results presented in this chapter on the LSE migration come from the use of this technique. A lot of this data has also been able to give insight on mortality reasons. Even the collision of an adult LSE during the spring migration with a small airplane could be meticulously documented using this method (Meyburg et al. 2018). How little was known about the transit in many countries before the era of satellite telemetry, in which practically almost the entire world population passes through, can be highlighted with the example of Eritrea, Ethiopia, Uganda and Sudan. For example, the LSE is called an uncommon to rare migrant in Ethiopia and Eritrea (Ash and Atkins 2009), while the entire world population of adults and also most of the younger individuals pass through here in spring (e.g., Meyburg et al. 1995, 2004b, 2007b, Meyburg and Meyburg 2009a, 2009b). The same is true for wintering areas. There are no records in Angola although some satellite tracked birds winter there even relatively far to the west (Fig. 1).

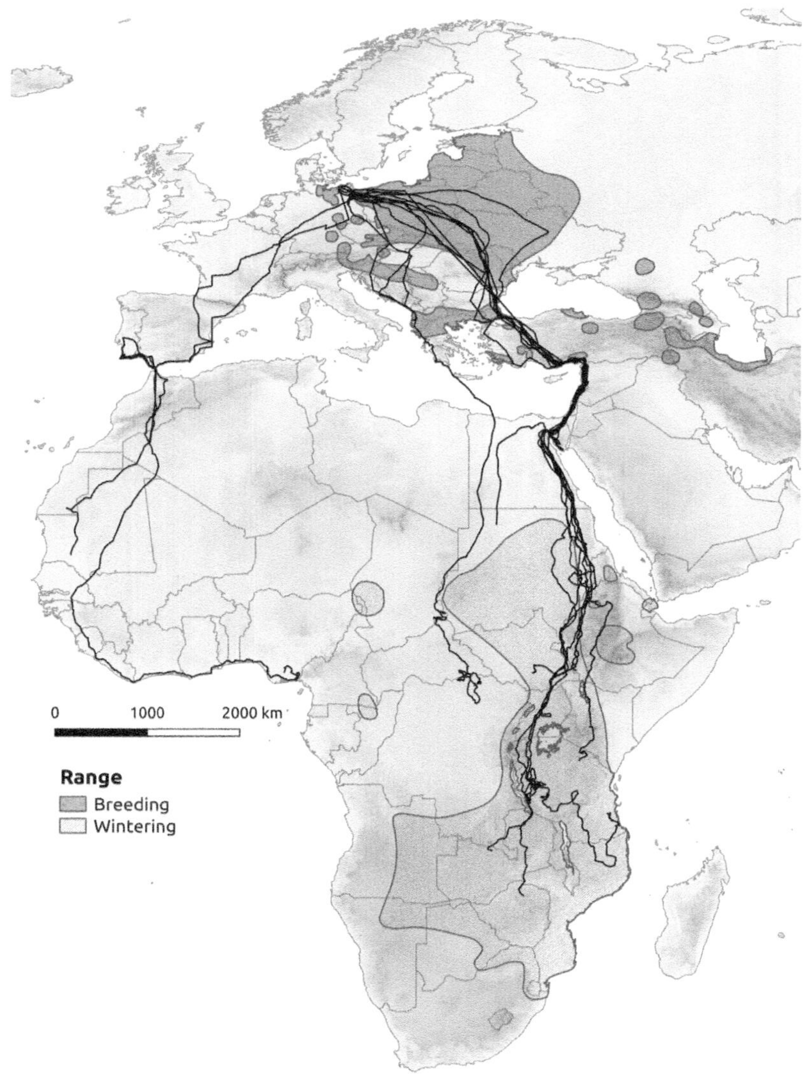

Figure 2. Autumn migration tracks of juvenile lesser spotted eagles.

Finally, other techniques to study migration used so far, mainly performed in Israel, are the tracking of LSEs and other migratory soaring birds by radar (Leshem and Yom-Tov 1996a, 1998, Spaar 1997) and following them by small planes (Leshem 1989, Leshem and Bahat 1999). This allows precise data to be obtained on, e.g., flight altitude, speed and daily activity, however only within the short migration stretch of Israel. The tracking of migratory birds using VHF telemetry has proved to be not successful.

Post-breeding Migration

After an average stay of 21 weeks in the breeding area (same value for both sexes), German tagged adults depart in mid-September. In most cases, it does not matter whether the pair has had breeding success or not. The females sometimes leave a little earlier than the males. Also immature individuals hardly leave earlier (Meyburg unp. data). Since all tracked adults wintered south of Lake Tanganyika (Meyburg unp. data), the corresponding latitude (8°52'36" S) was taken as a reference to compare the duration of all migrations, during both seasons. South of this latitude, migration was very slow and often interrupted by resting periods, making it difficult to distinguish between migration and resting. Therefore, the migration times given do not correspond to the whole journeys, but from the above mentioned latitude, for reasons of comparability. One of the future objectives is to determine whether there have been shifts in migration times or changes in duration of migration in the period 1994–2019, e.g., as a result of climate change and as a consequence of the El Niño phenomenon. Lake Tanganyika is reached by the end of October, with the journey lasting from six to seven weeks, thus just over 180 kilometres per day were covered on average, over a distance of approx. 8,340 km (Meyburg unp. data). The longest total distance covered on migration to Angola was 10,550 km, while 10,630 km to South Africa, clearly south of the Kruger National Park (Fig. 1). The three family members (male, female and fledgling), leave in all cases separate from each other as documented in detail in the greater spotted eagle (Meyburg et al. 2005, Meyburg unp. data). At the onset of autumn migration, juvenile eagles are without their parents and without hunting experience. This can cause them to take unfavourable and dangerous migration routes, e.g., to Greece, Italy and Spain, where the Mediterranean Sea is an almost insurmountable barrier (Fig. 2). They seldom succeed in overcoming these barriers, with many young eagles drowning. This may happen especially to young eagles at the western edge of the distribution area, but also to young eagles from the Baltic States that partly move, e.g., to Gibraltar (Meyburg et al. 2017). It is therefore vital to rapidly find the company of older, experienced conspecifics in order to learn from them the best migration route, but probably also to find food and good places for hunting.

All tracked adult eagles migrated over the Bosphorus (some also over the Dardanelles), where 58,327 individuals have been counted in the 2008 (Fülöp et al. 2014, Table 1). Birds that come from east of Germany also migrate across the Bosphorus. Whether there is a migration divide that separates the migration to the Bosphorus and the eastern edge of the Black Sea (Batumi, Georgia) remains unknown. Rather, a very wide transition area is expected (Meyburg unp. data). However, all LSEs head to the Eastern Mediterranean coast, in order to enter Africa. Satellite telemetry has shown that all adult LSEs pass through Lebanon both in spring and autumn, with no cases recorded east of it. Such flow includes also birds coming from the Eastern Black Sea flyway: at Batumi, an average of 7,715 birds has been counted in the period 2011–2018 (Wehrmann et al. 2019; Table 1). In this way, they concentrate in Israel, where an average of 100,000 individuals has been counted in recent years, peaking in early October (Krumenacker 2013; Table 1). The migratory flow here is very concentrated (Figs. 1 and 2), with a rough average of the 25% of the individuals passing the same day (Krumenacher 2013). The first-migrating juveniles, which migrate together with adult birds, follow their same route, in the very most case until at least Zambia or further into Southern Africa. However, some young eagles also winter much further north. For the first time, this could be documented in detail during the winter of 2007/2008 in a young eagle wintering mainly in South Sudan (Meyburg et al. 2008).

The migration speed of a couple of LSEs over the course of several years has already been documented (Meyburg and Meyburg 2009a). The duration of the migration to the southernmost point of wintering varied between 52 and 119 days (mean 81 days). There were noticeable differences in duration between the male (52–75 days) and female (64–119 days) mainly because the wintering areas were located at different distances to the south (Zambia for the male and South Africa for the female). In addition, for each of these two birds, the duration of the autumn migration (74–119 days) was markedly longer than its spring migration (52–64 days). The two fully tracked autumn journeys of this male differed only by one day. Overall, the female travelled much more slowly than the male during both autumn and spring migrations, which resulted in her autumn migration of 1997 taking more than twice as long as the male's spring migration in 1998.

The greatest daily flight distances recorded occurred during the autumn migration while crossing the Sahara Desert in Egypt and Northern Sudan, where steps lasting over 300 km/day were recorded, sometimes even extending to 350 or 400 km/day. The highest recorded daily flight distance for both birds on all routes was achieved by the female on 16 November 1998, when she covered 521 km in

Northern Sudan. Overall, between 14th and 17th November 1998, the bird flew 1,687 km, averaging 422 km/day over Egypt and Sudan (Meyburg and Meyburg 2009b).

Juveniles performing their first migration arrive to Africa via Turkey and Suez together with adult birds, and then follow their same route, in many cases until at least Zambia or further into Southern Africa (Meyburg unp. data). However, a few young eagles also winter much further north (Fig. 2). This was documented in detail for the first time in the winter of 2007/2008. And was the first LSE with a transmitter to winter north of the Equator. This bird wintered in South-Eastern Sudan and the neighbouring regions of Ethiopia, Uganda and Kenya, within an area extended north-south over 1,200 km and east-west over 600 km (Meyburg et al. 2008). Many juveniles do not manage to follow adults and learn the safe over land route to reach Africa via Suez. Indeed, mostly unsuccessful attempts to cross the Mediterranean Sea at wide, unsuitable spots, take place, with many first-autumn migrants drowning in the Mediterranean Sea. The fact that they require guidance by older migrants is also supported by translocation experiments of young eagles from Latvia and Eastern Poland to Germany, where they were reared and released into the wild by fostering and hacking (Meyburg et al. 2008, 2017a, Meyburg unp. data). If they do not catch up with older, more experienced conspecifics, there is very little chance that they will find the way to the Bosphorus. Often they try to cross the Mediterranean and die in the process. Out of 33 German tracked juveniles only less than half of them migrated to the Bosphorus, while the others headed south and some also southwest. Consequently, about half of all young eagles are lost due to an unfavourable direction of migration before reaching Africa. A similar process occurs in juveniles Egyptian vultures from the Balkans (Oppel et al. 2015, see Chapter 3). A few tracked LSEs reached Africa via the Strait of Gibraltar, where the sea crossing is shorter than in Greece and Italy, and a few individuals are also observed every year (Onrubia et al. 2011a). Two GPS-tracked young eagles migrating over Gibraltar reached as far as the Ivory Coast and Nigeria, with the latter individual unfortunately killed with a shotgun. It is striking that there are far fewer migrating birds in the Straits of Messina and Gibraltar in spring than in autumn, highlighting again that few juveniles using these routes during autumn manage to survive. In 2019/2020 a young GPS-tracked LSE flew on its first migration to Italy and overwintered in Sicily, making many attempts to cross the sea to Tunisia and Malta, but gave up after some time and returned to Central Europe in spring (Meyburg unp. data). On the opposite side of the Mediterranean Sea, in Tunisia, smaller numbers of LSEs have been observed in spring at Cap Bon in Tunisia (Thiollay 1975, 1977, Dejonghe 1980, Agostini 2005). In two cases, satellite-tracked first autumn migrants managed to cross the Mediterranean from Peloponnese (Greece) to Africa. One bird had already died afterwards during the migration through the Sahara. The other bird spent the winter in the Congo, but did not make the crossing from Libya to Europe in spring (Meyburg et al. 2017a, Meyburg unp. data).

Pre-breeding Migration

The complete spring migration could be assessed in 127 cases with adult LSEs by means of satellite telemetry from 1995–2020 during the German project. In many other cases there were not enough localizations to determine the beginning or the end of the migration with sufficient accuracy. These cases were not considered when calculating the length of migration. Some birds were lost during migration or the transmitters failed (Meyburg unp. data). The passage of the birds east of Lake Tanganyika took place between 9 February and 28 March. The males left the wintering areas later and crossed the corresponding latitude on average one week after the females. The peak transit throughout all years (1995–2020) was mid-March (Meyburg unp. data). During migration males moved on average faster than females, needing five days less, thus reaching the breeding sites after an average of slightly less than six weeks, during the second half of April. Females spent six and a half weeks migrating and arrived on average at about the same time. This results in an average spring migration speed of almost 200 km/day from the southern tip of Lake Tanganyika to the breeding sites in north-east Germany. The total migratory distance for the eagles wintering in Northern Namibia, for example, is about 10,800 km (Meyburg unp. data). There were strong individual and annual differences. For instance, in 2015, which was an exceptional year, migration took almost 20% longer than the average (Meyburg unp. data). Further analysis is needed to determine the reasons for the annual differences and the trend due to climate change over the entire study period. There were less telemetry data for the spring migration than for the autumn migration. Initially, in 1994 the early battery-powered transmitters were often unable to transmit long enough to transmit the return migration. Shortly afterwards, however, tags became available with solar charging with an almost unlimited duration of tracking and large numbers of fixes. Due to mortality and failure of tags, the number of tracked individuals was lower in spring, as all but two birds were marked in the breeding areas in Europe before autumn migration.

The spring departure from the wintering areas normally takes place quite suddenly and Lake Tanganyika is bypassed in the east, in the opposite direction as in autumn, but in contrast to autumn not so close to its eastern shore (Meyburg unp. data). Even the eagles which spend the winter in Namibia, do not shorten their migration route, but first move in almost eastern direction to the southeast of Lake Tanganyika, where the eagles come together similar to the autumn fan-shaped from all directions. Eagles wintering in Angola, Namibia and Western Botswana again make a long detour of several hundred kilometres to Lake Victoria on their route, bypassing Lake Taganyika, which in comparison would have been a direct line without this detour (Meyburg unp. data). The following part of the spring route mainly retraces the autumn one, but deviates very clearly in one area. In Africa, the eagles fly from South Sudan through Eastern Ethiopia into Eritrea getting closer to the Red Sea and then move north near the mountains parallel to the coast, thus not far inland as with the outward migration (Meyburg unp. data).

Even if, at a wide scale, the flyway is the same one followed during autumn migration, during spring standardized counts are very low in Israel (Table 1), probably because birds follow a more easterly route and on a broader front (Leshem and Yom-Tov 1998, Shirihai et al. 2000). Counts of thousands of individuals at Suez (Table 1) and an anecdotal daily count of 16,231 individuals over Quyrat Shemona (Israel) on 25 March 1991 confirm this scenario, besides, of course, individually tracked birds. In the Middle-East, the main migratory flow continues towards the Bosphorus, en route to Europe. Recent spring counts from Batumi confirm the importance of this watchsite also during spring migration (Table 1), even if lower numbers are recorded than during autumn. Satellite tracking shows that in Turkey, the majority of eagles follow the old route along the Mediterranean coast to the Göksu River area and then through the Taurus Mountains to the north-west. Some eagles, however, move from the Gulf of Iskenderun east of Adana in practically a straight line further to the Bosporus (Meyburg et al. 2000, 2004b, 2007b, Meyburg and Meyburg 2009a,b, Meyburg unp. data). All tagged 2nd calendar-year individuals left Africa in spring, but much later than the adults, and returned to Europe following the same pathway. Arrival in Europe is also much later than for the adults, although with a usually earlier arrival from year to year (Meyburg unp. data). Only very rarely one-year-old individuals were documented in the general area of their birthplace, but most frequently they spend the summer months far away. Eagles born in Germany spend their first summer hundreds of kilometres south and east of their birthplace, for example in Bulgaria, Belarus, Moldavia and Lithuania (Meyburg unp. data).

By the age of two years, they would have already returned to the vicinity of their birth place, but not before the second half of August. The departure from the breeding area, however, occurs at about the same time as the adults (Meyburg unp. data). Concerning migration duration and speed, the complete spring migration of a female adult LSE from winter quarters in the Kruger National Park (South Africa) to the breeding site in Germany, including all overnight stops, was documented in detail for the first time in 1998. On 21 February the eagle left the Kruger National Park and was observed directly arriving 64 days later at the nest site on 25 April. The bird migrated 51 days with an average daily distance of 211 km (range: 18–406). In-between it rested for 13 days, of which 4 days were a forced stopover in Eastern Hungary due to bad weather (Meyburg et al. 2007b). All other tracked adults behaved similarly, with slight individual differences.

Behavioural and Ecological Remarks

Sea crossings

For young, inexperienced LSEs, the Mediterranean and the Black Sea are dangerous ecological barriers, as they strictly avoid flying over larger water bodies as thermal gliders. The Mediterranean Sea in particular poses a very great danger to young birds on their first autumn migration. None of the 58 adult LSEs that we satellite-tracked since 1994 from European countries has ever tried to cross these two seas (Meyburg et al. 1995, 2000, 2004b, 2009b, 2017a). Even the migration from the southern tip of the Sinai Peninsula across the Gulf of Suez directly to Africa is avoided by most LSEs, if they arrive there deviating from the direct migration route—probably accidentally. They then travel almost to the north along the coast against the headwind to Suez bypassing the Gulf of Suez. This causes a loss of time of about one week, but occurs again and again in experienced adults (Meyburg et al. 2002, Meyburg unp. data). Some few satellite-tracked adult and juvenile individuals fly over the Gulf from the tip of the peninsula, however (Meyburg et al. 1995, 2008). Why also adults, so far only females, which take the correct short route in autumn to Suez for years and get there, fail to fly directly to Suez in the other years, remains a mystery. Since they move together in large flocks, they may rely on leading birds. Maybe they miss the connection in the morning and get lost alone or in small groups. In Zambia and South Africa, for example, large migrating and wintering groups were observed by the author, led by an individual (Meyburg et al. 2010).

Orientation and navigation

Successful bird migration can depend on individual learning, social learning, and innate navigation programs (Mueller et al. 2013). The navigation system of small birds is roughly known today. In contrast to many other bird groups, young eagles and other raptors do not migrate in family groups. Within the framework of a long-term population support project (since 2004) of the LSE, which is threatened with extinction in Germany, it was possible to gain insights into the migratory behaviour of migratory young eagles by means of translocation experiments (Meyburg et al. 2008, 2017a). In 2007–2011, 51 second hatched LSEs were transferred almost 1,000 km far in the WSW direction from Latvia. 18 Latvian young eagles were marked with satellite transmitters next to a larger number of German young eagles. A total of 51 young LSEs were satellite-tracked as part of the project, 33 German and 18 Latvian individuals, with non-experimental first-born nestlings also being equipped with transmitters as a control group, as well as adult birds. Only 45.4% (15 individuals) of the native unmanaged German juveniles migrated on the southeast route to the Bosphorus, corresponding to the simultaneously tracked adult eagles. Among the Latvian birds this proportion was 33.3% (6 individuals). 39.4% (13 individuals) of the German individuals and 50.0% (9 individuals) of the Latvian ones moved to the south (Greece, mostly Peloponnese) and Italy. 15.2% (5 German individuals) and 16.7% (3 Latvian animals) moved to the south-west, mainly to France and Spain, and in some cases crossed the Strait of Gibraltar. Thus, less than half of the young eagles of both groups migrated on the optimal route to the Bosphorus. The young eagles that travelled to the Bosphorus and then further along the main migratory route along the Eastern Mediterranean coast reached the African wintering areas like the adults. Most of the other young eagles were unable to fly over the Mediterranean Sea. They usually drowned only after they had covered a large part of the distance to the African coast. A young eagle flew from Sicily past Malta in the late evening and then drowned halfway to Africa. However, two individuals succeeded the crossing from Greece to North-Africa, which involved a non-stop water crossing of ca. 300 km, probably being the first proofs of such an event. Departure data of the young and adult eagles suggest that those young birds that left at the same time as the older adults accompanied them and thus learned the optimal migratory route, while the other young eagles were left on their own and only followed the apparently grossly innate migratory direction to the south. Some circumstances also suggest that they followed other species, such as ospreys, but could not cross the sea using the same route (Meyburg et al. 2017a, Meyburg unp. data). A large part of the adult Ospreys (*Pandion haliaetus*) tracked by myself crossed the Mediterranean Sea from Cabo de Gata (Andalusia, Spain) to Cape Viejo (Morocco). Also a young LSE tried to overcome these 150 km on exactly the same route and drowned only 5 km off the coast. Both translocated and unmanaged young eagles raised in their own eyries must learn the optimal migration route from unfamiliar, older eagles. On their own they have hardly any chance to survive the first autumn migration.

Observations carried out in autumn 2009 at Mount Olympus (Greece) fit well in this pattern: among 13 migrating LSEs, two adults were migrating towards the Bosphorus, while six juveniles were heading south-west, probably following the innate migration direction (Panuccio et al. 2013b).

Flight behaviour

Spaar (1997) shows that the LSE has very similar flight patterns to those of the steppe eagle (*Aquila nipalensis*), being both strongly dependent upon thermal updrafts: they spend the majority of time circling (36%) and gliding (53%), with just a 4% of flight time allocated to flapping. The climbing rate during thermal circling was almost 2 m/sec, while the cross-country speed was 10 m/sec.

The GSM-GPS transmitters I used from 2012 onwards made it possible to obtain many more locations than before, to determine flight altitudes without limit and at each point on the earth. On 21st March 2014 (490 locations on this day) for example an adult male started in Ethiopia 3 hours after sunrise. Within 12 minutes the eagle won 2,359 m height (climbing speed 3.27 m/sec), about three times the climbing speed in the breeding area. Between 12.37 and 15.00 the eagle flew five times at altitudes of 4,000 m and more, in one case 4,220 m above sea level or 3,302 m above the ground. After that he continued his migration until half an hour after Sunset. The highest altitude of 4,771 m above sea level was reached by a female on 29 March 2014 around noon in Northern Ethiopia, 2,710 meters above ground. Within nine minutes, the eagle had climbed 1,737 m. This gain in altitude corresponds to 3.2 m/second. In the time of the ascent the bird travelled 6.42 km in an easterly direction. Flight altitudes of about 5,000 do not seem necessary due to the ground level, there are probably other reasons (e.g., wind direction, lower air resistance) which needs further research (Meyburg and Meyburg in prep.).

At the Bosphorus, during autumn migration, more LSEs were observed with stronger and northerly winds. Moreover, they were flying lower in such wind conditions, and with higher air pressure (Panuccio et al. 2017). At Batumi, LSEs fly closer to the Black Sea coast especially when clouds develop on the hills inland, probably in order to find more favourable flight conditions (Vansteelant et al. 2014).

Population trends

Autumn counts in Israel offer a great possibility to monitor the global population of the LSE. From the late 1980s throughout the 1990s a steep decline is apparent (Krumenacker 2013). On the other hand, it seems that a positive trend can be seen in more recent years. Between 2004 and 2010, the number of birds recorded on average significantly increased by about the 35%. This is possibly a result of the strong expansion of the breeding range in Russia. The 30-year-average is 85,944 LSEs, while the average for the period 2007–2012 surpasses 96,000 individuals. The 2010 total of 124,000 birds represents the second highest count in history. However, there may be inaccuracies in the census protocol in Israel, and satellite tracking data suggest that Suez would be a better place, because the migration front is narrower. Vansteelant et al. 2020b analyzed a time series of LSEs counts from Batumi, which is much shorter than the Israel one, but takes into account different age classes. They found a lack of significant trends among non-juvenile birds, suggesting stable breeding population at least in the areas from which this migratory flow originates.

9

Greater spotted eagle *Clanga clanga*

Ülo Väli,[1,*] *Valery Dombrovski*[2] and *Paweł Mirski*[1]

ABSTRACT

The greater spotted eagle (*Clanga clanga*) is a flexible migrant raptor species exhibiting both short- and long-distance migration patterns. Its breeding range stretches across Eurasia from the Baltic Sea in the west to the Pacific Coast in the east; the winter distribution stretches along a wide belt across the entire southern Eurasia and extends to north-eastern Africa. Although this large soaring species migrates avoiding the crossing of large water surfaces it is only seldom recorded at watch-points, probably due to the low population size and relatively frequent use of short-distance migration. Results of telemetry indicate three main migration routes in Europe. The first leads through Eastern Europe to the Middle East, or further to North-Eastern Africa; the second through or by Carpathian Mountains to the Balkan Peninsula; the third through Central and Western Europe to South-Western Europe. The main wintering sites of the Asian populations are located in the Arabian Peninsula, the Indian Subcontinent, the Indochina Peninsula and South-Eastern China. To date, rather good knowledge exists on migration at the western margin of distribution range and further studies should focus more on populations in European Russia and Asia.

Introduction

Breeding biology across the distribution range

The breeding range of the greater spotted eagle (*Clanga clanga* Pallas, 1811) stretches along medium latitudes of Eurasia from the Kaliningrad Region (Russia) and Eastern Poland near the Baltic Sea in the West to Manchuria and the Pacific Coast in the East. In different parts of this vast distribution range, this opportunistic eagle has different stereotypes of nesting and foraging habitat selection. In the European part of the range, it is considered as a breeder of inaccessible swampy small-leaved forests close to open foraging grounds, such as fens, marshes and floodplains of rivers or lakes, where it feeds on voles (mainly water voles *Arvicola terrestris*) and medium-sized wetland birds (e.g., mallard

[1] Institute of Agricultural and Environmental Sciences, Estonian University of Life Sciences, Kreutzwaldi 5D, 51006, Tartu, Estonia.

[2] Laboratory of Molecular Zoology, Scientific Practical Centre for Biological Resources, National Academy of Sciences of Belarus, Academichnaia 27, 220072, Minsk, Belarus.

* Corresponding author: Ylo.Vali@emu.ee

Anas platyrhynchos, water rail *Rallus aquaticus*, spotted crake *Porzana porzana*, corn crake *Crex crex*; Cramp and Simmons 1980, Galushin 1962, Väli and Lõhmus 2000, Lõhmus and Väli 2005, Karyakin 2008a, Dombrovski 2010, Maciorowski et al. 2014). However, it is also able to forage on drained agricultural landscape where it feeds on insectivores, smaller voles (*Microtus agrestis*, *M. arvalis*) and medium-sized birds of drier habitats (e.g., grey partridge *Perdix perdix*, lapwing *Vanellus vanellus*, hooded crow *Corvus corone*; Väli and Lõhmus 2000, Väli and Lõhmus 2002, Dombrovski 2010). A hundred years ago, the greater spotted eagle was breeding also in the European forest-steppe zone, from where it was subsequently eradicated by humans (Belik and Vetrov 1998, Dombrovski and Panchenko 2009). In the diet of greater spotted eagle from this region, hamsters (*Cricetus cricetus*) and spotted ground squirrels (*Spermophilus suslicus*) were often recorded (Somov 1897, Portenko 1928, Kolesnikov 1960). In South of Siberia and the Far East, the array of nesting biotopes of the species becomes wider, from forest-steppe, dry terraces above the floodplain and mountain slopes to alder swamps and taiga forest, but usually with a significant predominance of open wetland habitats (Ryabtsev 1997, Karyakin 1998b, Berezovikov et al. 2008). Long-tailed squirrels (*Spermophilus undulatus*) and Daurian pikas (*Ochotona daurica*), in addition, water-associated small mammals, such as reed vole (*Microtus fortis*), play an important role in the diet of eastern greater spotted eagle populations (Karyakin 2008b, Melnikov 2008).

Greater spotted eagle builds its nest in a tree, which is usually located in forest interior. Often, it uses the dominant tree species of forest stand for nesting (Dombrovski and Ivanovski 2005). The height of the nest varies from 3 to 25 m (Cramp and Simmons 1980). It has one clutch of 1–2 eggs per year, and replacement clutch has been rarely recorded (Ü. Väli unpubl. data). Duration of incubation is 39–44 days (Cramp and Simmons 1980, Dombrovski 2019). In Belarus, mean date of the start of incubation is 22nd April (range 5.04–7.05); fledglings leave their nests on 6th August on average (range 20.07–26.08; Dombrovski and Ivanovski 2005).

Siblicide, or cainism (killing of younger chick at early age by the older one), significantly affects breeding success (Meyburg and Pielowski 1991). Interestingly, the frequency of cainism decreases in a gradient from West to East; whereas there have been no nests with two surviving chicks recorded in Poland (Maciorowski et al. 2014), the share of such nests is 3.6% in Belarus (Dombrovski 2005), 24% in the Volga-Ural region (Karyakin 2008a), 43.9% in Western Siberia and 64.3% in Altai, Eastern Russia (Karyakin 2008b). In line with that, fledging period decreases along the same gradient and is 60–65 days in the western part of the range (Cramp and Simmons 1980), 49–58 days in the Volga-Ural region (Karyakin 2008a) and 45–47 days in China (Gao Wei 1997 in BirdLife International 2001). After departing nests, fledglings are fed by their parents for about a month (Cramp and Simmons 1980), sometimes even up to two months.

Greater spotted eagle acquires full adult plumage only in fifth calendar year (Forsman 1999), but reproduction of birds having immature plumage characters has been recorded in spotted eagles (Meyburg et al. 1997, Väli 2010). Immatures tend to return to the breeding range and wander near occupied breeding territories or concentrate on places with high prey density (Dombrovski 2003). Information on natal dispersal is practically absent due to the high mortality rate of birds during their first year of life (Meyburg and Meyburg 2005, Maciorowski et al. 2014). In the only well-documented case, the chick marked with a GPS-transmitter began to breed at the age of no less than eight years, 60 km from its hatching site in Estonia (Ü. Väli and U. Sellis unpubl. data).

Numbers and threats

In the second half of the 20th century, numbers of the greater spotted eagle declined severely throughout the entire range. The species disappeared from many countries and regions, where it had previously nested—Finland, Slovakia, Hungary, Romania, Bulgaria, western Balkan region and the forest-steppe zone of Ukraine, for instance (Hagemeijer and Blair 1997). Currently, the species has remained as a breeder only in Russia, Belarus, Poland, Ukraine, Estonia, Kazakhstan, Mongolia and North-East China, as well as "in tiny numbers" in Pakistan and North-West India (Prakash 1988, Ferguson-Lees and Christie 2001, BirdLife International 2001). Single individuals, interbreeding with the sister species lesser spotted eagle *Clanga pomarina*, have been recently detected at the western margin of distribution range in Germany, Latvia and Lithuania (Väli et al. 2010). The species is considered as endangered (SPEC 1) in Europe and vulnerable (VU) in the global scale (BirdLife International 2017b).

The current size of the world population of the greater spotted eagle is estimated to be less than 10,000 mature individuals (BirdLife International 2017b). About 1,000 breeding pairs inhabit the European part of the range (BirdLife International 2017b). In Western Siberia and the Altai-Sayan region, which provides probably the most optimal habitat of the species in the Asian region, the number of greater spotted eagle is estimated at 1,200–1,445 breeding pairs (Karyakin 2008) while further to the East, the species is much rarer (Ryabtsev 1997, BirdLife International 2001).

Across its vast breeding range the greater spotted eagle distribution is rather patchy. The suitable habitat is fragmented due to anthropogenic transformation of natural habitats and this has caused declines in many regions (Cramp and Simmons 1980, Hagemeijer and Blair 1997, Maciorowski et al. 2014). The numbers decreased by 33% in transformed habitats, but remained stable in natural or weakly disturbed biotopes in Belarus in 2000–2010, for instance (Dombrovski 2013). In Estonia, where species has remained to breed almost only in transformed habitats, the decline in 2004–2010 was even more severe, twofold or 14% per year (Väli 2015a). The species is very sensitive to direct disturbance too. For example, greater spotted eagle prefers larger marshy forests for nesting near human-transformed biotopes in Belarus (Dombrovski and Ivanovski 2005). Accordingly, the problem of cutting suitable forest areas is particularly acute in such landscapes.

The eradication of rodents in the European part of Russia, the south of Western Siberia and Belarussian Polesie, widely practiced in these regions in the middle of the twentieth century, also had a negative impact (Belik 2000). In the Volga-Ural region and in Eastern Siberia, where a substantial proportion of greater spotted eagles nests in river floodplains, construction of hydroelectric stations and flooding of suitable habitats had a significant impact on its population (Ryabtsev 2000, Karyakin 2008a). Decline in the Baikal region has been associated with the ploughing of natural steppes and the disappearing of long-tailed ground squirrels (Melnikov 2008). Among other reasons behind the long-term decline in Eastern Siberia, unfavourable situation on wintering grounds in South-East Asia has been noted (Ryabtsev 2016). In some regions, the greater spotted eagle is threatened by electrocution on power lines (Maciorowski et al. 2014). In the Middle East and the Caucasus, one of the most significant threats to migratory raptors, including spotted eagles, is their illegal shooting by local people (Abuladze 2013, Sandor et al. 2017, Özbahar and Gül 2018). Deliberate killing is an important threat also during migration and wintering in Europe (Maciorowski et al. 2014).

A specific negative factor, involved in the decline of the greater spotted eagle, is the widespread hybridisation with more abundant lesser spotted eagle in the zone of sympatry in Central and Eastern Europe (Väli et al. 2010). This phenomenon is probably a direct consequence of recent anthropogenic transformations of the environment (Maciorowski and Mirski 2014, Dombrovski 2016), with the subsequent decline of the greater spotted eagle and also by the phenotypic plasticity of the two spotted eagle species (Lõhmus and Väli 2005). In Belarus, the percentage of hybrid individuals and mixed pairs is significantly higher in regions where the greater spotted eagle has a very low abundance due to the anthropogenic transformation of their natural habitats (Dombrovski 2016). Thus, the proportion of phenotypic hybrid individuals in the European museum collections was very low (about 4%) until the middle of the twentieth century, but was increased eightfold by the early 21st century (Dombrovski 2016). Gradual replacement of pure greater spotted eagle pairs with hybrids or mixed pairs, and thereafter with pure lesser spotted eagle pairs, has been recently recorded in suboptimal habitat with low breeding greater spotted eagle density (Väli et al. 2010, Maciorowski et al. 2015, Väli 2015).

Wintering

Similarly to the breeding range, the winter distribution of the greater spotted eagle stretches along a wide belt across entire Southern Eurasia and extends also to North-Eastern Africa. Birds from western populations migrate to a wide array of directions to winter in Southern Europe (mainly Greece), Anatolia, Israel and North-Eastern Africa (mainly the basin of river Nile; Meyburg and Meyburg 2005, Dan et al. 2005, Eagle Conservation Committee 2014, Dombrovski et al. 2018, Väli et al. 2018, Kotkaklubi and 5DVision 2019, Maciorowski et al. 2019). The main wintering sites of the Asian populations are located in the Indian Ocean basin and include the southern part of the Arabian Peninsula, the Indian Subcontinent, the Indochina Peninsula, and South-Eastern China (BirdLife International 2001). Birds winter solitarily or in small groups and usually no more than few wintering birds occur together at one site (Maciorowski et al. 2019).

Some birds could stay also northerly of the main wintering range. For example, in Central Europe wintering greater spotted eagles have been recorded thirteen times in Poland until 2016 (Stawarczyk et al. 2017) and there have been at least four sites with regularly wintering individuals in Hungary over the recent years (Juhász et al. 2018). Some individuals occur north of main range also in Asia, as indicated by winter observations from Uzbekistan, Turkmenistan, Northern Iran, and Transcaucasia (Dementiev and Gladkov 1951, Mitropolsky et al. 1987, Abuladze 2008). On the other hand, few individuals, including some tagged with satellite transmitters in Europe, have been observed in South Africa (mainly Zambia; Meyburg et al. 1998b, Meyburg and Meyburg 2005, Vallotton 2011, Maciorowski et al. 2014). In these extreme long-distance migrants mixed ancestry cannot be excluded, because interspecific hybrids are hard to identify and they tend to migrate further south than greater spotted eagles (Väli et al. 2018).

The wintering territory (home range) of the greater spotted eagle is characteristically small, compared to the lesser spotted eagle (*Clanga pomarina*), which moves over large areas in the course of winter (Väli et al. 2018). In most wintering areas, the greater spotted eagle chooses wetland habitats where it feeds on medium-sized birds and small mammals (Alivizatos et al. 2004, Pérez-García et al. 2014, Maciorowski et al. 2019). It is also periodically observed on landfills and dry fields rich in rodents (Wells 1999 in BirdLife International 2001). Sometimes, insects and carrion make up the winter diet (Ferguson-Lees and Christie 2001).

Knowledge Status

There is no known historical folklore related specifically to greater spotted eagle migration, probably due to difficulties in identification of large brown eagles, a group of raptors to which the greater spotted eagle belongs too. Indeed, telling apart greater spotted eagles from the lesser spotted and steppe eagles (*Aquila nipalensis*) is not always easy and the existence of hybrids between the two spotted eagle species does not make the task easier. This is an acknowledged problem in observation stations where migrating raptors are counted visually and distant large brown eagles (formerly known as *Aquila* eagles) not seldom remain unidentified (Alon et al. 2004, Hilgerloh 2009, Verhelst et al. 2011, Wehrmann et al. 2019). Identification of species and hybrids is easier on closer examination (Väli and Lõhmus 2004, Dombrovski 2006, Lontkowski and Maciorowski 2010) meaning that methods using marking techniques (e.g., ringing and telemetry) suffer less from identification problems. Next, we take a closer look at the knowledge status on migration of the greater spotted eagle resulting from the main sources of information.

Records from observation stations

The greater spotted eagle migrates using soaring-gliding flight over land (Hilgerloh 2009, Eagle Conservation Committee 2014, animation in Väli et al. 2018, Kotkaklubi and 5DVision 2019) and avoids crossing large water barriers similarly to other large soaring raptors (Bildstein 2006). Therefore migration of greater spotted eagle concentrates in narrow straits and could be watched visually in observation stations located at such places. At the same time, and for the same reason, this eagle is usually not recorded in islandic raptor watch-points.

As a northerly breeding species, the greater spotted eagle is recorded in observation stations in northern Europe, but only as a rarity. For example, only three migrating individuals were recorded in Falsterbo (Southern Sweden) during autumns 1986–1990 (Kjellen 1992). It is more often seen in raptor observation Stations in southern Europe but, as this species does not breed in Western or Central Europe and does not winter in West Africa, count lists of Western and Central Mediterranean observation stations expectedly lack the species completely or it is recorded extremely rarely, it is regularly observed in Eastern Mediterranean stations, but only in small numbers (Table 1). Higher numbers (up to few hundred; Wehrmann et al. 2019) have been recorded on the eastern coast of the Black Sea, i.e., on the flyway of birds probably mainly originated from European Russia, and in Israel where flyways from Central and Eastern Europe, as well as from Western Asia meet (Table 1). Individuals wintering in North-Eastern Africa are observed also in further south, at the western coast of the Red Sea (Table 1).

Notably, the proportion of greater spotted eagle is low in all observation stations (Table 1) and it is always outnumbered by other large brown migratory eagles. In Eastern Mediterranean and East Pontics, the lesser spotted eagle is hundreds or even thousands times more common (Leshem and Yov-Tov 1996, Alon et al. 2004, Abuladze 2013, Fülop et al. 2014, Panuccio et al. 2017) while in the Southern Middle-East and Southern Asia easterly breeding steppe eagles vastly outnumber the greater spotted eagle (Shirihai and Christie 1992, Leshem and Yov-Tov 1992, den Besten 2004). Both Lesser Spotted and Steppe Eagles form the bulk of migrating large eagles in North-Eastern Africa (Hilgerloh et al. 2009, Megalli and Hilgerloh 2013). Relatively high proportion of greater spotted eagles has been noticed only during a spring count in Bet-Shemesh (Israel) where it supposedly formed 15% of all raptors (i.e., 1,877 individuals! Frumkin et al. 1995). Given the general occurrence pattern, extremely low numbers and relatively low proportion among eagles, this record reflects probably a mistake.

Ring recoveries

Another traditional method to study migration is marking of birds and collation of information of their subsequent resightings or recoveries. To our knowledge wing-tagging has been used in the greater spotted eagle only for a short period in Belarus where 19 young were tagged in 2010–2011 resulting to two resightings (Dombrovski et al. 2011 and unpub. data). On the other hand, many individuals have been marked using common ornithological rings, but information from that type of tag is nearly unexceptionally possible to recover from dead birds only and thus the recovery rate is very low. In recent decades greater spotted eagles have been marked with colour rings too, mostly using various types of plastic rings. Special colour-ringing programmes started in Poland (valley of river Biebrza) in 2005, Estonia in 2006, Belarus in 2007 (Dravecky et al. 2008) and Ukraine in 2012. The recovery rate of colour rings is surprisingly high. For example, 11 out of 36 (31%) greater spotted eagle or hybrid nestlings colour-ringed in Estonia in 2006–2017 have been resighted later at least once, mostly during migration and wintering (Ü. Väli unpubl. data). In Belarus, 116 juvenile greater spotted eagles and 15 hybrids were ringed in 2007–2018 and 12 of them (9%) have been resighted (V. Dombrovski unpub. data). Seven recoveries (13%) have been made of 53 greater spotted eagles and hybrids (all age groups) ringed in 2005–2011 in Poland (Maciorowski 2012). The main reason for such a high recovery rate is probably the wintering of hundreds of greater spotted eagles in Southern Europe that is relatively densely manned by birdwatchers. Moreover, the rare greater spotted eagle is a target species for European birders who are highly interested to find this eagle, and possess good-quality equipment to record ring-codes from distance and keen to respond back on the information.

Table 1. Numbers of the greater spotted eagle (annual means of multiyear counts) and their proportion among all migrating raptors in migration observations stations across Southern Eurasia.

Location	Years	Season	Period	Mean no. of raptors in total	Mean no. of greater spotted eagles	Proportion of the greater spotted eagle (%)	Reference
Strait of Messina, Italy	1996–2000	Spring	01.04–27.05	26,172	0.2	0.001	Corso 2001a
Dadia, Greece	2003–2005	Spring		677	0.3	0.044	Schindler et al. 2014
Antikythira, Greece	2007–08	Spring	16.04.15.05	398	0.0	0.000	Lucia et al. 2011
Antikythira, Greece	2007–09	Autumn	21.08–20.10	1369	0.3	0.024	Lucia et al. 2011
Dobrogea, Romania	2003–2007	Autumn	Mid-August–2nd half of October	11,297	4.2	0.037	Fülöp et al. 2018
Dardanelles	2010	Autumn	22.08–01.10	1601	1.0	0.062	Panuccio et al 2017
Bosphorus	2010	Autumn	22.08–01.10	24,701	4.0	0.016	Panuccio et al 2017
Bosphorus	2008	Autumn	22.09–10.10	141,844	6.0	0.004	Fülop et al. in 2014
Batumi, Georgia	2008–2009	Autumn	21.08–31.10	829,085	74.5	0.009	Verhelst et al 2011
Batumi, Georgia	2011–2018	Autumn	17.08–16.10	1,039,330	464	0.004	Wehrmann et al. 2019
Northern Israel	1982–1990	Autumn		490,800	52	0.011	Leshem and Yov-Tov 1996
Northern Israel	1990–1999	Autumn	10.08–15.10	450,995	44	0.010	Alon et al. 2004
Eilat, Southern Israel	1980–1987	Autumn	15.10–15.11 or 25.08–30.11	16,967,	4.3	0.026	Shirihai and Christie 1992
Eilat, Southern Israel	1977–1988	Spring		795,265	6.3	0.001	Shirihai and Christie 1992
Eilat, Southern Israel	1977–1988	Spring	15.02–20.05	795,314	6	0.001	Leshem and Yov-Tov 1996
Iskenderun, Turkey	1976	Autumn		25,813	0	0.000	Sutherland and Brooks 1981
Iskenderun, Turkey	2000	Autumn		29,791	2	0.007	Can 2001
Iskenderun, Turkey	2013	Autumn	30.08–26.09	62,293	4	0.006	Oppel et al. 2014
Zait Bay, Egypt	2006	Autumn	20.08–29.10	9376	0	0.000	Hilgerloh et al. 2011a
Zait Bay, Egypt	2007	Spring	20.02–4.05	36,976	9	0.024	Hilgerloh 2009
Ayn Sokhna, Egypt	2012	Spring	01.03–02.05	154,276	57	0.037	Megalli and Hilgerloh 2013
Dharamsala, India	2001	Autumn	23.10–30.11		11		Den Besten 2004

Table 1 contd. ...

...Table 1 contd.

Location	Years	Season	Period	Mean no. of raptors in total	Mean no. of greater spotted eagles	Proportion of the greater spotted eagle (%)	Reference
Dharamsala, India	2002	Spring	19.02–29.03		15		Den Besten 2004
South Baikal, Russia	1996–2004	Autumn	25.08–14.10	6450	15	0.236	Alexeyenko and Fefelov 2018
Bedong, Malaysia	2012–13	Autumn	01.10–05.11	60,076	3.5	0.006	Eng et al. 2014
Chumphon, Thailand	2007–2008	Spring	Late February–early April	49,269	7	0.014	DeCandido and Nualsri 2009

Telemetry

Despite its rarity, or perhaps thanks to that, the greater spotted eagle was one of the first raptor species studied by means of satellite telemetry. The first bird was deployed by a transmitter in 1993 in the wintering ground in Saudi Arabia (Meyburg et al. 1995). Few years later, in 1996, the first bird from a breeding population in Biebrza Valley, Poland, was marked (Meyburg et al 1998, Meyburg et al 2005); the same population has been studied intensively until now (Maciorowski et al. 2014, Väli et al. 2018, Maciorowski et al. 2019). In 2005 telemetry studies started in Estonia (Väli et al. 2018) and in 2017 in Belarus (Dombrovski et al. 2018). Two birds have been deployed in Russia (Russian Raptor Research and Conservation Network 2017) and one more in wintering grounds in Arabian Peninsula (Strick et al. 2011).

Post-breeding Migration

Timing

At the western limit of the range, in Poland and Estonia, departing from breeding grounds takes place between mid-September and mid-October, significantly later than the lesser spotted eagle or interspecific hybrids (Väli et al. 2018). However, no such difference was recorded in Belarus where GPS-tracked spotted eagles left their territories between 13th and 28th September 2017 (Dombrovski et al. 2018). The last individuals, supposedly migrants from north, have been observed in Belarus on 10th October (Dombrovski and Ivanovsky 2005). Similarly, the peak of migration has been observed at the break of September and October in Ukraine (Domashevsky 2005). Timing of migration in eastern populations is not different. Greater spotted eagles depart Volga-Ural and Western Siberia in September (Karyakin 2008a,b), Baikal region between late August and early October, mostly in the second half of September (Ryabtsev 1997, Alexeyenko and Fefelov 2008, Elaev 2008, Alexeyenko et al. 2018), and North-Eastern China from early to mid-October (Gao Wei 1997 in BirdLife 2001). Notably, birds leaving the breeding ground, especially early-departing birds, not necessarily start migration, but may wander around breeding grounds or stop at a good foraging site instead (see below).

Data on migration phenology of the greater spotted eagle at bird observation stations is very scarce. In Batumi, at the eastern coast of the Black Sea, autumn migration was recorded between 12th September and 13th October in 2008–2009, with a median at 3rd October (Verhelst et al. 2011). In Northern Israel, peak passage was observed between 5th and 20th October (Alon et al. 2004) while in Eilat the species is recorded between the end of September and mid-November (Shirihai and Christie 1992). The greater spotted eagle is one of the latest migrants among raptors and its migration continues beyond the end of the normal survey period of raptor migration in autumn (Alon et al. 2004, Fülop et al. 2014) and thus species might be missed and phenology misdated. For example, in Antikythira, Greece, the greater spotted eagle was recorded only in years when observations lasted until the second half of October (Lucia et al. 2011). This is confirmed also by the telemetry results demonstrating that birds usually arrive their wintering grounds in October or November (Dombrovski et al. 2018, Väli et al. 2018, animation in supplementary materials).

There is a wide array of flight directions and distances used by migrating greater spotted eagles, considering both longitudinal and latitudinal patterns. For example, the individuals from a same breeding subpopulation are spread in winter along *ca* 20 latitudinal degrees (> 2200 km) and *ca* 30 longitudinal degrees (> 3300 km; Väli et al. 2018). This means that individuals may have up to two- or three-fold differences in their flight distances and this causes large variation in duration of migration.

As regards migration speed, birds flying from Belarus to Africa moved faster (170 km/day on average) than birds flying to Balkans and the Middle East (137 km/day) (Dombrovski et al. 2018). An eagle from Poland was reported to cover, on average, 217 km daily and even up to 490 km when flying over the desert. Conversely, a juvenile heading to Greece flew only 57 km per day (Meyburg et al. 2005). In addition to variation between individuals, same birds may have a very different mean migration speed over the years (Kotkaklubi and 5DVision 2019).

Main routes

Greater spotted eagles follow at least three main routes through Europe leading to different wintering sites (Maciorowski et al. 2014, Väli et al. 2018). The first leads through Eastern Europe to Middle East, or further to Africa; the second through or by Carpathian Mountains to Balkan Peninsula; the third through Central and Western Europe to South-Western Europe (Fig. 1). Since the greater spotted eagle strongly avoids crossing large water bodies, transcontinental migration routes concentrate on narrow straits, such as Bosphorus at the Mediterranean Sea. Although some greater spotted eagles have been observed in Crete or Sardinia, these were only inexperienced juveniles probably initially taking a risk to migrate over the sea but then remained on islands (Maciorowski et al. 2019).

Detailed migration routes have been recorded for more than 40 individuals deployed by GPS-tags in Poland, Estonia and Belarus (Meyburg and Meyburg 2005, Meyburg et al. 2005, Väli et al. 2018, Dombrovski et al. 2018) and for only few individuals from eastern population, Russia (Meyburg et al. 1995, Russian Raptor Research and Conservation Network 2017) and Kazakhstan (Strick et al. 2011). Individuals heading to the largest wintering site in Europe, the Balkan Peninsula, fly through Baltic States, Poland, Slovakia, Ukraine, Hungary, Romania and Balkan states or by-pass Carpathian Mountains from east through Romania and Bulgaria. Additionally, a bird from Estonia took a route to south-east and wintered east of the Black Sea (Kotkaklubi and 5DVision 2019). Birds leaving Europe mostly concentrate in Ukraine and continue along the straightest line through Romania, Bulgaria and Turkey across Bosphorus. Some birds remain in Anatolia, with larger wintering sites in South-West Turkey, e.g., in the Göksu Delta (Kirwan 2008, Maciorowski et al. 2019). Individuals that move further south may stay in Israel and Egypt, but most of them pass the desert zone and settle for the winter in the wetlands of White Nile at the border of Sudan and South Sudan or Blue Nile at the border of South Sudan and Ethiopia. Some individuals have found suitable wintering sites in deserts of Egypt and Chad. Movements of greater spotted eagles further south (e.g., Zambia) are recorded

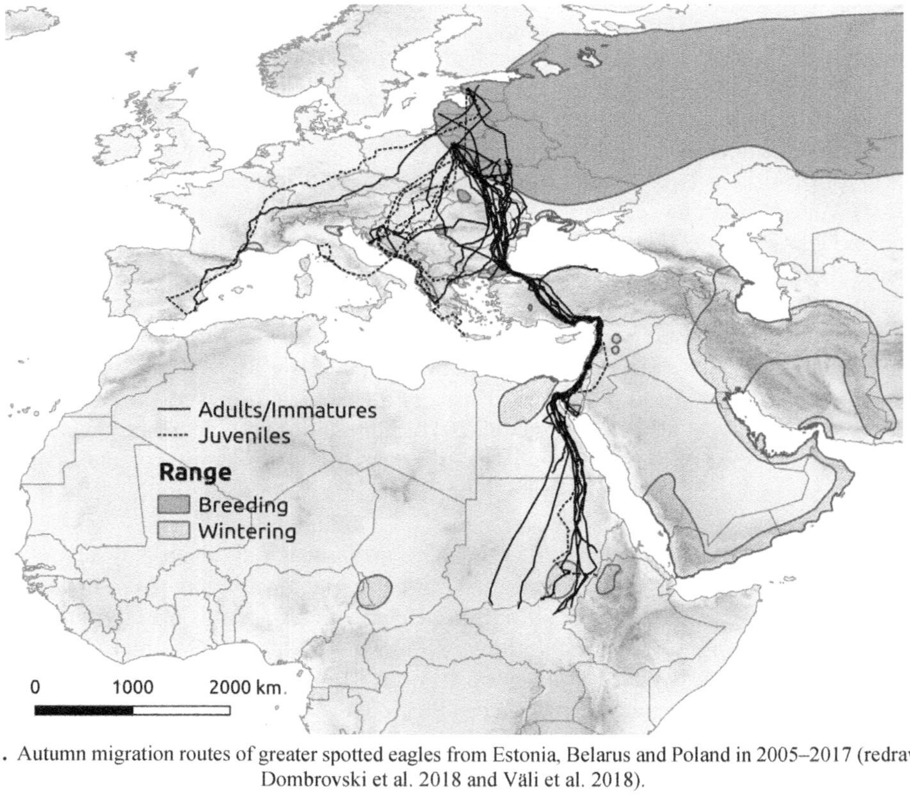

Figure 1. Autumn migration routes of greater spotted eagles from Estonia, Belarus and Poland in 2005–2017 (redrawn from Dombrovski et al. 2018 and Väli et al. 2018).

only seldom (Meyburg et al. 1998b, Dombrovski et al. 2018) and such birds may actually be hybrids between Greater and Lesser Spotted Eagles.

Only four Asian birds have been tracked. Two individuals were caught at wintering sites in Saudi Arabia and United Arab Emirates. The first one flew *ca* six thousand kilometres north-east to its breeding ground near Omsk (Russia), the other one through Turkmenistan and Uzbekistan to eastern part of Kazakhstan. This bird was in juvenile plumage, so it did not necessarily reached the species breeding range. Next autumn, the same bird headed almost directly southwards wintering in the Indus river delta. Recently, two juvenile greater spotted eagles, a male and a female, were also tagged in Altai, Russia. The male flew over three thousand kilometres and spent the winter in the Indus delta. Paths used by this individual in consecutive migrations were parallel, but spread up to about 600 km. The female headed almost directly southward for about four thousand kilometres to reach her wintering site in Central India. Large mountain massifs, such as Pamir and Himalaya are significant obstacles and probably most greater spotted eagles avoid these highland regions. However, the young female from Altai was able to cross the Kun-Lun Mountains flying at an altitude of more than 6,000 m (Russian Raptor Research and Conservation Network 2017). Birds from the Far East supposedly migrate along the coast of the Yellow and East-China Seas to reach wintering grounds in Indochina and China (Zalles and Bildstein 2000). The greater spotted eagle is mentioned as a migrant also in Mongolia, Southern China, Bhutan, Thailand, Laos, and Vietnam. Both migrants and small resident populations have been reported (BirdLife International 2001 in Pakistan, Nepal, northern India, Bangladesh and Myanmar), but the details of migration are not known there.

Migratory performance develops gradually over time with juvenile birds often taking suboptimal or more circuitous routes than adults. Interestingly, Western Balkans seems to be stronger attraction for juveniles whereas adults dominate among African migrants (Fig. 1). Large differences between autumn and spring routes may appear in unexperienced immature individuals and they may even switch to another flyway on their second autumn migration (Väli et al. 2018).

Pre-breeding Migration

Migration routes in spring are very similar (but less studied) to those in autumn and are described above. In this section we focus on timing of spring migration.

Data collected in telemetry studies indicate that birds from the western part of the range start their journeys to breeding grounds mostly in the first half of March, regardless if they are wintering in the Mediterranean Basin or in Africa (Meyburg and Meyburg 2005, Maciorowski et al. 2014, Dombrovski et al. 2018). Due to the low number of records, counts at observation sites provide only limited and thus highly variable information on phenology of migration. For example, in Egypt, at the northern coast of Gulf of Suez, migration ranged between 26th February and 28th March in 1982 and between 5th April and 1th May in 2012 (Megalli and Hilgerloh 2013), while at the south-eastern coast of Gulf of Suez, nine greater spotted eagles migrated north between 14th March and 16th April 2007, with the median on 11th April (Hilgerloh et al. 2019). A long-term study in Israel (Eilat) indicates that most greater spotted eagles pass there during mid-March to mid-April (Shirihai and Christie 1992).

Earliest regular arrival in breeding grounds, around 10th March, has been noticed in Ukraine (Domashevsky 2005). In Belarus, the first greater spotted eagles arrive between 9th and 25th March (Dombrovski and Ivanovsky 2005), much earlier than the co-breeding lesser spotted eagle (early April; Ivanovski 2012). GPS-tracking indicates that migrants from Africa arrive later than those from Europe; the difference could be a week or even more (Dombrovski et al. 2018). In Poland, greater spotted eagles that were wintering in Europe arrive at breeding sites between early March and the late April, while tagged African migrants only in mid-April (Maciorowski et al. 2014). In Estonia, the arrival of species has been observed in late March and in April (Väli and Lõhmus 2000) while tracked birds during the first half of April (Kotkaklubi and 5DVision 2019). Eastwards, eagles arrive at the break of March and April in Volga-Ural populations (Karyakin 2008a), but somewhat later, in the beginning of April, in Western Siberia (Karyakin 2008b). Arrival has been observed in mid-April in North-Eastern China and Eastern Russia (BirdLife International 2001) and in the second or third decade of April in Eastern Siberia (Elaev 2008).

Among eight birds breeding in Belarus, the average daily distance covered during spring migration was 151 km (range: 61–275 km); 150 km was the daily distance covered by a bird flying from Yemen to European Russia (Meyburg et al. 1995) and *ca* 260 km by an African migrant from Poland (Maciorowski et al. 2014). Interestingly, migrants from Africa to Belarus developed higher speed in the spring than in the autumn while the contrary pattern was observed among birds migrating within Europe (from Balkans; Dombrovski et al. 2018). Longer duration of spring migration was also observed in an individual wintering in Turkey (Meyburg and Meyburg 2005).

Also immatures move towards the breeding range in spring, but the onset of migration is considerably delayed and takes place in late March or early April (Eagle Conservation Committee 2014, Kotkaklubi and 5DVision 2019). However, immatures have been observed already in the first half of March in Belarus (Dombrovski 2004, but note the early arrival of adults in this region above). Some greater spotted eagles, possibly immatures too, are passing the migration bottleneck at Bosphorus in late May (Üner et al. 2010). Further north, in the Pomeranian region of Poland, at the southern coast of the Baltic Sea, immatures were most often recorded shortly at late spring (end of April, May), and they probably were still moving northwards (Sikora and Zieliński 2013). Although all tracked immatures have returned to north, there are some observations of immatures staying for the summer in the wintering grounds (Meyburg et al. 2016b).

Behavioural and Ecological Remarks

Although social influence may have a role in development of migration in large raptors (Meyburg et al. 2017), recent telemetry study focused on greater and lesser spotted eagles and their hybrids showed that migration strategy in this group is also under genetic control, which is consistent with the clock-and-compass mechanism known from passerines. Hence, migration-related traits such as onset, direction and location of wintering site depend in great extent on inherited "program" (Väli et al. 2018). Moreover, scarce data on the greater spotted eagle from counting points also confirm that this species does not fly in flocks where information could be shared by social interactions between conspecifics. Indeed, one of the first satellite telemetry studies on the species showed that greater spotted eagles families break up just before migration—a female was the first to leave the territory, followed 2–3 days later by an offspring and one week later by a male (Meyburg et al. 2005). Although the adults used the same flyway, their exact migration routes were not overlapping; the direction of juvenile (that perished on migration) remarkably differed from that of adults (Meyburg et al. 2005). Hence, although difference in migration routes taken by juveniles and adults/immatures may partly raise from learning, high mortality of juveniles (i.e., natural selection) is an important factor shaping the detected pattern (Fig. 1).

Not much is known about the post-breeding/pre-migratory movements. Some females may leave their home range well before migration and explore surrounding areas (see detailed descriptions of such behaviour in the lesser spotted eagle in Meyburg et al. 2007a) while others directly begin migrating. For example, a female breeding successfully in southern Estonia departed its breeding territory on 5th August, travelled through Latvia and Lithuania to Russia and stayed there, about 300 km far from its breeding territory, until the onset of southward migration on 15th October, whereas another female stayed over the years in its breeding territory until migration (Kotkaklubi and 5DVision 2019). At times, such pre-migratory movements are not easy to distinguish from stopovers made during migration. Moreover, some greater spotted eagles may fly to their wintering grounds without stopovers, especially when the total distance is short, while others make short stopovers to refuel and when facing adverse weather conditions (Kotkaklubi and 5DVision 2019). If stopovers are made, these last usually only few days, but seldom could amount up to two weeks; stopovers are more frequent and longer in autumn migration compared to spring migration (Ü. Väli and L. Puude unpub. data).

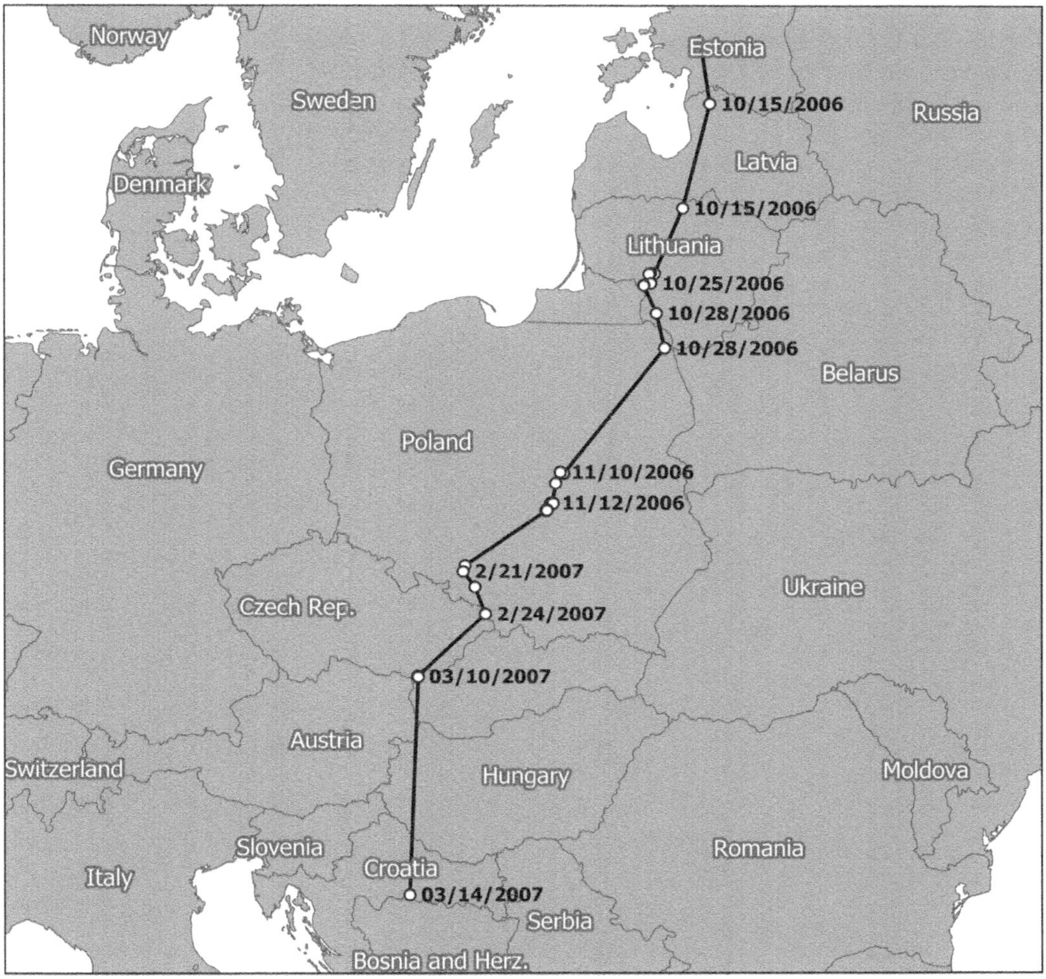

Figure 2. Intra-European movements during autumn and winter of an adult male GPS-tagged in Estonia (redrawn from Kotkaklubi and 5DVision 2019).

Sometimes it is hard to distinguish between stopover and wintering locations. For example, northernmost wintering birds, e.g., in Central Europe, may continue migration in the course of winter, probably due to worsening weather conditions. For example, a tracked male from Estonia moved in October ca 500 km south to the north-eastern corner of Poland; thereafter he was recorded in Central Poland in November and Southern Poland in February; in March he continued movement to south until Croatia (Fig. 2). Interestingly, two weeks later this male started its spring migration and was back in its breeding territory in ten days. On the other hand, late snow cover in breeding grounds may also stop spring migration. One should also note that registration of such intra-European movements during winter have been hindered due to technical limitations. At poor light conditions during winter in Europe solar-powered trackers are discharged and collection of data is hampered.

Apparently, females tend to make shorter migration flights than males. For instance, among GPS-tracked individuals breeding in Europe, all females have wintered in southern Europe, Turkey and Israel; although these areas have been used by males too, only known males have been recorded in Africa so far (Meyburg and Meyburg 2005, Dombrovski et al. 2018). This pattern, however needs further investigation and confirmation in a larger sample.

As mentioned above, immature greater spotted eagles usually return to the breeding range, but not necessarily to their natal or future breeding territory. For example, a male marked as a fledgling in Estonia in 2008, returned next spring to Estonia, but continued its journey northward to Finland where it spent the summer. Over the next eight years he always returned to Fennoscandia, which is out of the contemporary breeding range of the species, but only in 2016 was recorded as a breeder in Estonia again, 60 km far from its natal nest. Although there is no other published data on natal dispersal of the species, longer dispersal events probably exist since gene flow has maintained genetic diversity across populations (Väli et al. 2019).

10

Booted eagle *Hieraaetus pennatus*

Ugo Mellone[1,2]

ABSTRACT

The booted eagle (*Hieraaetus pennatus*) is mainly a summer resident in Eurasia, wintering in sub-Saharan Africa. It migrates by soaring-gliding flight, alone or in small, loose, flocks. In order to reach their wintering grounds, the majority of booted eagles concentrate at bottle-necks such as the Strait of Gibraltar, the Bosphorus and Batumi. The bulk of the European population breeds in the Iberian Peninsula and spend the winter in Sahel, even if some individuals remain in Spain. Others migrate towards Italy, using a route that is showing an increasing trend in the last years, albeit with a strong annual variation. Satellite tracking of Spanish birds show that these long-distance migrants need three to four weeks to reach the wintering grounds. In contrast with the majority of other migrant species, higher travel rates (hourly speed, daily distance, overall migration speed and overall straightness) are attained during autumn rather than during spring migration.

Introduction

The booted eagle (*Hieraaetus pennatus* Gmelin, 1788) is a medium sized raptor, with an average body masses of 709 g (male) and 975 g (female; Cramp and Simmons 1980) and a wing span of 110 cm (Bruderer and Boldt 2001), distributed in Eurasia, until Mongolia, and also breeding in Southern Africa with a different subspecies (Yosef et al. 2000, Ferguson-Lees and Christie 2001). It is a polymorphic species, with individuals showing either a pale or a dark plumage: the proportion in the western population is roughly 7:3 (Garcia-Dios 2016), while in Asia the percentage of dark individuals is higher (Shirihai et al. 2000). Its diet is quite generalist, consisting mainly of medium and small sized vertebrates, especially birds and rabbits, sometimes also insects (Cramp and Simmons 1980, Garcia-Dios 2016).

The majority of booted eagles breeding in Europe are trans-Saharan migrants, apart from the population breeding in the Balearic Islands which is completely sedentary. However, there are also tens of individuals wintering in Southern Europe (Baghino et al. 2007, Urios et al. 2017). The European population has been estimated in 4,400–8,900 pairs (Burfield 2008), with the majority occurring in Spain and France. According to migration counts at the Strait of Gibraltar the population is showing an increasing trend (Martin et al. 2016). On the other hand, data from Batumi suggest an overall stable trend for adults and a decreasing one for juveniles (Vansteelant et al. 2020b). Its conservation status is

[1] MEDRAPTORS (Mediterranean Raptor Migration Network), Rome, Italy.
[2] Vertebrates Zoology Research Group, University of Alicante, Alicante, Spain.
Email: u.mellone@gmail.com

"Least Concern", according to the IUCN (BirdLife International 2016b). The breeding season begins in early April: in Spain, one to three eggs are laid generally during the last week of this month, and the productivity may range from 0.7 to 1.88 juveniles per territorial pair (Garcia Dios 2016). The breeding habitat is a mosaic of forests with scrublands and herbaceous vegetation and the nests are normally located on trees, only sometimes on cliffs (Cramp and Simmons 1980). It seems that individuals wintering in Southern Europe concentrate close to wetlands (Baghino et al. 2007, Urios et al. 2017).

Given its morphology, the booted eagle is a typical soaring migrant, concentrating during migration in the main bottlenecks such as the Strait of Gibraltar and the Bosphorus (Cramp and Simmons 1980). Radar studies carried out in Israel did show that its flight parameters are similar to the ones of the European honey buzzard and Egyptian vulture (Spaar 1997). They are not gregarious migrants: at the Strait of Gibraltar half of the records involve solitary individuals, with the remaining ones being mainly related to loose groups of 2–3 individuals (30% of the total; Onrubia and Martin 2017).

Knowledge Status

The booted eagle is a regular sight in many raptor migration watchsites of Southern Europe: the highest counts come from the Strait of Gibraltar, which is the most important location where regular census are being carried out (Onrubia and Martin 2017). The species is also regularly observed in other migration hotspots of Italy and along the Eastern flyway. The first satellite tracked booted eagles have been two individuals tagged in Spain in the 2005 (Diaz-Ruiz and Cebollada-Baratas 2011). In the 2007, another individual was tagged in France, and followed for three seasons (Chevallier et al. 2010). Starting in the 2011, a large-scale study has been undertaken in Spain, with 18 adult individuals being tagged throughout the country (Urios et al. 2017). Almost five thousands booted eagles have been ringed in Spain: outside the Iberian Peninsula, the majority (21 out of 155) has been recovered in African countries of Maghreb and Sahel, but there are also two individuals collected in France and three in Italy that show the relative importance of

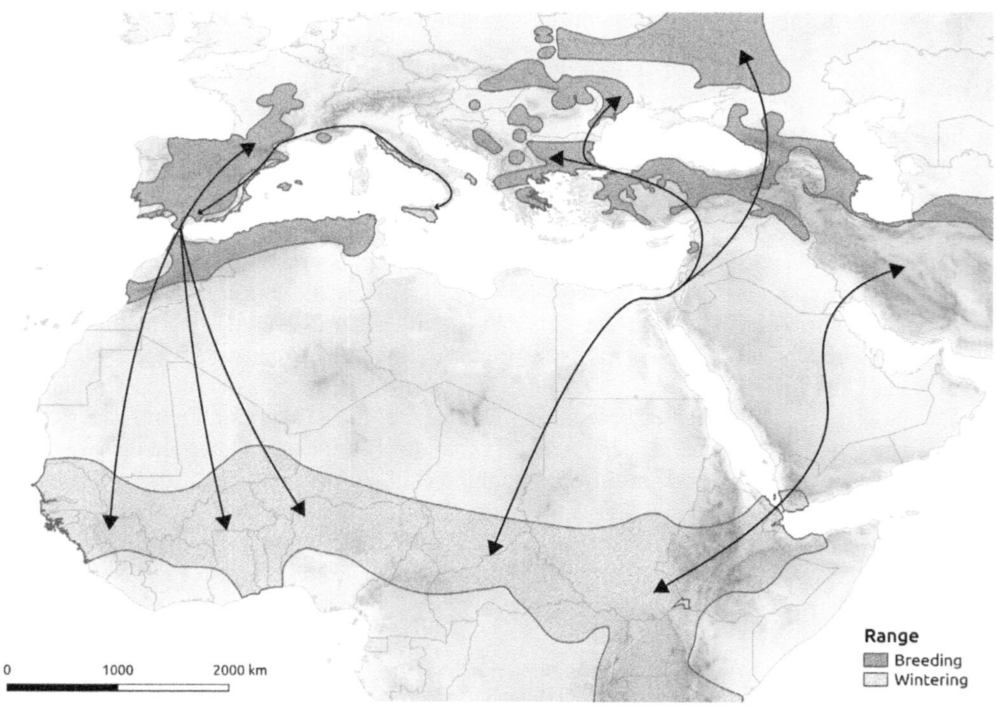

Figure 1. Schematic representations of the main migratory routes followed by booted eagles.

the flyway heading to Southern Italy (Fig. 1; Urios et al. 2017). Out of 59 booted eagles ringed in Israel, only one has been recovered, in Central Russia (Shirihai et al. 2000).

Post-breeding Migration

Western flyway

The western flyway shows the highest counts because it involves Iberian and French birds, which constitute the most conspicuous population of the species. Moreover, the great majority of them concentrate in just a single spot (Fig. 1). French birds enter the Iberian Peninsula by crossing the Pyrenees, mainly at Organbidexka, where an average of 67 individuals has been observed between 1981 and 2008 (Filippi-Codaccioni et al. 2010). The median departure date of satellite tracked individuals from Spain is the 14th of September. An average of 31,608 individuals has been reported at the Strait of Gibraltar in the period 2009–2012, mainly between August and October, peaking on the 19th of September, without any phenological difference among adults and juveniles (Onrubia and Martin 2017). In the last years, a tendency to advance the migration dates, probably in response to climate change, has been observed (Panuccio et al. 2016b). They concentrate in the narrowest area of the Strait, trying to limit the water-crossing as much as possible. When the weather conditions are too risky to cross the sea, eagles may stop in this region for up to six days, moving inland for even 100 km, as shown by satellite tracked individuals (Mellone 2017). Autumn migration lasts about three weeks (23.4 ± 11.4, N = 24; Mellone 2017), with the eagles crossing the Sahara Desert on a broad front, until they settle in their wintering grounds in the Sahel, located mainly between 13° and 16°N.

The flyway that leads Iberian booted eagles to the Italian Peninsula involves a disproportionate lower number of individuals, but is perhaps more intriguing, becoming increasingly more conspicuous in the last years (Fig. 1). While before the species was only an irregular winter visitor in Italy, nowadays tens of individuals are regularly observed during the cold months especially in the southern part of the peninsula and in Sicily (Baghino et al. 2007). Visual observations targeting at this species are carried out, although not regularly, at several watchsites located along this flyway, in Eastern Spain, Southern France and Liguria/Tuscany in Italy (Premuda et al. 2007b). The annual variation in these data show strong bouts during "invasion" years, for example in the 1985, 2004, 2011, 2017 and 2018, generally during the first half of October. For example, 629 booted eagles were counted heading North in Castellón (Spain) and then 568 in Arenzano (Liguria, Italy) in the 2004 (Premuda et al. 2007b), 516 in Bellet (France) in the 2011 (Kern et al. 2011) and 494 at Capriglia (Tuscany, Italy) in 2017 (Premuda 2017a). The 2018 has been by far the year showing the most conspicuous flux on this route, with 1,233 individuals counted from Capriglia, peaking on the 7th of October (385 individuals; Premuda 2018). In the 2019, 556 individuals have been observed from the same watchsite. During peak years, the age ratio is strongly biased towards juveniles (Premuda et al. 2007b), with their proportion decreasing in the following years, and the peak taking place earlier, even in September (Premuda 2019b), suggesting that the same individuals (the surviving ones) are involved, returning to the previous wintering grounds (Baghino et al. 2007). The Spanish origin of these booted eagles is confirmed also by ringing recoveries (Premuda and Baghino 2004, Mellone 2017). They spend the winter in Italy, especially in Sicily: in "good" years, tens of individuals may cross the Strait of Messina (e.g., 209 individuals in the 2018, Cento et al. 2018). Probably only a few attempt to cross the Mediterranean sea en route to Africa: the highest reports are of 23 individuals in the 2007 in Marettimo (Panuccio et al. 2011) and 38 individuals in Malta in the 2016, unfortunately the last ones probably all killed by poachers (Bonavia and Vella 2016).

Even more puzzling than booted eagles reaching Italy are those observed in the Balearic Islands during autumn. No standardized counts have been carried out, but the phenomenon has apparently increased in the last five years. A total of 249 individuals have been observed from the southernmost point of Mallorca disappearing towards the South on the 30th of September 2017, including a group of 70 individuals (L. Ventoso pers. comm.). During the same season, a flock of 210 individuals has been recorded on the Illa de l'Aire (Menorca) on the 18th of October (Catchot 2017). It seems that they were mainly juveniles and remained in the area for several days with no evident migration direction, taking advantage of the high abundance of rabbits and lizards to replenish their energy stores, which, according to the capture of a few individuals, were very low (Catchot pers. comm.). The above mentioned flock including 210 individuals is, according to the available knowledge, the largest ever recorded for the species and, not surprisingly, is associated with a route involving a long sea-crossing (at least 80 km). Gregariousness is important for the crossing of large surfaces of water, because an increase in flock size increases the probability that the flock will contain one or more individuals with a pronounced tendency to continue migrating (Agostini et al. 1994). Agostini et al. (2015a) suggested that, on the basis of its morphology, the booted eagle might be able to cross large sea stretches at least as similar species as the honey buzzard and the black kite do but is limited by the fact that does not migrate in large flocks as the latter species. Probably those booted eagles are "trapped" in Menorca like the southbound juveniles short-toed snake eagles (*Circaetus gallicus*) in Sicily, and show a conflictive behaviour when facing the water barrier (Agostini et al. 2009, Mellone et al. 2016). The recovery of a wing-tagged individual (Garcia-Dios 2017) in Mallorca suggests that these individuals do not belong to the resident Balearic population (Urios et al. 2017) and that they come from the mainland, but is not clear where they are heading to, because the different behaviours observed in the two islands do not fit in a coherent pattern. There is no evidence suggesting that they spend the winter in the islands (Catchot pers. comm.), given also that they would then have to compete with the dense local population.

Still, it is unknown why hundreds of juveniles from the Iberian Peninsula in some years head north-east instead of crossing the Strait of Gibraltar: other than the influence of anomalous weather conditions (Premuda et al. 2007b), perhaps the emergence of odd behaviours ("mirror-image" orientation, see De Sante 1983) may be more evident in an increasing population as the Spanish one, even more in a climate change scenario in which new strategies may be rewarding from a fitness point of view. This phenomenon may lead birds normally heading to the south-west switching to a north-east bearing. However, this would not explain why booted eagles change their flight direction to the south once they have entered the Italian Peninsula. Perhaps also the use of the coastline as a leading line may play an important role: visual observations carried out in Batumi (Georgia), on the Black Sea coast, show that, for unknown reasons, booted eagles adhere to the coastlines more markedly than other soaring migrants (Vansteelant et al. 2014).

None of the 16 individuals followed by satellite tracking at least during one migration took the Italian route, but two showed a short-distance migration strategy, travelling for only 2–4 days and spending the winter in the wetlands South of Valencia (Mellone 2017), where up to 12 individuals have been observed during winter (Palomino and Molina 2012).

Eastern flyway

Since the breeding population of booted eagles in Eastern Europe is less dense than in the Western countries, the counts concerning birds coming from this region are much lower than in the Strait of Gibraltar. Birds coming from European Russia concentrate mainly at the eastern coast of the Black Sea: an average of 6,475 ±665 (SD) have been counted between the 2011 and 2018 at Batumi (Georgia), peaking around mid-September; here, the average annual count of non-juvenile is equivalent to ~ 3% of the global breeding population, while juveniles represent about one fifth of the totals and are decreasing at a pace of 10% per year, even if is difficult to draw firm conclusions on this trend (Vansteelant et al. 2020b). Much fewer individuals are observed on the opposite shore of the Black sea, at Burgas Bay in Bulgaria: between 2004 and 2017 the totals have been ranging from 49 to 466 individuals, with a tendency to increase (Michev et al. 2018). These individuals join the Balkan population at the Bosphorus: here, 497 individuals have been reported in 2010, with only 20 observed in the same year at the Dardanelles Strait (Panuccio et al. 2017). A maximum of 32 booted eagles has been observed in the island of Antikythira and all aged birds (19) were juveniles (2009; Lucia et al. 2011): this small fraction may either spend the winter in Crete (Londei 1991) or try the long sea-crossing to Libya. Another important watchsite in Turkey is the Belen Pass, where 1,028 individuals have been counted in 2013 (Oppel et al. 2014). In Israel, the average count collected in Kfar Qassem in the time frame 1982–1987 is of 1,391 individuals, peaking around the 23rd of September (Shirihai et al. 2000), while between the 1990 and the 1999 it dropped to 653 individuals (Alon et al. 2004) counted in the Northern Valleys. However, it is difficult to draw any conclusion on population trends, due to methodological differences among the surveys. The maximum count from Suez is 1,104 in the 1984 (Shirihai et al. 2000). A consistent number of booted eagles reaches the African continent following a different flyway and crossing at Bab-el-Mandeb, where 1,123 individuals have been reported in the 1987 (Welch and Welch 1988).

Pre-breeding Migration

Western flyway

From the Spanish shore of the Strait of Gibraltar, an average of 6,662 booted eagles have been counted between 2009 and 2012, from February to June, peaking on the 31th of March. During this season immatures migrate later than adults, peaking in early May (Onrubia and Martin 2017). The stunning difference between autumn and spring counts at this location may be due to the lack of many immatures which die during winter or spend the summer in Africa, and possibly because in autumn eagles stop in the Strait area for several days, showing conflicting behaviour such as aborted sea-crossings and retreat flights before undertaking the final crossing, thus making a precise census very difficult. The same behaviour occurs also in spring, but obviously only on the Moroccan side of the Strait (Mellone et al. 2013a), keeping therefore the spring observations in the European side less susceptible of double counts. Concerning satellite tracked individuals, the median date of departure is the 12th of March, with the entire journey lasting 28.4 ± 8 (SD) days (N = 24; Mellone 2017). Therefore, unexpectedly, the pre-breeding migration takes more time than the post-breeding one, as I will also discuss below.

Spring counts on the flyway connecting Italy with Spain fundamentally mirror those taken during the preceding autumn: for example, after the "invasion" recorded in autumn 2018, 296 individuals were counted at the Strait of Messina in 2019 (Adami et al. 2019). As occurs also in other broad-winged migrants during spring, some individuals decide to by-pass the Strait using the island of Panarea (Aeolian archipelago) as a stepping stone towards the Italian Peninsula—here, 30 individuals have been reported in the 2005 (Baghino et al. 2007) and 60 in the 2012 (Gustin 2012). These individuals spent the winter in Sicily—the migratory flow from Africa across the Central Mediterranean may be considered as very minor (Agostini 2005). These booted eagles, joined by those that spent the winter in the Italian Peninsula, continue their travel *en route* to France and Spain (Fig. 1); during this season the coverage in the watchsites of Tuscany and Liguria is more irregular, being aimed at other species, but it is worthwhile to report the total of 423 individuals in 2019 in Capriglia (Premuda 2019a). The fact that the species is quite rare in the migration hot-spots of the Adriatic coast, where thousands of raptors of other species directed to Eastern Europe are observed (Gustin and Sorace 2004, Premuda et al. 2004), confirms that these eagles belong to the Western Europe population. Finally, concerning the Balearic archipelago, up to 50 individuals all together have been observed in the western tip of Mallorca, leaving the island towards north-west (9/4/2018, M. Mc Minn. pers. comm.), highlighting again how the role of this archipelago within the migration routes of the booted eagle is far from being clarified.

Eastern flyway

Booted eagles may leave Africa either at Bab-el-Mandeb (953 on 5–7/3/1990, Welch and Welch 1998; 1,773 on 2–10/3/2013, Rayaleh et al. 2013) or at Suez (Fig. 1). Here, during March and April

2012, 1,177 individuals have been counted from the Ayn Sokhna watchsite (Megalli and Hilgerloh 2013). In Israel the highest total was 605 individuals at Beer Sheva in the 1985 (Shirihai et al. 2000). Spring raptor migration at the Bosphorus has been much more rarely studied than autumn one. The only available comprehensive study, made by Uner et al. (2010) in the period 18/3-31/5 2006 reports 161 booted eagles, with a daily peak of 25 on the 6th of April.

Behavioural and Ecological Remarks

Perhaps the biggest peculiarity of booted eagle migration behaviour is the fact that according to satellite tracking of Spanish birds, they migrate significantly more quickly during autumn rather than during spring (ca. 23 vs. 28 days, see previous paragraphs). Such pattern is striking in contrast with the majority of migrating birds worldwide (Nilsson et al. 2013). With the aim of disentangling the reasons of such differences, Mellone et al. (2015a) analyzed a sample of 15 autumn journeys and 10 spring ones. It came out that the migration speed was significantly faster in autumn than in spring (106.5 ± 34.5 (SD) km/day in spring vs. 146.8 ± 46.7 (SD) in autumn), migration routes were straighter during autumn (straightness index = 0.76 ± 0.13 (SD) in spring vs. 0.85 ± 0.09 (SD) in autumn), and non-travelling days (days with less than 50 km between the two roosting sites) were less frequent in autumn (10 ± 6 days in spring vs. 6 ± 8 in autumn). The daily flight time lasts ca. 8 hours, with no seasonal differences. Making within and among seasonal comparisons of the migration performance in two regions (within and outside the Sahara Desert) the lowest speeds occur in spring outside the desert, both at the hourly (21.5 km/h vs. 26.2–27.7) and daily scale (145.3 km/day vs. 193.2–208.8). Linear mixed models also testing variables such as sex and day length highlighted that the highest daily distances are attained during autumn in the desert, favored by tailwinds and thermal uplift (Mellone et al. 2015a). The maximum known daily distance is 541 km, being recorded on an individual flying out of the desert in spring (between 27° and 32°N), and the highest hourly speed 72 km/h (Mellone 2017). The seasonal differences are significant even when controlling for weather conditions, which are slightly more favourable during spring (Mellone et al. 2015a). Still, it remains unknown why they migrate more quickly during autumn: may be the competition for optimal winter territories is stronger than for breeding ones. Or, conversely, they deliberately choose to migrate at a slower speed during spring, discarding a time-minimization strategy, in order to avoid the depletion of their energy reserves, being in better condition to ensure and defend a good site for reproduction.

Vidal-Mateo et al. (2016) analyzed the effect of forward and perpendicular winds on three species of raptors along the Western flyway, including the booted eagle. They found that booted eagles took advantage of tailwinds both in autumn and spring, increasing their rate of forward movement on average by 8.98 (autumn) and 10.35 km/day (spring), being displaced by crosswinds by 6.77 (autumn) and 8.07 km/day (spring), for every meter per second of wind. The proportion of compensation segments was higher at the onset of autumn migration and at the end of the spring one, in order to reach very precise goal areas as the Strait of Gibraltar and the breeding area. Wind effects have been studied also at a local scale during autumn migration along the northern side of the Bosphorus by Panuccio et al. (2017), showing that booted eagles compensate for wind drift at the site when crosswind speed increases. At the same site the flight altitude of booted eagle decreases when tailwind is blowing (Panuccio et al. 2017). At the Strait of Gibraltar, during the same season, counts were highest on days with locally northerly (tail) and westerly (lateral) winds and high pressure following days with local easterly winds and lower pressure, and before the middle of the count season (Miller et al. 2016). On the basis of 1.8 hours of radar observations carried out in Israel during both seasons, Spaar (1997) reports that booted eagles spend 34% of their flight time circling in thermals and 53% gliding, with the remaining time frames spent soaring in straight line (6%), circling and flapping (3%) and flapping and gliding (3%). Their inter-thermal gliding air speed was positively affected by the climbing rate in thermal circling and by the altitude above the ground.

Concerning foraging behaviour during migration, the booted eagle, as other soaring migrants, is supposed to rely mainly on fat reserves gained before the journey (Smith et al. 1986, Panuccio et al.

2006). As they may forage in a huge array of environments (Lopez-Lopez et al. 2016), it is difficult to understand from available satellite tracking data if days with lower migration speed and less directed movements were actually dedicated to stopover or were due to bad weather conditions (Mellone 2017). The latter case is more probable close to ecological barriers such as the Strait of Gibraltar or the Sahara Desert. However, this kind of movement pattern was not infrequent also in more benign habitat, especially in Southern Spain during spring migration. A striking exception comes from a bird that stopped during both migrations in a restricted area located in Morocco, 250 km south of the Strait. It concentrated its movements there for 27 days in autumn and 17 days during spring, mainly within a radius of ca. 12 km from the roosting area, showing movement patterns that resemble those of the breeding and the wintering areas (Mellone et al. 2013a, Mellone 2017). Even if there was no visual confirmation, it is possible to assess that this individual interrupted migration to perform true and not opportunistic foraging sessions, a behaviour that has been seldom observed among soaring migrants. Finally, some satellite tracks show unusual undirected movements with criss-crossing paths within the Sahara Desert, whose reasons are still unknown and will be the subject of future studies (SEO/Birdlife unpub. data, Mellone et al. in prep.).

11

Steppe eagle *Aquila nipalensis*

Michael McGrady,[1,*] *Evgeny Bragin*,[2,†] *Igor Karyakin*,[3] *Nyambayar Batbayar*[4] and *Todd Katzner*[5]

ABSTRACT

The steppe eagle (*Aquila nipalensis*) is a globally endangered, full migrant raptor that breeds in the southern temperate zone from European Russia in the west to eastern Mongolia, Dauria and adjacent north-eastern China in the east. It winters in Africa, the Middle East and Southern and South-Eastern Asia, and migrations can sometimes entail journeys > 10,000 km in length. Kazakhstan, Russia and Mongolia are the breeding strongholds. Declines in the breeding population, which are estimated to total 50–60%, are most obvious in Europe. Migration occurs during August–October, and again during February–April. For some populations, migration paths form notable clockwise loops, as steppe eagles use different seasonal flyways to avoid barriers such as the Red Sea, the Gobi Desert or the Himalayas. Threats to the population include loss of habitat, persecution, inadvertent poisoning, and electrocution. Very large aggregations settle at anthropogenically-generated waste disposal sites, and these sometimes provide potential for mass poisonings and increased risk of electrocution. Creation of new waste disposal sites may have caused changes in the wintering distribution of steppe eagles in recent decades. Robust regional and range-wide population estimates are lacking, as are important details about food availability and risk of poisoning and electrocution, and these gaps undermine effective management.

Introduction

The steppe eagle (*Aquila nipalensis* Hodgson, 1833) is a large, open-country Palearctic raptor. Traditionally, two subspecies have been identified, *A. n. orientalis* in the western part of the breeding range, *A. n. nipalensis* in the east, but morphological and genetic studies suggest it to be monotypic (Clark 2005, L. Zinevich unpub. data). Until the end of the 20th century the steppe eagle was considered conspecific with the tawny eagle *A. rapax* (Cramp and Simmons 1980, del Hoyo et al. 1994, Clark 2005). Steppe eagle populations are completely migratory (Cramp and Simmons 1980, del Hoyo

[1] International Avian Research, Am Rosenhügel 59, 3500 Krems, Austria.
[2] Naurzum Nature Reserve, Kostanay, Kazakhstan.
[3] Sibecocenter LLC, Novosibirsk, Russia.
[4] Wildlife Science and Conservation Center of Mongolia, Union Bldg B701, UNESCO Str., Ulaanbaatar 14210, Mongolia.
[5] U.S. Geological Survey, Boise, Idaho, USA.
† Deceased
* Corresponding author: mikejmcgrady@aol.com

et al. 1994), and migration distances can be intermediate or very long, sometimes exceeding 10,000 km (Meyburg et al. 2003, Meyburg and Boesman 2013, Javed et al. 2014, Batbayar and Lee 2017, Karyakin et al. 2018). Steppe eagles are solitary breeders that nest in flat, hilly, or mountainous (up to 2,300 m) terrain, typically in semi-arid or arid grasslands, and grassy-mosaic landscapes on the fringes of more arid areas or forest edges (del Hoyo et al. 1994, Snow and Perrins 1998, Bold and Boldbaatar 1999, Gombobaatar et al. 2012). Nests are large and made mostly of sticks, although they frequently contain substantial human and animal refuse—wire, fencing, clothing, dung, and even scrap iron. Nests are built on flat ground, on or at the top of steep slopes or raised ground, on small rock outcrops in open landscapes, and occasionally in small trees or bushes or on buildings, derelict farm equipment, or power poles (Cramp and Simmons 1980, Shevchenko 1986, del Hoyo et al. 1994, Bold and Boldbaatar 1999, Medzhidov et al. 2011, Gombobaatar et al. 2012, Karyakin et al. 2013).

Breeding commences in early to mid-April in the western part of the distribution (Shevchenko 1986, Amosov 2011), early May in the east (Daurian steppe; Peshkov 1976, Goroshko 2012) and the end of May in the Altai mountains (Mitrofanov 2015). Steppe eagles lay 1–5 eggs, most commonly 2–3. Incubation lasts about 45 days, and fledging occurs at 58–65 days. Fledglings become fully independent at about 11 weeks (Agafonov et al. 1957, Cramp and Simmons 1980, del Hoyo et al. 1994).

The steppe eagle was once considered the most abundant large eagle in the world, but it is currently globally Endangered (BirdLife International 2017c, 2019d). Karyakin (2018) estimates that 25,500–46,065 pairs exist.

The most important populations are in western and central Kazakhstan, transboundary Russia-Kazakhstan-Mongolia and central Mongolia (Karyakin 2013, 2018). Declines are most obvious in Europe, but numerical declines and range contractions in central and eastern Asia may be underappreciated (Karyakin et al. 2013). Combined totals from across the species' range suggest a decline of ~ 50–60% in recent decades (Karyakin 2015, BirdLife International 2019d), although the population in Mongolia appears to be rather stable and its range has not contracted (N.B. unpub. data).

The current breeding distribution of steppe eagle is huge, and mostly within a latitudinal band in the southern temperate zone (*c.* 43°N to 53°N) that extends from Southern European Russia (about 43°E) in the west across Kazakhstan and Mongolia to north-eastern parts of inland China. Isolated breeding populations occur in Turkey, north-central China, and in parts of Russia north of 53°N (Karyakin 2013, 2018, Karyakin et al. 2016a, 2016b, BirdLife International 2019d). Breeding range contraction has occurred in recent decades, especially along the western (e.g., Bulgaria, Ukraine, European Russia), northern, and southern edges (Karyakin et al. 2013, Karyakin 2018) of the species' distribution.

Wintering occurs in four areas: Africa, the Middle East, south Asia and Southeast Asia. In Africa, wintering areas extend south from Sudan and Ethiopia, with a few individuals reaching South Africa (Meyburg et al. 2003). In the Middle East, most steppe eagles winter in Arabia, especially Saudi Arabia, Yemen and Oman. On the basis of the small proportion of 2nd calendar-year birds among migrating steppe eagles at Eilat (Israel) during spring, Weiss (2018) suggested that many juveniles spend the winter in Arabia and do not migrate to Africa. More research is needed to confirm this. Meyburg et al. (2003) reported large concentrations of individuals of different age classes wintering in Tanzania. The Southern Asian wintering area extends from south Iran in the west, across the Northern Indian subcontinent to Myanmar. In Southeast Asia the wintering grounds are mostly in south China, but they also extend to peninsular Malaysia (Cramp and Simmons 1980, Helbig and Wells 1990, del Hoyo et al. 1994, BirdLife International 2019d).

Birds from European Russia, west and central Kazakhstan, and Turkey typically winter in the Middle East, Southern Iran, and Africa (Meyburg and Boesman 2013, Bekmansurov et al. 2018). Eagles from Altai, Siberia, north-eastern Kazakhstan, and central and eastern Mongolia winter mainly in Southern China, Northern India, Pakistan, South and Southeast Asia (Ellis et al. 2001, Batbayar and Lee 2017, BirdLife International 2019d). However, even if north-south connectivity is generally strong, some "crossover" occurs (I.K. unpub. tracking data).

Steppe eagles face a wide range of threats that vary in importance seasonally and across their global distribution. On breeding grounds, a major cause of range contractions and population declines has been ascribed to conversion of steppe grasslands into agricultural land (Dementiev et al. 1951, Cramp and Simmons 1980), and concomitant declines in food resources for eagles (Peshkov 1976, Belik 2004). An important direct cause of mortality of birds of all ages is electrocution (Pererva and Blokhin 1981, Lanovenko and Abdunazarov 1983, Pererva 1986, Medzhidov et al. 2011, Amartuvshin and Gombobaatar 2012, Matsyna et al. 2012, Dixon et al. 2013a,b, Levin and Kurkin 2013, Karyakin et al. 2016a, 2016b), mostly on medium-voltage distribution lines. Negative direct and indirect effects of agrochemicals have been a threat since the middle of the 20th century. The fact that nests are usually easily accessible can increase risk of predation, nest destruction, disturbance and persecution (Dzhamirzoev and Bukreev 2009, Karyakin et al. 2013, BirdLife International 2019). In places, fire destroys perhaps 5–15% of nests, and kills nestlings (Karyakin et al. 2016a, 2016b). The effects of climate change on steppe eagles are poorly understood, but this process could affect the phenology of their mammalian prey, their migration, or the frequency of dangerous extreme weather events throughout the annual cycle.

Threats on migration include shooting and trapping, electrocution, collisions with wind turbines and powerlines, and poisoning. The threats of electrocution and inadvertent poisoning seem ubiquitous across steppe eagle migratory routes, though their severity varies geographically.

Many threats to steppe eagles that occur on the breeding grounds are also relevant in wintering areas, although winter-time threats are relatively poorly studied. The winter shift to a more scavenging lifestyle exposes eagles to a heightened threat of poisoning. Electrocution is a threat, especially at wintering aggregations (Keijmel et al. 2020). Diclofenac, and presumably other non-steroid anti-inflammatory drugs (NSAIDs), are toxic to eagles (Sharma et al. 2014); this threat is perceived to be greatest on the Indian sub-continent. Although details are scarce, distribution and abundance of food is almost certainly declining in some areas due to habitat destruction and climate change, and this will affect eagles. Finally, shooting of eagles, and capture and sale in markets may also occur with enough frequency that they are demographically relevant (the authors, unpub. data).

Knowledge Status

Steppe eagle ecology on the breeding grounds is rather well-known, either through direct field studies or by implication from what is known about other *Aquila* eagles. Visual observations from bottleneck sites (e.g., Eilat, Bab-el-Mandeb, western Caspian coast, Himalayan slopes) have provided, and continue to provide, initial insight about the migratory habits of this species. In recent decades knowledge gaps about migration and distribution in remote locations and away from bottlenecks have been partially filled by data from tracking via satellite. Most tracking has been of juvenile birds marked at their nests in Kazakhstan, Russia, Mongolia and Turkey (e.g., I.K., E.B., T.K., N.B., R. Efrat, M. Horvarth unpub. data). Tags also have been deployed on birds along the migration route and in wintering areas, mostly in Arabia (e.g., Meyburg et al. 2003, Javed et al. 2014, M.M., B. Meyburg unpub. data). Historical data gaps exist, especially in Africa and India, in part because of the earlier lumping of tawny and steppe eagles.

Post-breeding Migration

Autumn migration of steppe eagles in the steppe zone of European Russia and Kazakhstan begins at the end of August or in early September (Fig. 1). In Southern Siberia and Western Mongolia migration begins at the end of September or in early October. The last migrants from Southern Siberia depart at the end of October and pass through Kazakhstan during the beginning of November. Thus, autumn migration of steppe eagles through Kazakhstan appears to occur in two waves (Karyakin et al. 2019). As is the case for other *Aquila* eagles, timing of autumn migration likely is affected by age and breeding status (Bragin 1983, 1987). Subadults (2nd–3rd calendar year) are

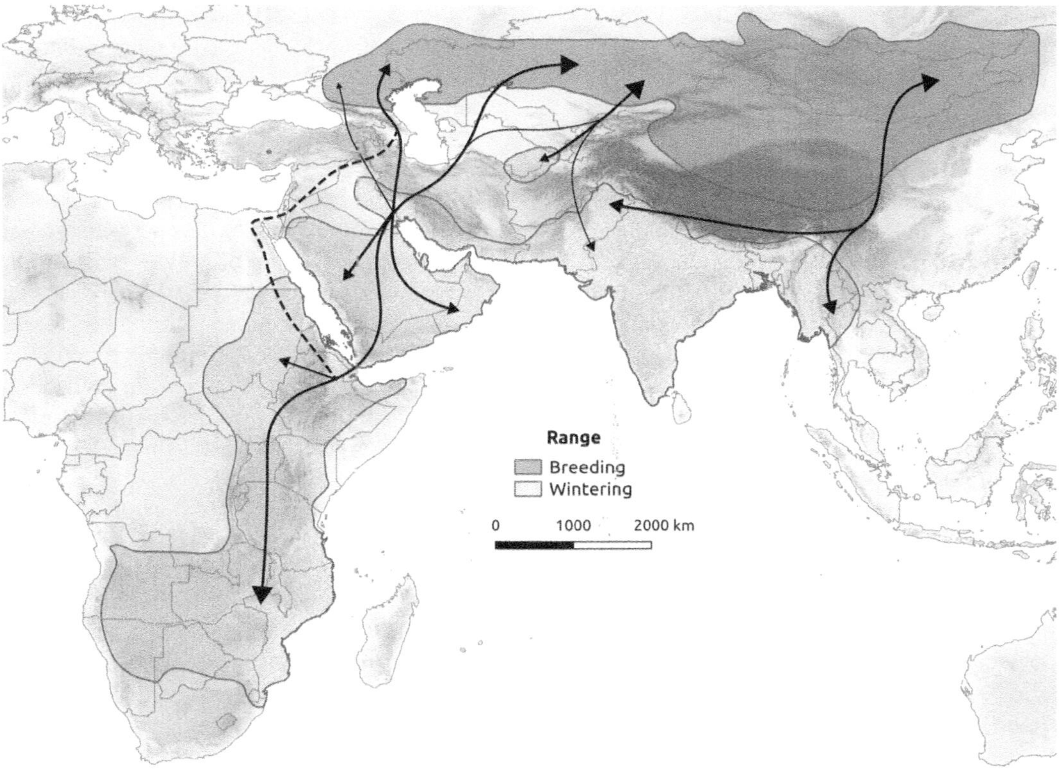

Figure 1. Schematic representation of migration routes used by steppe eagles as deduced from the published information (see references), and unpub. data from tracking via satellite (see text). The route through Suez and Israel (dashed arrow) represents a route followed mainly, but not exclusively, during spring.

typically the first to migrate, sometimes even in the first days of August. They are then followed by juveniles and adults that have not bred successfully. Adults that have bred successfully often are the last to leave the summering areas.

Upon leaving the breeding grounds migration occurs across a rather broad front, although geographical features concentrate the passage (Welch and Welch 1991b). At least 420 birds were recorded at Chokpak Pass, Kazakhstan, during 4 September–22 October 2002 (Kovalenko et al. 2002). 80 eagles were recorded at the Murgab River delta, Turkmenistan, east of the Kopetdag Range on 20 October 1976 (Rustamov and Kurbanov 2011). In 1990, intensive autumn migration was observed along the middle reaches of the Amu Darya River, near Seydi, Turkmenistan; on 16 October more than 140 eagles were seen during a single half-hour observation period (Saparmuradov and Eminov 1994). Autumn migration from the western populations skirts around the Aral and Caspian Seas, and the Arabian Gulf (Welch and Welch 1991b, M.M., E.B., I.K., T.K., unpub. data from tracked birds). Panuccio et al. (2018c) estimated that 350 steppe eagles passed during just 25 hours of observation in October 2017 in the Alborz Mountains, Iran (southeastern corner of the Caspian Sea), and Heiss and Gauger (2011) reported daily maximum counts of 1,400 eagles during autumn 2008 at Shirvan National Park, Azerbaijan. Numbers observed passing through the west and central Caucasus and west of the Caspian Sea are relatively low and declining (Abuladze 2013), reflecting, in part, the low density of breeders in European parts of Russia. On average, 568 steppe eagles/year have been counted in Batumi, Georgia, during 2011–2018 (Wehrmann et al. 2019, see also Abuladze 1999,

Abuladze and Edisherashvili 2003, Verhelst et al. 2011). Passage via the Bosphorus is more or less non-existent (Cramp and Simmons 1980, Fülöp et al. 2014, Panuccio et al. 2017), again likely a result of declines in the eastern European population. Juvenile steppe eagles tracked from a small population in Anatolia, Turkey, migrated via the Levant and Sinai to the eastern regions of Sahel (M. Horvath and T. Burak, pers. comm.). Steppe eagle migration in autumn across Iran occurs mostly on a relatively broad front from the SE corner of the Caspian Sea to the head of the Arabian Gulf. Most birds that migrate through western Iran and Iraq are destined for wintering areas in Arabia or Africa. Passage into Africa occurs mainly via the Bab-el-Mandeb Straits between Yemen and Djibouti. There, Welch and Welch (1988, 1991b) reported a passage of 60,987 birds during 15 October–1 November 1985, and 76,586 during 3 October–9 November 1987. Relatively recent data suggest that the land bridge through Israel and Northern Egypt is less important during this season (Shirihai et al. 2000, Hilgerloh et al. 2011a), though earlier counts recorded strong migration there (64,800 in 1981; Bijlsma 1983). Arrival dates at the more northern wintering areas in the Middle East are in October. However, the flow of migrants that ultimately winter farther south continues, mostly within the Rift Valley. Birds settle into wintering home ranges all along this route, with some moving as far south as South Africa (Simmons 1997, 2005).

Data obtained from 19 satellite tracked individuals suggest that populations of steppe eagles from central Kazakhstan, the eastern portion of its Russian distribution, Mongolia and western China, use two major flyways, skirting west and east of the Pamir-Alay mountain system (Karyakin et al. 2019). Eagles from the western Mongolia, central and eastern Kazakhstan, and north-west China pass to the west of this mountain system, along Lake Balkhash (Kazakhstan) from the north, or along the Dzhungarian Alatau and Trans-Ili Alatau mountain ranges. They then cross the Karatau (Kazakhstan) and Nuratau (Uzbekistan) Ranges and fly along the western foothills of the Zarafshan and Hisor Ranges to reach the southeastern corner of the Karakum Desert. From there, some eagles cross the Kopet Dag Range and head to Arabia, others cross the Iranian Plateau and migrate to Southern Iran, and still others head south along the border between Iran and Afghanistan to the lower reaches of the Indus River in Pakistan (Bari et al. 2020) and further into India (e.g., Gujarat and Rajasthan; Sharma and Sundar 2009, Karyakin et al. 2019). Birds that go to the east of the Pamir-Alay Mountains are from Mongolia, Northern China, the Baikal Region of Russia and Dauria, and they pass through Gobi-Altai and the northern part of the Gobi Desert. These birds mostly seem to avoid the large desert area of the Tarim Basin in western China, although some migrate through the Karakoram Mountains. From Mongolia, eagles cross the Eastern and Southern Qinghai-Tibetan Plateau, then move toward the Hengduan Mountains. Eagles cross the Himalayas using a rather small number of mountain passes and move along their southern slopes before fanning out to wintering locations in India and Pakistan.

Counts from south of the Himalayas are somewhat confusing, the result perhaps of both population fluctuations and counts being made at different locations over relatively few years (DeCandido et al. 2013). De Roder (1989) reported 7,852 steppe eagles (approx. 44% adults, 28% immatures and 28% juveniles) during 20 October–7 November 1985 passing a watch site south of Annapurna, Nepal. During autumn migrations 2012–2014, Subedi et al. (2017) report 6,166–8,684 eagles passing at Thoolakharka, Nepal; in 2019, 5,127 steppe eagles were counted at that site in autumn (Gurung et al. 2019). Age composition during migration near Annapurna changes as the fall season progresses (De Roder 1989), with juveniles becoming a larger proportion of the passage as the migration advances. Age composition at different watch sites led Subedi et al. (2017) to suggest that migration routes might vary with age. Most eagles that cross the eastern Himalayas subsequently migrate west along the southern slopes into (mostly) Northern and Western India and Pakistan (I.K. and N.B. unpub. data, See also: Ellis et al. 2001, Den Besten 2004, DeCandido et al. 2013, Subedi et al. 2017, 2018, Batbayar and Lee 2017), and return along this path in the spring. However, some of the eagles crossing the Eastern Himalayas head south to Myanmar, some occasionally reaching Thailand. These individuals spend the winter in west India, detouring around the Himalaya (I.K. and N.B. unpub. data, see also: Ellis et al. 2001, Den Besten 2004, DeCandido et al. 2013, Subedi et al. 2017, 2018).

Some steppe eagles from Eastern Mongolia and north-central China migrate to wintering areas in Southern China and Myanmar. Details of this far eastern migration are scarce; no birds have been tracked along this route. MaMing and Zhao (2013) state that this route is bipartite, with birds migrating along either an inland or a near-coastal path. The two paths later join up with the migratory flow from central Mongolia that skirts desert areas of west-central China. Regular wintering areas are mostly below 29° N latitude, though some eagles regularly winter in Eastern Turkmenistan (between 35° and 38° N latitude; Karyakin et al. 2019). The northernmost known regular wintering location in central Asia is at a rubbish dump in South-Eastern Kazakhstan (43.52°N 76.75°E). There also are a few records of steppe eagles wintering in Mongolia in areas with high densities of rodents (Bold and Boldbaatar 1999).

Pre-breeding Migration

Observations at Eilat indicate that some eagles may leave wintering areas as early as mid- to late December, but that most leave during January–late March (Shirihai et al. 2000). This generalization is supported by data from individuals tracked via satellite (Meyburg 2003, the authors, unpub. data). Weather may be a factor in the timing of migration out of some winter localities, in that rain influences the availability of termites, a primary food source for some wintering eagles (Brown et al. 1982). Observations from Eilat, Israel and central Saudi Arabia suggest adults typically leave wintering locations before juveniles (Shirihai et al. 2000, M. Keijmel, pers. comm.); this is consistent with data from other migratory *Aquila* eagles (Miller et al. 2016, Rus et al. 2017). Indeed, Weiss (2018) reports that 95% of the individuals that could be aged in February in Eilat were adults, but that dropped to around 40% during March. The main migration route out of Africa follows the African Red Sea coast (Fig. 1; Meyburg et al. 2003, 2012, Javed et al. 2014), or just inland of it (M. Habib pers. comm.). Well-known spring migration observation sites are at Suez and Ayn Sokhna, Egypt and Eilat, Israel (Shirihai et al. 2000, Megalli and Hilgerloh 2013, Weiss 2018), where the passage is concentrated at the northernmost extent of the Red Sea. In Eilat, an annual average of 18,821 steppe eagles passed in spring during five survey years between 1977 and 1988; during 2014–2018 an annual spring mean of 15,039 eagles was counted (Weiss 2018). At Ayn Sokhna (Egypt), 8,837 eagles were counted during 1 March–2 May 2012 (Megalli and Hilgerloh 2013); at Suez 15,778 were counted in 1982 (in Shirihai et al. 2000). Some studies suggest that the number of spring migrant eagles and the proportion of juveniles has decreased in Israel in recent decades (Yosef and Fornasari 2004, Zduniak et al. 2010), but Weiss (2018) is less supportive of this conclusion. Satellite tracking and observations of steppe eagle aggregations in Arabia (Keijmel et al. 2020) suggest that for at least 20 years an increasing proportion of the global population has been wintering there, rather than migrating to Africa; this change may, to some extent, account for the changes observed in Israel (see Weiss 2018). Apparently, only a small number of steppe eagles exit Africa over the Bab-el-Mandeb Straits, supporting the conclusion of Meyburg et al. (2003) from tracking data that most steppe eagles perform a loop migration when in Africa (see also behavioural remarks). McGrady et al. (2014) recorded only 21 eagles during 9 days of observations on the Djiboutian side of the Bab-el-Mandeb. Those few that make the crossing join eagles that wintered in (mostly western) Yemen. Some move up the Saudi Arabian Red Sea coast and converge with the birds crossing Southern Israel and Lebanon. Others cross Saudi Arabia, passing mostly north of Riyadh, then on towards the Kuwait-Saudi Arabia-Iraq border. From the head of the Gulf, the main flow of migration out of Arabia heads toward the SE coast of the Caspian Sea, crossing Iran, where it meets up with the migrants that have come across from the head of the Red Sea, Southern Israel and Jordan. From Southern Israel, most of the spring migration is thought to head east towards the Caspian Sea (Shirihai et al. 2000). The small proportion of eagles destined for the depleted Eastern European breeding grounds skirt around the eastern coast of the Black Sea (Karyakin et al. 2018, 2019). Estimates of spring passage through Georgia are 4,000–7,000 per year during the 1970s, 3,000–5,000 per year during the 1980's, and 2,000–3,000 per year in the 1990s and 2000s (Abuladze 2013). Presumably these birds pass mostly through Central and Eastern Georgia, as only a small number migrate through the well-known migration bottleneck near Batumi (during spring 2020 only 120 individuals were counted; https://www.trektellen.nl/site/totals/1047/2020). Meanwhile, the main migration flow along this route heads towards central Asia, splitting at the Caspian Sea with some eagles heading north along the western coast of the sea, and others crossing through Northern Iran along the Southern Caspian coast. In Dagestan, to the northeast of the Caucasus and along the western edge of the Caspian Sea, in the late 2000s, at least 2,000–3,000 individuals were recorded during spring migration (Dzhamirzoev and Bukreev 2009), and individuals tracked via satellite have also used this route in spring (R. Efrat, Kuwait Environmental Lens unpub. data). In an area between the Caspian Sea and the Western Kopetdag, Turkmenistan, in the 1980s and 1990s, about 15,000 steppe eagles were annually recorded in spring (Bukreev 1998). In Iran and Turkmenistan, this branch of migrants is joined by those departing from Arabia, Southern Iran and Pakistan. The migration then proceeds northward, fanning out into breeding areas in Western and Central Kazakhstan and Southern Russia. Upon passing the Caspian Sea, eagles arriving from Africa and Arabia spread out to areas in the western part of the Central Asian summer range. Tracking data suggest that birds following the Western Caspian route generally settle in summering areas west of those used by the eagles passing south and east of the Caspian (the authors, unpub. data). Even at these higher latitudes, visible passage east of the Caspian Sea (i.e., in west Kazakhstan) can be heavy (Shevchenko et al. 1978, Shevchenko 1986, Klimov 1986, Berezovikov et al. 1992). Tracking of subadult, non-breeding, steppe eagles reared and tagged in Southern Siberia, and of unknown provenance tagged in the wintering grounds, revealed that most of them spend the summer in Western Kazakhstan (Karyakin et al. 2019, M.M., B. Meyburg unpub. data).

Less is known about the migration of birds that winter from Southern Iran across to the east as far as Pakistan and North-Western India, though they seem to migrate in spring along much the same route as in autumn. Eagles that winter in the Indus River basin cross the Pamir-Alay from the west, usually skirting the mountains at a distance of 100–200 km. Eagles that winter in Northern India are thought to fly along the southern slopes of the Himalayas to the western part of those mountains (De Candido et al. 2013). That said, there are few observations of migration in this area; Den Besten (2004) reports 8,194 birds counted during 29 days of observations in spring 2001. From there birds avoid crossing the highest areas of Himalaya, fly through the Hengduan Mountains, and migrate to summering areas in Mongolia and the Baikal Region of Southern Siberia, Russia. Spring migration through Altai occurs in April (Petrov 2006, Irisov and Irisova 1984). The springtime migration of steppe eagles that winter in Southern China is a matter of speculation. They are thought to migrate in spring along the eastern edge of the Tibetan Plateau toward summering areas; no evidence exists for an Asian east coastal flyway in spring (MaMing and Zhao 2013).

Eagles start arriving on the breeding range as early as late February, but mostly in March, extending into mid-May; arrivals are generally earlier at more southerly locations (Bragin 1983, 1987, Gavrilov and Gistsov 1985, Kovshar et al. 2004). In April, eagles appear in all the basins and hollows where they breed in the Altai-Sayan Region and the Minusinsk Basin (Khakassia and south of Krasnoyarsk Kray, Russia). In Dauria, eagles appear in the second half of March (Goroshko 2012). In Mongolia, steppe eagles first arrive during mid-March–early-April, with the earliest arrivals occurring in the central part of the country, where Brandt's voles (*Lasiopodomys brandtii*) are numerous.

Behavioural and Ecological Remarks

The steppe eagle is among the largest raptors performing long-distance migrations, and a notable feature of its migratory behaviour is the loop migration that occurs both in Africa and Eastern Asia (e.g., China, Mongolia). Meyburg et al. (2003) suggested that the loop migration at the Bab-el-Mandeb Straits between Africa and Arabia is shaped by prevailing east winds, which favour the sea-crossing during autumn but hinder it during spring. Barriers and weather are also likely factors in the loop migration by steppe eagles in Eastern Asia. Steppe eagles also circumvent other water bodies, including the Caspian, Black, and Aral Seas, and the Arabian Gulf (there is no evidence of significant numbers crossing at the Straits of Hormuz).

Steppe eagle migration also flows along terrain features that benefit soaring. The relative benefit of following those "leading lines" may vary with weather, especially thermal and wind conditions. Research in Israel using radar indicated that flight tactics that combine soaring and inter-thermal gliding with straight-line gliding along rather large areas of thermal uplift result in efficient migration (Spaar and Bruderer 1996). Specifically, that research showed that: (1) steppe eagles increased their gliding airspeed when exploiting strong thermal convection while flying against the wind, resulting in an overall increase in their cross-country speed; (2) climbing rate in thermals was highest around noon (mean 1.9 m/s, maximum 5 m/s); (3) mean cross-country speed was 12.4 m/s and the mean gliding airspeed was 15.6 m/s, and; (4) 90% of steppe eagles flew below 1,000 m above ground, 50% flew below 400 m. When crossing mountain ranges, steppe eagles use passes, although they have been recorded flying over the Himalayas at 7,200 m asl, the highest recorded flight for any *Aquila* species (Batbayar and Lee 2017).

Information from many of the steppe eagles recently tracked via satellite has yet to make it into the published literature. Both published (e.g., Meyburg et al. 2012, Karyakin et al. 2018) and unpubl. data (the authors) indicate that pace of migration varies amongst individuals, age classes, season and along different stretches of their migration path. For example, an Altai-reared juvenile eagle travelled about 1,000 km in 23 days in autumn (Karyakin et al. 2018), birds that wintered in Southern Africa migrated about 200 km/day in spring (Meyburg et al. 2012), and birds wintering in Arabia covered about 100 km/day in spring (M.M. and B. Meyburg unpub. data). The routes travelled by individual birds can vary from year to year. In autumn the eagle population includes juveniles that have never migrated, and their inexperience is an additional factor that might contribute to variation in migration and vagrancy. Some birds migrate long distances without breaks (either north or south), others will stopover for days or even weeks (Karyakin et al. 2018, 2019, M.M., B. Meyburg unpub. data, R. Efrat et al. pers. comm.). Stopover locations can be anywhere along the migration route. Data from tracked birds suggest stopover sites are locations where eagles can prey on colonial rodents or scavenge, especially at anthropogenic waste disposal sites. Migrant eagles may stop for several days at sites on the Tibetan Plateau that support high densities of plateau pika (*Ochotona curzoniae*; N.B. unpub. data). On fall migration and on wintering grounds, most steppe eagles are less dependent on small mammals for food than they are during summer. They often scavenge at carcasses of animals and at anthropogenic waste sites, where huge numbers can sometimes be seen (e.g., more than 6,500 near Sharqah, Saudi Arabia, Keijmel et al. 2020). In Africa, swarming termites (e.g., *Hodotermes mossambicus*) are an important food source for some steppe eagles. In winter, steppe eagles can range over larger areas than they do in summer, but they apparently do not do this if food is concentrated and consistently available. Some eagles are thought to settle into extremely small winter home ranges focused on *Quelea quelea* colonies (Brown et al. 1982, del Hoyo et al. 1994).

Tracked eagles show some degree of fidelity to wintering and summering places, and this contributes to the annual similarity of migration routes for some individuals (the authors unpub. data). However, there are exceptions: one juvenile eagle tracked from its natal location in Russia (53.85°N, 91.53°E) spent the first winter in Pakistan-India (24.39°N, 68.29°E – 23.57°N, 69.33°E), the second summer in central Kazakhstan (47.16°N, 57.09°E) and the second winter in Southwestern Saudi Arabia, Eastern Yemen and Western Oman (17.5°N, 50.65°E – 17.00°N, 53.95°E). Also, two young eagles tracked from Central Mongolia, spent their second summers in Northern and Central Kazakhstan (N. B. unpub. data). Some tracked juveniles reared in Turkey returned to their natal area in subsequent years (M. Horvath and T. Burak, pers. comm.). Social interactions, with less experienced individuals switching routes according to decisions taken by more experienced ones, may have a role in flexibility of route selection by steppe eagles, as occurs for other flocking migrants (see Chapters 7 and 8). There are few data on flocking behaviour and in particular on the occurrence of mixed age flocks during autumn migration; De Roder (1989) reported that in Nepal 58% steppe eagles migrated in small groups (< 5 birds), 30% in flocks ranging between 5 and 20 individuals, and 12% in larger flocks.

Finally, an analysis of migration phenology in Israel during spring revealed a rapid advance in the date of first passage of steppe eagles (Zduniak et al. 2010). The authors of that report attribute this change to changing climate in wintering areas. They also suggest that increasing competition for breeding territories due to dwindling habitats may be an additional factor.

12

Eastern imperial eagle *Aquila heliaca*

Evgeny Bragin,[1,†] *Márton Horváth*,[2] *Sharon Poessel*[3]
and *Todd Katzner*[3,*]

ABSTRACT

Eastern imperial eagles are a short-, medium-distance, partially-migratory, or even non-migratory, raptor that breeds at the forest-steppe interface in Eurasia and winters in Northern Africa, the Middle East or South Asia. Migratory strategies of imperial eagles are diverse. Eagles breeding in Central and Southeast Europe and south of the Black Sea usually are year-round residents or partial- or short- distance migrants that winter in the Balkan Peninsula, Northern Africa, or western parts of the Middle East. Eagles that summer to the east of the Black Sea are usually medium-distance migrants that winter in the Middle East or south Asia. Migration tends to follow topographic features, avoids water-crossings and, especially for young birds, may be intermittent and indirect. Populations of imperial eagles are small, in decline in some parts of the distribution, and the species faces a large number of threats including electrocution, persecution, and capture for sale in markets.

Introduction

The eastern imperial eagle (*Aquila heliaca* Savigny, 1809) is a large raptor of the Eurasian forest-steppe interface. Adult imperial eagles are large and brown, often with white epaulets (shoulder markings), and a bright yellow-gold on the top of their head and the back of their neck. Nestlings have dark wings and their breast is ochre with brown streaking. Sub-adults (between 1 and 5 years of age) are intermediate in plumage, becoming mottled and progressively darker as they age. Imperial eagles reach their final adult plumage only in about their 6th calendar year. The species shows no known plumage dimorphism, although males generally are 10–25% smaller than females in weight and other morphological measurements.

The present-day breeding distribution of imperial eagles extends across Eurasia, from Austria and Czechia in the west, into Bulgaria, Turkey, Ukraine, and Russia, to Northern Kazakhstan, Mongolia and the region of Lake Baikal in Russia. Some birds breed in southern parts of Central Asia, including Uzbekistan, Kyrgyzstan, and extreme Western China. To the west of this distribution, in Spain and

[1] Faculty of Natural Science, Kostanay State Pedagogical Institute, Kostanay, Kazakhstan.
[2] MME BirdLife Hungary, Budapest, Hungary.
[3] U.S. Geological Survey, Boise, Idaho, USA.
† Deceased
* Corresponding author: tkatzner@usgs.gov

Portugal, this species is replaced by the closely related Spanish imperial eagle (*A. adalberti*), a non-migratory bird, once considered a subspecies of the eastern imperial eagle.

Populations of imperial eagles are residents or short-distance, medium/long-distance or partially-migratory. During the non-breeding season, individuals from some populations stay on or near their breeding territories (especially in Central Europe, Turkey and the Balkans; Horváth and Kovács 2009, Stoychev et al. 2018). In contrast, other birds from those same regions, as well as birds from colder and more northerly areas—Ukraine, Russia, Kazakhstan—make short-, medium-, or, occasionally, long-distance migratory movements. These movements take birds to southeast Europe, the Middle East, North Africa, and south- and southeast Asia. Linkages (i.e., migratory connectivity) between northern breeding areas and southern wintering areas are not well known.

For this species in particular, it is sometimes difficult to distinguish long-distance dispersal displacements of juveniles from true migratory movements. This is for two reasons, both associated with sampling bias. First, the majority of satellite tracked individuals have been tagged as juveniles and it is not well known if the long-distance journeys performed during their first autumn are repeated later in life. Second, the majority of birds telemetered have been tagged in putatively non-migratory populations in Western Europe. Other data are inconclusive in this regard. For example, in Israel, adults constitute just 8% of the observed birds in migration watchsites (Shirihai et al. 2000), suggesting therefore that long-distance movements are performed mainly by juveniles. However, in Central Asia, all adults leave breeding grounds in autumn and large numbers are observed wintering in the Middle East. Despite this lack of clarity, in this chapter we will use the term "migration" for all medium- and long-distance movements.

There have been many studies on ecology and breeding biology of eastern imperial eagles conducted in the former Soviet Union, but few appear in the English-language scientific literature. More recently there have been a number of breeding ecology studies from farther west in the species' distribution—Hungary, Serbia, Slovakia, Czechia and Bulgaria (Chavko et al. 2014, Demerdzhiev et al. 2014, Horváth et al. 2014).

As mentioned above, throughout its range, this species is almost universally associated with the forest-steppe interface. However, in parts of Hungary, imperial eagle nests are built in dense forests on mountains high above their grassland foraging areas. Likewise, near Lake Baikal, nesting habitat tends more towards forest than steppe, although even there foraging occurs in open habitats. Throughout their range, eastern imperial eagles are generally associated with, and feed upon, colonial birds and mammals, especially *Spermophilus* spp. ground squirrels and members of the Family Corvidae. Nevertheless, at times their diet can be highly diverse, and in Central Asia and Central Europe, well over 100 species have been recorded among their prey (Bragin 1989, Katzner 2003, Horváth et al. 2018b).

Populations of eastern imperial eagles are generally considered small and appear to be declining throughout much of their range (BirdLife International 2019e). This is especially true in the extreme eastern parts of the species range. In western parts of the range, these declines appear to have been stemmed, in part because of extensive research and conservation actions (especially in Hungary, Bulgaria, and the Balkans; Demerdzhiev et al. 2011). In contrast, populations in Eastern Russia appear to have declined dramatically (Ryabstev and Katzner 2007). Conversely, in some central parts of the range, for example Kazakhstan and the Russian Altai, populations appear to be stable. The trend is less clear in other parts of this central distribution, with reports of 20th century increases in Eastern Ukraine, the Volga Basin and the Southern Urals, but then 21st century declines in some regions of the Volga Basin and Southern Urals (Belik 1998, Belik and Galushin 1999, Borodin 2003, Borodin and Smirnova 2006, Galushin 2006, Vetrov and Milobog 2008, Bakka et al. 2010, Viter 2013, Karyakin et al. 2014).

The size of the global imperial eagle population is poorly known. In Europe, population estimates suggest that there are < 4,000 individuals (BirdLife International 2019e). However, the stronghold of the population is in Russia and Kazakhstan. Population estimates vary widely, ranging from 1,650 to 7,800 breeding pairs for the two countries together (BirdLife International 2019e). Given this uncertainty, the global population is considered to be in the range from 3,500 to 15,000 individuals.

Imperial eagles are reported as one of the least common species at a watchsite in Thailand (DeCandido et al. 2008) and as occurring in small numbers on migration and during winter in Greece (Sakoulis 2001), Lebanon (Nielsen and Christensen 1970), Israel (Alon et al. 2004), Bulgaria (Michev et al. 2011), Turkey (Fülöp et al. 2014, Oppel et al. 2014, Panuccio et al. 2017), Georgia (Verhelst et al. 2011), Djbouti (Welch and Welch 1988) and Italy (Corso 2001). That said, others have cited Bulgaria, Turkey and the Eastern Mediterranean as important migration bottlenecks (Heredia 1996), and some sources suggest that, historically, up to 100 eagles could be counted on migration in both spring and fall through Israel (Shirihai and Christie 1992, Leshem and Yom Tov 1996) and possibly in Georgia (van Maanen et al. 2001).

Knowledge Status

In contrast to the large number of nesting ecology studies, there are relatively fewer migration studies of this species. Migration data for imperial eagles generally come from three sources. First are migration watchsite data; imperial eagles are regularly observed at watchsites throughout Eurasia. That said, because the counted numbers are low (Meyburg and Kirwan 2016), the global population is small, many populations are partially migratory, and several important hawkwatch sites are south of primary eagle wintering grounds, there is only limited information gathered by migration counts. Second, other important data come from marking studies. About 200 birds were wing-tagged in Central Kazakhstan, and many thousands have been ringed in Hungary and the former Soviet Union. Finally, and most recently, there have been a number of telemetry studies, nearly all of nestlings tagged on breeding grounds. Birds tracked have been from across the range of this species, but especially in the western part of their distribution (Horváth et al. 2018a). In particular, studies have been conducted of migration of eagles originating on nesting grounds in the Baikal region, Eastern Russia, the Russian Altai, northern Kazakhstan, the Central Volga region in European Russia, Central and Southeast Europe, the Caucasus, and Turkey. A few birds have also been captured and tagged on wintering grounds in the Middle East.

There are, however, gaps in knowledge of migration of this species. In particular, from a geographic perspective, there is limited information on movements of individuals from Central and Western Russia and western states in the former Soviet Union (Ukraine, Belarus, Moldova). Likewise, because count numbers are low and tagging data are incomplete, there is limited understanding of issues such as migratory connectivity, habitat use on migration, stopover behaviour, and age-related and among- and within-individual variation in migration behaviour.

Post-breeding Movements

Timing of migration—departure and arrival

Observational data are central to understanding timing of autumn migration. Within the former Soviet Union, autumn migration takes place from about the end of August to the middle of November. However, there is considerable regional variation in rates of passage and times of peak passage. These likely correspond to how far north an observational site is, which populations are observed migrating past that site, and the age of the birds. In general, observational data from sites farther to the south suggest a broader temporal window of migration than do data from farther north. This suggests that some eagles from northern latitudes may engage in long stopovers *en route*.

In the Northern Caucasus, the most intensive flight occurs at the end of September (Polivanov et al. 1985, Polivanov et al. 1999). Similarly, on the lower Don River, migration of imperial eagles was noted throughout September (Belik 1999). Farther east, in the Urals, more birds are seen in October and into early November (Karjakin 1998), with local breeders leaving territories at the end of September or in October.

Timing of migration in Central Asia appears to be later than it is at sites farther north. At Naurzum, in North-Central Kazakhstan, local nesting birds depart breeding territories at approximately the end of September. Young-of-the-year birds depart 3–5 weeks earlier, and non-territorial or non-breeding birds leave even earlier. However, migratory individuals are observed as early as late August and migration can persist until the end of October or, in some years, the beginning of November (Bragin 1983, 1987). Farther south, in the Northern Kyzylkum desert, nesting birds leave territories in mid-late September (Krivosheev 1960).

Passage dates are consistent with these departure times. In the Northern Aral region, average date of passage is ~16 September, with peak migration between 27 August and 24 September, and birds observed until the end of October (Warshawskii 1957, Berezovsky 1983). In the eastern parts of Central Asia, at Chokpak Pass between Shimkent and Taraz in Kazakhstan, imperial eagles are generally seen between late August and late October (Gavrilov and Gistsov 1985). Likewise, in the nearby foothills of the Talas Alatau in Kyrgyzstan, the flight starts around the second half of September, it peaks in late September, and it continues until the end of October or early November (Kovshar 1966).

Despite these flights, substantial numbers of eagles sometimes are observed in November between the Caspian Sea and the Kopetdag Mountains of Turkmenistan (Isakov and Vorobyev 1940). Similarly, in the Central Kyzylkum, on the Ustyurt Plateau, and in the Golodnaya and Karshinskya steppes, migrants are regularly seen in October and November, often concentrating in groups in areas where saiga antelope (*Saiga tatarica*) are harvested, or near aggregations of livestock. The dispersed nature of these observations may provide insight into why imperial eagles are seen in low numbers at migration watchsites and why migration is observed over such a large window of time.

The eastern part of the range of this species extends to the Lake Baikal region of Russia. On the southwest coast of the lake, migratory birds are observed in late September through mid-October, with a peak in October (Ryabstev 1999, Krasnoshtanova 2001). Numbers of eagles are not extremely high and seasonal counts between 1996–2003 fluctuated from 24 to 41 individuals (Fefelov et al. 2004).

There also are some telemetry data from which timing of migration can be inferred. For birds marked as nestlings in Russia's Middle Volga Region (Ulyanovsk Oblast and Tatarstan), Northern Kazakhstan, the Southern Ural Mountains, the Altai Region of Russia, and near Lake Baikal, Russia, departure dates were in late September or early or mid-October (Ueta and Ryabstev 2001, RRRCN 2014, 2016, AquilaSystem 2019, Korepov et al. 2019). For those birds that lived into a second year, their second migration began later than did the first migration. When the end of migration is recorded, it has occurred in mid-late November. Finally, a single adult tagged in Northern Kazakhstan departed on 12 October (SAP, EAB and TEK unpublished observations).

Routes of migration

Nestling imperial eagles from eight principal areas have been tracked on migration. These are (1) Central Europe (Hungary, Austria, Slovakia, Czech Republic and Romania), (2) the Balkans (Bulgaria, North Macedonia and European Turkey), (3) Anatolia, (4) Georgia, (5) the Middle Volga region of Central Russia, (6) Northern Kazakhstan and the nearby Southern Ural Mountains of Russia, (7) the Altai Region of Russia, and (8) the Baikal Region of Russia. Each region has its own migratory route.

Birds from west of the Black Sea are only rarely migratory and the telemetry work on these birds has focused mostly on nestlings. As such, it can be difficult to distinguish between "dispersive" and "migratory" individuals. Data from 57 nestlings tagged in Hungary and Austria show that the majority of them (50) performed only dispersal movements, up to Latvia, Ukraine, or even Russia, although most travel shorter distances and remain within Eastern Europe. Five of them reached Greece, in some cases spending the first winter there. Two travelled as far away as Syria or even Ghana (with the latter case involving remarkable sea-crossings, see details below; MME – BirdLife Hungary, 2020).

Nestling eagles marked in the Middle Volga Region (Ulyanovsk Oblast and Tartarstan) have migrated either to the east or to the west of the Caspian Sea. Of those that migrated west of the Caspian Sea, the majority (5 of 6) travelled not far from the coast, through Eastern Azerbaijan, to ultimate

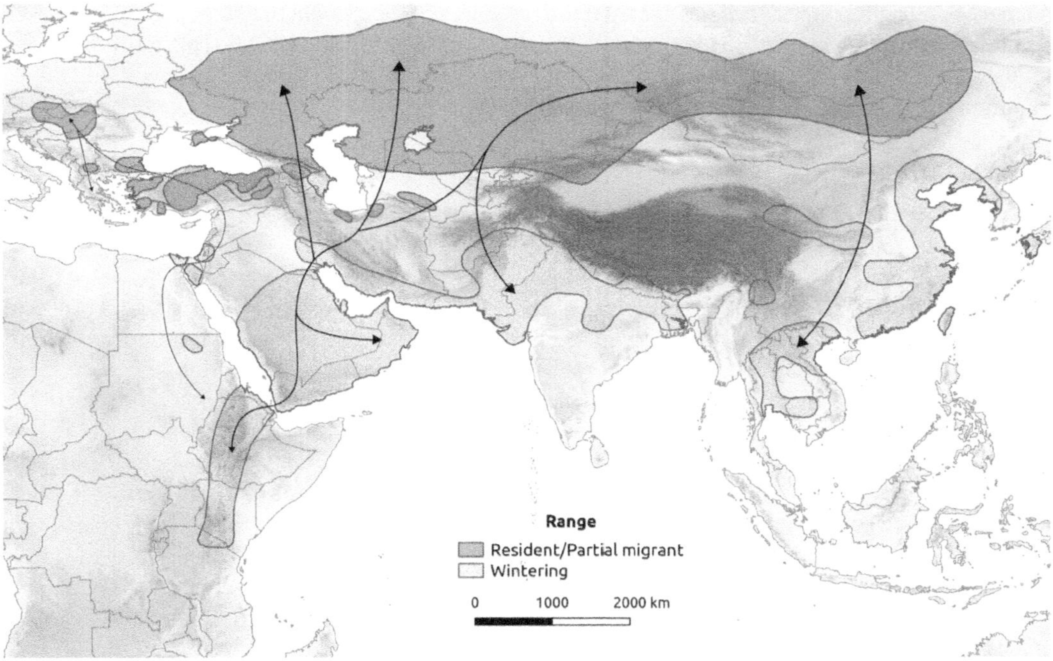

Figure 1. Schematic representation of the long-distance movements performed by eastern imperial eagles.

wintering grounds in Iran, Iraq or the Arabian Peninsula. The sixth bird in this group passed through the Caucasus Mountains in Georgia, and then passed through Turkey, Syria and Jordan. Subsequently it migrated along the Red Sea coast to an ultimate wintering area in Yemen (Fig. 1).

Several of the birds marked in this same region, as well as all of those marked in Northern Kazakhstan and the Southern Ural Mountains of Russia, migrated east of the Caspian Sea (RRRCN 2016, Poessel et al. 2018, Karyakin et al. 2018b, AquilaSystem 2019). All migrated close to the Aral Sea, through the Ustyurt region of Western Kazakhstan and Turkmenistan, sometimes also crossing Uzbekistan. These birds generally wintered either in Iran, Saudi Arabia or Yemen. One individual marked in the Middle Volga Region was atypical, in that after travelling through Saudi Arabia and Yemen, it subsequently crossed into Africa at Bab-el-Mandeb and wintered in Central Ethiopia (Fig. 1).

Nestling imperial eagles also have been tagged in the Russian Altai region (RRRCN 2014). Most of these birds migrated through Eastern Kazakhstan (n = 4 of 5 tagged) to wintering grounds in Uzbekistan (n = 1) or Pakistan (n = 2; although one of these two birds did not send data for a complete migration). The fourth bird from this group migrated through Kazakhstan on a more westerly route, ultimately passing through Iran and Iraq. Finally, a single imperial eagle ringed in the Altai region was recovered alive in the Thar Desert, Rajasthan, India in March 2015 (Sangha and Poonia 2016).

Among three nestlings tagged in Georgia, two moved to the Middle East (Turkey and Iraq), while the other one showed shorter-distance movements to Azerbaijan (MME—Sabuko, 2020). A small number of nestlings have also been outfitted with transmitters in the region near Lake Baikal in Russia (Ueta and Ryabstev 2001). Migration of these birds began along the shore of Lake Baikal and then continued south through Central Mongolia and Central China. The three that survived migration established wintering areas in Southern China, Northern Burma and Northern Thailand.

Other information on migratory routes comes again from limited observational records of birds seen at watchsites. The species is regularly observed at Batumi, Georgia, where between 11 and 62 individuals have been reported every year from 2011 to 2018 (Wehrmann et al. 2019). At Bab-el-Mandeb, 70 individuals were counted during autumn 1987, mainly 1st and 2nd year birds, thus consistent with the age proportion reported in Israel (Shirihai et al. 2000). The largest reported counts of migrating imperial eagles are from the Suez area of Egypt, where 556 imperial eagles were counted in autumn 1981, with a maximum of 97 on a single day in late October (Bijlsma 1983). Such high counts are inconsistent with those from all other sites within the same flyway (e.g., in Israel, a maximum count of 64 in 1980; Shirihai et al. 2000).

Pre-breeding Movements

Timing of migration—departure and arrival

There are substantial observational data on timing of spring migration and on arrival at nesting territory by imperial eagles. In general, imperial eagles are observed on migration in March and April, with arrival on southern nesting areas occurring early during this period and arrival on more northerly areas at the end of March or early April. That said, there is considerable variation in observational data and young birds regularly migrate well into May.

Migration is observed along both the east and west coasts of the Caspian Sea. On the eastern side, in Western Kazakhstan and Uzbekistan, migration is observed from early March to mid-April (Salikhbaev et al. 1970, Antipov and Shubenkin 1984). In Western Turkmenistan, migration has been observed as late as 28 April 1939 (Isakov and Vorobyev 1940). Finally, on the western coast of the Caspian Sea, in Dagestan, Russia, migration is observed until the end of April (Butyev et al. 1989).

Information on timing of imperial eagle migration also can be gleaned from a large number of observations from desert regions of Central Asia. In Central Uzbekistan, eagles may be observed in early March (Mitropolsky et al. 1987), but most are seen in the first half of April (Ostapenko et al. 1976, Lanovenko and Abdunazarov 1983). North of there, in Kazakhstan, migration takes place from the middle of March to the middle of April (Spangenberg and Feigin 1936, Krivosheev 1960, Kovshar 1966, Gavrilov and Gissov 1985, Poslavsky 1986). This timing appears to coincide with the arousal from hibernation of yellow ground squirrels (*Spermophilus fulvus*; Warshawskii 1957). Farther east, migration is heavier in April. For example, in the Zalarinsky District of the Irkutsk region in Russia, there are observations of migrating eagles from 10–12 April (Sonin and Lipin 1980).

Imperial eagles arrive in the southern part of the breeding range (Ukraine, Georgia, Uzbekistan) in early March (Mitropolsky et al. 1987, Pilyuga 1991, Abuladze 1996). In the Northern Caucasus, eagles appear usually in mid-late March (Dinnik 1881, Khokhlov 1995, Ilyukh and Khokhlov 1999, Komarov and Lipkovich 1999, Parfenov 2006). Similarly, in the steppe zone that makes up the core of this species range, imperial eagles arrive on territory in the second half of March (Korelov 1962, Zubarovsky 1977, Bragin 1980, Korovin 1983, Khokhlov 1995, Karjakin 1998). Timing is generally linked to latitude and thus, at Naurzum in Northern Kazakhstan, arrival averages ~ 14–16 March (Bragin 1983), but farther north in the steppe regions of the Urals, arrival is generally at the end of March or beginning of April (Karyakin 1999). Finally, in the Baikal area of Eastern Russia, the first imperial eagles appear between late March and early April (Elaev 1997, Ryabstev 1999).

Telemetry studies also provide insight into the timing of imperial eagle migration. However, it is difficult to identify a general pattern, since sample sizes are small and tagged birds belong to different populations. Imperial eagles tagged the previous year as nestlings in the Altai Region of Russia left wintering areas near Karachi, Pakistan, in mid-late March and arrived in the region where they hatched in late April and early May (RRRCN 2014). Another bird tagged in the same area in 2016 left a wintering area near Chimkent, Kazakhstan in late February and returned to its natal area in late May. Birds tagged as nestlings in Kazakhstan and the Volga Region of Russia began their spring migration from the Arabian Peninsula (Iran, Saudi Arabia), in late March, April or May (Poessel et al. 2018, Korepov et al. 2019). In the case of these late-season migrations, it is often noted by authors that adults were already in breeding territory while these younger birds were still migrating.

Routes of migration

Tracking data on spring migration are sparse, as many birds are tagged as nestlings and die before spring migration is attempted or completed. In addition, the region with the greatest number of tagged birds, west of the Black Sea, including Hungary and Bulgaria, is an area where populations are partially migratory. Tagging of these often non-migratory birds contributes a great deal to our understanding of the ecology of this species, but comparatively less to our understanding of its migratory behaviour.

Imperial eagles trapped on wintering grounds in Saudi Arabia have been tracked along migration routes to summer home ranges spanning a large portion of Eurasia (Meyburg and Meyburg 2010). Three of these birds (one adult male, two adult females), migrated on the western side of the Caspian Sea, through the Caucasus, to summer ranges in the middle-Volga and south-Ural region. Others (two adult males, one subadult female) migrated east of the Caspian Sea, to nesting grounds in Central Kazakhstan and the Ural region of Russia, or to summering (non-breeding) grounds in Xinjiang, China, not far from the Kazakhstan border. Other birds tagged as nestlings in Russia and Kazakhstan travelled similar migration routes, east or west of the Caspian Sea, after wintering in the Middle East (RRRCN 2016, Poessel et al. 2018).

Two imperial eagles tagged as nestlings in the Altai Region of Russia wintered near Karachi, Pakistan (RRRCN 2014). Both of these birds migrated north and to the west of the Hindu Kush Mountains and then travelled into southeast Kazakhstan. These birds passed between Balkhash and Lake the Tien Shan Mountains, with final destinations in Eastern Kazakhstan and the Altai region of Russia.

Other information on migratory routes comes from limited observational records of imperial eagles from a variety of watchsites, mainly in the Middle East. Eight individuals were counted at Bab-el-Mandeb during 9 days in March 2013 (Rayaleh et al. 2013), confirming that few birds spend the winter in tropical Africa. Between 12 and 95 individuals were reported across 8 seasons at a watch site in Israel between 1977 and 1997 (Shirihai et al. 2000), and 32 individuals were observed crossing the Bosphorus during spring 2006 (Üner et al. 2010). The latter count may be of individuals from the small and partially migratory population in Eastern Europe.

Behavioural and Ecological Remarks

Little is known of the behavioural and ecological correlates of migration of imperial eagles. That said, some inference can be drawn from the observations noted above. First, as is the case for other *Aquila* eagles, it appears that this species avoids crossing large water bodies. Imperial eagles are known to cross Balkhash Lake, Kazakhstan, but primarily at a bottleneck about 10 km wide. Likewise, visual observation and telemetry data confirm that the species crosses at Bab-el-Mandeb strait, between Yemen and Africa; at its narrowest, this requires a ~ 25 km water-crossing.

Some individuals tagged in Hungary attempt crossing the Mediterranean Sea, but the majority of those stop at the southern tip of Peloponnese. In one exceptional case, an individual made a ~ 400 km crossing between Greece and Libya, spent the winter in Ghana, and then returned to Hungary crossing the open water between Tunisia and Italy.

One telemetry study associated GPS locations from a small number of birds with habitat variables (Poessel et al. 2018). In that study, during autumn migration, imperial eagles selected for forested and mixed vegetation cover types and topographically rough habitats with greater frequency than would be expected based on their availability on the landscape. Spring migration was also associated with stronger thermal updraft than was autumn migration. Observational data corroborate these sparse telemetry data. Imperial eagles are observed at migratory bottlenecks where topography funnels them together. They avoid water-crossings and, when near water, tend to follow lowland, often wooded, areas.

Little is known about characteristics of migration, such as stopover behaviour, migratory connectivity, feeding strategies or daily distances travelled. Imperial eagles are not thought to be a flocking migrant. Daily distances travelled have been reported to be about 130 km/day in autumn and 80 km/day in spring (Poessel et al. 2018), and there is evidence of long stopovers during migration of pre-adult birds (RRRCN 2014). There is also some evidence of patterns of migratory connectivity. For example, when birds from west of the Black Sea make long-distance movements, they go either towards the Middle East or even to Africa. These birds have been recorded wintering or found dead in Bulgaria, Greece, Israel, Turkey, Syria, Saudi Arabia, Sudan, Ethiopia, and Ghana (Stoychev et al. 2014, Horváth et al. 2018a). Likewise, birds from east of the Black Sea and west of the Ural Mountains have been tracked to Iran, Saudi Arabia, and neighboring states (see citations above and Parchami-Araghi et al. 2015). Birds tagged from the Altai Region winter in Central Asia, from Kazakhstan to Pakistan, and birds tagged near Lake Baikal winter in southeast Asia. Shirihai et al. (2000) report the case of a bird that may be an exception to the rule: a tracked individual breeding in Northern China and wintering in Saudi Arabia for two consecutive years.

Threats to imperial eagles on migration are not well understood. However, known causes of death of eagles include shooting, electrocution, capture for sale in markets, and capture in leg-hold traps (EAB, TEK, I. Karyakin, personal observations). Other likely causes of death on migration include collision with power lines, disease, and poisoning (Heredia 1996, Stoychev et al. 2014).

13

Levant sparrowhawk *Accipiter brevipes*

Herbert Stark[1,*] and *Felix Liechti*[2]

ABSTRACT

The Levant sparrowhawk (*Accipiter brevipes*) is a monotypic, fully migratory raptor with a small breeding range in SE Europe and an estimated breeding population of 3,500 to 6,900 breeding pairs. They winter mainly in the east Sahel zone of sub-Saharan Africa, although the extent of their winter quarters is not well known. Levant sparrowhawks migrate in large numbers over a narrow front through the Middle East. This chapter presents an overview of the current knowledge, the timing of migration at major bottlenecks, and highlights the strategies and routes of Levant sparrowhawks.

Introduction

The Levant sparrowhawk (*Accipiter brevipes* Severtsov, 1850) is a medium sized raptor with a wing span of 69–74 cm. Males weigh on average 175 gr and females have a weight of 220 gr (Mebs and Schmidt 2006). Morphologically, the Levant sparrowhawk is very closely related with the Eurasian sparrowhawk (*Accipiter nisus*). In contrast to the Eurasian sparrowhawk the Levant sparrowhawk appears slimmer with more pointed wings and as a specific characteristic it has black wingtips. Its name originates from its breeding range, the Levante on the easterly edge of the Mediterranean Sea (Mebs and Schmidt 2006).

The Levant sparrowhawk has a small breeding range which is concentrated on Southeastern Europe, including large parts of the Balkan Peninsula respectively, the easterly parts of the Ukraine, Southwestern Russia up to the Ural river further on to the Caspian Sea into westerly Kazakhstan. Additionally it ranges from Asia Minor through Transkaukasien into northwesterly Iran. The European breeding population is estimated at 3,500 to 6,900 pairs with an unknown trend (Birdlife International 2017a). The new European Breeding Bird Atlas (Keller et al. 2020) shows that there have been losses in some parts of the breeding range but overall the current trend appears to be one of gradual expansion. However, the size of many national populations probably continues to be underestimated, owing to identification difficulties and the species' secretive nature (Gorman

[1] Landgasse 2, 78479 Reichenau, Germany.
[2] Swiss Ornithological Institute, Seerose 1, 6204 Sempach, Switzerland.
* Corresponding author: herbert_stark@t-online.de

1998). For example, European Russia was until recently thought to hold 700–1,000 pairs but is now estimated to host 2,000–2,800 pairs (Keller et al. 2020). It is likely that the species has benefited from the development of human settlements with trees and from the establishment of plantations, orchards and vineyards, all of them providing suitable nesting habitats (Gorman 1998). The large difference between the European population size estimate and the numbers recorded during migration counts in Israel, where more than 60,000 birds have been recorded (Alon et al. 2004) shows the challenges of monitoring this species (Keller et al. 2020). Levant sparrowhawks breed in fragmented deciduous forests in river basins but also occur in broadleaved forests and mountain slopes usually below 1,000 m asl. They feed on a variety of small birds, mammals, reptiles and insects (Hagemeijer and Blair 1997).

Knowledge Status

Knowledge about the migration of Levant sparrowhawks is deeply routed in many traditional survey programs established in several countries (Fig. 1) over many years, reaching from the Black Sea countries (Bulgaria, Georgia, Turkey) down to East Africa (Kenya) with a hotspot area in Israel, where a large proportion of migrating Levant sparrowhawks are passing through the narrow area between the Arabian desert and the Mediterranean Sea. This regular visual observation time series allows to delineate the phenology of this species at a given location by recording the first and last migrants and the days with peak migration intensity. Additionally, such data can also be used to compare flyway population sizes and assess population trends over time (Martin et al. 2016).

The Levant sparrowhawk is a long distance migrant using both strategies of soaring/gliding and flapping flight. During daytime large flocks of Levant sparrowhawks share updrafts with typical soarer species gaining height above ground and then glide to cover large distances for a high proportion of their migration. This strategy of migration is energy efficient and allows large birds like raptors, vultures, storks and cranes to travel long distances at low transport costs (Alerstam et al. 2003). During night Levant sparrowhawks use the flapping flight, with a typical regular pattern of flapping phases with a given wingbeat-frequency (5–6 Hz) interrupted by pauses (Stark and Liechti 1993, Spaar 1997). At the end of the post-breeding migration season the proportion of night migrants increases, showing that time constraints to reach their wintering grounds dictate their migration behaviour (Stark and Liechti 1993).

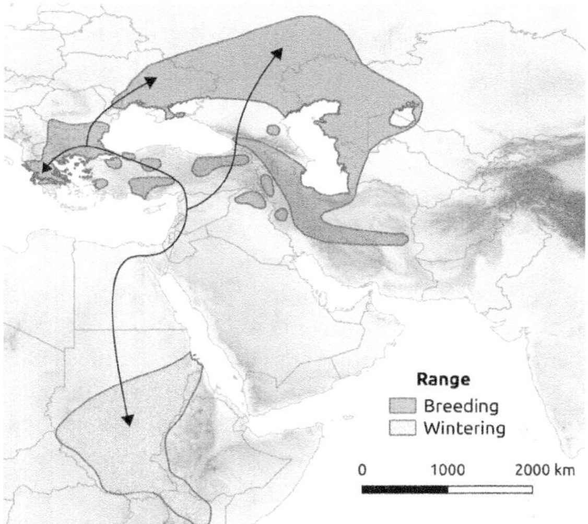

Figure 1. Schematic representation of the migratory routes followed by Levant sparrowhawks.

Migration observation locations where data of Levant sparrowhawks are available consistently over at least one migration period are in Bulgaria (Michev et al. 2018), Georgia (Wehrmann et al. 2019, Vansteelant et al. 2014), the Caucasus region (Verhelst et al. 2011), the Bosphorus (Fülöp et al. 2014), Iskenderun (southern Turkey; Oppel et al. 2014), Israel (Alon et al. 2004), Egypt (Hilgerloh 2009, Hilgerloh et al. 2011a, Megalli and Hilgerloh 2013) and Kenya (Pearson et al. 2014, pers. sightings 2013). Some of these survey campaigns have accumulated long time series over many years, whereas others have only provided data for one season. However the phenology for Levant sparrowhawk can be summarised quite reliably for the East Mediterranean Flyway. The species winters from Southern Arabian Peninsula to the East Sahel of sub-Saharan Africa down to South Africa, covering travelling distances of 5,000 to 6,000 km. Principal migration routes lie entirely within the Middle East (Frumkin et al. 1995, Shirihai et al. 2000).

Post-breeding Migration

Starting in the most northern part of the East African-Palearctic migration system in Bulgaria, raptors are surveyed at the famous site of Burgas Bay at the Western Black Sea coast, also known as the Via Pontica (Michev et al. 2018). At this hotspot Levant sparrowhaks numbers were increasing from 1979 to 2017. During the years 2004 to 2017 maximum counts of 600 to 800 birds were reached in the second half of September (Michev et al. 2018).

At the same latitude, but on the opposite side of the Black Sea, near the city of Batumi in the Republic of Georgia there is also an established longterm raptor survey. Since 2008, ornithologists of the Batumi Raptor Count have been monitoring the autumn raptor passage daily from mid-August until mid-October (Wehrmann et al. 2019). From 2011 to 2018 the numbers of Levant sparrowhawks reached average numbers of 3,508 individuals per year and represent 18–35% of the estimated world population (Birdlife International 2017a).

1,000–2,000 pairs are estimated to breed in Greece, mostly in the Eastern part of the country (Handrinos and Akriotis 1997). As the species is uncommon during both migrations on the island of Antikythira (Lucia et al. 2011), Panuccio et al. (2013b) suggested that Greek birds concentrate mainly at the Bosphorus, with some of them performing a detour to avoid the crossing through the Central-Eastern Mediterranean, although they are able to use the flapping flight with relatively low energetic costs (Agostini et al. 2015a).

On the Bosphorus several studies have been conducted in the last decades. The most important one has been carried out in 2008 by Fülöp et al. (2014), that reported the passage of 2,455 individuals using four survey stations located on the Asian shore.

At the bay of Iskenderun, in the Southeast of Turkey. Oppel et al. (2014) reported for 2013 (mid-August to mid-October) totally 15,700 individuals. There, Levant sparrowhawks were seen passing in large flocks (max. 770 individuals in a single one) peaking at the end of September.

The best observed and documented part of the East African-Palearctic migration system is when raptors are passing over Israel. Over many years Israel authorities had to engage huge efforts to minimize air strikes of military aircrafts with migrating birds, especially large birds like raptors, storks, cranes and pelicans. For this and other reasons many studies have been conducted over many years providing a huge data set at several sites (Bruderer et al. 1997, Leshem and Yom-Tov 1996a, 1996b, Frumkin et al. 1995, Shirihai et al. 2000, Yosef 1995).

Extreme passage dates in Israel extend from 1st September to 20th November (Shirihai et al. 1996). In 1994, during the Northern Valleys Counts, 60,390 Levant sparrowhawks were recorded (Alon et al. 2004). 90% of the annual total was reported between 16th and 29th September.

Pre-breeding Migration

Pre-breeding migration of Levant sparrowhawks starts in early to mid-March from their wintering grounds in East Africa (Pearson et al. 2014). Birds arrive at their breeding grounds in late April to mid-May after travelling 5,000–6,000 km from Central and East Africa, along routes that don´t differ from their post-breeding ones. At Eilat, Israel, large concentrations of Levant sparrowhawks were found during spring (Shirihai and Christie 1992, Yousef 1995). About 90% of the world population travels through Israel in a short period of a fortnight, mostly concentrating in late april, with adult males moving before females, followed by immatures (Shirihai et al. 2000). Yosef et al. (2003) reported that as many as 50.000 Levant sparrowhawks were counted each spring at the end of the Gulf of Aqaba over Eilat, Israel. Besides Israel, other significant concentrations during spring are reported in Batumi (Georgia), where in 2019 and 2020 1,166 and 1,561 Levant sparrowhawks were counted with peak numbers between the end of April and the beginning of May (max. one day counts 453 on the 5th May 2019 and 633 on the 29th April 2020).

Behavioural and Ecological Remarks

For the Eastern Black Sea flyway, Vansteelant et al. (2014) describe the influence of wind conditions on the behaviour of migrating birds and found that Levant sparrowhawks are detected along the coast mostly during easterly winds, perhaps due to the fact that they tend to fly at lower altitude to reduce the drift effect of winds blowing them offshore. Hilgerloh et al. (2011a) report that for the Isthmus of Suez the main flight direction in spring is influenced by crosswinds and is mainly directed to NE.

Yosef et al. (2003) reported that Levant sparrowhawk in better body condition according to body mass and wing chord length migrated earlier in spring. Moreover, the median date of passage for adults preceded that of juveniles by 2.5 days. They believe that juveniles on their first spring passage migrate slower than adults and they are more likely to be later and in poorer body condition, confirming earlier findings by Gorney et al. (1999). The body mass and index of physical condition of migrating Levant Sparrowhawks trapped in baited traps was lower than birds trapped in mist nets (Gorney et al. 1999). The extent to which biases occur may be different even for closely related species. For comparison, in this study Eurasian sparrowhawks (*Accipiter nisus*) did not show this pattern.

Spaar et al. (1998) presented a comprehensive study of migration behavior of the Levant sparrowhawk carried out by radar tracking. They showed that the optimal migratory strategy with respect to time and energy depends on feeding conditions en route: in poor conditions, both time and energy are minimized by a pure soaring-gliding flight strategy. If food is available en route, soaring-gliding flight should be combined with flapping flight when no thermals are available, as this will minimize time spent on migration. Levant sparrowhawks flew at similar altitudes as other migrating raptors in Israel during the day, even if they prove to be able to reach such altitudes also under poor thermal conditions, using flapping fight. Soaring-gliding birds maximized cross-country airspeed according to optimal flight theory and, thus, minimized time needed per distance. In flapping-gliding flight, they adjusted their airspeed with respect to the wind to fly at the maximum range speed, suggesting that they minimized energy consumption per distance.

Information concerning the wintering and breeding areas and population size and structure of Levant sparrowhawk is extremely limited. Further studies to bring light in the secretive life of this species should be initiated, possibly by tracking individuals with GPS devices.

14

Eurasian sparrowhawk *Accipiter nisus*

Paolo Becciu[1,2]

ABSTRACT

The Eurasian sparrowhawk (*Accipiter nisus*) is a partial migrant, travelling mainly on short to medium distances. It is sedentary in Europe, apart the Scandinavian region, where it is a summer visitor. It migrates using a mixed flight strategy of soaring-gliding and flapping flight modes. It is a fly-and-forage migrant, hunting small resident and migratory birds. The most remarkable counts are localized in North Europe, as well as in Central- and South-European mountain passes and at barrier-crossing sites, such as the Alps, the Pyrenees, the Strait of Gibraltar and the Bosphorus. Furthermore, ring recoveries data helped us to understand the main migratory directions from Scandinavia and North Europe towards Central, South-West Europe, and Africa. Some individuals migrate to Northern Africa and a very small proportion even to sub-Saharan Africa. Future studies with the support of satellite tracking are needed to expand our knowledge of the migration of the Eurasian sparrowhawk.

Introduction

The Eurasian sparrowhawk (*Accipiter nisus* Linnaeus, 1758 - "sparrowhawk" hereafter) is a small falcon-looking hawk. Adult males have bluish slate-grey upperparts, rufous cheeks, and rufous-barred underparts; adult females have slate-grey upperparts with brown-grey barring below. Juveniles have dark-brown upperparts and the underparts' barring often coarse, broken up and irregular on breast. The sparrowhawk is a sexual size dimorphic species, with females being up to 25% larger than males; one of the greatest size differences between the sexes in any bird species (Newton 1986). Sparrowhawks weigh on average 200 g with a wingspan ranging between 62 and 74 cm (Newton 1986, Bruderer and Boldt 2001). It is a predator specialised in catching woodland birds (Newton 1986). Males tend to take smaller birds, including tits, finches, and sparrows; females catch primarily thrushes and starlings, but can kill birds weighing 500 g or more (Newton 1986). It is widely distributed in different types of woodlands, among temperate and subtropical regions of the Palearctic (Ferguson-Lees and Christie 2001). The sparrowhawk is a partial migrant breeding almost everywhere in Europe and Western North-Africa. In Asia it is a summer visitor from Western Russia to the Kamchatkan peninsula and

[1] Animal Flight Laboratory, Department of Evolutionary and Environmental Biology, University of Haifa, Israel.
[2] Department of Migration, Max Planck Institute of Animal Behavior, Radolfzell, Germany.
Email: pbecciu89@gmail.com

Japan. Wintering grounds range from Central and South Europe, Africa to the Middle East, Indian subcontinent, and South-Eastern China. In the Western Palearctic, the resident populations are mostly in East, Central and South Europe, and North Africa. Scandinavian populations are mainly migratory. Following a longitudinal gradient towards East, the breeding and wintering grounds of this species become more separated, consequently increasing the migration distance. For instance, wintering grounds in Asia span from Iran to India and the South-Eastern part of the continent, with several resident populations in Japan and the Korean peninsula.

The European population has been estimated in 340,000–450,000 pairs (Burfield 2008), and its conservation status is "Least Concern", according to the IUCN (BirdLife International 2020). The breeding season is between May and August in habitats of well-grown, extensive areas of woodland, often coniferous or mixed.

The sparrowhawk can use mixed flight strategies of soaring-gliding and flapping flight depending on environmental conditions and internal motivation. The species mainly uses active flapping flight, and usually migrates singly or in small flocks (Ferguson-Lees and Christie 2000). During migration they could probably use soaring-gliding flight to minimize energy expenditure covering large distances. The sparrowhawk is also considered a fly-and-forage migrant, hunting on local or migratory small birds during migration flights.

Knowledge Status and Migration Patterns

Most of the information about the migration of this species comes from observational studies and reports from bottlenecks. In addition, ringing recoveries provide general information on migration distances and directions. To our knowledge, the species has never been studied by any tracking means, except for few radar studies which recorded its flight at local scale (Malmiga et al. 2014, Meyer et al. 2000).

The analysis of ring recoveries of birds ringed in Central and Southern Finland suggests that the migration likely occurs on the NE-SW axis (Fig. 1), with wintering grounds spanning from South France, Northern Italy to Germany, South Denmark, Poland and Czechia (Saurola et al. 2013). Trajectories do not seem to differ between first-year and older sparrowhawks ringed as nestlings. The migrating behaviour of sparrowhawks breeding in North Europe is known for more than a century and inspired a pioneering study on navigation in birds. In 1938, Drost published a study showing that translocated juveniles move along an inherited NE-SW axis during their first migration, while translocated adults adjusted their direction trying to locate the axis along which they had migrated in previous years to successfully reach their goal destination.

The breeding population in Scandinavia migrates in high numbers, as recorded in Southern Sweden, Finland, and on the Polish Baltic coast. Long-term trends of counts carried out at Falsterbo (Sweden) during the post-breeding migration show a rough average of 20,000 sparrowhawks passing each year from 1973 to 2019. The counts progressively increased from the 1970s (decadal average of 9,680 birds) to the 2010s (28,554) (source: www.falsterbofagelstation.se, accessed on 13/06/2020). At this location, the sparrowhawk is often the most numerous raptor species recorded in autumn, with juvenile birds migrating before than adult birds, with a median date of 17 September and 17 October, respectively (Kjellén 1992). Post-breeding counts on the Baltic coast of Poland (Vistula Spit, Gulf of Gdansk) recorded an average passage of 6,000 sparrowhawks per year between 2008 and 2011, with a maximum of 8,320 in 2011 (Polakowski et al. 2014).

Migratory sparrowhawks are passing the Alps in autumn in substantial numbers between the Jura Mountains and the Lake Geneva, with a rough yearly mean of 1,100 counted between 2000 and 2016 (LPO Haute-Savoie 2016). The maximum registered number was ca. 2,500 birds in 2013. Following a South-West route to the Pyrenees many sparrowhawks were observed on the East side of the Ardeche mountains, often between 500 and 900 individuals per year between 1984 and 2019 (source: www.migraction.net). At Col d'Organbidexka (Pyrenees) around 500 individuals per year were observed

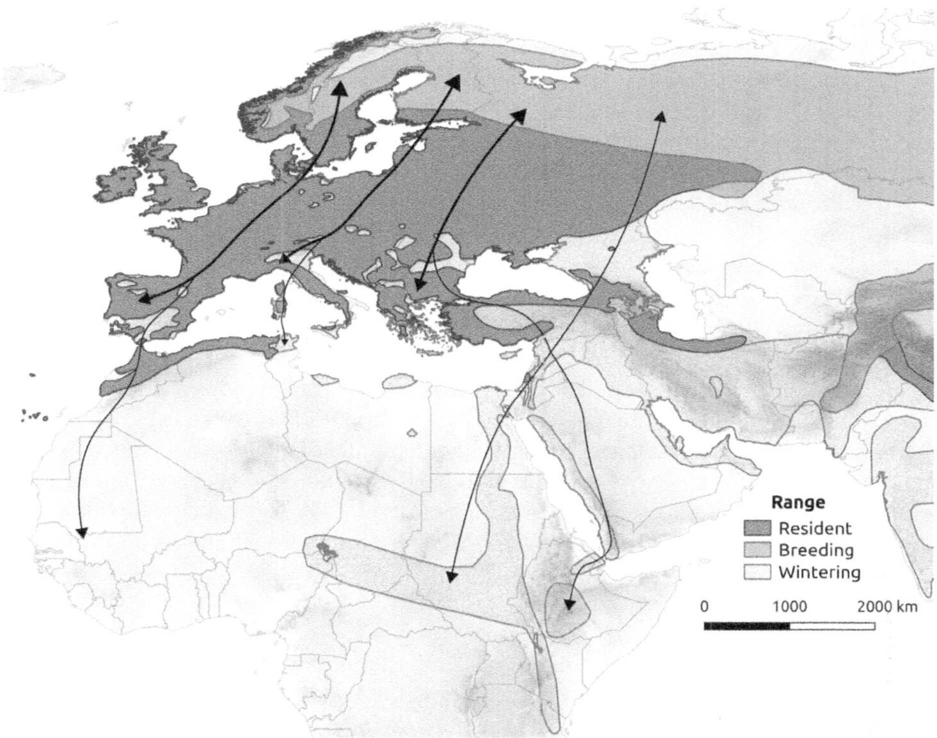

Figure 1. Schematic representation of the long-distance movements performed by Eurasian sparrowhawks breeding in Europe.

in the last 38 years (1982–2019) (source: www.migraction.net, accessed on 09/06/2020). At the Strait of Gibraltar, average autumn counts, that peak after mid-September, increased from 610 (1976–1977) to 2,220 (2011–2012; Onrubia 2016). It is not clear if such increase is real or is just a reflection of an effort improvement. However, these counts highlight that, even if it involves a small portion of the European population, the migratory flow towards North Africa is consistent and may include even trans-Saharan individuals (A. Onrubia, personal communication), as shown also by sporadic observations in the Sahara desert (Salewski et al. 2005, U. Mellone personal communication; Fig. 1). Hundreds of sparrowhawks from Northern and Central Europe also cross the Alps in order to spend the winter in Italy, as shown by ringing recoveries (Spina and Volponi 2008) and by some migration counts: for example, 296 individuals have been counted at Cima Comer (Lombardy, Italy) during spring, peaking in late February (Leo et al. 2006). Numbers of migrating sparrowhawks across Italy and Central Mediterranean islands are relatively low, suggesting that only a few reach Africa through this flyway (Agostini et al. 2005a, Sammut and Bonavia 2004): for example 61 have been counted from Malta in autumn 2019 (Bonavia 2019). In this scenario, counts reported from Elba island are quite remarkable, for example 296 individuals en route to Corsica in autumn 2006 (Paesani and Vanni 2006; Fig. 1). At the major bottleneck of the region, the Strait of Messina, the most recent maximum counts are 76 during autumn (Cento et al. 2018) and 158 during spring (Giordano et al. 2016).

In Eastern Europe, sparrowhawks are regularly observed in migration bottlenecks, such as the Bosphorus (Turkey) and Batumi (Georgia). In the first location, 1,500–2,000 individuals are observed during each migration season (Üner et al. 2010, Fülöp et al. 2014, Arslangündoğdu et al. 2018). Larger numbers are recorded in Batumi, where a maximum of 7,452 sparrowhawks were observed in autumn 2009 (Verhelst et al. 2011), and 15,703 in spring 2020 (source: https://trektellen.nl/species/graph/0/1047/100/0, accessed on 09/06/2020). Over Greek islands, the sparrowhawk was observed mostly in autumn migration but in small numbers (Panuccio et al. 2013b). On Antikythira island, a maximum of 145 individuals were observed on the autumn of 2009, while only 23 birds were recorded during the previous two autumns, with spring passage being virtually absent (Lucia et al. 2011). In any case, those birds probably spend the winter in Crete (Panuccio et al. 2013b). Observations in the Middle East suggest that the desert crossing might not be unusual. At the observation point in Eilat, South Israel, a yearly mean of 183 sparrowhawks in spring and 51 in autumn was recorded between 1977 and 1998, with a peak in spring 1997 with 366 birds (Shirihai et al. 2000). In Northern Israel, the counts increased to a yearly mean of 784 between 1982 and 1998 (only autumn data), with a peak on autumn 1986 with 1761 birds (Shirihai et al 2000). These counts are in agreement with those carried out at Ayn Sokhna (Egypt) in spring 2012, when 512 sparrowhawks were observed heading north (Megalli and Hilgerloh 2013). Finally, 2,135 individuals have been counted during the autumn migration of 1987 at Bab-el-Mandeb (Yemen), peaking on the 16th of October (Welch and Welch 1988). Such significant count suggests that probably the species is a more widespread winter visitor to East Africa than previously thought (Ash and Atkins 2009, Zimmerman et al. 2020; Fig. 1).

Overall, these data suggest that migratory individuals of the European population mainly migrate on short and medium distances, with some individuals spending the non-breeding period in North Africa and a lower proportion south of the Sahara desert. The individuals migrating the longest distances use the Western flyway and cross the Mediterranean through the Strait of Gibraltar.

Behavioural and Ecological Remarks

The lack of tracking studies on the sparrowhawk limits our understanding of its behavioural ecology during migration greatly. However, local observational and radar tracking studies help us understand the migratory behaviour of sparrowhawk at specific sites along their migration routes and long-term changes, likely related to the climate change.

In contrast to larger raptor species, the sparrowhawk seems to migrate at lower altitudes, not least to minimize the risks when winds are strong (Malmiga et al. 2014). This strategy seems to be related to their behaviour of hunting during their trips, which might also require an increase of time spent in flapping flight (Malmiga et al. 2014). In addition, sparrowhawks were found to be less inclined to exploit thermal updrafts (Rudebeck 1950) but they adjusted their airspeed and groundspeed to tailwinds and headwinds (Malmiga et al. 2014), likely to reduce the cost of transport and maintain a steady cross-country speed. Sparrowhawks were found to aggregate on the coastal side of a migration bottleneck near Batumi, Georgia, during periods when winds blew offshore, suggesting aggregation caused by wind drift (Vansteelant et al. 2014). Their flight capabilities likely favour the crossing of a mountain chain. For instance, they were found more prone to cross the Alps (Schmid et al. 1986) than the larger common buzzards, *Buteo buteo*.

Recent global climate change has been shown to affect the phenology of the sparrowhawk. Lehikoinen et al. (2010) found that Finnish sparrowhawks have been advancing spring arrival, breeding, and autumn departure considerably since the late 1970s, with no changes in migration distance (assessed with ring recoveries). Furthermore, early migrants, who are the ones with the highest reproductive success, show the strongest advance in the timing of spring migration. In addition, Lehikoinen et al. (2010) found that the breeding success of Finnish sparrowhawks has increased significantly over the study period (1973–2007), likely due to a reduced exposure to organochlorine pollutants, rather than factors related to climate change. The advance of autumn migration departure of Finnish birds seems to be most evident in young individuals (Lehikoinen et al. 2010). However, a significant and consistent advance in autumn migration dates was found for sparrowhawks migrating at the Strait of Gibraltar between 1999 and 2014 (Panuccio et al. 2016b). A similar trend was also found in the Pyrenees, between 1981 and 2008 (Filippi-Codaccioni et al. 2010). This pattern is in line with other long-distance migrant species, but not with short-distance migrants. This could be because the sparrowhawk is a predator specialized on passerines, in contrast to the delayed short-distance species (hen harrier, common kestrel and common buzzard) that feed on small mammals (Filippi-Codaccioni et al. 2010). Small mammals do not migrate but occur all over the Europe year round, while passerines are mainly migratory. Sparrowhawks are possibly aligning their migration with passerines, while mammal eaters do not need to do so (Lehikoinen 2009). Future studies are needed to understand the migratory behaviour and movement ecology of the sparrowhawk in a changing climate, possibly tagging with satellite transmitters individuals from Northern Europe.

15

Western marsh harrier *Circus aeruginosus*

Nicolantonio Agostini

ABSTRACT

The western marsh harrier (*Circus aeruginosus*) is a partial migratory species showing a strong sexual dimorphism, with populations breeding in Northern, Eastern Europe and in most of Asia wintering in the Mediterranean region, tropical Africa and the Indian subcontinent. This species migrates on a broad front, largely using flapping flight when thermal currents are not available, and moving along more or less parallel flyways undertaking long sea-crossings. As a result, unlike typical soarers, rarely seasonal counts of more than 1,000 western marsh harriers are recorded at traditional bottlenecks, even if hundreds of birds are regularly observed at several watchsites, especially in the Mediterranean region. Combining results obtained through visual observations, satellite telemetry and ringing recoveries reveals a differential migration pattern in which juveniles tend to migrate on shorter distances and slower than adults and, among adults, females on shorter distances than males. The female-biased sex ratio observed among adult birds wintering at higher latitudes has been explained on the basis of the Bergmann's rule. The higher proportion of adult males among birds at migration watchsites fits in the same scenario, even if it may be related also to a male-biased sex ratio in the source populations. Despite unfavourable winds in Africa during spring migration, western marsh harriers cover longer distances per day in this period, probably as a result of a time minimization strategy, to reach earlier their breeding grounds.

Introduction

The western marsh harrier (*Circus aeruginosus* Linnaeus, 1758), hereafter "marsh harrier", is a medium-sized raptor with an average body mass of 0.54 Kg (male: 0.40–0.73 kg; female: 0.54–0.96 kg) and a wing span of 1.28 m. It is distributed in the Palearctic from Atlantic to central Asia and, during the winter, in the Afro-tropical region and in the Indian subcontinent (Clarke 1995, Bruderer and Boldt 2001, Ferguson-Lees and Christie 2001, Naoroji 2011). The most recent estimates reported a world population of 500,000–999,999 breeding individuals (Status: Least Concern; BirdLife International 2020). This species is mainly resident in Western, Southern Europe and North Africa, although some birds show dispersive, nomadic or short-distance movements. It is chiefly migratory

MEDRAPTORS (Mediterranean Raptor Migration Network), Rome, Italy.
Email: nicolantonioagostini@gmail.com

in Northern and Eastern Europe such as in most of Asia, with some remaining in Asia Minor, Iraq and Middle East (Ferguson-Lees and Christie 2001, AlSheikhly and AlAzawi 2019).

The European population has been estimated in 99,300–184,000 pairs (BirdLife International 2017a) with an increasing trend mostly in eastern countries since 1980s. In Western Russia, it has been estimated about the half of the whole European population (BirdLife International 2004). This species shows strong breeding site fidelity (Strandberg et al. 2008b) and the nest is built usually in dense, tall vegetation growing in water. The breeding season lasts April-July or June-August depending on latitude, and the clutch consists of four-five eggs with a breeding success of 1.6–3.3 chicks fledged per pair (Clarke 1995). Some studies reported male-biased sex ratio in broods of birds breeding in The Netherlands and Poland (% of males 54.8 and 54.2, respectively) with the proportion of males produced increasing during the season and not concentrated during the period of good food conditions (Zijlstra et al. 1992, Witkowsky 1989). As male marsh harriers are about 20% lighter than females and therefore the cheaper sex to rear, requiring the 20% less energy as nestlings (Riedstra et al. 1998), it has been suggested that their proportion could increase in denser populations in which competition for food increases (Simmons 2000, Agostini et al. 2021). Polygyny, nearly always bigamy, has been widely reported for this species, being more prevalent in certain regions and strictly related to the abundance of food (Clarke 1995, Simmons 2000). In populations showing higher frequency of polygyny, higher proportions of females are expected (Kjellén 1992, Clarke 1995). As regards the age of first breeding, some immatures can colonize unoccupied and newly developed breeding habitats or fill vacancies among adult breeders, and since males show a slower maturation (Clarke 1995, Simmons 2000) a higher proportion of females is expected among younger breeders, especially among yearlings. However, some immatures, arriving earlier in the season and taking a territory, probably because migrating on shorter distances (Agostini 2001), are ousted at a later date by adults (Newton 1979). Among adults, older males return earlier to the breeding grounds and have higher success than the younger ones (Altenburg et al. 1987), while primary females have more success than secondary ones because nestling mortality increases in secondary nests (Altenburg et al. 1982). Marsh harriers usually hunt in reed beds and often over farmland, preying on marshland birds but also insects and other birds (Clarke 1995). After the end of their postfledging dependency, juveniles can faster improve their hunting skills by social learning, selecting habitat patches where prey can be caught successfully (Kitowsky 2009).

The marsh harrier migrates on a broad front with relatively small concentrations at traditional migration bottlenecks (Zalles and Bildstein 2000, Agostini and Panuccio 2010). They mostly migrate solitary or in small loose flocks, although flocks of few dozens of individuals have been reported, sometimes roosting on trees (Sammut 2005, Panuccio and Agostini 2006b). Because of its size and morphology, the energy consumption of the flapping flight in the marsh harrier is relatively low if compared with that required in large obliged soaring raptors (Panuccio et al. 2013c, Agostini et al. 2015a). As a result, this species shows an intermediate behaviour between typical soarers and flappers, largely using flapping flight during migration. In so doing, marsh harriers start to move early in the morning using the whole day for migration, undertake long sea crossings and sometimes fly also during the night (Clarke 1995, Spaar 1997, Spaar and Bruderer 1997, Meyer et al. 2000, Panuccio et al. 2002, Mellone et al. 2012). When flying through large water bodies, marsh harriers use several small islands to enable soaring flight or as stopover sites for roosting and/or hunting (Thake 1983, Frost 1994, Rebassa 1995, Agostini and Logozzo 1998, Jonzén and Pettersson 1999, Agostini et al. 2003, Panuccio et al. 2004, Sammut 2005, Sammut and Bonavia 2004, Panuccio and Agostini 2006b).

Differences in the plumage allow to determine the age and sex of migrants during the fieldwork, making possible the study of differential migration by direct visual observation. However, because of the easier identification of adult males (Forsman 1999, 2016), some studies focusing on this topic reported the passage of birds labeled female/juvenile, meaning that they were either adult females or juveniles, to estimate the total numbers of birds belonging to different age and sex classes limiting a possible male-biased distortion (Kjellén 1992, Agostini and Logozzo 2000, Agostini et al. 2003, 2017a, 2017b, 2021). Interestingly, a study on a resident population in west-central France revealed a substantial proportion (34%) of adult males showing a permanent female mimicry, and only 10% looking like the typical field-guide male mostly observed in other populations and at migration watchsites, so much so that this population appears as an outlier (Blanc et al. 2013, Wehrmann et al. 2019, Agostini et al. 2021).

Knowledge Status

At the beginning of the second half of the last century, the European population of the marsh harrier decreased dramatically as a result of shooting, poisoning by pesticides, and habitat loss (Fenech 1992, Clarke 1995, del Hoyo et al. 1994). However, since 1990s' the population is increasing (status: Least Concern; online datazone BirdLife 2020), especially in Eastern and Northern Europe (BirdLife International 2004), probably because of the adoption of protective measures (del Hoyo et al. 1994, Clarke 1995, BirdLife International 2004). This trend has been confirmed by migration counts made in the Central Mediterranean (Beaman and Galea 1974, Agostini et al. 2003, 2017b, Sammut and Bonavia 2004, Sammut et al. 2013).

Studies on the marsh harrier migration have been made by direct visual observation, ringing recoveries, satellite and radar tracking. The passage of hundreds of birds occurs at several watchsites mostly in the Mediterranean area during both spring and autumn (Zalles and Bildstein 2000, Agostini 2001, Agostini et al. 2003, 2017a, Roth and Corso 2007, Corso et al. 2009). Conversely, substantial counts (> 1,000 individuals) have been recorded at few watchsites, in France (Gruissan, autumn; Leucate, spring: Zalles and Bildstein 2000), Italy (Strait of Messina, spring: Corso 2001a; autumn: Agostini et al. 2017b; Ustica, spring: Panuccio et al. 2014; Conero, spring: Brichetti and Fracasso 2018), Georgia (Batumi, autumn: Vansteelant et al. 20120b), Israel (autumn: Kfar Qassem, Northern Valleys; Shirihai et al. 2000) and over the island of Malta (autumn: Sammut et al. 2013). A total of 23 individuals, 20 adults and three juveniles, belonging to the Swedish (N = 17) and Low Countries (N = 6) populations, were studied using satellite telemetry (Strandberg et al. 2008, Vansteelant et al. 2020a). Some of these birds (all juveniles and four adults) showed west-oriented pre-migratory movements well before the onset of migration. Radar studies were carried out in the Negev Highlands and in the Arava Valley (Israel; Spaar 1997, Spaar and Bruderer 1997) and at the Strait of Messina (southern Italy; Mateos-Rodríguez and Liechti 2012, Agostini et al. 2017b) during both spring and autumn, along the southern coast of Spain, 140 km north-east of Gibraltar, in autumn (Meyer et al. 2000). Finally, migration patterns in relation to age and sex classes have been studied by analyzing ringing recoveries data of birds belonging to the European population (Panuccio et al. 2013d). Ringing recoveries also show that the Central Mediterranean flyway is used mainly by individuals coming from central and Eastern Europe, including Finland, too (Spina and Volponi 2008, Saurola et al. 2013).

Post-breeding Migration

Marsh harriers breeding in Northern and Eastern Europe migrate heading SW-SSW along more or less parallel flyways showing a leapfrog migration and wintering in the Mediterranean basin and in Central-West Africa north of the Equator: birds breeding furthest north winter furthest south, birds that breed furthest west tend to winter furthest west, while a large overlap in the wintering grounds was reported in birds breeding in Sweden and the Netherlands (Fig. 1; Agostini et al. 2001, Ferguson-Lees and Christie 2001, Panuccio et al. 2005a, Strandberg et al. 2008b, Agostini and Panuccio 2010, Panuccio et al. 2013d, Saurola et al. 2013, Vansteelant et al. 2020a). As regards migration counts, larger concentrations have been reported at Batumi (Georgia), located between the Black Sea to the west and the Pontic Mountains to the east, and at the Strait of Messina, located between Southern continental Italy and Eastern Sicily. In particular, at Batumi, on average 6,489 individuals were counted each autumn from two observation posts in the period 2011–2018 (Vansteelant et al. 2020b). At the Strait of Messina on average 2,055 marsh harriers were counted by a single observation post in the period 2011–2017 (Agostini et al. 2021), a maximum of 5,146 birds in 2015 but using three observation

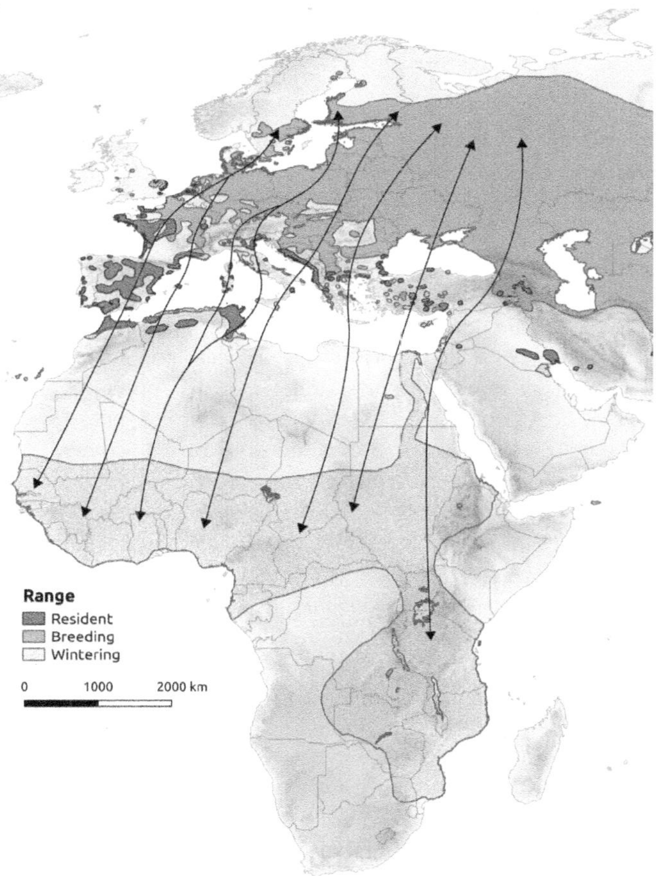

Figure 1. Schematic representation of the migratory routes followed by long-distance migratory populations of western marsh harriers.

posts (Brichetti and Fracasso 2018). Over Malta, on average 1,105 birds each autumn were counted in the period 2005–2012 with a maximum of 2,227 individuals in 2010 seen by a single observation post (Sammut et al. 2013). In Israel, a mean annual count of 1,176 birds was reported during the Northern Valleys surveys in the period 1988–1999 (Shirihai et al. 2000). Moreover, at Gruissan, along the south-eastern coast of France, on average 1,080 individuals were observed each autumn season in the period 1983–1993 (max. 3,160 in 1984; Zalles and Bildstein 2000). Among other watchsites where long term counts have been conducted during autumn, a significant concentration of marsh harriers occurs at Falsterbo, in Southern Sweden (average of 842 in the period 2010–2019; https://www.falsterbofagelstation.se/) while fewer are observed at the Strait of Gibraltar (average of 440 in the period 2009–2012; Onrubia 2016) and at the Bosphorus/Dardanelles (Zalles and Bildstein 2000, Panuccio et al. 2017). Hundreds of individuals are recorded on several Mediterranean islands during September (Agostini and Panuccio 2010, Lucia et al. 2011).

As regards migration phenology, at Falsterbo the peak occurs during the last ten-days of August, at Batumi and in the Mediterranean region in mid September, in Israel in late-September, probably as a result of a latitudinal pattern (Kjellén 1992, Shirihai et al. 2000, Agostini and Panuccio 2010, Vansteelant et al. 2020b, Agostini et al. 2021). As mentioned above, some studies focused on the differential migration in relation to sex and age classes. Surveys made at Falsterbo, showed that juveniles migrate earlier than adult females which, in turn, migrate earlier than adult males. Moreover, juveniles largely outnumbered adults, perhaps as a consequence of the latter being less inclined to follow leading lines, while, among adults, females outnumbered males, perhaps as a result of a tendency towards polygyny in the Swedish population (Kjellén 1992). Conversely, at Batumi and at several Mediterranean watchsites located in Italy (Circeo promontory, Marcellinara Isthmus, Strait of Messina), Malta and Greece (island of Antikythira), adults outnumber juveniles while, among adults, males outnumber females (Agostini and Logozzo 2000, Agostini et al. 2003, Panuccio et al. 2013c, Vansteelant et al. 2019, Agostini et al. 2021). In addition, both at the Strait of Messina and at Batumi, adult males migrate earlier than females, and at the first site adult females slightly earlier than juveniles (Agostini et al. 2021).

The study made by satellite telemetry on birds belonging to the Swedish (eight adult females, six adult males and three juveniles) and Low Countries populations (six adult males), showed that nearly all migrated via France, Spain and North-Western Africa during both spring and autumn. As regards the autumn, juveniles did not cross the Sahara Desert and two wintered in southern Europe. All adults, with the exception of a female wintering in southern Europe, reached Tropical Africa visiting several post-migratory stopover areas. Among Swedish birds, the autumn migration lasted and covered on average 28 days and 4,243 km (excluding pre-migratory movements), with juveniles migrating shorter distance (on average 1,476 km) and slower (on average 136 km day^{-1}) than adults (males = 5,252 km and 167 km day^{-1}; females (= 4,426 km and 173 km day^{-1}), and wintering at higher latitudes. However, among adults, although females migrated shorter distance than males (a female wintered in southern Europe) the difference was not significant neither in length and duration (Standberg et al. 2008). Also recoveries of ringed birds showed the same pattern in relation to the length among adults, while recovery data from Finnish birds, unlike those from the Swedish ones (Strandberg et al. 2009d), did not reveal any difference between adults and juveniles in wintering areas (Panuccio et al. 2013d, Saurola et al. 2013). As regards the stopover behaviour of Swedish tracked birds, juveniles did not stop during migration while adults showed some days of stopover in Europe and North Africa just before or after water-crossings.

Although male-biased counts reported in the Central-Eastern Mediterranean region and at Batumi could reflect male-biased populations in Central-Eastern Europe and in Western Russia, in those surveys juveniles appear underrepresented, especially in the Mediterranean region (Agostini et al. 2021), probably because they migrate shorter distances. Observations made in Central and Southern Europe wetlands showed that most wintering birds are juveniles while, among adults, nearly all surveys reported that females outnumbered males (Zijlstra 1987, Gonzàlez 1991, Clarke et al. 1993, Handrinos and Akriotis 1997, Agostini and Logozzo 2000, Underhill-Day 2002, Panuccio et al. 2005b, Castelijns and Castelijns 2008). Apparently, almost exclusively adult males belonging to resident populations winter in Europe. To explain these results it has been suggested that a substantial number of females, mostly juveniles, might cover shorter distances than males wintering at higher latitudes on the basis of the Bergmann's rule (Agostini and Logozzo 2000, Panuccio et al. 2005b). As a matter of fact, larger females capture larger prey and, probably, are able to tolerate colder temperatures and fast longer than males, better surviving at higher latitudes during the boreal winter. Among adults, this strategy could allow these females to reach their breeding grounds earlier the following spring thus improving their reproductive success mating with an older male and/or being the primary female of a polygynous one. On the other hand, among juveniles, also yearling females arriving earlier in the season could have more chances to mate during their first reproductive season. Although, in agreement with this time minimization hypothesis, the satellite study showed that the spring migration of adult females lasted less than that of adult males (Strandberg et al. 2008b), future research should verify if the same occurs among juveniles also investigating their sex ratio in populations wintering at higher latitudes. Wide wintering areas of this species are also located in Eastern Europe, around North-Western Black Sea such as in the Balkan Peninsula (Ferguson-Lees and Christie 2001, Abdulaze 2013), but there is lack of information about both numbers and proportion of age and sex classes of birds wintering there.

Pre-breeding Migration

Marsh harriers leave their wintering grounds in early-mid March, crossing the Mediterranean between late-March and early April (Strandberg et al. 2008b, Agostini and Panuccio 2010). In this period larger concentrations were reported in the Central Mediterranean region, where they move along more or less parallel flyways such as during autumn (Fig. 1). In particular, at the Strait of Messina on average 2,462 and 2,218 individuals were counted in the periods 1996–2003 and 2006–2016, respectively (max. annual count 3,414 in 2000; Corso 2001a, Brichetti and Fracasso 2018). Hundreds of marsh harriers have been reported at the Cap Bon promontory (Tunisia), over the islands of Marettimo, Ustica, Panarea and Malta (Thiollay 1975, 1977, Agostini 2001, Panuccio et al. 2002, Sultana and Gauci 1982, Mellone et al. 2007, Agostini et al. 2017a) and along the Adriatic coast, en route to the Balkans, at Capo d'Otranto (Gustin and Pizzari 1998, Premuda et al. 2004) and on the Conero promontory (max. 3,465 at the Conero in 2009; Brichetti and Fracasso 2018). At the Strait of Gibraltar migration counts did not report higher numbers of marsh harriers in spring (Finlayson 1992), with an average of only 210 individuals in the period 2009–2012 (Onrubia 2016). In north-western Italy, hundreds of marsh harriers migrate along the Ligurian Appennines apparently using the western flyway, being probably part of the same flow observed at Leucate, located in the southeastern coast of France (Zalles and Bildstein 2000). Among birds tracked by satellite telemetry belonging to the Swedish and Low Countries populations, only the spring migration of adults was considered, since just one juvenile survived during its first winter (Vansteelant et al. 2020a). Although the spring migration patterns concerning duration, travel speed and distance were very similar to that recorded in autumn, these birds showed a narrower front when crossing the Mediterranean between Morocco and Spain, more concentrated in the area of the Strait of Gibraltar, even though visual observations did not show a greater spring than autumn passage at this site (Onrubia 2016). In addition, they showed several stopover in Northern Morocco and Southern Europe just before or after barrier-crossings such as during autumn. On the other hand, longer water-crossings were observed in spring: one via the Balearic Islands, and another one between southern Morocco and Portugal. As mentioned above, in this period the migration of females lasted less than the males' one among Swedish birds (27 vs 46 days) although no significant differences was found both in the distance (4,865 vs 6,166 km) and migration speed (193 vs 139 km day^{-1}; Strandberg et al. 2008b). Nevertheless, females began migrating some days later than males and reached their breeding grounds some days earlier. However, surveys made both in the Central Mediterranean (Gustin and Pizzari 1998, Pandolfi and Sonet 2003, Panuccio et al. 2004, Agostini and Panuccio 2010) and at the Strait of Gibraltar (Finlayson 1992), showed that adult males tend to migrate earlier than females in spring. In addition, in this period a partial overlap in the migration of adults and juveniles has been reported, perhaps the result of the contemporaneous passage of yearlings wintering in the Mediterranean basin and adults wintering in tropical Africa but beginning earlier their northward migration (Agostini 2001).

Behavioural and Ecological Remarks

Studies made by satellite telemetry have shown that migrating marsh harriers change their behaviour in relation to crosswinds showing drift, compensation and overcompensation during different legs of their journeys such as in relation to different crosswind directions, especially in order to avoid being drifted over the ocean (Klaassen et al. 2011, Vansteelant et al. 2020a). In addition, at individual level, they show a stronger repeatability in time than in space due perhaps to a stronger endogenous control of timing and a more flexible choice in routes (Vardanis et al. 2011, 2016). As regards seasonal migration patterns, tracked birds belonging to the Low Countries and Swedish populations showed a clockwise loop migration when migrating through Africa and Southern Europe, leaving a drift effect in response to crosswinds (westerly winds in Europe and easterly winds in Africa) and following a more westerly route in spring than in autumn when the tail wind support is stronger (Klaassen et al. 2010, Vansteelant et al. 2020a). Such loop migration arises in Africa at the onset of spring, when the harriers are markedly drifted to the west by strong easterly winds. As mentioned above, in the Mediterranean region they showed a wider front of migration in autumn, more restricted in the area of the Strait of Gibraltar during spring. Moreover, their migration speed was lower in autumn due to stopover behaviour, shown almost exclusively in Europe and North Africa. Interestingly, during the passage through the Sahara Desert, marsh harriers covered longer daily distances during spring, although migrating under less favourable wind conditions, spending more time flying and more energy probably as a result of a time minimization strategy in this period (Mellone et al. 2012a). Because of the loop migration between Africa and Southern Europe, the total migration distance in autumn among Swedish birds was on average only 1.6% longer than the straight line between breeding and wintering locations, while the spring migration was on average 4.3% longer (Strandberg et al. 2010). Interestingly, Low Countries (six adult males) and Swedish birds (two adult males and two adult females) followed distinct autumn migration corridors, but used the same western corridor to cross the Sahara in spring. Swedish harriers departed and arrived two weeks later than Low Countries in autumn. Conversely, they arrived to their breeding areas around the same time in spring, although Low Countries harriers initiated their spring migration on average three weeks before Swedish harriers. In particular, harriers showed high flexibility concerning departure timing and stopover duration during spring migration, with Low Countries birds making long stopovers in the Sahara and North Africa perhaps due to worst foraging conditions at their non-breeding sites towards the end of the non-breeding season. Once reached Europe, marsh harriers minimized the remaining migration distance with Low Countries and Swedish harriers crossing the Pyrenees on the western and eastern side, respectively (Vansteelant et al. 2020a).

The radar study carried out in Southern Israel during both spring and autumn, showed that in soaring-gliding flight marsh harriers react to different thermal conditions by adjusting their gliding airspeed to the actual climbing rate in thermal circling. As a result cross-country speed is related to climbing rate, and maximized in soaring-gliding flight such as in typical soarers. As regards the flapping flight, it was regularly used predominately soon after sunrise and before sunset, when thermal updrafts are scarce or weak, and more often in opposing than in tail winds. Moreover, their flight altitude above the ground was higher in autumn, probably due to better thermal conditions in this period, and increasing during the afternoon (Spaar and Bruderer 1997).

The radar study made along the coast of Southern Spain in autumn, 140 km north-east of Gibraltar, showed that marsh harriers undertook the long sea-crossing towards Africa (150 km) when they could exploit a good tail wind support, maintaining higher altitudes after the crossing of a mountain chain. Conversely, they tended to follow the coast using soaring-gliding flight with weaker tailwinds, headwinds and side-winds which did not provide a tail support. Unlike birds following the coast, those crossing the sea did not reduce the airspeed with increasing tailwind-support, probably to shorten the crossing time lowering the risk of unpredictable weather changes. When crossing in lateral winds, harriers increased airspeed with increasing sidewind-component to reduce drift and consequently the crossing distance over water (Meyer et al. 2000).

At the Strait of Messina, radar studies were made along the continental slope during both spring (along the coast) and autumn (inland). During spring, marsh harriers flew below the optimal flight altitude with respect to tailwind assistance during the sea-crossing such as European honey buzzards, but passing at higher altitudes. Probably, unlike the latter species, they did not decrease their altitude when flying over water by using the flapping flight to a larger extent (Mateos-Rodriguez and Liechti 2012). A negative relationship between flock size and flight altitude was found during the study made in autumn inland of Strait. It suggested that marsh harriers, unlike more obligatory soaring species (Kerlinger 1989), do not use flocking to locate thermals during migration (Agostini et al. 2017b). This is not surprising when considering that, as mentioned above, harriers are more prone to use flapping flight during migration, showing irregular flocking and mostly flying singly or in small flocks. In this period, the average flight altitude of birds was lower of that reported in southern Israel in the same period but very similar to the altitude reported there in March, suggesting a similar development of thermal convention in the two areas in September (Strait of Messina) and March (southern Israel).

A recent study made at the Strait of Messina has shown that this species suspends moult during autumn migration (Ramirez and Panuccio 2019), with females showing a more advanced moult stage in agreement with the conclusion that in this species they commence moult earlier than males (Kjellén 1992). In his study in southern Sweden, Kjellén suggested that this would explain why adult females also migrated earlier than adult males at Falsterbo. However, as mentioned above, unlike Falsterbo, at the Strait of Messina and Batumi adult males migrated slightly earlier (Agostini et al. 2021). Future research should verify if eventual intersexual differences concerning migration length and speed among marsh harriers belonging to easternmost populations might affect their migration phenology at lower latitudes.

As mentioned above, in a resident population in West-Central France, a substantial proportion of adult males shows a permanent female mimicry, while only 10% look like the typical field-guide male, so much so that this population appears as an outlier. In particular, the large number of adult males showing the grey plumage seen at migration watchsites is in agreement with the conclusion that adult males showing a permanent female-like plumage are very rare elsewhere out of France (Agostini et al. 2021). Therefore, sexual mimicry is rare or absent in completely migratory populations, probably because its adaptive value emerges mainly in males wintering in cold climates, e.g., resident ones, which do not segregate from females and have to compete with them. Interestingly, a study made during the breeding season, revealed that the agonistic behaviour of female-like males is more directed towards females, while they are less attacked by typical males, thus avoiding costly fights and spending more time to access to feed resources (Sternalsky et al. 2012). Moreover, female-like males appear more cooperative than typical ones during the cold season, being more aggressive towards predators on winter roosts (Sternalsky et al. 2010, 2012). These personality traits, perhaps associated with the colour morph (Sternalski et al. 2012), probably make the sexual mimicry more advantageous in cold climates, as those experienced by resident populations during the boreal winter, and this would explain the rarity or the absence of female-like males in long-distance migratory populations, which spend the winter in more favourable conditions.

16

Hen harrier *Circus cyaneus*

Nicolantonio Agostini

ABSTRACT

The hen harrier (*Circus cyaneus*) is a partial migrant, being entirely migratory in Northern, Eastern Europe and Asia, and resident, partial or short-distance migrant in Western Europe. Its wintering grounds are located in Europe, North Africa and in a relatively narrow band between Turkey and Japan. This species regularly hunts during migration, moving on a broad front singly or in very small loose flocks, mostly using the flapping flight. As a result, few birds are detected at raptor migration watchsites. During post-breeding migration, the highest counts have been reported on the Vistula Spit, along the southern coast of the Baltic Sea (Poland), and also slightly lower at Falsterbo, in Southern Sweden. Adults were seen migrating earlier than juveniles, probably under the pressure to get good winter territories. Ring recoveries of Scandinavian birds revealed that they move SSW to reach their wintering grounds mostly located in Central-Western Europe. Few British birds winter in mainland Europe, but some individuals tracked by satellite reached Northern Spain undertaking long water-crossings through the Bay of Biscay. Very few birds cross the Mediterranean to overwinter in North Africa. During pre-breeding movements counts of the same magnitude of those of Falsterbo were reported at Skagen (Denmark). In this period, males migrate earlier than females, probably to acquire the best nesting territories.

Introduction

The hen harrier (*Circus cyaneus* Linnaeus, 1766) is a medium-sized slender raptor showing a high degree of reversed size dimorphism, with an average body mass of 0.43 kg (male: 0.35 kg; female 0.53 kg) and a wing span ranging between 114 and 121 cm (Forsman 1999, Simmons 2000). It shows the widest distribution among Old World harriers, from Atlantic to eastern Asia, nesting above the Arctic Circle until the Northern Kirghiz Steppes, North Mongolia, Manchuria, Ussuriland, Sakhalin, Kamchatka and westernmost Chinese Turkestan (Ferguson-Lees and Christie 2001). Recent population estimates reported a world population of 100,000–499,999 breeding individuals (Status: Least Concern; BirdLife International 2020). This species is entirely migratory in Northern, Eastern Europe and Asia, and is a resident, partial or short-distance migrant in Western Europe. Wintering areas are widely distributed through Europe, North Africa, Turkey, Syria, Lebanon, Israel, Iraq, Iran, Indomalayan region, Southern China, North Laos, Japan and Korea (Ferguson-Lees and Christie 2001).

MEDRAPTORS (Mediterranean Raptor Migration Network), Rome, Italy.
Email: nicolantonioagostini@gmail.com

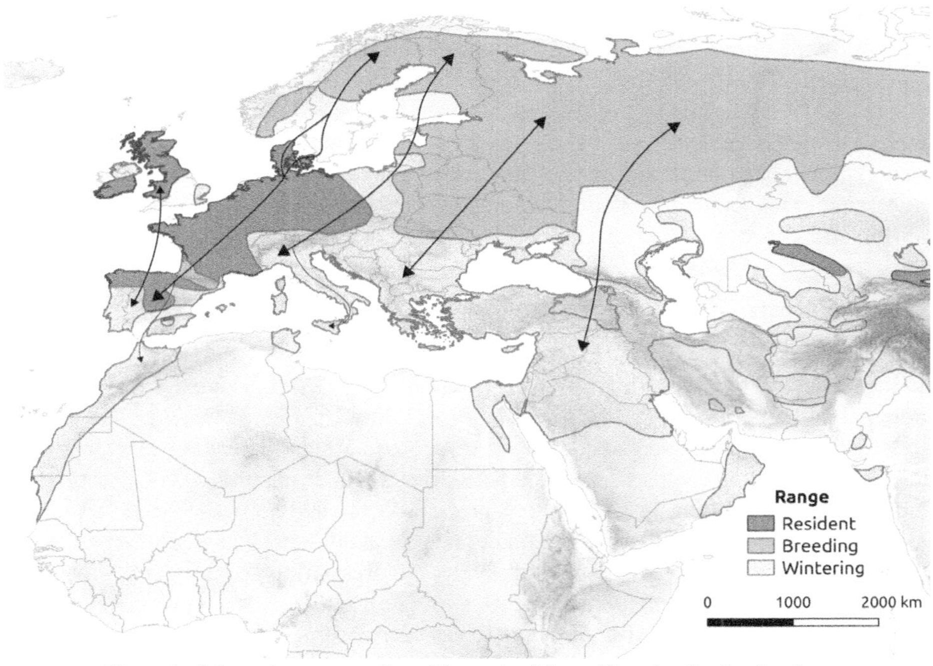

Figure 1. Schematic representation of the routes followed by migrating hen harriers.

The European population is estimated at 30,000–54,400 breeding females, which equates to 60,000–109,000 mature individuals, showing rapid declines in 2000s, mostly due to the transformation of habitat, after the previous one in the DDT era (Bildstein 2006, BirdLife International 2017a). This species does not show strong breeding site fidelity, needing open areas with low ground cover to hunt, catching mostly small mammals and birds, but also frogs, snakes, lizards and insects. The breeding season lasts from early April to mid July, but sometimes beginning late February and ending in late August, with clutches usually of four or five eggs. The nest is built on the ground by using dead vegetation, but sometimes it has been recorded nesting in trees. Although the hen harrier nests usually in natural habitats, a recent increase in the number of pairs nesting in cereal crops has been reported. Hen harriers defend their nesting territory, but with males having larger home ranges than females, feeding up to 10 km from the nest (Arroyo et al. 2009). Only the female incubates (29–37 days for each egg), consequently the male provides the food in this period such as during the first few weeks of hatching. The young remain in the nest for 31–35 days. Polygyny is quite common in this species, mostly related to the abundance of food, especially to the vole cycle, with the dominant female generally rearing a larger brood. Although a polygynous male usually pairs with two or three females, there are cases of males pairing with up to six. The harem often forms a loose colony (Watson 1977, Balfour and Cadbury 1979, Gensbøl 2008, Hardey et al. 2013).

The hen harrier migrates on a broad front (Fig. 1), and very low numbers are recorded at traditional watchsites (Zalles and Bildstein 2000). It regularly hunts during migration and thus avoids intraspecific competition for food, migrating singly or sometimes in very small loose flocks. In so doing hen harriers mostly use flapping flight during migration moving low to the ground and passing often undetected. Conversely, they usually form communal roosts during the winter (Bildstein 2006).

Knowledge Status

Studies on the hen harrier migration have been carried out by direct visual observation, radio tracking, satellite telemetry and ringing recoveries. As mentioned above, because of their migration habits, very low numbers are recorded at raptor migration watchsites. Highest counts during post-breeding migration have been reported along the Vistula Spit (Poland) and at Falsterbo (Sweden), while during return movements Skagen (Denmark) and Batumi (Georgia) are the most important watchsites (Kjellén 1992, 1998, Bildstein 2006, https://trektellen.nl/site/yeartotals/1047/2020/-1). The movements of tens of British birds have been studied by radio tracking and satellite telemetry, but detailed analyses have not yet been published (https://www.gov.uk/government/publications/hen-harriers-tracking-programme-update/hen-harrier-annual-tracking-update). As regards ringing recoveries, studies involved mostly birds belonging to the population of Northern Europe, in particular those of Finland, Sweden and Britain (Etheridge and Summers 2006, Saurola et al. 2013).

Migration Patterns

During post-breeding movements, significant numbers of hen harriers migrate along the southern coast of the Baltic Sea passing along the Vistula Spit (Poland), where on average 633 birds per season were counted in the period 2008–2011 (maximum 915, median date 24 October; Polakowsky et al. 2014). A weaker passage has been recorded at Falsterbo, in Southern Sweden. Here, on average 239 individuals were counted each season in the period 1998–2003 (maximum 379), seen from August until November, and peaking in late September early October (Kjellén 1992, Bildstein 2006). At the site, adult females migrate earlier than adult males, and adult males earlier than juveniles. Thus, adults migrate earlier than juveniles probably under the pressure to acquire good winter territories. Among adults, females outnumber males, perhaps as a result of a tendency towards polygyny in this species (Kjellén 1992). A female-biased sex ratio has been recorded in the breeding population of the Orkney Islands (Balfour and Cadbury 1977). At Falsterbo, the proportion of juveniles greatly varies among years, probably as a consequence of rodent numbers in the northern breeding range of the species (Watson 1977, Kjellén 1998). As expected, in years of poor breeding success an earlier passage has been recorded (Kjellén 1992). In addition, a significant passage, averaging 200 birds, has been recorded in Southern Finland, at the Porkkala Peninsula (Zalles and Bildstein 2000). At watchsites located at lower latitudes, numbers of hen harriers rarely reach a few dozen (Zalles and Bildstein 2000, Shirihai et al. 2000), while a substantial passage has been reported at the three watchpoints in the Western Pyrenees, Organbidexka, Lindux and Lizarrieta, where on average a total of 332 birds were counted in the period 2009–2011 (de Juana and Garcia 2015). Conversely, very few (< 10 per season) have been observed at the Strait of Gibraltar in recent surveys (Onrubia 2016), although a maximum of 114 birds was reported in 1974 (de Juana and Garcia 2015). Since its migratory flow is virtually non-existent through the Central Mediterranean (Agostini 2005, Sammut et al. 2013), very few are expected to winter in North Africa. A long term survey at Organbidexka, made between 1981 and 2008, showed a delayed migration of this species probably due to climate change, while ringing recoveries data suggest a significant decrease in distances covered (Filippi-Codaccioni et al. 2010).

The pre-breeding movements largely retrace post-breeding ones. As a matter of fact, higher counts involve birds breeding in Scandinavia, which concentrate at Skagen (Northern Denmark); there, on average 201 hen harriers were counted each season in the period 1972–1998 (maximum 548 in 1975), peaking mid April-early May (Zalles and Bildstein 2000, Bildstein 2006). In Denmark, a significant pre-breeding passage has also been reported along a southernmost flyway, at Rørvig (maximum 181 in 1991) and Hellebæk (maximum 163 in 1990). Finally, at Batumi (Georgia) 218 hen harriers were counted in spring 2020 (https://trektellen.nl/site/yeartotals/1047/2020/-1). In the Mediterranean region, a few dozen are observed along the south-eastern coast of France (e.g., at Leucate; Zalles and Bildstein 2000) and at the Strait of Messina, between eastern Sicily and the "toe" of the Italian Peninsula, the latter being mostly birds wintering in Sicily (Corso 2001a, Brichetti and Fracasso 2018, Agostini unpub. data). These birds are expected to reach the island in late autumn, since observations made at the Strait of Messina between mid August and mid October did not report any passage (Cento et al. 2018, 2019). Unlike post-breeding migration, males precede females in spring perhaps under the pressure to acquire the best nesting territories (Bernis 1980, Finlayson 1992, Bildstein 2006).

Studies based on ringing recoveries, mostly involving nestlings, revealed that birds belonging to the populations breeding in Sweden and Finland migrate heading SSW along more or less parallel pathways, with some birds recovered at over 2,000 km (Fig. 1; Saurola et al. 2013). They mostly winter in Central-Western Europe, with birds found in the Netherlands, France, Northern Spain and Northern Italy. This pattern of migration has been confirmed by ringing recoveries in Spain, where the wintering population includes birds from France, Britain, Germany and Finland (de Juana and Garcia 2015). In addition, birds ringed in Finland, Germany and Czech Republic have been recovered in Italy (Spina and Volponi 2008). Some recoveries of Finnish birds in Poland, Belarus, Russia and Latvia suggest that some of them winter also in Eastern Europe. Both ringing recoveries and satellite tracking data of British birds, show that a minority crosses the English Channel to winter in the Continent (Etheridge and Summers 2006). Interestingly, ringing recoveries showed that a higher percentage of first-year males in the East Highlands covered longer distance than females, and the records in mainland Europe involved mainly first-year males. In autumn, the shortest recovery distances were belonging to adult males. Finally, some Dutch birds were recovered in England (Clarke and Watson 1990). Some satellite tracked British individuals spent the winter in the Iberian Peninsula, undertaking long sea-crossings *via* northwestern France through the Bay of Biscay, confirming that for these birds, such as for other harriers, water surfaces are not insurmountable ecological barriers during migration (https://www.gov.uk/government/publications/hen-harriers-tracking-programme-update/hen-harrier-annual-tracking-update).

Behavioural and Ecological Remarks

A few hen harriers breeding in Scandinavia winter in Southern Sweden. Kjellén (1994) compared age and sex ratio in birds migrating and wintering there, showing that the proportion of adults was higher among wintering birds than among migrants. In addition, among adults, more females were seen among wintering birds. These results are in agreement with the social-dominance hypothesis which suggests that the dominant adults and the larger sex (in raptors the female) generally winter furthest north. They are also in agreement with the Bergmann's rule since larger females are able to tolerate colder temperatures and fast longer than males, better surviving at higher latitudes during the boreal winter. However, since this species does not overwinter in tropical Africa, both males and females have to cope with stressful environments during the cold season in large areas of their wintering ground.

As expected, in this species the proportion of juveniles recorded among migrants at Falsterbo is correlated with the annual fluctuations in rodents availability in the breeding grounds (Kjellén 1994, 1998).

17

Pallid harrier *Circus macrourus*

Julien Terraube,[1,*] *Francois Mougeot*,[2] *Ari-Pekka Auvinen*[3] and *Beatriz Arroyo*[2]

ABSTRACT

Pallid harriers (*Circus macrourus*) are long distant migrants that breed from Europe to Central Asia. This chapter reviews and discusses current knowledge on the migration of this species. Information coming from satellite tracked birds breeding in Central Asia (core breeding populations located in Kazakhstan) revealed a migratory divide, as birds migrated either to India or to Africa, over a strikingly broad wintering range. Spring migration is faster and more direct than autumn migration and important migration bottlenecks have been identified along the Black Sea coast (Georgia) and in Middle East countries. For more western breeding populations, important migration bottlenecks include the Middle East, and the Strait of Messina, with a possible loop migration. Increasing reports of migrating pallid harriers in Western Europe coincide with the recent establishment of a new breeding population in Finland that migrates to Africa through Western Europe (France, Spain).

Introduction

The pallid harrier (*Circus macrourus* Gmelin, 1770) is a long-distance migrant whose breeding populations extend from Western Europe to China (del Hoyo et al. 1994). Central Asia (North-Central Kazakhstan and South-Western Siberia) represents the core of the breeding distribution of pallid harriers worldwide (Fig. 1). Its main wintering grounds are open grasslands and agricultural areas throughout the Indian subcontinent, the Sahel belt in Africa, and the eastern African savannahs (Limiñana et al. 2015). The pallid harrier is included in Annex I of the European Birds Directive (2009/147/CE) and is categorized as Near Threatened on the IUCN Red List (BirdLife International 2018a) as a result of population declines during the twentieth century. It is a medium-size raptor whose flight pattern is characterized by flapping-gliding flight, with a high proportion of flapping and low gliding performance, which allows them to migrate under unfavourable thermal and wind conditions, and have broad-front migration (Spaar 1997, Spaar and Bruderer 1997).

[1] Global Change and Conservation Lab, Organismal and Evolutionary Biology Research Program, Faculty of Biological and Environmental Sciences, University of Helsinki.

[2] Instituto de Investigación en Recursos Cinegéticos (IREC, CSIC-UCLM-JCCM), Ronda de Toledo 12, 13005 Ciudad Real, Spain.

[3] Finnish Environment Institute, Oulu Office, P.O. Box 413, 90014 University of Oulu, Finland.

* Corresponding author: jterraube@gmail.com

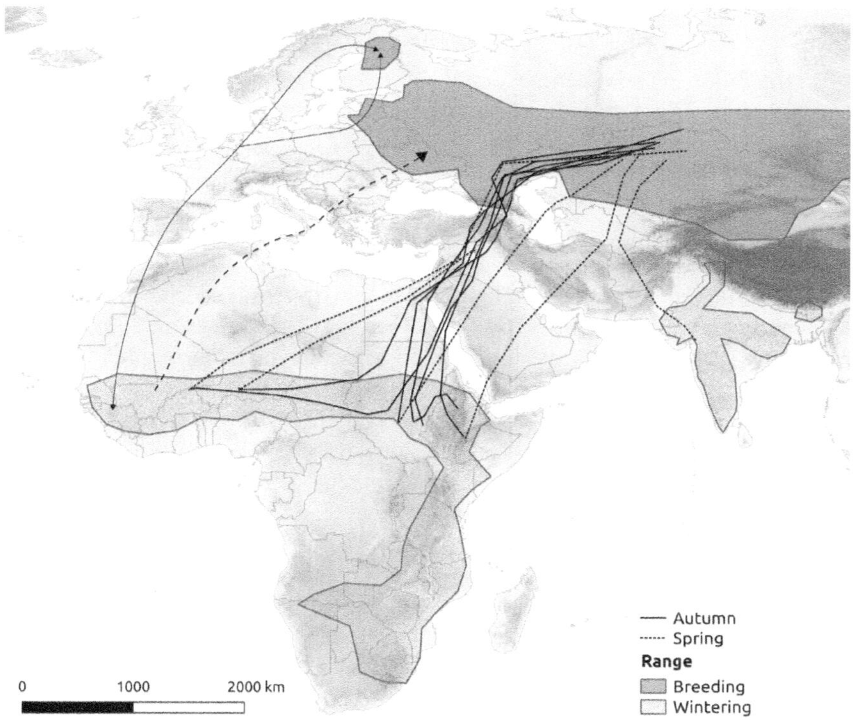

Figure 1. Post-breeding (solid lines) and pre-breeding (dotted lines) migration routes of pallid harriers identified using satellite-tracking data (taken from Terraube et al. 2012a). The dashed line shows the schematized pre-breeding route through the Central Mediterranean, and the thin line shows the approximate routes of birds trapped in Finland and marked with GPS tags (http://www.luomus.fi/en/pallid-harriers).

Knowledge Status

Knowledge on pallid harrier migration to date is based on observations at bottleneck migration points (Corso and Cardelli 2004, Panuccio and Agostini 2006c), reports by birdwatchers during migration movements in France/Spain (Ollé et al. 2015, Réglade et al. 2015), ringing recovery data (Terraube et al. 2012a) and satellite tracked birds from populations breeding in Kazakhstan (Terraube et al. 2012a) and Finland (unpublished data; http://www.luomus.fi/en/pallid-harriers).

In the following sections, we first summarize available information on the migration ecology of Asian pallid harrier populations. We then highlight the existence of Central Mediterranean migratory routes associated to westernmost Russian breeding populations and provide preliminary evidence on a new migration route arising from the establishment of a breeding population of pallid harrier in Northern Europe.

Migration Patterns of Asian Populations

An analysis of ringing and recovery data of pallid harriers in and from Kazakhstan, obtained from the Institute of Zoology, Almaty (KZ), suggested that central Asian pallid harriers used two alternative migration routes: a 'western' route to the African wintering grounds, which was evidenced by three ring recoveries in Central Kazakhstan from birds ringed in Tunisia and France, and an 'eastern' migratory route linking Central Kazakhstan to the Indian wintering grounds, which was evidenced by two ring recoveries in South-Eastern KZ, and one in Northern India (see Fig. 1 in Terraube et al. 2012a).

This inter-continental migratory divide was confirmed by remote sensing using data from 6 birds tagged with satellite transmitters on their breeding ground in North-Central Kazakhstan in 2007–2008 and one first-year male tagged on its wintering area in India in 2008 (Terraube et al. 2012a). The six individuals tagged in North-Central Kazakhstan migrated towards Africa, whereas the bird tagged in India spent the following summer in Central Kazakhstan, ca. 300 km east of the breeding sites of the other tagged birds. These data therefore suggest the coexistence of distinct migratory strategies within the core breeding population, and that birds from one breeding area may winter over a strikingly broad range not only across continents (i.e., India or Africa), but also within continents: the wintering areas of the 6 birds migrating to Africa had a longitudinal spread of c. 4,500 km, as two of them spent the winter in Ethiopia, two others in Sudan, and the last two other birds arrived to Sudan but then shifted to Central-Western Africa and settled at the Burkina Faso—Niger border and in Niger (Fig. 1). Pallid harriers seem to be able to shift between different wintering sites to track local variations in resource availability (small passerines, rodents or locusts) (Limiñana et al. 2015).

In terms of migratory routes, the bird tagged in winter in India migrated through Pakistan, Afghanistan, Turkmenistan and Uzbekistan, to spend the summer in Central Kazakhstan (Fig. 1). The six individuals tagged in North-Central Kazakhstan migrated west of the Caspian Sea, through Russia, Georgia, Armenia, and then crossed countries of the Middle East (Turkey, Syria and Iraq). Two to three of these birds flew a route near the Batumi region, along the Georgian (east) coast of the Black Sea when crossing the Caucasus area. In autumn, tagged birds travelled an average distance of 7,340 ± 1,378 (SD) km, at an average speed of 108 ± 67 km. day^{-1}. A pre-migratory staging area, used during up to a month, located north to the Kazakh breeding areas, was detected in various birds (Terraube et al. 2012a). The mean duration of the autumn migration including this period was 83 ± 31 days. Subsequently, some birds used a more "terrestrial" route crossing Israel, Jordan and Egypt, whereas others crossed Saudi Arabia, then the Red Sea, Egypt and Eritrea (Fig. 1).

Spring migration routes towards Kazakhstan were more direct, and marked inter-individual variations were observed, mainly dependent on the location of wintering areas (see Fig. 1, and Terraube et al. 2012a). The two birds wintering in Western-Central Africa came back via Egypt after a northward trip through the Sahara Desert, thereafter using a similar route to those taken during the autumn. The two birds that wintered in Eastern Africa (Sudan and Ethiopia) crossed the Red Sea and Saudi Arabia, one crossing the Caspian Sea and the other passing east of it (and not west, as the birds did during the autumn migration), coming close to the migration route of the bird wintering in India (Fig. 1, and Terraube et al. 2012a). Spring migration duration was therefore shorter and faster: the average distance travelled was 6,888 ± 1,878 (SD) km and tagged pallid harriers showed an average speed of 258 ± 105 km day^{-1}. Spring migration was also more direct with no stopover recorded, and lasted an average of 28 ± 5 days (Terraube et al. 2012a).

An interesting point arising from the satellite-tracked birds in Asian countries is that half of the signal losses (either because of failure or bird mortality) occurred on the wintering areas and during migration. A large proportion of the tagged birds from central Kazakhstan transit through migration corridors in Georgia, along the Black Sea coast and in other Middle East countries (van Maanen et al. 2001), including the Batumi area in Georgia where an average of 1,491 ± 612 (SD) pallid harriers have been counted each year during autumn migration during the period 2011–2018 (Vansteelant et al. in press). At this migration bottleneck area, a widespread capture and shooting of raptors, including harriers, have been reported (Sándor et al. 2017, Sándor and Anthony 2018). The pallid harrier migration count data point out the critical importance of this watch site for Central Asian pallid harrier populations wintering in Africa, but the extremely high variability of annual counts does not allow to detect demographic trends, as they migrate on a broad front and cross the Caucasus as far east as Azerbaijan (Heiss and Gauger 2011, Terraube et al. 2012a). Recent research estimated that 100–400 pallid harriers are killed every year in Georgia (Brochet et al. 2017). These authors also highlighted pallid harrier as one of the six globally threatened bird species being most strongly impacted by illegal killing in Europe as a result of widespread poaching in Caucasian countries. Although comprehensive information on current shooting and trapping levels on migratory birds in Middle East countries is currently lacking, a report also indicated that migratory birds face considerable hunting and trapping pressure in Syria (Attar 2005), where the signal of one of the pallid harriers tagged in North-Central

Kazakhstan was also lost (Terraube et al. 2012a). Priority should be given to assessing the sensitivity of the pallid harrier and other raptors' populations to illegal hunting along these migration routes. Analysis of contaminants in breeding pallid harrier from North-Central Kazakhstan highlight a potential exposure to DDTs on wintering areas in Africa or India, although levels were overall low and below harmful levels (see Espin et al. 2018 for more details).

Migration Patterns of European Populations

The Central Mediterranean is one of the most important spring European flyway for this species. Birds cross the Sicily Channel on a broad front between March and May, reaching the Italian Peninsula through the Tyrrhenian Sea (Panuccio and Agostini 2006c), or via the Strait of Messina where the highest concentration was reported with more than 100 individuals per year since 1998 (Corso and Cardelli 2004). Subsequently, pallid harriers mainly cross the Adriatic Sea, especially from Capo d'Otranto, en route to the Balkans and East European countries (Premuda et al. 2004). Despite polygyny is not known in this species (Simmons 2000), adult males, which migrate earlier, are greatly outnumbered by females (roughly 1:2; Corso and Cardelli 1994, Panuccio and Agostini 2006c). Immatures migrate later in the season, showing that unlike other long-distance migratory raptors they come back to the breeding range already during their second calendar-year. Other birds chose to reach Europe through the Aegean Sea and the Bosphorus (eBird 2019). Autumn passage in the Central Mediterranean is virtually absent (Agostini and Logozzo 1997, Panuccio et al. 2005a, Sammut et al. 2013). This suggests that this species might perform a loop migration in certain areas of its distribution. It is probably linked to shorter spring migration routes and large-scale movements in African wintering areas due to spatial variations in food abundance. The individuals passing through the Central Mediterranean area are probably those wintering on the westernmost area of the wintering range (Burkina-Faso, Mali, Niger), and may be those breeding on the westernmost area of the traditional breeding range (European Russia, Ukraine), although for a species with such great breeding dispersal, they could include also birds breeding further East.

Israel is a minor migration bottleneck for pallid harriers in the Middle East, with dozens of individuals recorded every spring in Eilat (Shirihai and Christie 1992). The species is scarcer in autumn at Eilat but an average of 35 birds per year has been recorded in Northern Israel, from the Northern Valleys/Kfar Qassem surveys (Alon et al. 2004). Annual totals for Israel have not changed significantly between 1990 and 2014 (Perlman 2015).

In addition to the Central Mediterranean migratory routes, increasing numbers of pallid harriers have been observed during migration and wintering periods in Western Europe in the last 15 years. In France, records of pallid harriers increased abruptly since the early 2000's (Réglade et al. 2015). Ollé et al. (2015) highlighted a similar increase in the total number of pallid harrier records in Spain since 2008. Figure 2 shows these data, together with data updated for Spain using the rare bird reports for the years 2010–2014 (Gutiérrez et al. 2011, 2013, Copete et al. 2015, Gil Velasco et al. 2017). This suggests an increasing breeding population of pallid harriers in Europe, or a shift in the migration routes from the Asian populations.

The first scenario seems to be more likely given that, at least for Spanish observations in the time period 2010–2014, the sharpest increase in observations relates to birds seen during autumn migration (Fig. 3), discarding potential Asian migrants performing loop migrations in spring (Corso and Cardelli 2004). Additionally, an increasing number of birds has been reported overwintering in France and Spain in the last decade (Fig. 3). These changes are concomitant with the expansion, since 2011, of a small population of pallid harriers breeding in Finland (Saurola 2012). First breeding events have also been recorded in the Netherlands in 2017 (Henry 2018) and in Spain in 2019 (Jubete and Mougeot 2019).

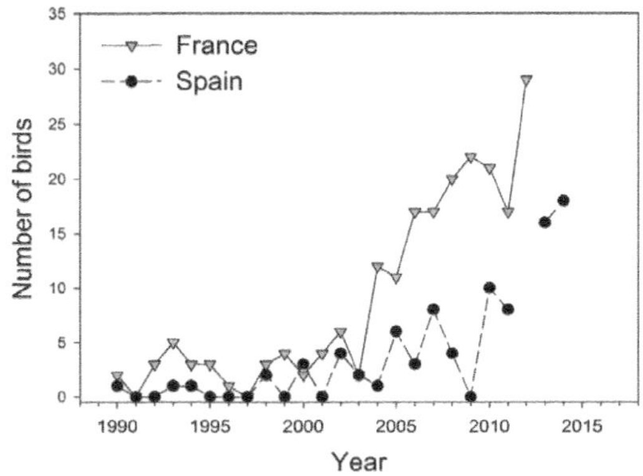

Figure 2. Observations of pallid harriers in France and Spain certified by the rare bird committees of France in 1990–2012 (Réglade et al. 2015, Ollé et al. 2015) and of Spain in 1990–2014 (Ollé et al. 2015, Gutiérrez et al. 2011, 2013, Copete et al. 2015, Gil Velasco et al. 2017).

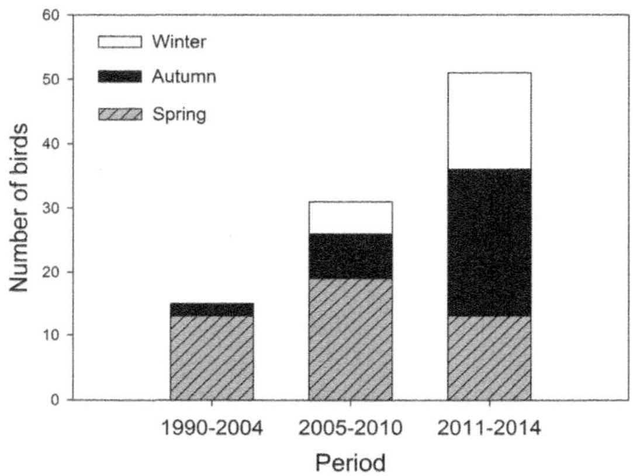

Figure 3. Changes over time in the number of pallid harriers observed in Spain during the spring migration (April–May; grey striped), autumn migration (August–November; black bars) and during winter (December–January; pale white bars). Data based on the rare bird reports for Spain of the Ardeola journal (Gutiérrez et al. 2011, 2013, Copete et al. 2015, Gil Velasco et al. 2017).

In Finland, most breeding events have been recorded in the north-west of the country (Oulu region). Local numbers of breeding pairs fluctuate in relation to small rodent abundance in the same way as Asian populations (Terraube et al. 2012b). However, contrary to the core Asian populations, Finnish pallid harriers settle mainly in clear-cut areas located within patches of boreal forest fragments and agricultural fields. Satellite tracking of Finnish pallid harriers indicates that individuals from this northern population follow a migration route connecting Northern Europe to Western Africa through France and the Iberian Peninsula (Auvinen et al. in prep.), similar to the migration route of Montagu's harriers breeding in Western Europe (Trierweiler et al. 2014; see Fig. 1). Autumn records of birds observed in France and Spain suggest the existence of two migratory routes for European birds, an 'eastern' route through the Rhône Valley, Catalonia and following the Mediterranean coast, and a more 'western' route, following the French Atlantic coast, Navarra and crossing the whole Iberian plateau down to Gibraltar (Ollé et al. 2015, Réglade et al. 2015). Further satellite tracking data should confirm this pattern.

Behavioural and Ecological Remarks

Recent studies conducted throughout the distribution range of the pallid harrier suggest the existence of diverse individual migratory strategies within and between populations and highlight the plasticity of this nomadic species: (1) migratory connectivity was reported between breeding populations from Central Asia and wintering areas in India and Africa; (2) a population recently settled in Northern Europe seems to use a 'classical' western migration route connecting Northern Europe to Western Africa. Further research is needed to understand how the use of different migration routes and wintering areas might impact large-scale population dynamics in this species and explain the recent colonization of new breeding territories at the western limit of the pallid harrier distribution range.

18

Montagu's harrier *Circus pygargus*

Almut E. Schlaich[1,*] and *Raymond H.G. Klaassen*[1,2]

ABSTRACT

The Montagu's harrier (*Circus pygargus*) is a medium-sized long-distance migratory raptor with a southwest-Palearctic breeding distribution and an Afrotropical/Indomalayan wintering distribution. Their long and slender wings allow them to alternate between flapping-gliding and soaring-gliding flight. This makes Montagu's harriers less dependent on thermals compared to obligate soaring migrants, which has several consequences for their migration patterns. (1) Being a facultative soaring migrant allows Montagu's harriers to cross larger water bodies, resulting in broad front migration. Consequently, they are observed in comparatively small numbers at migration bottlenecks and watchpoints. (2) Furthermore, Montagu's harriers can use a relatively broad daily time window for travelling, and hence achieve high daily travel distances. Montagu's harriers, however, do not fly the whole day but frequently interrupt their flights for foraging and resting. Tailwinds have a large effect on daily travel distances not only because of the wind support itself but also because Montagu's harriers interrupt their flights less often, and continue to fly for a longer time period on days with tailwinds. Variation in weather accounts for almost all variation in travel distances among regions and between seasons.

A large satellite telemetry and GPS-tracking dataset is available for Montagu's harriers breeding in Europe. This revealed three main migration routes: a western route via Spain, a central route via Italy and an eastern route via Greece. Montagu's harriers generally follow the same route they used in autumn back in spring, although spring migration routes tend to lay (slightly) west of autumn routes. Only Eastern European breeding birds follow a clockwise loop by taking the eastern route via Greece in autumn and the central route via Italy in spring. These loop migration patterns are caused by dominating eastern winds in the Sahara Desert. Migration speeds are similar between routes but are lower in spring (114–162 km/d) compared to autumn (173–187 km/d) because of more and longer stopovers in north-west Africa during spring.

Introduction

The Montagu's harrier (*Circus pygargus* Linnaeus, 1758) is a medium-sized long-distance migratory raptor with a southwest-Palearctic breeding distribution and an Afrotropical/Indomalayan wintering

[1] Dutch Montagu's Harrier Foundation, PO Box 46, 9679ZG Scheemda, The Netherlands.
[2] Conservation Ecology Group, Groningen Institute for Evolutionary Life Sciences, University of Groningen, PO Box 72, 9700AB Groningen, The Netherlands.
* Corresponding Author: almut.schlaich@grauwekiekendief.nl

distribution (Ferguson-Lees and Christie 2001). The European population is estimated by BirdLife International at 54,500–92,200 breeding pairs which represent only 11–35% of the global population (BirdLife International 2017a). The global estimate of 266,000 to 499,000 mature individuals (BirdLife International 2017a) is thus largely based on highly uncertain numbers of breeding pairs in Russia. Ferguson-Lees and Christie (2001), however, estimated the global population at 60,000–71,000 breeding pairs (thus 120,000–142,000 mature individuals which is half of the BirdLife International estimate) and the European population at 9,800–15,000 breeding pairs. Based on the BirdLife International estimates, the IUCN categorizes the species as Least Concern (BirdLife International 2017a), even though Montagu's harriers are declining and red-listed in many European countries including key strongholds of the species in France and Spain.

In recent times, Montagu's harriers have shifted from breeding in natural habitats like dunes and moors towards breeding in agricultural crops all over Europe. This makes the species dependent on protection measures, as chicks often do not fledge before harvesting of the crop (Arroyo et al. 2002). In addition, food abundance might limit breeding populations which can be seen for example in the extreme dependence of some populations on a single main prey species. For example, in the Netherlands, Montagu's harriers strongly rely on the common vole (*Microtus arvalis*) with laying date, clutch size and annual population growth rate positively related to vole abundance (Koks et al. 2007). Population fluctuations in France have also been shown to be related to vole abundance (Millon and Bretagnolle 2008). This dependence on small mammals is related to the low abundance of alternative prey species, which is mainly the case in modern intensive European agricultural landscapes (Terraube and Arroyo 2011). In general, birds are the main prey of Montagu's harriers in most of its distribution range and diet is much more diverse in landscapes with higher availability of alternative prey (Terraube and Arroyo 2011).

During winter, European Montagu's harriers spend about six months in the Sahel (Trierweiler and Koks 2009). Satellite tracking has revealed that they are itinerant and use on average four distinct wintering sites that are located progressively further south and to which individuals are site-faithful in consecutive years (Schlaich 2019, Trierweiler et al. 2013). Consecutive wintering sites are on average 200 km apart and Trierweiler et al. (2013) proposed that these mid-winter movements are related to grasshopper availability. Montagu's harriers prefer open landscapes with mosaics of savannah and cropland (Augiron et al. 2015, Limiñana et al. 2012b, Trierweiler et al. 2013). In winter, Montagu's harriers are acridivorous and mainly feed on resident grasshopper species (Mullié 2009, Mullié and Guèye 2010, Trierweiler and Koks 2009, Trierweiler et al. 2013). They aggregate on communal roosts with several up to hundreds or even thousands of individuals (Cormier and Baillon 1991, Trierweiler and Koks 2009). Very large communal roosts have also been reported from India (Clarke et al. 1998, Ganesh et al. 2000, Verma 2005). Montagu's harriers declined significantly between 1969–73 and 2003/2004 in their Sahelian wintering range by 73% outside protected areas and 56% inside protected areas (Thiollay 2006b). Even though the decline is less steep in protected areas, the surface area of these areas is small and their protection status in most cases not guaranteed. The fact that there are few protected areas, and that Montagu's harriers often prefer agricultural habitats, results in most of the harriers residing outside protected areas (Limiñana et al. 2012b). Ongoing habitat deterioration and destruction due to the ever-increasing human population pressure diminishes the value of protected and unprotected areas ever more in the future. Even though there are some regions in West Africa that seem to be used by greater numbers of individuals, the vast extent of the wintering range of European Montagu's harriers seems to make the implication of protection measures nearly impossible (Limiñana et al. 2012b).

Knowledge Status

The Montagu's harrier is a broad-front migrant which is consequently observed in comparatively small numbers at migration bottlenecks and watchpoints. Therefore, migration counts, coming mainly from locations around the Mediterranean and the Black Sea, provide only limited information on the migratory behaviour of this species. In the Western Mediterranean, where many raptor species concentrate at the Strait of Gibraltar, Spain, to cross at the narrowest point, only ~ 1,300 Montagu's harriers are estimated to pass during autumn migration compared to ~ 59,000 European Honey-buzzards (Onrubia 2016). In general, more Montagu's harriers are counted during autumn than during spring migration at the Strait of Gibraltar (García and Arroyo 1998). Between 2009 and 2012, on average 498 Montagu's harriers were recorded in autumn and 178 in spring (Onrubia 2016). In the Central Mediterranean however, significant passage is observed during spring (Corso 2001a, Panuccio and Agostini 2006c). Corso (2001a) observed in 1996–2000 on average 463 individuals with peak passage in mid-April at the Strait of Messina, Italy. During spring counts in 2002 and 2004, 126 individuals were counted over Ustica and 143 at the Strait of Messina, respectively (Panuccio and Agostini 2006c). At Capo d'Otranto, Italy, 88 Montagu's harriers were observed in April 2003, mainly en route to the Balkans (Premuda et al. 2004). In autumn however, fewer individuals pass at the Strait of Messina, e.g. 49 and 38 Montagu's harriers were counted in 2018 and 2019, respectively (Cento et al. 2018, 2019). Based on these observations, it was suggested that Montagu's harriers perform a counter-clockwise loop migration, concentrating at the Strait of Gibraltar in autumn and crossing the Central Mediterranean during spring migration (Agostini and Logozzo 1997, García and Arroyo 1998).

This hypothesis could only be contradicted when small enough tracking devices became available to track individual Montagu's harriers throughout their annual cycle. From 2005 onwards, satellite tracking revealed the most important routes for several European breeding populations and their wintering areas (Limiñana et al. 2007, Limiñana et al. 2012c, 2013, Trierweiler et al. 2007, 2014; Fig. 1). Spanish individuals crossed the Mediterranean Sea in the Strait of Gibraltar area in autumn and returned via the same route in spring to their breeding areas, only moving slightly more westerly above Africa in spring (Limiñana et al. 2012c). Individuals from North-Western European breeding populations (Netherlands, Denmark, North-Western Germany) migrated mainly on a western route via Spain (78%) and to a smaller extent on a central route via Italy (22%). In spring they also returned along the same routes. Individuals from central breeding populations (North-Eastern Germany) used the central route or a more eastern route via Greece in autumn and spring. Individuals from Eastern European breeding areas (Eastern Poland, Belarus) migrated along the eastern route in autumn and returned mostly via the central route in spring (Trierweiler et al. 2014). These individuals thus make a clockwise loop migration which is the most likely explanation of the higher numbers observed in the Central Mediterranean in spring. However, very few have been reported from visual counts in Greece, during autumn (Lucia et al. 2011), probably because the migratory flow is quite broad (Fig. 1). The likely reason of this large-scale loop are prevailing easterly winds over the Sahara Desert.

Much less is known for Montagu's harriers breeding further east and wintering in Eastern and Southern Africa. In the Middle East during counts in the 1970s till 90s, only 73 ± 69.7 (SD) male Montagu's harrier were observed at Kfar Qassem/Northern Valleys in autumn and 6 ± 9 and 18 ± 15.5 at Eilat in autumn and spring, respectively (Shirihai et al. 2000). However, substantial numbers of Montagu's harriers are counted every autumn near Batumi, Georgia. On average, yearly, 4436 non-juveniles and 2491 juveniles passed through this bottleneck between 2011–2018 (Vansteelant et al. 2020b). These are supposed to be Russian breeding birds which winter in Eastern and Southern Africa. Interestingly, much less individuals are observed during spring migration (180 in 2020; https://trektellen.nl/site/yeartotals/1047/2020/-1), suggesting that the birds seen during autumn use another route in spring.

In the meantime, a large dataset on the movements of Montagu's harriers tracked by satellite telemetry and GPS-tracking devices has accumulated (Figs. 1 and 2). This includes tracks from Montagu's harriers from Europe (UK, Netherlands, Denmark, Germany, France, Poland and Belarus) as well as from India. The map clearly shows that we still lack tracking data from the birds breeding in Russia, which are presumed to spend the winter in Eastern and Southern Africa. First tracks from Montagu's harriers wintering in India show that these birds originate from breeding areas in Kazakhstan. These birds circumvent the Himalaya mountain ranges in a large westerly loop in autumn as well as in spring (Figs. 1 and 2). Since tracking data is lacking from Russia and Central-Western

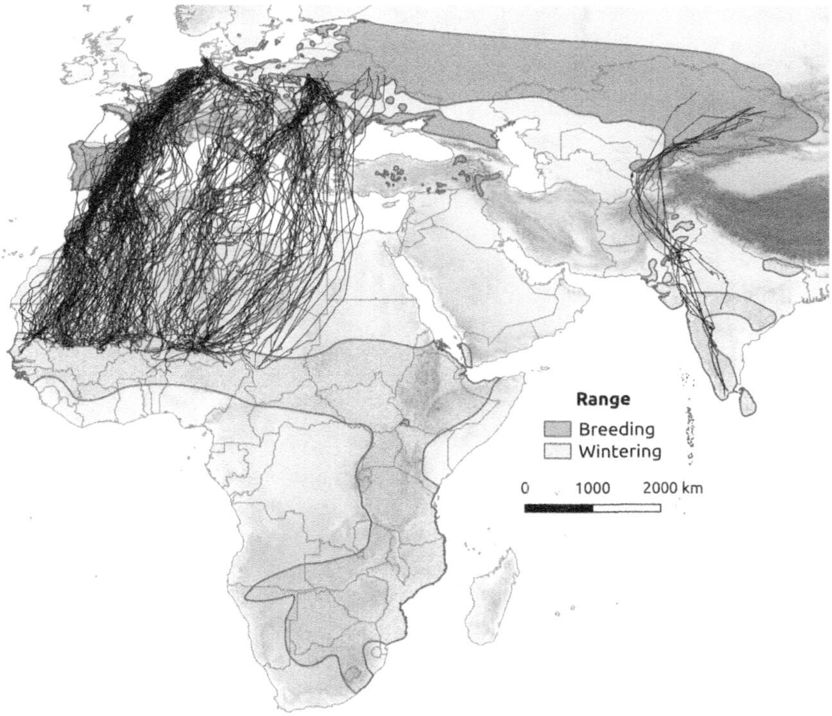

Figure 1. Autumn migration tracks belonging to 145 adult Montagu's harriers tracked by satellite and GPS telemetry between 2005 and 2020 (N = 247 tracks).

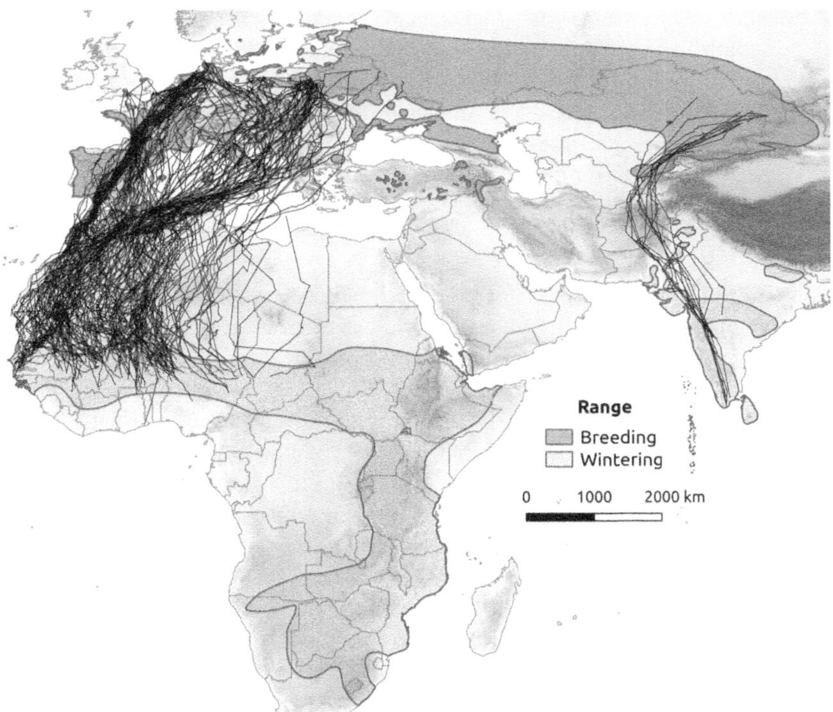

Figure 2. Spring migration tracks belonging to 145 adult Montagu's harriers tracked by satellite and GPS telemetry between 2005 and 2020 (N = 200 tracks).

Kazakhstan, we do not know where the migratory divide is situated that separates populations migrating to Africa and populations migrating to India. At the scale of the whole breeding population we observe clear migratory connectivity with population specific migration routes.

When zooming in on Europe, for which most tracking data is available, we see that breeding birds from Western European populations winter in Western West Africa whereas birds from Eastern European populations winter in Eastern West Africa (Limiñana et al. 2012c, Trierweiler et al. 2014). However, wintering ranges largely overlap since individuals from one population do not all winter in the same region but spread out over a vast longitudinal range. Breeding birds from the same area or even the same colony were found to winter more than 1,200 km apart (Limiñana et al. 2012c, Trierweiler et al. 2014). Still we observe the pattern that western breeding birds tend to winter further to the west than eastern breeding birds, and given the large overlap in wintering range between western and eastern population we conclude that migratory connectivity is weak at this scale (Webster et al. 2002).

Post-breeding Migration

Montagu's harriers, especially failed breeders, occasionally make long pre-migratory movements during which they often visit other breeding populations, presumably prospecting for future breeding events (Limiñana et al. 2008, Trierweiler et al. 2007, 2014). Autumn migration starts at the end of August, and takes 16 days for Spanish breeding birds (Limiñana et al. 2012c) and one month for north-western breeding birds (Trierweiler et al. 2014). Birds arrive at the wintering area around the 10th of September (Spanish birds) and end of September (north-western breeding birds). Distance covered during migration is about twice as long for north-western birds (4,800 km) compared to Spanish ones (2,600 km). The overall autumn migration speed was comparable for Spanish (187 km/d) and north-western breeding birds (173 km/d).

Trierweiler et al. (2014) compared the timing and migratory performance between migration routes (western route via Spain, central route via Italy, and eastern route via Greece), but found no dramatic differences in timing, duration, distances and migration speed, despite differences in landscape, topography and habitat composition. This suggests that the behaviour of these migrants is to some extent 'hard-wired' and that differences in landscape play a minor role. Spanish breeding birds migrate much shorter distances, and thus their migrations take much less time (cf. above), but the fact that their migration speed is similar to north-western breeding birds corroborates the idea that migratory performance is similar between populations.

Pre-breeding Migration

Spanish Montagu's harriers leave the wintering area about two weeks earlier (half March) compared to their northern conspecifics (end of March). Spring migration takes 27 days for Spanish birds and 35 days for north-western breeding birds, which means that they arrive in the breeding area at on average the 11th of April (Spanish birds) and the 5th of May (north-western breeding birds; Limiñana et al. 2012c, Trierweiler et al. 2007, 2014). Total migration distance is slightly longer compared to autumn, 2,930 km for Spanish birds and 5,306 km for north-western breeding birds. Notably, the resulting overall migration speed is lower for spring migration compared to autumn migration (114 km/d for Spanish birds and 162 km/d for north-western breeding bird; Limiñana et al. 2012c, Trierweiler et al. 2007, 2014). As daily travel distances do not vary between seasons (see Section 'behavioural remarks'), this must be contributed to the birds making more or longer stopovers. The stopover areas in North-Western Africa play a central role here. The fact that migration speed is lower in spring compared to autumn makes the Montagu's harrier one of the exceptions to the general pattern of faster spring migrations (Nilsson et al. 2013; see also Chapter 10).

Montagu's harriers generally follow the same route back in spring, although spring migration routes tend to lay (slightly) west of autumn routes. As mentioned above, only Eastern European breeding birds seem to make a larger scale loop migration taking the eastern route via Greece in autumn and the central route via Italy in spring (Trierweiler et al. 2014). The westward shift in tracks is caused by dominant easterly winds over the Sahara Desert (Limiñana et al. 2013), as shown for western marsh harrier (*Circus aeruginosus*) and osprey (*Pandion haliaetus*) (Klaassen et al. 2010, Klaassen et al. 2011). For Montagu's harriers migrating via the western route via Spain, this means that they get pushed towards the coast of the Atlantic Ocean, and subsequently birds congregate along the coast of Morocco to cross the Mediterranean Sea west of Gibraltar (Fig. 2). Birds from the central and eastern breeding populations drift westwards during the crossing of the Sahara Desert, for which they compensate in Northern Africa where they travel eastward (Fig. 2). One of the consequences of these wind drift-induced detours is that the birds generally travel longer distances in spring compared to autumn (Limiñana et al. 2012c, Trierweiler et al. 2007, 2014).

Behavioural and Ecological Remarks

Montagu's harriers have relatively long and slender wings (and thus a high wing aspect ratio), which allows them to travel by both flapping-gliding and soaring-gliding flight (Spaar and Bruderer 1997). This flexibility in flight behaviour makes Montagu's harriers less dependent on thermals compared to obligate soaring migrants, allowing them to migrate also under less favourable thermal and wind conditions (Spaar and Bruderer 1997). Montagu's harriers migrating over Southern Israel showed a higher percentage of flapping-and-gliding flight in spring (27% of observation time) than in autumn (14%). The proportion of soaring was 25% in spring and 35% in autumn (Spaar and Bruderer 1997). Flapping-gliding flight was observed during the whole day in spring, but in autumn most birds migrated mainly by soaring-gliding flight. Only shortly after sunrise, when thermals have not yet developed, flapping-gliding flight was dominant in autumn (Spaar and Bruderer 1997).

As Montagu's harriers depend less strongly on thermals for migration, they can use a relatively broad daily time window for travelling. A time budget analysis based on GPS-tracking data of Montagu's harriers from the Netherlands revealed that daily flights usually were initiated within 2 hours after sunrise, and daily flights ended within 1–2 h before sunset, resulting in an average daily migration time of 10.4 h (Klaassen et al. 2017). This is more than the daily migration time of obligate soaring migrants. A peculiarity of migrating Montagu's harriers is that they do not travel the whole time but frequently interrupt their flights, suggesting fly-and-forage migration behaviour (Klaassen et al. 2008). Observations from the Middle East confirm that they hunt for small birds and insects while migrating (Shirihai et al. 2000).

Interruptions occurred on 78% of all travel days, and comprised, on average, 14% of the average daily migration time. Close inspection of the GPS tracks revealed that birds were moving (suggesting foraging behaviour) during 41% of the interruptions, and stationary (suggesting resting behaviour) during 32% of the interruptions (27% of the interruptions were too short to enable classification). The balance between interruptions for foraging and interruptions for resting varied between regions, with relatively more interruptions for foraging in Europe and relatively more interruptions for resting in Africa (Klaassen et al. 2017). The fact that Montagu's harriers still relatively frequently interrupted their flights for foraging in Africa, also in the middle of the Sahara Desert, suggests that this region is less hostile for migrating harriers than previously thought (or, alternatively, that harriers are efficient in exploiting scarce foraging opportunities). Daily time budgets were surprisingly similar among regions and between seasons. Daily flight times were only marginally longer in Africa (9.4 and 9.0 h for autumn and spring, respectively) than in Europe (8.6 and 8.7 h, respectively), and did not differ between seasons (Klaassen et al. 2017).

The travel performance of migrating Montagu's harriers varies notably between days; the distance harriers travel on a day varies between 20 and 790 km (Klaassen et al. 2017; excluding overnight flights associated with sea crossings—see below). This variation is largely explained by variation in the time flying per day: the more hours per day the flight is interrupted the fewer kilometres are covered (Klaassen et al. 2017). Weather has a profound additional effect on travel performance in several ways. Montagu's harriers are less affected by precipitation and crosswinds than obligate soaring migrants, but at the same time do not seem to profit to the same extent as obligate soaring migrants from favourable soaring conditions (Vansteelant et al. 2015). Tailwind has a positive effect on hourly travel speeds and thus daily travel distance (Klaassen et al. 2017, Limiñana et al. 2013, Vansteelant et al. 2015). In addition, Montagu's harriers tend to travel more hours per day, and make less interruptions, when wind conditions are favourable, in extreme cases even extending their flights into the night (Klaassen et al. 2017). Due to this direct effect of wind on daily travel time, a strong positive effect of tailwind on daily travel distance is found (Klaassen et al. 2017).

Daily travel distance varies only marginally among regions and between seasons, in which longer distances are covered in Africa (270 km/d in autumn and 234 km/d in spring) than in Europe (228 km/d in autumn and 217 km/d in spring), and longer distances in autumn (228 km/d in Europe and 270 km/d in Africa) than in spring (217 km/d in Europe and 234 km/d in Africa). These differences in migration performance can be accounted for by variation in weather conditions between regions and seasons. I.e. when accounting for effects of weather (which explain 30–40% of the variation in travel performance), region and season themselves explained less than 1% of the variation in daily travel distances (Vansteelant et al. 2015).

The Montagu's harrier is a diurnal migrant, like most raptors, although the results from satellite telemetry suggested that they occasionally extend their diurnal travels into the night (Trierweiler et al. 2014). More accurate GPS-tracking data showed that nocturnal flights are not extremely uncommon in Montagu's harriers (about 3% of all flights; Klaassen et al. 2017). Nocturnal flights were associated with sea crossings (especially the crossing of the Mediterranean Sea in spring) but also two overnight flights over land were recorded. When wind conditions are favourable (i.e., strong tailwinds), Montagu's harriers sometimes extend their daily flights into the night achieving very high travel distances (on average 917 km with a maximum of 1488 km; Klaassen et al. 2017; note that these flights continued after midnight and thus took more than one day).

Another consequence of Montagu's harriers being facultative soaring migrants is that they do not congregate at narrow sea crossing points but migrate in a broad front, crossing large distances of open sea if weather conditions are favourable. This is illustrated by the observation that migrating harriers at Dardanelles, Turkey, show a much lower tendency to follow the peninsula compared to obligate soaring migrants, but cross the strait directly, independently of wind conditions (Panuccio et al. 2017). Many Montagu's harriers migrate along the east coast of Spain and cross the Mediterranean Sea from Cabo de Gata in a 4–6 h flight, rather than making the detour via the shortest crossing point at Gibraltar (Fig. 1). In agreement with this, near Malaga (between the two above mentioned locations), 74% of Montagu's harriers cross the sea instead of continuing along the coast (Meyer et al. 2003). The longest sea crossings are observed in spring for Montagu's harriers travelling through the eastern part of the Mediterranean Sea, using the islands as stepping stones.

During autumn migration, Northern European Montagu's harriers make stopovers both in Europe and Northern Africa (25% of all stopovers) whereas they mainly stopover in Northern Africa during spring migration (45%; Trierweiler et al. 2014). A major stopover area in Morocco/Algeria was predominantly used by birds travelling via the western route, both in autumn and spring. Of those birds travelling via Spain, 41% made stopovers in this region in autumn and 76% in spring (Schlaich 2011). Birds stayed there on average for 9.7 days in spring and 8.7 days in autumn, respectively. In spring, they feed on passerines such as larks and their eggs and nestlings as was discovered during field expeditions (Schlaich 2011).

In very rare cases, adult Montagu's harriers over-summer in Africa. A Danish male, over-summering after an exceptionally demanding breeding season, showed no intention to migrate during onset of spring migration but spent more than 17 months in West Africa, part of it outside the normal wintering range of the species (Sørensen et al. 2017).

Satellite tracking data is not only useful to discover and describe routes, stopover areas and wintering ranges, but can also be used to determine seasonal mortality. Mortality of adult Montagu's harriers during the breeding and wintering season seems to be generally low, whereas it is higher during migration, peaking especially during spring migration (Klaassen et al. 2014, Schlaich 2019).

Even though migration is a relatively short period in the annual cycle, daily mortality is higher than during the stationary periods. However, Schlaich (2019) found an alarming increase in mortality during winter and spring migration between 2005–2010 and 2012–2017: total mortality in winter doubled and mortality during spring migration increased 1.5-fold. The remarkably high mortality during spring migration is believed to be related to adverse environmental conditions, in particular headwinds over the Sahara Desert. But it cannot be excluded that part of these birds die due to carry-over effects of harsh conditions and intensified foraging at the end of the winter (Schlaich et al. 2016).

Acknowledgements

Additional data for figures 1 and 2 was very kindly made available by the following persons and organizations: UK—M. Thomas, B. Image, J. Scott (Royal Society for the Protection of Birds) and P. Castle (Wiltshire Ornithological Society); Germany (Saxony-Anhalt)—R. Fonger; Germany (North Rhine-Westphalia)—H. Illner (Arbeitsgemeinschaft Biologischer Umweltschutz/Biologische Station Soest); Germany (Diepholzer Moorniederung)—BUND Diepholzer Moorniederung; Denmark—Dansk Ornitologisk Forening (BirdLife Denmark); Poland—D. Krupinski (Towarzystwo Przyrodnicze "Bocian"); France— Groupe Ornithologique des Deux-Sèvres and Centre d'Etudes Biologiques de Chizé (CEBC-CNRS); India—T. Ganesh (ATREE). We are very grateful to Tonio Schaub for help with data preparation.

19

Red kite *Milvus milvus*

Javier Vidal-Mateo,[1,*] *Javier de la Puente,*[2] *Ana Bermejo*[2] and *Vicente Urios*[1]

ABSTRACT

The red kite (*Milvus milvus*) is endemic to Western Palearctic. It is a partial short-distance migrant, with most of the individuals in North-Eastern Europe migrating in a SW direction to spend the wintering season mainly in Southern France and the Iberian Peninsula. However, resident birds occur in different areas of its distribution range. Satellite tracked individuals belonging to populations from Central Europe show that autumn migration begins between late September and early November, taking between two and three weeks to reach the Iberian Peninsula, although some individuals can made long stops along the route, prolonging the migration period. A lower proportion of individuals migrate to Italy and the Balkans. In spring, birds leave the wintering quarters mainly between mid-February and mid-March, with a wider range of dates for the immatures. Red kites enter Spain mainly through the western part of Pyrenees. Birds travel mainly during the hottest hours of day and usually migrate in small groups of less than 10 individuals.

Introduction

The red kite (*Milvus milvus* Linnaeus, 1758) is a medium-sized raptor, with 60–72 cm long and a wingspan of 143–171 cm; males weigh 757–1,221 g and females 960–1,600 g (Orta et al. 2018). This species is endemic to the Western Palearctic, from Southern Sweden through Central Europe to Ukraine and the Mediterranean basin, and also Great Britain (del Hoyo et al. 1994, BirdLife International 2017d). The three largest populations are located in Germany, France and Spain, and although it declined globally in the last decades due to persecution, poisoning and agricultural changes, many populations have increased the number of individuals in recent years (Aebischer 2010a, Molina 2015, BirdLife International 2017d). Therefore, currently the species is considered as "Near threatened" by the International Union for Conservation of Nature and Natural Resources (IUCN 2018), with a European population of 25.200–33.400 pairs (BirdLife International 2017d).

Most red kites from Central and Eastern Europe are migratory and spend the winter mainly in Southern France and the Iberian Peninsula (del Hoyo et al. 1994). Instead, populations from southern areas of its distribution range and that of Great Britain are resident. A few migratory individuals can

[1] Vertebrates Zoology Research Group, University of Alicante, Apdo. 99, E-03080 Alicante, Spain.
[2] SEO/BirdLife, Bird Monitoring Unit, C/ Melquiades Biencinto, 34, E-28053, Madrid, Spain.
* Corresponding author: javier.vidalm90@gmail.com

reach Africa through the Strait of Gibraltar (Onrubia 2016). Birds start the migration towards their wintering grounds between August and November, and they return to their breeding areas between February and April (Snow and Perrins 1998, Pfeiffer and Meyburg 2009).

Laying usually takes place in March or April (Mammen and Stubbe 1995, Mougeot and Bretagnolle 2006) and clutch size is one to three eggs (Sergio et al. 2005, Mougeot and Bretagnolle 2006). It nests in trees, inhabiting open wooded lands mixed with farmland, pastures or heathlands during the breeding period; in winter it prefers more open lands, such as pastures, crops, wasteland or scrub, where it roosts in groups in clumps of trees (del Hoyo et al. 1994). The red kite's diet is varied, and although it tends to a scavenger feeding visiting vulture restaurants or dumps, it also feeds on small mammals, birds, reptiles and other small preys, being able to exploit different food resources according to local availability (Cramp and Simmons 1980, Blanco et al. 1990, García et al. 1998).

Knowledge Status

Tens of red kites have been tagged with satellite transmitters in different countries, such as Austria, Czech Republic, France, Germany, Poland, Slovakia, Spain and Switzerland (Pfeiffer and Meyburg 2009, Aebischer 2010b, Scherler et al. 2015, Raab et al. 2017, Literak et al. 2018, Maciorowski et al. 2019, Vidal-Mateo et al. 2015), but for some of those studies detailed analyses are not yet available.

Concerning visual observations at migration bottle-necks, the most important watchsites for the species are located in France (Fort l'Écluse and Pyrenees; Urcun and Bried 1998a, Mission Migration 2018, LPO Haute-Savoie 2016) and Sweden (Falsterbo; Kjellén and Roos 2000). In these bottle necks, censuses have been performed for several years, sometimes also through different decades.

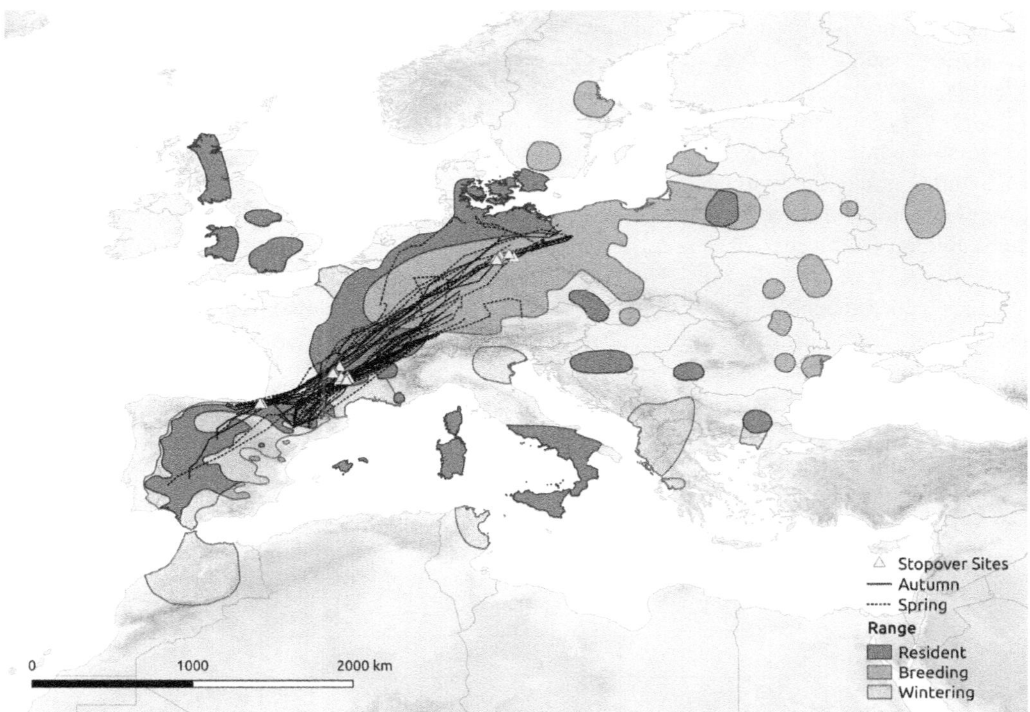

Figure 1. Migration routes of GPS tracked red kites between the breeding/summering areas in Central Europe and wintering quarters in Spain (sources: SEO/BirdLife unpub. data, Vidal-Mateo 2019). White triangles indicate long stop-overs (17–45 days).

Post-breeding Migration

Most of red kites that breed in Central and North-Eastern Europe migrate to the southwest to spend the winter, mainly in Southern France and in the Iberian Peninsula (Vidal-Mateo et al. 2015), but also to Southern Italy and the Balkans (Fulco et al. 2017, Raab et al. 2017). Falsterbo is the main bottle neck for Swedish migratory individuals, and counts show a marked increase: 227 individuals on average were reported in the period 1980–1989, 817 in 1990–1999, 1,807 in 2000–2009 and 3,178 in 2010–2017; finally, 4,574 individuals were observed in 2017 (Kjellén and Roos 2000, Kjellén 2017). Here, juveniles strongly outnumber adults, which on the contrary are mainly resident (Kjellén 1992, 1994).

There is also a growing number of resident birds in Central Europe (Aebischer 2010a, BirdLife International 2017d): according to 21 winter recoveries of Swiss ringed red kites, 29% remained in Switzerland (Aebischer 2010b) while most migrated following a South-West direction.

In route towards the main wintering areas, from the watchsite of Fort l'Écluse (Rhone valley, France) 8,000–10,000 individuals have been counted between 2011 and 2016, showing a marked increase in comparison with the previous decade (LPO Haute Savoie 2016).

Red kites enter in to Spain crossing the Pyrenees, mainly through the western part (Urcun and Bried 1998a). Along this flyway, the watchsite where more individuals can be counted is the Organbidexka pass, although there are other spots where substantial numbers of individuals can also be observed, according to the weather conditions (Urcun and Bried 1998a); the number of birds that cross by the eastern part is given below (Table 1). In particular, at Organbidexka pass, on average 3080 individuals/year were observed in the period 1981–2008 (Filippi-Codaccioni et al. 2010). In 2017, 8796 individuals were counted at this watchsite, with the average peak date being 15th October, taking into account all available years (Mission Migration 2018). Spain has an estimated population in winter of more than 50,000 individuals, while the breeding pairs are less than 2,500 (Molina 2015): this suggests that, albeit those watchsites concentrate substantial numbers of red kites crossing the Pyrenees, the real figures are even higher.

Migratory birds that cross the Strait of Gibraltar towards Africa are very few with a mean of 27 individuals counted in 2009–2012 (Onrubia 2016).

Satellite tracked individuals indicate that autumn migration began in late September and early November, mainly the first half of October, covering distances of a few hundred kilometers to more than 2,300 km (Aebischer 2010b, Peiffer and Meyburg 2009, Vidal-Mateo 2019). At Falsterbo, an earlier departure of juveniles was observed (Kjellen 1992, 1994), although Nachtigall (2008) using wing tags recorded conflicting evidences, and Urcun and Bried (1998a) reported that juveniles would migrate later than adult, in the Pyrenees.

Migration schedules of adult red kites are best known thanks to birds tagged in Germany, and during winter in Spain (Pfeiffer and Meyburg 2009, Vidal-Mateo 2019). German birds (7 individuals) did not perform stopovers and required between 12 to 28 days to accomplish the journey. Birds tagged in Spain belong to populations from Poland, Germany and Switzerland (4 individuals, 14 migrations, including also three 2nd cy birds in their 2nd autumn migration). They

Table 1. Number of red kites counted at four passes of the Pyrenees during autumn migration. Organbidexka, Lindux and Lizarrieta are at the western part of the chain, and Eyne at the eastern part (Source: Mission Migration 2017, https://www.migraction.net).

	2013	2014	2015	2016	2017
Organbidexka	5,518	3,528	6,752	5,931	8,796
Lindux	3,162	2,162	4,724	4,705	6,548
Lizarrieta	2,705	1,304	1,244	3,492	4,367
Eyne	27	123	203	183	552

had as mean departure date the 16th October. The majority of these birds made direct migration journeys to their wintering areas (Fig. 1), taking 18 ± 8 (SD) days to complete the migration and showing only a few stopover days (0–5). However, two individuals exhibited a different behaviour: they stopped for longer periods (between 17 and 45 days) in one or several zones where they made short displacements until they resumed the travel (Fig. 1). This prolongation of migration delayed the arrival date to the final destination until the third week of January. One of these two birds started the migration from Switzerland on 30th October and stopped 45 days in Lozère (Southern France) and 18 days in Navarra (Northern Spain), arriving on 23th January to the wintering area where it had been captured the previous season. The other bird behaved in the same way for three consecutive years: after starting the migration from Lubusz (Western Poland), it remained in Eastern Germany for several weeks, until it resumed the journey to Spain; during one year it stopped in Cantal and Gers (France) without arriving to Spain (Vidal-Mateo 2019).

Among 15 juveniles tracked from Poland, the median departure date was the 16th August, with the journey lasting 91.5 ± 45.2 days. Such variation was due to different stopover strategies (37.5 ± 10.7 days per individual), with birds starting later spending less days on stopovers (Maciorowski et al. 2019). All these individuals migrated towards France and the Iberian Peninsula, apart from one that went to the Balkans.

Migration routes of 15 red kites (all juveniles except for two adults), fitted with transmitters in Austria, the Czech Republic and Slovakia show a much larger variation in migratory behaviour and orientation. Only three went to the Iberian peninsula, three to Italy (by crossing the Adriatic Sea), one adult went to Corsica, and five to the Balkans. Three individuals, including an adult, remained in Central Europe (Raab et al. 2017).

Pre-breeding Migration

The main bottlenecks where red kites are recorded during autumn migration do not show significant concentrations during spring. A maximum of 500 individuals has been recorded at Col de l'Escrinet in the 2015 (France, Mission Migration 2018). Therefore our knowledge is based mostly on satellite tracking studies.

According to tracked individuals, birds leave the wintering quarters in Spain mainly between mid-February and mid-March (Aebischer 2010b, Pfeiffer and Meyburg 2009, Vidal-Mateo 2019), taking between one to three weeks to complete the journey (Fig. 1).

Although there are not significant differences in the migration duration, immatures (2nd cy) start the spring migration later than adults (Fig. 2), ending also later: for 21 recorded migrations of adults tagged in Spain, the mean onset date was the 27th February, while for 8 immatures the 29th March (Vidal-Mateo 2019). The number of stopover days for these birds ranged between none and 11, with an average of 2 days.

Behavioural and Ecological Remarks

The proximity between the breeding and wintering areas could explain the variability in movement ecology of this partially migratory raptor, with short-distance migrants showing a large variation in migration speed, while other individuals are sedentary, mainly in the southern portion of its distribution range. Such variation occurs even in the same individuals throughout their life cycle. Wintering site fidelity seems to be the commonest pattern but, for example, an adult female tracked over five migration periods showed strong annual variation in the selection of the wintering areas (albeit during two years used the same one), even avoiding to migrate during one year (Pfeiffer and Meyburg 2009).

Spring migration seems to be faster than autumn migration, being completed in less days and with higher speeds (Pfeiffer and Meyburg 2009): in spring the distance covered each day is around 130 km/day while in autumn it is around 95 km/day, making this difference statistically significant (Vidal-Mateo 2019). Moreover, long stop-overs have been observed only during the autumn migration

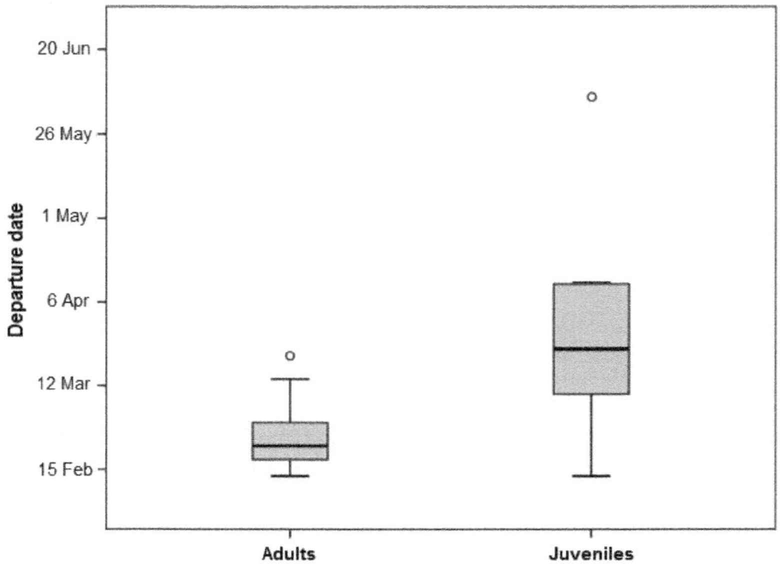

Figure 2. Comparison of departure dates for spring migration between adults and immatures. The median, standard deviation, range and extreme values are shown.

(even lasting weeks). As in several other species, these seasonal differences could be related with a time minimization strategy due to the pressure of having a good breeding site after the pre-breeding migration (Kokko 1999). The competition for the breeding resources can also explain why immatures begin to migrate later than adults and with a wider range of dates in spring. Significant differences between age classes have been only found in the migration dates, but not for daily distances (Vidal-Mateo 2019).

Unlike trans-Saharan migratory raptors, which cover daily distances around 200–250 km/day (Mellone et al. 2012a, 2015a, López-López et al. 2014), the red kite cover lower distances which are more similar to other short-distance migrants, such as the Common Buzzard (Strandberg et al. 2009a), with an average speed of 84 km/day.

The "stepping-stone" migration pattern observed in two satellite tracked adults (Vidal-Mateo 2019) has not previously been recorded in raptors, to our knowledge, but is known for other taxa (see Stach et al. 2012, McKnight et al. 2013). This strategy possibly allows the red kites to exploit different food resources in suitable areas before resuming the journey.

Birds travel mainly during the hottest hours of day, between 9AM and 3PM (Urcun and Bried 1998a). Although they can form flocks of dozens of individuals, they mostly migrate in smaller groups of less than 10 birds (Urcun and Bried 1998a, Gatter 2000).

20

Black kite *Milvus migrans*

*Alejandro Onrubia** and *Beatriz Martín*

ABSTRACT

The black kite (*Milvus migrans*) is one of the most abundant raptors in the world and one of the commonest migratory raptor in the Palaearctic flyways. The species has a very broad distribution, with resident populations in Africa, Southern Asia and Australia, and long-distance migratory populations in Europe, the Middle East and Central Asia. Black kites generally migrate in lax groups and over a broad front, except where detours around major water bodies and higher mountains become necessary. The European population is divided into three major flyways. The population breeding in Western Europe has a migratory route across the Strait of Gibraltar to wintering areas in West Africa. Black kites from Italy and, to a lesser extent, Central Europe tend to use the Central Mediterranean route crossing mostly through the Sicily Channel and wintering in Western Africa. The eastern population, breeding in East Europe and Middle East, migrates around the eastern coast of the Mediterranean and Black Sea towards the Red Sea, which is either crossed near Suez in the north, or via the Bab-el-Mandeb Straits in the south, and spends the winter in the Middle East and Eastern Africa. Satellite telemetry studies revealed a complex age-dependent pattern of migration performance shaped by a gradual improvement, more evident during the pre-breeding journey, with the earliest individuals being advantaged as regards their survival and fitness.

Introduction

The black kite (*Milvus migrans* Boddaert, 1783) is a medium-sized raptor (44–66 cm; 600–1200 g; wingspan 120–153 cm; Ferguson-Lees and Christie 2001, Palomino 2016), widely distributed across the Palearctic as well as in the Afro-Malagasy, Indomalayan and Australasian regions (Orta et al. 2020). This species, however, is absent in the Neartic and Neotropic regions although vagrants have been recorded in America (e.g., Brazil, Tavares-Nunes et al. 2015). At least seven subspecies are currently recognized, but more work is needed to help to clarify sister relationships within the group (Orta et al. 2020). The nominal subspecies (*Milvus migrans migrans,* hereafter black kite), which is the subject of this chapter, is a breeding raptor in the Western Palearctic and Central Asia, almost entirely migratory, that usually winters south of the Sahara Desert in Africa (Sahel belt) and also in the Middle East (Shirihai et al. 2000, Ferguson-Lees and Christie 2001). Specifically, wintering grounds of black kites breeding in Central-Western Europe are commonly located in Western Africa

Fundación Migres, CIMA, Tarifa, Cádiz, Spain.
* Corresponding author: aonrubia@fundacionmigres.org

Table 1. Overview of the magnitude and timing of spring and autumn migration of black kite from ground observations at selected migration bottlenecks along his distribution range.

Country	Location	Spring migration				Autumn migration				Source
		Magnitude	Range	Peak	Range	Magnitude	Range	Peak	Range	
Western Mediterranean Flyway										
Sweden	Falsterbo					15	0–82	16-Sep	20 Aug–20 Sep	Kjellen 1992, 2004, https://www.falsterbofagelstation.se
Germany	Helgoland	< 5	0–2	15-Apr	10 Apr–25 May	< 5	0–1	5-Oct		Dierschke et al. 2011
Poland	Gulf of Gdansk					< 5				Polakowski et al. 2014
France	Wetern Pyrenees					28,000	12,000–47,000	5-Aug	20 Jul–30 Aug	Urcun and Bried 1998b, www.migraction.net
France	Eastern Pyrenees (Eyne)					3,800	2,650–5,200	17-Aug		www.migraction.net
Portugal	Sagres					200	125–270	5-Sep	10 Aug–15 Oct	Palma and Reja 1994, Tome et al. 1998
Spain	Gibraltar	169,000		15-Mar	20 Feb–10 May	224,600		16-Aug	29 Jul–10 Sep	Bernis 1980, Finlayson 1992, Garcia and Bensusan 2006, Onrubia 2015
Central Mediterranean Flyway										
Spain	Balearic Islands	< 50		5-Apr	20 Mar–25 May	< 50		25-Aug	10 Aug–20 Sep	Rebassa 1995, Rebassa and Ventoso 2015
Italy	Messina/ Sicilian Channel	1,000	540-1,010	8-Apr	10 Mar–20 May	3,500		30-Aug	20 Aug–15 Sep	Hein and Kisling 1991, Agostini 2002, 2005, Agostini et al. 2000, 2004, Corso 2001
Eastern Mediterranean and Middle East Flyway										
Romania	Dobrogea					36	10–50	4-Sep	17 Aug–27 Sep	Fulop et al. 2018
Bulgaria	Bourgas	< 50				250	35–800	21-Aug	10 Aug–2 Oct	Michev et al. 2011, Shurulinkov et al. 2011
Greece	Antikythira	12	10–15			75	15–80			Lucia et al. 2011
Georgia	Batumi	250,000	220,000–283,000			136,900	80,000–170,000	17-Sep	3 Sep–5 Oct	Verhelst et al. 2011, Abuladze 2012, Vansteelant et al. 2019, Werham et al. 2019, https://trektellen.nl/site/yeartotals/1047/2020/-1
Azerbaijan	Besh Barmag					< 5				Heiss and Gauger 2011, Heiss 2013, Heiss 2016
Turkey	Bosphorus	225	120–260	30-Mar	15 Mar–31 May	250	40–2,800	31-Aug	25 Jul–25 Sep	Collmait and Croxall 1967, Porter and Willis 1968, Beaman et al. 1975, Shirihai et al. 2000, Üner et al. 2010, Fülöp et al. 2014, Panuccio et al. 2017

Turkey	Dardanelles	125				40				Panuccio et al. 2017, Uysal and Tosunoğlu 2018
Turkey	Gulf of Iskenderun/ Belen Pass					550			25 Aug–15 Sep	Sutherland and Brooks 1981, Shirihai et al. 2000, Oppel et al. 2014
Lebanon	Harissa					3,800				Shirihai et al. 2000, Meyburg et al. 2020
Israel	Kfar Qassem/ Northern Valleys					1,700	1,000–2,700	9-Sep	25 Aug–10 Oct	Leshem and Yom-Tov 1996, Shirihai et al. 2000, Alon et al. 2004
Israel	Eilat	28,000	12,000–37,000	29-Mar	17 Mar–10 Apr					Leshem and Yom-Tov 1996, Shirihai and Crhistie 1992, Shirihai et al. 2000
Egypt	Gulf of Suez	10,000	2,300–11,000	5-Apr	10 Mar–5 May	100	75–110		13 Sep–3 Nov	Bijlsma 1983, Shirihai et al. 2000, Megalli and Hilgerloh 2013
Egypt	Zait Bay	1,600		31-Mar	20 Mar–25 Apr	60		15-Sep	20 Aug–15 Oct	Hilgerloh 2009, Hilgerloh et al. 2009, Hilgerloh et al. 2011, Megalli and Hilgerloh 2013
Djibouti/Yemen	Bab Al Mandeb					580	55–600			Welch and Welch 1988, 1989, 1991, 1998
Ethiopia	Lake Langano	> 1,000				10,000			20 Sep–30 Oct	Vittery 1983, Zalles and Bildstein 2000
Asian Flyway										
Kazakhastan	Chokpak Pass	110		2-Apr		2,700		15-Sep		Gavrilov and Gistov 1985
Russia	Chergiskyi// Slyudyanka	710				150		20-Sep		Zalles and Bildstein 2000
Iran	Elborz Mountains	> 400		15-Apr						Ullman and Ullman 2010, Panuccio et al. 2018
Pakistan	Haleji Lake	50		15-Feb						Roberts 1991
India	McLeod Ganj	80			20 Feb–30 Mar					Juhant and Bildstein 2017
Nepal	Khare					100			20 Sep–05 Nov	Juhant and Bildstein 2017
Nepal	Thoolakharka					550	100–600		15 Sep–13 Nov	Subedi and De Candido 2013, 2014, Subedi and Gurung 2018, Gurung et al. 2019
China	Guncang	> 170								Juhant and Bildstein 2017
Thailand	Chumphon// Khao Dinsor	< 20	8–13	23-Mar		170	125–170	10-Oct	27 Sep–9 Nov	DeCandido et al. 2004, 2009, Nualsri et al. 2010, Pajunen et al. 2012, Siponen et al. 2013
Malaysia	Bedong Kedah					< 20				Eng et al. 2014

(Cramp and Simmons 1980, Meyburg and Meyburg 2009a). The other black kite subspecies include *M.m. govinda* (E Pakistan, India, China, Indochina and Malay Peninsula), *M.m. affinis* (Sulawesi, New Guinea, Australia), the black-eared black kite—*Milvus migrans lineatus/formosanus* (Ural Mountains, Siberia, Japan, India, Myanmar, China; wintering south to S Iraq, S India and SE Asia), and the yellow-billed black kite—*Milvus migrans aegyptius/parasitus*—(Egypt, SW Arabia and coastal E Africa S to Kenya, Africa S of Sahara and Madagascar). These subspecies are mainly residents, short-distance migrants (*M.m. govinda, M.m. formosanus, M.m. parasitus*), or display continental scale movements (*M.m. lineatus*) (Cramp and Simmons 1980, Ferguson-Lees and Christie 2001, Orta et al. 2020).

The results of ringing and tagged birds show that the most important wintering areas for migratory black kites (*M.m. migrans*) are located in sub-Saharan Africa throughout the Sahel belt from Senegal to Ethiopia (Sergio et al. 2014, Ovčiariková et al. 2020), and also recently increasing in the Middle East and Turkey (Shirihai et al. 2000, Biricik and Karakaş 2011, Panuccio et al. 2014, Vansteelant et al. 2019, Meyburg et al. 2020). Recently, black kites have also begun to winter regularly in Spain, Sicily, Ukraine and Greece (Tsvelykh and Panjushkin 2002, Sarà 2003, Palomino 2012, Literák et al. 2017). On the other hand, black-eared kites (*M.m. lineatus*) spend the winter mainly in Eastern Africa and the Indian subcontinent, but increasing their winter occurrence is also in the Middle East (Meyburg et al. 2020). The wintering distribution of black kites in the Sahel is highly dependent on food sources that are highly unpredictable, such as the locust outbreaks (Sánchez-Zapata et al. 2007). Although most individuals winter south of the Sahara Desert in Africa and also in the Middle East, numbers of black kites that are recently shortening their migratory routes and wintering in the Mediterranean basin are increasing (Panuccio et al. 2014). Similarly, remarkable numbers of wintering black kites have been recently observed in southeast Turkey (Biricik and Karakaş 2011). Both shifts in the wintering range of black kites in the Mediterranean and the Middle East regions seems to be related to climate change (Panuccio et al. 2014). Regarding the migratory phenological patterns, according to counts of migrating birds at the Strait of Gibraltar, passing dates of black kites during autumn are both advancing (in the earlier phases of the migration distribution) and delaying (in the later phases), leading to an enlargement of the total migratory passage time in the last 40 years (1970s–2000s; Scholer et al. 2016, Panuccio et al. 2016b). Tagged birds also show that black kites spend their first 1 or 2 years in Africa and usually start breeding in Europe when they are 3–6 years old. Thus, most of the juveniles tend to stay in tropical Africa during their first summer, many also move north but to substantially lower latitudes than natal areas (in Morocco, Algeria, Spain and Turkey). During the second year, they winter again in the tropical part of sub-Saharan Africa, but subsequent summer quarters were located north from the winter quarters, some staying throughout the summer in Africa, some in south to Central Europe, some in Turkey and even Northwestern Russia. Many 3-year-old birds (fourth calendar year) attempt to nest usually close to their natal nest (Sergio et al. 2014, Ovčiariková et al. 2020).

Black kites are able of both soaring and flapping flight thanks to their relatively narrow, long wings, compared to typical soarer species (Spaar 1997). For this reason, the flight behaviour of this species during migration is halfway between broad-winged raptors like buzzards, eagles and vultures and relatively long-winged raptors like harriers. As a result, they mostly use soaring-gliding flight over land during migration to minimize energetic costs and concentrate at straits and isthmuses, but they are also able to cross large bodies of water using long flapping flight (Panuccio et al. 2014, Santos et al. 2017, 2020, Mellone 2020). Being a gregarious species, birds often occur in communal roostings and usually breed in lax colonies (Cramp and Simmons 1980, De Juana and Garcia 2015, Palomino 2016). Similarly, they can migrate solitarily, in loose flocks (del Hoyo et al. 1994, Ferguson-Lees and Christie 2001), or even in groups including several hundreds of individuals (Bernis 1980, Panuccio et al. 2014). Regarding the habitat, black kites are ubiquitous throughout habitats, feed on almost any type of prey, including carrions, and exploit concentrations of food that are spatially and temporally unpredictable (Viñuela and Veiga 1992, Cortes-Avizanda et al. 2011). They are also frequently found in rubbish dumps, where they eat the human refuses (Blanco 1994, 1997, Palomino 2012).

Overall, nominal black kite shows certain nomadic or dispersive tendency after breeding but generally leave their breeding grounds in Eurasia between July and October, returning between January and June from the wintering grounds (Ferguson-Lees and Christie 2001, De Juana and Garcia 2015, Palomino 2016). As mentioned above, black kites largely relay on thermal updrafts over land and thus they tend to avoid the crossing of wide extensions of water where a lack of such vertical air currents requires flapping (Santos et al. 2017, 2020). This fact leads to large concentrations of birds along well-defined narrow fronts during migration. Specifically, along their migratory routes, large numbers of black kites can be observed at so-called migratory 'bottlenecks' (see below), where birds can be easily monitored by visual observation (Martín et al. 2016). This species has an extremely large distribution range all over the world, and its population size (> 1 million of mature individuals) ultimately makes it one of the most abundant raptors on Earth (BirdLife International 2020). The available information, points out to an overall decreasing trend in this species due to different human-made causes (BirdLife International 2020). However, black kite is globally considered as "Least Concern" due to population and range size criteria (IUCN 2016, Birdlife International 2020). In Europe, there are 81,200–109,000 breeding pairs estimated (i.e., about 162,000–218,000 mature individuals; BirdLife International 2015) and inference from migrating counts indicates that the European breeding population is increasing (8–9% annual rate of change during 1999–2013; Martín et al. 2016), although for populations beyond Europe the trend is mostly unknown (BirdLife International 2020). According to migration counts, black kite numbers in Europe exhibit one of the largest increases among the most common long-distance migratory raptor species (Martín et al. 2016, Vansteelant et al. 2019).

Knowledge Status

The black kite is one of the most well studied diurnal raptors of the Western Palaearctic, but knowledge about the migratory routes and the connection with the African wintering grounds are not yet completely understood (Panuccio et al. 2014). Data collected at fixed watchsites where black kites concentrate during migration have been used to delineate the phenology and to compare the importance of flyways in terms of population size, and assess population trends over time (Martín et al. 2016). The Strait of Gibraltar, between Spain and Morocco, and Batumi (Georgia), between the Black Sea and the Pontic Mountains, are the most important bottlenecks for the species, as more than 100,000 individuals are observed every season (Onrubia 2015, Wehrmann et al. 2019).

The marking with metal rings or specific marks (wing-tags, plastic rings) has provided complementary information about origins and destinations of black kites migrating through Eurasia and Africa. The first ringing campaigns of black kites in Europe started in the 1940s and continue today, summing up more than 25,000 ringed birds who produce more than two thousands of recoveries or resightings (SEO/BirdLife 2012, Bairlein et al. 2014). These recoveries show the role of the Iberian peninsula as a corridor for the breeding populations of Central Europe (France, Switzerland, Germany, Poland, even Italy and Czech Republic) and their connections with Western/Central Africa wintering areas (SEO/BirdLife 2012, Bairlein et al. 2014). However, Eastern Germany, Slovenia and Czech Republic may be situated in a zone of migratory divide, as some marked breeding birds were recovered on routes suggesting the crossing of the Mediterranean via Italy or the Balkans, with some recoveries in Uganda and Chad (Cepák et al. 2008, Kralj et al. 2013, Bairlein et al. 2014). The smaller populations from Scandinavia and the Baltic countries produce recoveries both through the western route (via Denmark or France) and the eastern route (Lebanon, Jordan) (Fransson and Pettersson 2001, Saurola et al. 2013). Black kites ringed in Western Russia (Volga Basin) produced recoveries southward in the northwestern shores of Caspian Sea (Volga Delta) and also in Israel (Northern Valleys, Eilat), while black kites from Altai (black-eared Kites *M.m. lineatus*) are recovered southward in India and Pakistan (Russian Ringing Scheme - http://rrrcn.ru/en/ringing/bd-). Finally, black kites ringed in Mongolia (*M.m. lineatus*) produce recoveries in Eastern India and Myanmar (Balachandran et al. 2018).

Black kites migration strategies have been described also through satellite telemetry. The most comprehensive study has been carried out in the Doñana population (Southern Spain), from which 92 individuals with different experience levels (aged 1–27 years) have been GPS tracked (Tanferna et al. 2012, Sergio et al. 2014), in one of the highest sampling efforts ever carried out for a raptor tracking research (total = 346 tracks). Other individuals have been tagged in Bulgaria, Ukraine, Czech Republic, Germany, Slovakia and Italy (Meyburg and Meyburg 2009a, Ovčiariková et al. 2020, Panuccio et al. 2018b, http://www.birdtelemetry.cz/en/maps/?druh=black-kite). Outside Europe, 13 black-eared kites (*M.m. lineatus*) have been equipped with GPS/GSM tags in Altai Territory (Southern Russia; http://www.birdtelemetry.cz/en/maps/?druh=black-eared-kite). Finally, 73 black kites tagged in Southern Spain with high-frequency-GPS devices with accelerometers gave new insights related to flight dynamics during the sea crossing at the Strait of Gibraltar (Santos et al. 2017, 2020, Marques et al. 2019). Finally, the flight behaviour of migrating black kites has been also studied by means of tracking radar, in different locations (Spaar 1997, Meyer et al. 2000, Mateos-Rodriguez and Liechti 2012, Horvitz et al. 2014).

According to these different sources, the main migratory routes of black kites can be summarized in this way: Western Mediterranean, Central Mediterranean, Eastern Mediterranean-Middle East, Central Asia and Eastern Asia (see Fig. 1):

(1) the Western Mediterranean flyway crosses the Iberian peninsula and funnels through the Strait of Gibraltar to reach wintering quarters in sub-Saharan Western Africa, mostly from Senegal to Nigeria. This route is followed mainly by birds breeding in Western and Central Europe

(2) the Central Mediterranean flyway involves the longest sea crossings and is followed mainly by birds breeding in Italy and, to a lesser extent, Central-Eastern Europe. This route shows strong seasonal differences, which will be discussed in the chapter. A much smaller migratory flow, links

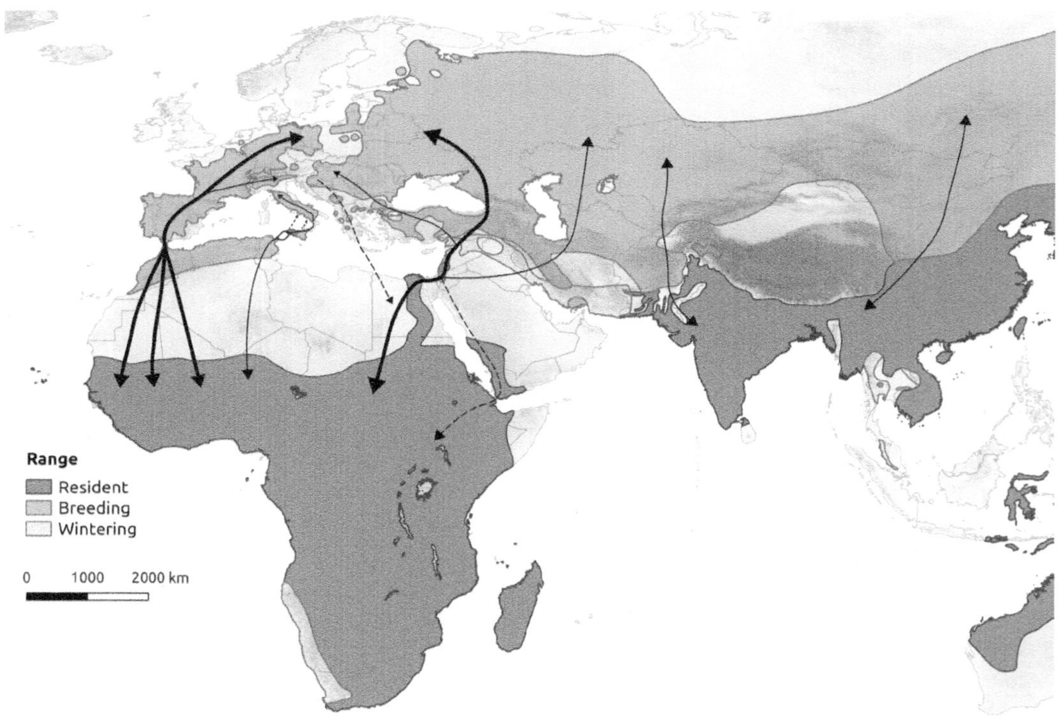

Figure 1. Schematic representation of the main routes followed by black kites. Dashed arrows (Eastern Mediterranean and Arabian Peninsula) represent routes followed mainly during autumn, while dotted ones (Tyrrhenian Sea) only during spring. See text for details.

Greece to North Africa. Once in Africa, black kites cross the Central Sahara to reach wintering quarters in the Sahel belt, from Senegal to Cameroon.

(3) Black kites from Eastern and Northern Europe tend to follow the Eastern Mediterranean-Middle East flyway, circumnavigating the Black Sea from both sides, and concentrating in the Bosphorus or mainly in Batumi. Then they cross Turkey on a broad front, but tending to concentrate in the Eastern coast of Mediterranean (Syria, Lebanon, Jordan, and Israel). From there, part of black kites cross the Sinai peninsula, to move southward following the corridor of the Nile or the coasts of the Red Sea, reaching mountains of Eritrea and Northern Ethiopia and the savannas of Sudan. Another flow is channeled through the mountainous corridor of the eastern shores of the Red Sea in Saudi Arabia, until reaching Yemen, and then crosses the Bab el Mandeb Straits. Most of these kites overwinter in the savannas of Chad, Central African Republic, Sudan and Ethiopia, but to lesser extent further south (Kenya, South Africa) or even in Egypt and the Arabian peninsula. This flyway is also used by black-eared kites (*M.m. lineatus*) coming from Central Asia (W Kazakhstan, Turkmenistan Uzbekistan).

(4) Black-eared kites from Central Asia (S Russia, E Kazakhstan, Turkmenistan and Uzbekistan) tend to fly around, or to cross, the higher mountains of Tian Shan, Hindu Kush and Himalayas to overwinter mostly in the Indus Valley in Pakistan and western parts of India.

(5) Black-eared kites from Eastern Asia (E Russia, Mongolia) tend to bypass the higher mountains of Himalayas on the east, to overwinter in the Indomalayan region (E India, Bangladesh, Myanmar).

Post-breeding Migration

For black kites breeding in Europe, autumn departure occurs during July-early September, with southern populations moving on average earlier than northern ones (mid-July vs early-August, respectively), and arriving to their wintering grounds in a variable period between September–December, till their final settlement in their wintering grounds in Tropical Africa (Table 2). During this post-breeding migration, they follow a journey of ~ 3.000–5.000 km along a broadly similar direct route as during their spring migration (Sergio et al. 2014, 2017, 2019, Ovčiariková et al. 2020), with similar pattern of dates and distances in black kites following different flyways, but tending to have, over the Sahara desert, more eastern-ward routes than spring ones (Table 2, Fig. 1). The Western Mediterranean route gathers mostly black kites from all Western Europe, which cross the Iberian Peninsula in a broad front, but funneling through the Strait of Gibraltar to cross in large numbers to North Africa and the Sahara until reaching the savannas of the Western Sahel (SEO/BirdLife 2012, Bairlein et al. 2014). Coming mainly from the forests of France, Germany and Switzerland, and to a lesser extent, from Poland, Czech Republic, Slovakia, Austria or even Northern Italy, thousands of black kites migrate on a broad front to reach the Pyrenees, that they try to cross avoiding the highest peaks of Central Pyrenees. Thus, more than 30,000 black kites concentrate in the mountain passes of Western Pyrenees (Lindux, Organbidexka, Lizarrieta), and to a lesser extent by the more eastern passes (2,500–5,500 in Eyne) (Urcun and Bried 1998b, www.migraction.net). In Organbidexka, black kite migration extends from July 15th to September 15th, with 90% of passage between July 25th and August 25th, and peak of migration at August 5th. The migration extends during all the daytime (7:00–18:00), but the number of migrants and the mean flock size were highest during the warmest moments of day, ranging 6–15 individuals (Urcun and Bried 1998b). The transit of the Iberian Peninsula occurs in broad front although mostly following a southwest direction (Fig. 1; Meyburg and Meyburg 2009a, Ovčiariková et al. 2020). Daily monitoring in the Strait of Gibraltar between 1999–2019 ("Migres programme") indicates that autumn migration extends for about 4.5 months (139 days), from July 5th (first contact crossing) to October 20th (last contact), although 80% of the black kites cross in 32 days, between August 1th and September 1th, peaking around August 16th (Table 3, Fig. 2). The maximum number of black kites crossing in one day was recorded on 15-August-2004, when 33.030 kites crossed the Strait after seven days of bad weather.

Table 2. Overview of spring and autumn migration distances and speeds of black kites along the Western, Central and Eastern Mediterranean, and Central Asia flyways based on satellite or GPS tracking data (Ovčiariková et al. 2020, Russian Raptor Research and Conservation Network 2020). Includes repeated journeys from the same individuals. Data have been pooled irrespectively of age classes. Spring routes from Central Mediterranean include also birds spending the summer in North Africa (Argelia, Tunisia).

Migration parameter	**Spring migration**		**Autumn migration**	
	Mean	**Range**	**Mean**	**Range**
Western Mediterranean (n = 15)				
Departure date	25-Mar	3 Jan–04 May	20-Aug	15 Jul–25 Sep
Arrival date	19-May	20 Apr–15 Jun	30-Sep	5 Sep–20 Nov
Travel duration (days)	51	30–125	42	15–70
Travel distance (km)	4,190	2,800–5,300	4,620	2,800–6,500
Speed (km/day)	101	63–177	125	90–210
Central Mediterranean (n = 9)				
Departure date	30-Mar	25 Mar–3 Apr	7-Aug	2 Jul–31 Aug
Arrival date	18-May	30 Apr–1 Jun	16-Oct	9 Sep–17 Dec
Travel duration (days)	45	25–60	51	15–80
Travel distance (km)	4,020	3,500–5,000	4,725	3,500–5,400
Speed (km/day)	109	58–200	120	70–233
Eastern Mediterranean-Middle East (n = 14)				
Departure date	12-Apr	20 Mar–20 May	9-Aug	25 Jul–25 Aug
Arrival date	5-May	15 Apr–9 Jun	24-Sep	12 Sep–27 Oct
Travel duration (days)	20	15–25	43	20–65
Travel distance (km)	3,480	2,520–4,130	4,905	4,100–5,520
Speed (km/day)	177	138–237	122	74–205
Central Asia (n = 18) (M.m. lineatus)				
Departure date	14-Apr	1Apr–1 May	4-Aug	5 Jul–11 Sep
Arrival date	17-May	23 Apr–30 May	16-Oct	3 Sep–3 Dec
Travel duration (days)	31	19–45	67	45–140
Travel distance (km)	3,475	2,950–5,210	3,795	2,815–5,415
Speed (km/day)	121	84–153	72	23–205

Table 3. Timing of black kite during autumn passage through the Strait of Gibraltar (Spain), Batumi (Georgia) and Eilat (Israel) (Shirihai et al. 2000, Verhelst et al. 2011, Onrubia 2015).

	Gibraltar (Spain)	**Batumi (Georgia)**	**Northern Valleys (Israel)**
Duration (days)	139		74
Duration 90% (days)	32	34	45
First record	05-July		26-Aug
5%	29-Jul	02-Sep	27-Aug
10%	01-Aug		31-Aug
50%	**16-Aug**	**19-Sep**	**13-Sep**
90%	1-Sep		4-Oct
95%	7-Sep	06-Oct	10-Oct
Last record	20-Oct		28-Oct

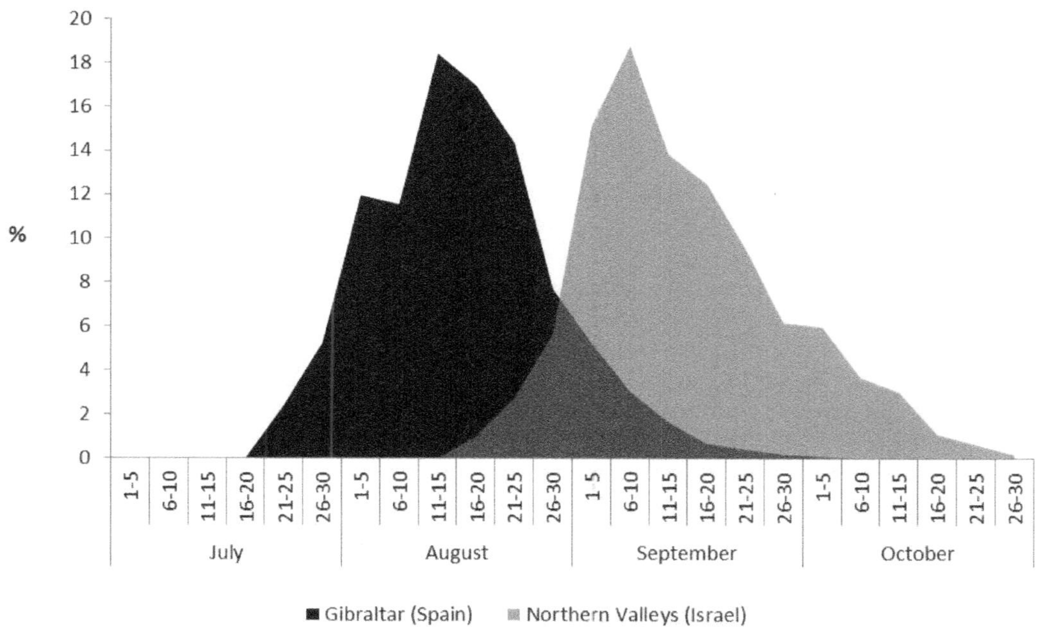

Figure 2. Timing of black kite during autumn passage through the Strait of Gibraltar (Spain) and Northern Valleys (Israel) (percentage daily passage) (Migres database, Shirihai et al. 2000).

About the time of passing, they were detected crossing the sea between 7:00 to 19:00 (12 hours), but almost 80% of passage concentrated between 10:00 to 16:00 (6 hours), when the stronger thermals occurred. When passing through this bottleneck, the black kite appears travelling in groups of variable size, from a few individuals to several hundreds. During post-breeding migration, the average size of the groups is 53.1 ± 0.84 (SE) individuals (ranging from 1 to 7,440 birds), with 80% of the groups containing less than 70 black kites. In general, the size of the flock is markedly greater than during the pre-breeding movement. About the age of black kites crossing, on a sample of 3.560 birds 31% were juveniles and no differences on timing of passage was detected depending on age. Regarding the magnitude of the passage, more than 224,000 black kites are estimated to cross every autumn season (Onrubia 2015), and the available data in the period 1970–2019 indicate a clear increase in the passage in recent years as the source population has burgeoned.

The Central Mediterranean route is used mostly by black kites belonging to the Italian population and, to a lesser extent, by some individuals, probably mainly juveniles, from Central-Eastern Europe (Spina and Volponi 2008, Ovčiariková et al. 2020). The migration peak occurs in a short period between the end of August and the beginning of September (peak August 30th), with a large overlap in the migration period of adults and juveniles (Agostini et al. 2000, 2004c). Flying southward, they tend to follow the Italian Peninsula, avoiding crossing the Tyrrhenian Sea. Upon reaching the Strait of Messina, located between the "toe" of the Italian peninsula and Eastern Sicily, black kites head towards the western side of the island, crossing the Mediterranean Sea via the Channel of Sicily and flying over the islands of Pantelleria and Marettimo, sometimes forming huge flocks (max. 1100 seen together; Agostini et al. 2000) that enable juveniles to learn the safest over-water routes from adults (Agostini et al. 2004c, Panuccio et al. 2018). The higher numbers recorded during autumn counts along this migration corridor (ca. 3.500 vs 1,000 birds), suggest that likely more black kites use the Central Mediterranean flyway during southbound movements (Corso 2001a, Panuccio and Agostini 2010b). By contrast, in Northwestern Italy, only a few hundred black kites are observed migrating through the Maritime Alps, mostly coming from Northern Italy, and probably Austria and

Czech Republic, primarily en route to the Strait of Gibraltar (Toffoli and Bellone 1996, Giraudo and Toffoli 2003, Ovčiariková et al. 2020).

Through the Eastern Mediterranean and Middle-East flyway there are black kites from the Balkans and other eastern countries (Hungary, Romania, Bulgaria, Ukraine), which are channeled through the Bosporus and Dardanelles, before crossing in wide front through Western Turkey, mixing with the local populations. However, part of the Balkans population crosses the Eastern Mediterranean between the Peloponnesus, Crete and Libya-Egypt (Lucia et al. 2011, Ovčiariková et al. 2020). In the 1950s and 1960s, few thousands black kites were estimated crossing through the Bosphorus (Porter and Willis 1968, Porter and Beaman 1985), but their numbers have dramatically decreased in surveys made in recent years (Fülöp et al. 2014, Panuccio et al. 2017), in agreement with the negative trend in populations of Eastern Europe recorded since the end of the last century (BirdLife International 2004). From there, black kites tend to fly along the shores of the Eastern Mediterranean, concentrating on the Gulf of Iskenderum (Belen Pass) to continue following the coasts of Syria, Lebanon, Israel (peak of passage September 9–13th), Jordan and Egypt (peak of passage September 13–15th), where birds from the small local populations are also incorporated (Fig. 2 and Table 3; Shirihai et al. 2000, Zalles and Bildstein 2000, Alon et al. 2004). In Israel, in most years, peak numbers occurred on, or around, the 9th September, and a secondary peak, often involving hundreds of birds, occurred in late September, and may reflect, at least in part, the arrival of locally wintering birds (Alon et al. 2004). Black Kites migrate throughout the area in a broad front and in loose small flocks (< 10), sometimes in loose drawn-out flocks of 20–30, up to 56, frequently flying at low speeds and very low altitudes (50–100 m above ground), but also up to 300–400 m and more (Meyburg et al. 2020).

Black kites coming from the important populations of Western Russia are funneled between the Caucasus and the Black Sea and concentrate in Batumi, where the passage of more than 160,000 individuals has been reported in 2017 (Georgia), mostly in September (peak September 17th) (Table 3; Verhelst et al. 2011, Abuladze 2012, Vansteelant et al. 2019, Werham et al. 2019). They move southwestward on a broad front, up to the Northern Valleys, the Dead Sea and the Negev, to join the main migratory flow and tending to converge in Sinai peninsula and Red Sea (Ovčiariková et al. 2020). However, the numbers recorded in Sinai (Eilat, Suez) are comparatively much lower than in spring, pointing out to a different route followed by black kites during post-breeding migration (Shirihai et al. 2000). Given the negligible numbers observed at the Bosphorus and the gulf of Iskenderun in Turkey (Fülop et al. 2014, Oppel et al. 2014), it is suspected that most of the black kites observed in the Lebanon corridor and Negev came from the Eastern Black Sea flyway (Wehrmann et al. 2019, Vanstelant et al. 2019). Moreover an important proportion of these kites were not *M.m. migrans*, but the eastern subspecies *M.m. lineatus*, which tend to migrate later and, although their routes are still poorly understood, appear to overwinter in increasing numbers in Middle East (Turkey, Syria, Israel), and probably in Eastern Africa, perhaps due to food availability provided by new rubbish dumps (Forsman 2016, Vansteelant et al. 2019, Meyburg et al. 2020, Ferguson-Lees and Christie 2001). However, information concerning migration of black kites entering Africa from Asia is incomplete and further research should verify if they move on a broad or a narrow front through the Middle East and the Red Sea.

Hundreds of black kites cross the Sinai peninsula during the second-half of September and October, moving southward following the corridor of the Nile or the coasts of the Red Sea, until reaching the mountains of Eritrea and north from Ethiopia and the savannas of Sudan. Another flow of black kites is channeled through the mountainous corridor of the eastern shores of the Red Sea (Hejaz and Asir mountains) in Saudi Arabia, until reaching Yemen, where a few dozen were seen passing through the Bab el Mandeb Straits, between Yemen and Djibouti (Welch and Welch 1988, 1989). In Sudan and Ethiopia, tens of thousands of black kites arrive at the end of September and early November, for overwinter in the region, while some black kites reach Congo basin in Central Africa and rarely Southern Africa (Zalles and Bildstein 2000).

On the Central Asian flyway, black-eared kites (*M.m. lineatus*) coming from Eastern Caspian Sea and Altai and Ural mountains (S Russia, E Kazakhstan, Turkmenistan, Uzbekistan) have to negotiate the crossing of the mountains of Tian Shan, Hindu Kush and Himalayas to overwinter mostly in the Indus Valley in Pakistan and western parts of India. The passage of hundreds or thousands of black kites has been recorded from mid-September to mid-November in the Tian Shan and Pamir Mountains in the great valleys of Wakhan river or on the Indus river, as well as in the foothills of the Himalayas mountains, up to Nepal and Bhutan (Zalles and Bilstein 2000, Subedi and DeCandido 2013, 2014, Juhant and Bildstein 2017, Subedi and Gurung 2018). Moreover, recent tracking results show that black-eared kites may also cross the Western Himalaya, flying for more than 500 km at altitudes > 3,500 m a.s.l., sometimes even of up to more than 6,500 m a.s.l. (Kumar et al. 2020).

Pre-breeding Migration

For black kites wintering in West Africa, spring migration starts during February with adult breeders arriving to the Iberian Peninsula at the end of March and to Western-Central Europe in mid-April, while immatures or non-breeders usually migrate later (Sergio et al. 2014, 2017). During this pre-breeding migration, they follow a journey of ~ 3.000–4.000 km along a broadly similar direct route as their autumn migration (Meyburg and Meyburg 2009, Sergio et al. 2014, 2017, 2019). Birds wintering in East Africa and Middle East start migration a bit later (early April) but arrive in Eastern Europe and Caucasus mostly during May, after travelling 2.500–4.000 km from Central and East Africa, along routes that don't differ substantially from autumn ones (Table 2, Fig. 1). According to birds breeding in Southwestern Spain (data from GPS-satellite tracked birds, n = 92, Sergio et al. 2014), kites departed from Africa over 5 months (23 January–23 June). This wide range was dictated by the sequential departure of different age contingents (Sergio et al. 2014), and the departure date during spring movements is probably the most important factor in terms of the overall migration performance of black kites (Sergio et al. 2014). The main corridors used during spring migration in the Mediterranean are shown in Fig. 1.

Similarly to other raptor species, the diffuse spread of migration paths across the Sahara in West Africa are usually slightly further west than the autumn ones as a result of seasonally varying winds over the Sahara (Klaassen et al. 2010, Vansteelant et al. 2017, Vidal-Mateo et al. 2016), converging towards a corridor in the Atlas and Rif mountains in Morocco and Algeria before crossing the Mediterranean Sea at the Strait of Gibraltar. Once the birds reach the Iberian Peninsula they continue in a broad front to their ultimate destinations, usually avoiding flights over higher mountains in the main ranges (Sierra Nevada, Central Pyrenees, Alps; Ovčiariková et al. 2020).

Tracked black kites from wintering areas in Western Africa to Spain show a very similar pattern of migration when compared to their autumn performance: they needed 18 days to complete the spring journey from Senegal to the Iberian peninsula (≈ 3,100 km) flying a daily average of 180 km/day (usually 6–8 hours/day), with some days of stopover, and spending 8 days for the Sahara crossing and 6 days for passing north-African mountains in Morocco (Sergio et al. 2014). The main migratory passage occurs at the Strait of Gibraltar, where more than 165,000 kites are estimated to cross every year (Onrubia 2015). Data obtained in the daily monitoring program in 2009–2012 (Migres Program) showed that spring migration in the Strait of Gibraltar extends for about 4 months (129 days), from February 2th (first contact crossing) to June 10th (last contact), although 80% of the black kites cross in 70 days, between February 23th and May 3th, peaking around March 15th (Table 4 and Fig. 3). The duration of spring passage doubles that of the autumn (70 days vs 32), caused by the more vague and delayed arrival of immatures or non-breeding individuals (Onrubia et al. 2011b). The maximum number of black kites crossing in one day was recorded at 3-March-2009, when 15,827 kites were detected (Migres Programme database). About time of passing, black kites cross the sea between 7:00 to 21:00, but more than 95% of passage concentrated on 6 hours, between 10:00 to 16:00, with maximums at 12:00–14:00, when stronger thermals occur. When passing through this bottleneck, they fly in groups of variable size, from a few individuals to several hundreds of birds. During spring migration, the average size of these groups is 14.0 ± 0.3 (SE) individuals (range of 1 to 902 birds), and 80% of the groups are made up of less than 20 black kites. About the age of black kites crossing,

Table 4. Timing of black kite during spring passage through the Strait of Gibraltar (Spain) and Eilat (Israel).

	Gibraltar (Spain)	**Eilat (Israel)**
Duration (days)	129	121
Duration 90% (days)	80	18
First record	02-Feb	20-Feb
5%	20-Feb	18-Mar
10%	23-Feb	21-Mar
50%	**15-Mar**	**28-Mar**
90%	02-May	6-Apr
95%	10-May	10-Apr
Last record	10-Jun	20-Jun

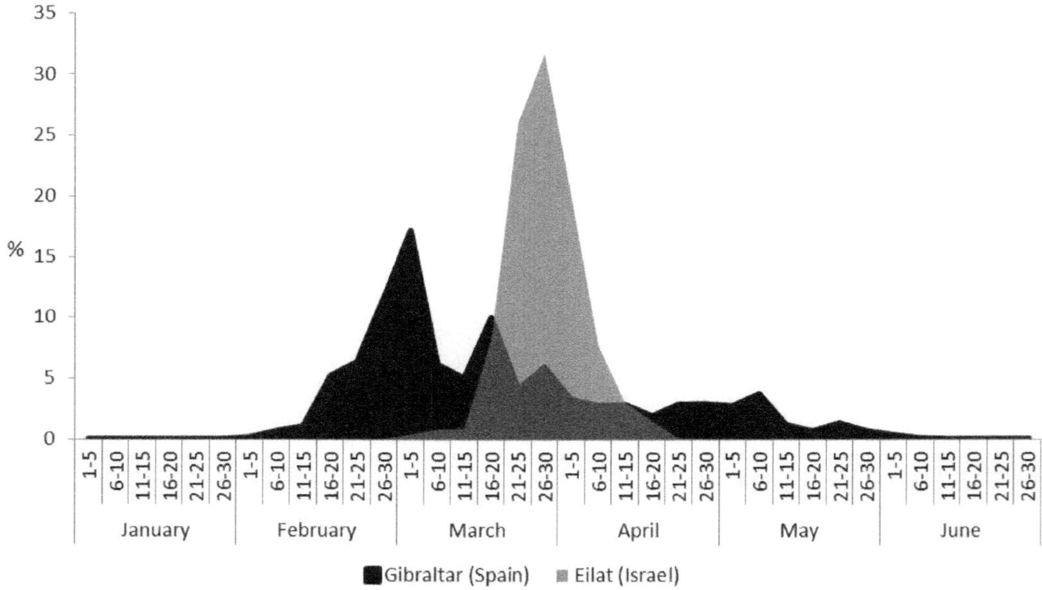

Figure 3. Timing of black kite during spring passage through the Strait of Gibraltar (Spain) and Eilat (Israel) (percentage daily passage) (Migres database, Shirihai et al. 2000).

on a sample of 737 birds, 67% were adults and 32% were immatures. In agreement with satellite tracking data (Sergio et al. 2014), adults tended to cross earlier than immatures, with their peak of migration at February 27th while immatures peak at May 20th.

In the Central Mediterranean, black kites coming from Western-Central Africa, cross the Sahara on a broad front and concentrate at the Cap Bon peninsula (NE Tunisia) before crossing the Mediterranean Sea (150–200 km) through the Sicilian Channel. Few individuals have been observed migrating outside this area, as shown by research made at the islands of Malta, Lampedusa and along the Sardinia-Corsica corridor (Beaman and Galea 1974, Premuda et al. 2007a, Corso et al. 2009). Upon reaching Western Sicily, black kites tend to migrate on a broader front with a passage of hundreds of birds through the Tyrrhenian Sea, in particular via the islands of Ustica and Panarea (Panuccio et al. 2004, Mellone et al. 2007). In the Strait of Messina, black kites summing up around one thousand birds, with peak of passage of adults between late March- early April (Corso 2001a, Panuccio and Agostini 2010b), while through the Tyrrhenian Sea a maximum of 302 birds were counted at the island of Ustica in 2002 and 109 at Panarea, in 2006 (Corso 2001a, Panuccio et al.

2004, Mellone et al. 2007). In this area no individuals were recorded migrating during the first half of March (Agostini and Malara 1997), but hundreds of birds were reported migrating later in the season, between late April and early May, including a significant numbers of immatures (Agostini and Duchi 1994, Agostini and Logozzo 1998, Panuccio et al. 2004, Panuccio and Agostini 2010b). Unlike during autumn, birds migrate singly or in small flocks, rarely comprising tens of individuals (Panuccio and Agostini 2010b). After crossing the Strait of Messina, the species is virtually absent at watch-sites located on the Adriatic coast of Italy (Gustin and Sorace 2004, Premuda et al. 2004, 2008), suggesting that the individuals migrating through the Central Mediterranean in spring belong mainly to the population of the Italian peninsula and do not cross the sea en route to the Balkans, differently from other species (Premuda et al. 2004, 2008). Interestingly, some tens of black kites are observed in March in Northwestern Italy (Arenzano on the Ligurian Apennines and the Apuan Alps) moving towards the Northern Apennines (Premuda and Baghino 2012), probably coming from the Strait of Gibraltar. Overall, the pre-breeding migration through the Central Mediterranean, in comparison to the post-breeding one, involves much less individuals, occurs on a broader front and, unexpectedly, is less concentrated in time (Panuccio et al. 2014). The phenology is delayed in comparison with what occurs in the Western flyway, probably in response to the different timing of breeding of the populations from the Italian peninsula (Panuccio and Agostini 2010b). Finally, the seasonal difference in the amount of migrating black kites may be connected both with the lack of juveniles summering in Africa and with the fact that during spring, hundreds of black kites breeding in Northern Italy do not cross the Central Mediterranean as they probably do in autumn and instead enter in Europe from the Strait of Gibraltar, shaping thus a loop migration pattern, which however needs be confirmed through satellite telemetry.

Black kites wintering in East Africa (Sudan, Ethiopia) commences their movements in January–March, flying north-east until they reach the Red Sea, and then follow its western coastline northwards until Suez in Egypt and Eilat in Israel, where 10,000–30,000 are recorded every year from late January to mid-June, peaking in the last week of March and first week of April (Fig. 3 and Table 4) (Shirihai et al. 2000, Megalli and Hilgerloh 2013). Passage at Eilat, in some years, consists of two waves, the principal movement occurring in the second half of March, and a smaller one in the second week of April: until the end of March, most appear to be adults, but in late April-early May, second-year birds predominate (Shirihai et al. 2000). On the other hand, birds crossing through Bab-El Mandeb Straits or wintering in the Arabian Peninsula progress north following the Red Sea coastal plain, before eventually dispersing into a broad front across the Arabian desert, Jordan, Iraq, and Syria (Fig. 1) (Shirihai et al. 2000). In North-Eastern Egypt, Jordan and Southern Israel, the flyways begin to separate depending on whether birds head to breeding areas in Central Asia, the Middle East, or Western Turkey and the Balkans. Black kites from the Balkans, and presumably also from Central and Western Turkish breeding grounds, generally follow the eastern coastline of the Mediterranean north through Jordan, Israel, Lebanon and Syria reaching the Gulf of Iskenderun, from where they turn north-westward towards their breeding grounds. Most birds using this flyway reach the European continent via the Bosphorus or Dardanelles, where a few hundred are recorded between mid-March to end of May, peaking on March 30th (Collmait and Croxall 1967, Üner et al. 2010, Uysal and Tosunoğlu 2018). However, return passage in spring principally bypasses the Red Sea, through Sinai and the Levant, and thence via the Eastern Black Sea route, where more than 220,000 black kites are registered in Batumi (https://trektellen.nl/site/yeartotals/1047/2020/-1), with comparatively few entering Europe via the Bosphorus (Shirihai et al. 2000, Alon et al. 2004).

Behavioural and Ecological Remarks

The black kite is a medium-sized raptor with relatively high wing loading (3.1 kg/m^2; Agostini et al. 2015a), favourable for obtaining high inter-thermal gliding speed. As a result, they usually rely on soaring-gliding flight exploiting thermal and orographic uplift, where they can ascend at relatively high speed (1.5 m/s). In so doing, black kites reach variable heights above the ground, mostly between 200–600 meters, occasionally above 1,200 m, then gliding at moderate speed (9.9–13.6 m/s), resulting in a low-energy flight that leads them to make long daily movements with little energy consumption (Mateos-Rodriguez and Liechti 2012, Horvitz et al. 2014, Santos et al. 2017, 2020). However, comparing with obliged soaring raptors like vultures and eagles, black kites frequently employ flapping flight, being also capable to use efficiently weaker thermals, migrating for more hours throughout the day (Horvitz et al. 2014, Sergio et al. 2014, Santos et al. 2017). Indeed, detailed radar tracking of individuals on active migration over land revealed that the majority of time is spent on circling and gliding, but with an higher proportion of flapping in comparison with obligate soarers (8%), situating therefore the black kite in the same cluster of the marsh harrier (*Circus aeruginosus*; Spaar 1997). During migration, the temporal pattern of black kites obtained by direct observations and especially by tracked birds shows daily periods of intense activity of about 6–10 hours (average 8.5 h, typically from 8:00 a.m. to 3:00 p.m.), with flight speeds of 40–80 km/h, allowing them average displacements of about 180–220 km/day (maximum 663 km/day; Meyburg and Meyburg 2009a, Sergio et al. 2014, Santos et al. 2017). As the soaring-gliding flight consumes little energy, these birds do not need to forage en route, although occasionally they can make stops to feed, if they find suitable opportunities, for example rubbish dumps (Panuccio et al. 2014). In general, adult black kites show better migratory performance compared with juveniles and the development of migratory behaviour follows a consistent trajectory, gradual and prolonged, which is promoted by both individual improvements and selective mortality, mainly operating on juveniles during the post-breeding migration. During pre-breeding migration, early arrival to the breeding grounds is mandatory to ensure a high quality breeding site enabling a higher fitness, and it has been shown that early departure date is the most important variable in this scenario: even if some late migrating individuals were flying quicker than early ones, they could not compensate their delay (Sergio et al. 2014). Another interesting result of the survival analysis in Sergio et al. (2019) is that mortality rates are significantly higher also for kites that are 7–11 years old, probably because of a strong effort during the breeding season.

The dependence of these soaring raptors on the existence of updrafts, poses serious limitations for performing long movements over large bodies of water where there are no thermals. The longest water crossing performed regularly during spring migration, and thus involving experienced individuals which perform it deliberately, is the one across the Tyrrhenian Sea (Panuccio and Agostini 2010b), which entails at least 220 km. However, longer oversea flights are extremely rare and performed mainly by inexperienced juveniles (Ovčiariková et al. 2020). As a matter of fact, observations made in both Spain and Southern Italy confirm the reluctance of black kites to undertake water crossings; at the island of Marettimo, during autumn migration, they hesitated when facing the open sea by soaring over the island for many hours (Agostini et al. 2000), such as at the Cap Bon promontory during spring migration, especially when migrating in smaller flocks (Agostini and Duchi 1994). At the Strait of Gibraltar, where the opposite coastline is 14 km away in the narrowest point, they followed the shortest sea crossing. When weather conditions are not ideal for this short flight over water, especially on days with strong crosswinds, kites stop on the Spanish side of the Strait, sometimes for several days, accumulating by the hundreds or thousands (Bernis 1980, Finlayson 1992, Miller et al. 2016). In particular, black kite sea crossing performance have been specifically studied using GPS and triaxial accelerometer for 73 birds while passing the Strait of Gibraltar (Santos et al. 2020): most crossing attempts were made at the narrowest stretch of the Strait and many individuals attempting to cross turned back upon reaching the sea. With optimal weather conditions, black kites needed 20–30 minutes to cover the 14 km across the Strait, with speeds of 36–50 km/h, starting to 600–1,200 m.a.s.l and arriving on the Maroccan coast at an altitude of 20–100 m.a.s.l., soaring and climbing rapidly as soon as they reach the coastline (Santos et al. 2020). Sea crossing performance was driven by their age, the wind conditions, the starting altitude and distance to Morocco, with young birds flying longer paths and reaching lower altitude above the sea than adults, thus assuming more risks. Strong crosswinds during sea crossings forced birds to extend the flapping flight at low altitude, and the way in which birds start the sea crossing was crucial for their success, particularly the starting altitude, which dictated how far birds can reach with reduced flapping effort.

The current state of research shows that migration timing of the black kite largely varies across the Mediterranean basin (see Table 1) thanks to the highly plasticity in migration behaviour of this species (Scholer et al. 2016, Panuccio et al. 2017). Compared to the Strait of Gibraltar, the migration occurs later in the Central Mediterranean and around the Black Sea (Kirwan et al. 2008, Sergio et al. 2009, Panuccio and Agostini 2010b). Such delayed spring migration is likely due to different factors as climate, population density, the seasonal timing of migration and the distance between breeding and wintering grounds (Panuccio et al. 2014). In general, among birds migration is more time-selected during spring (Nilsson et al. 2013) and, therefore, such a pattern is quite unusual also from this point of view. In particular, birds are expected to migrate more quickly during spring in a scenario of intra-specific competition during the breeding season (Kokko 1999). In agreement with research made by the visual observations that we reviewed, different authors (Meyburg and Meyburg 2009, Sergio et al. 2014) showed that black kites breeding in Germany and Spain covered the migratory distance during autumn migration more rapidly than during spring movements, unlike other raptor species (Mellone et al. 2012a, Nilsson et al. 2013). This is evident also in the Central Mediterranean flyway, where the migratory flow is much more concentrated in autumn rather than during spring. Some authors speculated on this pattern, suggesting that it may be related to higher competition for resources in the African wintering areas where migratory European birds meet resident individuals (Agostini 2002, Panuccio et al. 2014). As a result, black kites should minimize time during autumn migration, and some birds crossing the Mediterranean at the Strait of Gibraltar in spring could use loop migration passing through the Central Mediterranean in autumn (for instance, birds belonging to the important population at the foothills of the Italian Alps; see also Premuda and Baghino 2012).

21

Rough-legged buzzard *Buteo lagopus*

Nicolantonio Agostini

ABSTRACT

The rough-legged buzzard (*Buteo lagopus*) is a Holarctic migratory species breeding in the Arctic and subarctic regions and wintering in a band from 40° to 55°N. The nominate subspecies *B. l. lagopus* breeds in Scandinavia and eastwards to regions between Ob and Yenisey, and winters mainly in Central and Eastern Europe until Central Asia, including a few in Southern Scandinavia and East England. During both post- and pre-breeding movements highest concentrations occur in the Baltic region, peaking in late September to mid-October and in mid-April to early-May. Substantial numbers undertake sea-crossings through the South Baltic, the Gulf of Bothnia and the Gulf of Finland. However, larger numbers are expected to leave and enter Scandinavia via SE Finland, some passing through the Karelian Isthmus and along the Vistula Spit. An adult female trapped in the wintering ground in The Netherlands and tracked by satellite during pre-breeding migration used this pathway, reaching the breeding ground in the Kola Peninsula after two months. A total of 112 individuals were tracked during pre-breeding migration in Eurasia and North America, showing that their movements respond to snowmelt progression. During post-breeding movements, adults were seen migrating slightly earlier than juveniles, and adult females slightly earlier than adult males. Observations made on rough-legged buzzards wintering in Southern Sweden suggest that in this species adults tend to migrate further than juveniles and, among adults, males further than females.

Introduction

The rough-legged buzzard (*Buteo lagopus* Pontoppidan, 1763) is a medium-sized broad-winged raptor with an average body mass of 1.0 kg (male: 0.85 kg; female 1.22 kg) and a wing span ranging between 127 and 148 cm. It is a Holarctic migratory species that breeds in the Arctic and subarctic regions and winters in a band from 40° to 55°N. It also shows nomadic and irruptive movements, mainly in years of extensive snow cover in their usual wintering grounds (Yamaguchi et al. 2017). Three subspecies live in Eurasia (*B. l. lagopus*; *B. l. menzbieri*; *B. l. kamtschatkensis*), one in North America (*B. l. sanctijohannis*), with a world population of 300,000–1,000,000 mature individuals (Status: Least Concern; BirdLife International 2020). This chapter refers to the movements of the nominate subspecies *B. l. lagopus* which breeds in Scandinavia and eastwards to regions between Ob and Yenisey, and winters in Central and Eastern Europe until Central Asia, with a few in Southern Scandinavia and East England

MEDRAPTORS (Mediterranean Raptor Migration Network), Rome, Italy.
Email: nicolantonioagostini@gmail.com

(Ferguson-Lees and Christie 2001). The European population has been estimated in 37,200–79,200 pairs showing wide annual fluctuations (Saurola et al. 2013, BirdLife International 2017a). As a matter of fact, the breeding success of this species is strictly related to the lemming and vole cycle, these small mammals forming 80–90% of its diet. It mostly hunts birds when rodents are scarce (Pokrovsky et al. 2014). Breeding success is also negatively affected by extreme weather events which may become more common with climate change (Pokrovsky et al. 2014). The rough-legged buzzard breeds in the tundra, needing open country with low rough vegetation to hunt, but extending into the northernmost taiga. The breeding season lasts from mid-April to early-September, and nests are built on rock, low hummock in tundra or occasionally on trees. Clutches consist of 2–5 eggs, up to seven in top vole year, but it does not breed at all when rodents are scarce (Hagen 1969). The density of breeding pairs in the Siberian tundra was found to be correlated with the small mammal density, but, interestingly, that study did not report a similar effect on clutch size (Potapov 1977). The rough-legged buzzard winters in more varied habitat, such as farmland, coastal marshes, wetlands, steppes and prairies, forming communal roosts (Ferguson-Lees and Christie 2001).

Knowledge Status

Studies on the migratory movements have been made by direct visual observation, ringing recoveries data and satellite tracking. Survey on both post- and pre-breeding migration were made mostly in the Baltic region, where at several watchsites in Denmark, Sweden, Finland and Poland a significant passage has been recorded (Zalles and Bildstein 2000). At Falsterbo, autumn movements have been studied also in relation to sex and age classes (Kjellén 1992, 1998). A total of 112 rough-legged buzzards have been tracked during pre-breeding migration in a large scale study both in North America and Eurasia (Curk et al. 2020), while a bird has been tracked during pre-breeding migration from The Netherland to the Kola Peninsula (Schaub et al. 2018). Finally, thousands of rough-legged buzzards

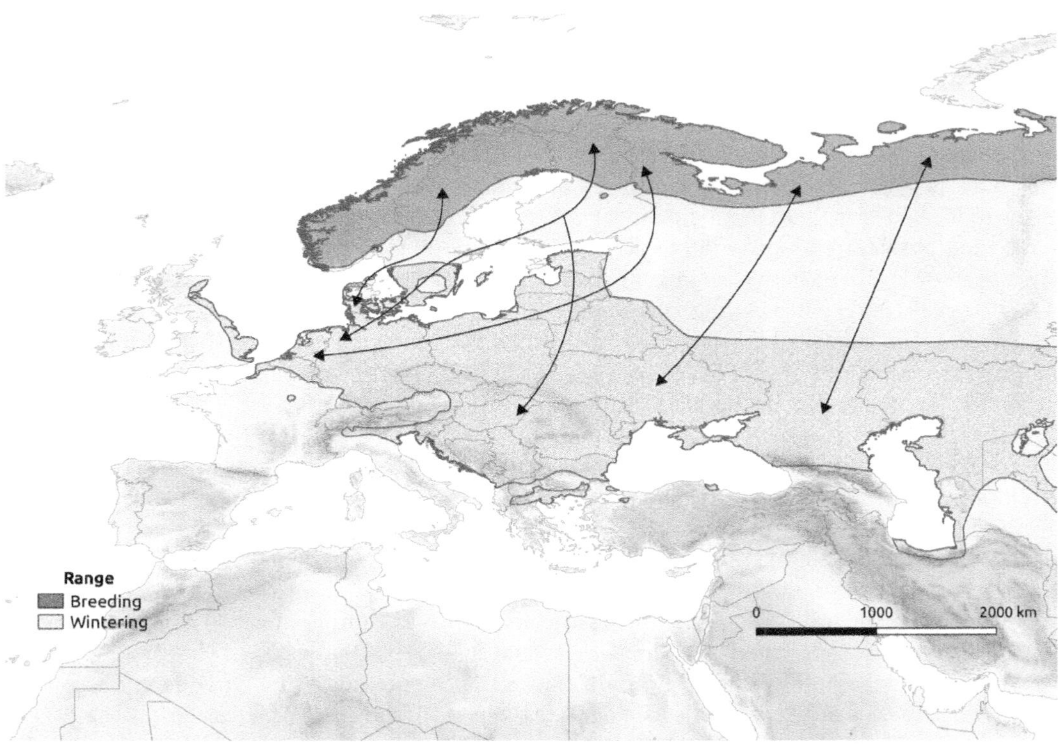

Figure 1. Schematic representation of the migratory routes followed by rough-legged buzzards breeding in Europe.

have been ringed in Finland with recovered birds providing data on their wintering distribution (Saurola et al. 2013).

Migration Patterns

Post-breeding movements start in late-August, peaking in late-September to mid-October, and occur on a broad front (Fig. 1). Substantial numbers undertake water-crossings through the Gulf of Bothnia, via the Åland Archipelago (between Southwestern Finland end Eastern Sweden), and through the Finland Gulf concentrating at the Porkkala Peninsula, where on average 400 birds per year were counted in the period 1980–1993 (Zalles and Bildstein 2000). There, rough-legged buzzards probably use the Estonian island of Naissaar (located 40 km S of the peninsula) as a stepping stone. However, larger numbers are expected to avoid the water crossing, moving through SE Finland into Northwestern Russia (e.g., 3,222 counted between 9–12 October 1982; Ferguson-Lees and Christie 2001), some funneling through the Karelian Isthmus and recorded at Kellovuori, where on average 716 birds were counted per season in the period 1993–1997 (maximum one day count 1,155 on 9 October 1982; Zalles and Bildstein 2000). In addition, a few hundreds were seen following the southern coast of the Baltic Sea passing along the Vistula Spit (Poland); there, on average 280 rough-legged buzzards were counted in the period 2008–2011 (maximum 415 in 2008; Polakowski et al. 2014). In Sweden, significant numbers of birds head SW crossing south Baltic Sea and concentrating at Falsterbo (average 889 in the period 1973–2019; maximum 2,733 in 1978), Bornholm (1,151 between 17–19 October 1978) and Blekinge (Zalles and Bildstein 2000, Ferguson-Lees and Christie 2001, Gensbøl, B. 2008; https://www.falsterbofagelstation.se/). Telemetry data of Swedish common buzzards (*Buteo buteo*; Strandberg et al. 2009b) showed that adults prefer to cross the Baltic Sea close to Helsingborg (north of Falsterbo) where the distance over water is even over four times shorter; it is likely that this occurs also among rough-legged buzzards. As regards migration phenology, at Falsterbo adults were seen migrating slightly earlier than juveniles, and adult females slightly earlier than adult males. Second-year birds were the first passing, probably because they complete moult before older birds (Kjellén 1992). However, some adults detected at this watchsite were still moulting flight feathers, suggesting that they urge to reach their wintering grounds especially in year of poor rodents' population in their breeding grounds. Some irruptive movements of juveniles on the southernmost areas of the wintering range occur in year of good breeding success.

Also during pre-breeding movements highest counts have been reported in the Baltic region. At Skagen, located at the northern tip of Jutland (Denmark), on average 1,153 rough-legged buzzards have been counted per season in the period 1972–1998 (maximum 2,718 in 1978) peaking mid-April–early-May. In Eastern Denmark, a substantial passage was also observed at Rørvig (maximum 1,443 in 1989) and Hellebæk (maximum 600 in 1987; Zalles and Bildstein 2000). Such as during post-breeding movements, substantial numbers of birds cross both the Bay of Bothnia and the Gulf of Finland concentrating at Valsörarna (Southwestern Finland; average 2,121 per season in the period 1974–1993) and at the Porkkala Peninsula (average 600 per season in the period 1980–1993), respectively. Significant numbers have been also detected along the Bay of Bothnia at the Hailuoto island (Zalles and Bildstein 2000). Large numbers are expected to reach Scandinavia passing through SE Finland on a broad front. There, hundreds of rough-legged buzzards were seen concentrating at Simpele (average 551 per season in the period 1980–1993), located north of the Karelian Isthmus (Zalles and Bildstein 2000). An adult female trapped in her wintering ground in The Netherlands and tracked by satellite during spring migration used just this pathway, flying around the Gulf of Finland and passing through the Karelian Isthmus, but reaching the Kola Peninsula after two months performing unexpected detours in Northern Scandinavia (Schaub et al. 2018). Snow directly influences resource availability, and the movements of 112 birds tracked during pre-breeding migration in Eurasia and North America showed that rough-legged buzzards respond to snowmelt progression, probably using snow presence as a cue to time their arrival at their breeding grounds (Curk et al. 2020).

Thousands of rough-legged buzzards have been ringed in Northern Finland, nearly all as nestlings. Ring recoveries data showed that Finnish birds winter in a wide area between Denmark and Western Russia, with the furthest recovery 2,796 km from the natal site. More than half of ringed nestlings reported dead have been recovered during the first year of life (Saurola et al. 2013).

Behavioural and Ecological Remarks

In a study made in Southern Sweden, Kjellén (1994) compared the proportion of rough-legged buzzards among migrants and wintering birds. Unlike other species of raptors wintering there, the proportion of juveniles was significantly higher among wintering birds than migrant ones, suggesting that in this species adults tend to migrate further than juveniles. Among adults, such difference has been found also among females, but it was not significant. These results partially agree with studies made in North America where, in agreement with the social-dominance hypothesis and the Bergmann's rule, juvenile rough-legged buzzards clearly migrate further than adults and, among adults, males further than females (Gauthreaux 1985, Palmer 1988). A further evidence that juvenile migrate shorter distances than adults in Europe has been given by an influx of rough-legged buzzards in Southern Germany in January 1987, caused by extensive snow cover in Central Europe. There, only 25% of wintering birds were juveniles, compared to 68% of wintering birds in Southern Sweden the same winter (Dobler et al. 1991, Kjellén 1994). A study made in Central-Eastern Poland has shown that the number of rough-legged buzzards wintering there greatly varies among years, being higher in years with lower daytime temperature but, interestingly, being not affected by the depth of snow. As regards the age and sex structure, adults outnumbered immatures and among adults females outnumbered males (Kasprzykowski and Cieśluk 2011).

As expected for a vole specialist, the proportion of juveniles among both migrant and wintering birds in Southern Sweden is correlated with the annual fluctuations in the quantity of rodents in the breeding grounds (Kjellén 1994, 1998).

22

Long-legged buzzard *Buteo rufinus*

Guilad Friedemann,[1,*] *Yossi Leshem*,[1] *Gil Bohrer*,[2]
Avi Bar-Massada[3] and *Ido Izhaki*[4]

ABSTRACT

The long-legged buzzard (*Buteo rufinus*) is a medium-sized, diurnal, raptor species with a Southern Palearctic distribution. Although the classic literature indicates that some of its populations are migratory, partially migratory or resident in different areas, its actual status in various areas and different populations is not fully clear. This uncertainty is also manifested by its weak passages, at very small numbers, at the main raptors bottlenecks. Recently, the movement ecology of long-legged buzzard has been studied in the Judean Israeli population, and has revealed an exceptional movement pattern. This population performs a poleward migration from their southern breeding grounds in Israel to their northern non-breeding grounds in Syria, Turkey, and Russia, where long-legged buzzards spend boreal summer and autumn. Therefore, they show an opposite migratory direction in comparison with the majority of other avian species.

Introduction

The long-legged buzzard (*Buteo rufinus* Cretzschmar, 1829) is a medium-sized and broad-winged diurnal raptor species (weight: 1,100–1,600 g; wing span: 115–170 cm) which belongs to the *Accipitridae* family and to the *Accipitriformes* order (Ferguson-Lees and Christie 2001). It is a polymorphic species, with individuals that can be divided into four main plumage types: pale; intermediate; rufous; and much less frequently, dark morph (Forsman 1999). The long-legged buzzard has a Southern Palearctic distribution, from Southeast Europe to Northern Africa, and from Asia Minor to Northwest China. It has two subspecies: *rufinus* and *cirtensis*. The first breeds in Southern Turkey, Greece, Syria, Iraq, Iran, the Balkan peninsula, and around the coast of the Caspian Sea to Central Asia. The second breeds in Northern Africa and has a more southern distribution. Interestingly, a recent molecular phylogenetic study suggests that *cirtensis* may be taxonomically closer to the Eurasian buzzard *Buteo buteo* rather than to *B. rufinus* (Jowers et al. 2019). In this chapter, we will

[1] Department of Zoology, George S. Wise Faculty of Life Sciences, Tel Aviv University, Israel.
[2] Department of Civil, Environmental and Geodetic Engineering, The Ohio State University, Columbus, OH, USA.
[3] Department of Biology and Environment, University of Haifa at Oranim, Kiryat Tivon, Israel.
[4] Department of Evolutionary and Environmental Biology, Faculty of Natural Sciences, University of Haifa, Israel.
* Corresponding author: Gilad.fridman@gmail.com

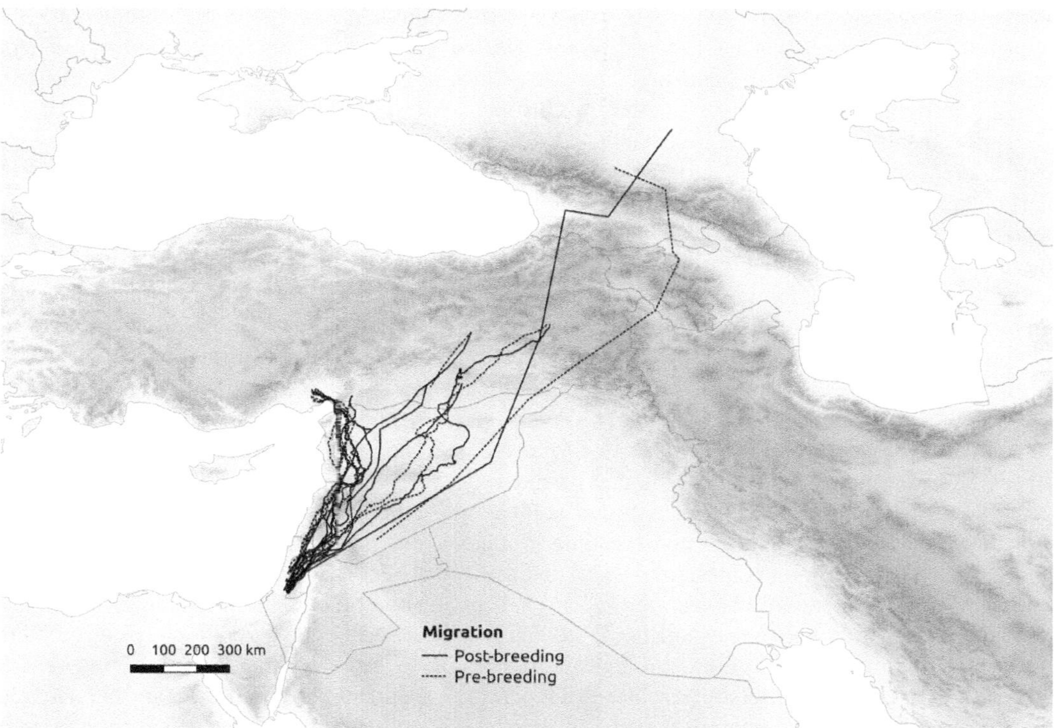

Figure 1. Migratory tracks followed by GPS tracked long-legged buzzards breeding in Israel and moving northward during the post-breeding migration.

refer exclusively to the nominal *rufinus* subspecies, which its preliminary global population size is estimated as ~ 69,500–113,000 breeding pairs (BirdLife International 2019a).

Long-legged buzzard is known to nest on cliffs and only occasionally on trees (Friedemann et al. 2011). While pairs are monogamous and return to the same territories each year (Friedemann et al. 2016) it is known as a territorial species, which aggressively defends its territory from interspecific and intraspecific intruders (Friedemann et al. 2017). It forages mainly in Mediterranean garrigue habitats and its diverse diet comprises mainly reptiles, birds and mammals (Friedemann et al. 2016).

Knowledge Status and Migration Patterns

Although the classic literature indicates that some long-legged buzzard populations are migratory, partially migratory or resident in different areas (Cramp and Simmons 1980, Shirihai 1996, Snow and Perrins 1998), its actual status in various areas and different populations is not fully clear (Cramp and Simmons 1980), and its movement ecology has been studied only recently in the Israeli population, through GPS telemetry (Friedemann et al. 2020). The species is regularly observed at major migration watchsites in Eastern Europe and in the Middle East, but never in consistent numbers. Moreover, all these various bottlenecks' records should be listed carefully as some observations need confirmation in view of possible confusion with similar species, such as the steppe buzzard (*Buteo buteo vulpinus*). For example, Shirihai et al. (2000) have dismissed high counts reported in the early 1980s from Suez (Egypt) during spring. Counts from Cap Bon (Tunisia; Thiollay 1975, 1977) were related to double counts of local, non-migrating, *cirtensis* individuals (Agostini 2005).

The most widespread bird migration pattern in the northern hemisphere involves annually repeated seasonal movements between northern breeding grounds and southern non-breeding grounds.

However, Friedemann et al. (2020) followed 8 breeding adults in Israel with high-frequency GPS transmitters, and showed that the strategies followed by these individuals contradicted this usual pattern by revealing that, once finished the breeding duties, they migrated northward to their non-breeding grounds in Syria, Turkey, and Russia, thus performing a poleward migration (Fig. 1).

Friedemann et al. (2020) suggest that the reverse pattern of poleward movements has an ecological explanation, and is likely related to food availability in their breeding and non-breeding grounds. In particular, long-legged buzzards occupy rich environments and track specific ecological productivity levels along latitude gradients during their breeding and non-breeding seasons. This is manifested by the maintenance of a relatively consistent NDVI level across their annual cycle. These low NDVI levels are linked to the foraging ecology of long-legged buzzards, which forage in relatively open areas characterized by low vegetation (Friedemann et al. 2016). Thus, this poleward movement allows these long-legged buzzards to exploit productive environments in northern latitudes, where resource availability peaks after the Israeli breeding season ends, and the availability of resources in the breeding territories decreases. Hence, this reverse movement direction is driven by a combination of conditions, which involves early breeding followed by a food-shortage in the breeding areas, and a specific flush of prey in summer at higher latitudes, not fully exploited by local breeders.

The tagged long-legged buzzards nested in their Judean breeding grounds between December-July, and migrated towards their non-breeding grounds in the northern latitudes of Syria, Turkey, and Russia. All migration routes followed a relatively well-defined NNE-SSW corridor, and the outbound migration started between the second half of June and the beginning of July. Two long-legged buzzards migrated to the same non-breeding locations during two and four consecutive years and used a roughly similar migration route each year. Long-legged buzzards spent an average of 3.8 (± 2.3 SD) days within the outbound routes. All individuals spent their non-breeding time at relatively limited areas, from mid-June until the end of November. The average duration of return migration was 4.8 (± 5.5 SD) days.

Visual observations from raptor-migration bottlenecks add a degree of uncertainty concerning the long-legged buzzard's actual migratory behaviour in different populations. For example, Shirihai (1996) indicated that long-legged buzzards passed over Israel in small numbers, probably involving birds wintering in Africa (Ferguson-Lees and Christie 2001); 62 individuals passed over the Northern Valleys during the 1989 autumn migration (mostly in October), and 105 individuals passed over Eilat during the 1989 spring migration (mostly in March). Similar numbers were recorded during 1990–2018 as part of the long-term migration surveys of the Israel Ornithological Center (The Society of the Protection of Nature in Israel - SPNI); an average of 33.2 (± 18.2) individuals passed over the Northern Valleys during the autumn migrations 1990–2013 and an average of 86.5 (± 36.4) individuals passed over Eilat during the spring migrations 2015–2018.

Other migration watchsites recorded similar numbers: during autumn migration, in 2008 and 2009, 88 and 109 long-legged buzzards, respectively, were recorded passing over Batumi (South Western Georgia; Verhelst et al. 2011), where 70 individuals have been counted during spring 2020 (https://www.batumiraptorcount.org/migration-count-data). 24 long-legged buzzards were recorded passing over the Bosphorus (Turkey) during autumn migration in 2008 (Fülöp et al. 2014), and 28 during spring migration in 2006 (Üner et al. 2006). Moreover, during the 2010 autumn migration, 7 and 15 long-legged buzzards were recorded passing over the Bosphorus and the Dardanelles (Turkey), respectively (Panuccio et al. 2017), while 37 individuals were recorded around the gulf of Iskenderum (Turkey) during autumn 2013 (Oppel et al. 2014). In Ayn Sokhna (Egypt), a more southern bottleneck, a total of 322 long-legged buzzards were counted during the 2012 spring migration survey (Megalli and Hilgerloh 2013), and at Bab el Mandab Straits 131 individuals were recorded in autumn 1987 (Shirihai et al. 2020).

Finally, the species is regularly observed, but in very small numbers, even along the Central Mediterranean flyway during spring migration, especially in the Strait of Messina and at Otranto Cape (Corso 2001a, Premuda et al. 2004). This weak passage (5–10 individuals per year), occurs mainly in the peak period of European honey buzzards (*Pernis apivorus*; late April–May), and probably involves individuals from Eastern Europe. Overall, such small numbers may reflect the low breeding densities of long-legged buzzards (Snow and Perrins 1998) or the fact that some populations are probably resident ones. Therefore, additional research, and especially additional GPS tracking across different areas of the species distribution, are critical in order to better understand its movement ecology, and by doing so, to better understand those necessary steps that should be taken for its conservation.

23

Eurasian buzzard *Buteo buteo*

*Ülo Väli** and *Paweł Mirski*

ABSTRACT

The Eurasian buzzard (*Buteo buteo*) is the one of the commonest Western Palaearctic raptors and an obligate soaring migrant avoiding crossing of large water surfaces. This flexible raptor species exhibits various migration strategies, which are largely attributed to subspecies. Westerly distributed Common buzzard *B. b. buteo* is partial, short-distance migrant wintering mostly in Western and Central Europe. The main migration direction is south-west along several minor migration corridors over land. The Common buzzard is a late and slow migrant in the autumn; spring migration is faster and it arrives early in breeding quarters. Easterly distributed steppe buzzard *B. b. vulpinus* conducts long distance migrations to sub-Saharan Africa. After the funnelling of birds from a large area from Eastern Europe and Western Asia, the main migration flight enters Africa in Djibouti south of the Red Sea (Bab al Mandab Strait); spring migration circumvents the Red Sea north and crosses Sinai. Autumn migration of steppe buzzard is earlier and spring migration later than those of the nominate form. Temporal changes in migration patterns of the Eurasian buzzard have been observed, being associated with climate change.

Introduction

The Eurasian buzzard (*Buteo buteo* Linnaeus, 1758) is a medium-sized diurnal raptor. Its breeding range covers nearly all of Europe, being absent only in its very northern (Northern Fennoscandia) and western margins (most of Ireland), and extends to Asia easterly up to Mongolia and China and southerly into South-Western Asia (southern coasts of the Black and the Caspian Sea). The Eurasian buzzard is the most numerous species among Western Palaearctic raptors (Ferguson-Lees and Christie 2001) and its population in Eurasia accounts for 2.1–3.7 million mature individuals, of which 1.6–2.8 million breed in Europe (BirdLife International 2017e). The success of Eurasian buzzard is based on its plasticity in habitat use. It occupies a wide range of habitats from managed woodlands interspersed with clearcuts to agricultural landscape with only small patches of woods (Melde 1971, Walls and Kenward 2020). The adaptivity of the species is further illustrated by its prey choice. Although the Eurasian buzzard prefers to forage on small mammals in general, it opportunistically chooses from wide array of prey species available in a particular area (Melde 1971). In various European regions, mice, voles, rabbits, birds, reptiles and amphibians form its staple diet (Walls and Kenward 2020).

Institute of Agricultural and Environmental Sciences, Estonian University of Life Sciences, Kreutzwaldi 5D, 51006, Tartu, Estonia.

* Corresponding author: Ylo.Vali@emu.ee

The Eurasian buzzard is divided into a number of subspecies (7–16, depending on author), many of them having an insular or otherwise restricted range. The subspecies fall into two main groups, which are named according to the numerously and spatially dominant subspecies *B. b. buteo* and *B. b. vulpinus*, respectively. Here we focus on these two subspecies, and refer to these with common names—common buzzard (*B. b. buteo*) and steppe buzzard (*B. b. vulpinus*), respectively.

Migratory behaviour of the two subspecies is strikingly different. Populations of westerly distributed nominate form are sedentary, partial migrant or short-distance migrant, while the easterly distributed Steppe buzzard is a long-distance migrant, performing one of the longest migrations of any bird of prey, up to 14,200 km (Newton 2008). Indeed, nearly a century ago, Grote (1930) pointed out that buzzards in the western range margin (Britain) are obligate residents, Scandinavian populations migrate to Western Europe and Siberian buzzards have their winter quarters in Southern Africa. However, the location and dimensions of the subspecies contact zone are still unclear and in the Baltic region, for instance, various authors have determined it differently (reviewed by Väli and Vainu 2015). Also the subspecies dividing in Fennoscandia has raised different views. Earlier references suggested that Southern Sweden falls within the range of nominate form whereas the northern and central parts are inhabited by the steppe buzzard (Rudebeck 1950). However, Ulfstrand (1977) did not find any bimodality in morphology of Swedish buzzards and concluded that the range border of the steppe buzzard lies east of the Baltic Sea. The same is actually suggested by the south-westward autumn migration directions of most buzzards throughout Sweden and in South-Western Finland, while only some birds in Northern Sweden and in most of Finland head south-east (Saurola 1977, Alerstam 1990, Saurola et al. 2013). Indeed, only few steppe buzzards are noticed as migrants in Falsterbo (Rudebeck 1950) and the bulk of buzzards is formed of nominate form conducting their short-distance migration to Western Europe (Alerstam 1990). On the other hand, all but one recoveries of birds ringed in Southern Africa have been made in Russia east of 30°E (Melde 1982, Alerstam 1990, Shmeleva and Haritonov 2016) confirming the link between migration pattern and subspecies status.

Why the two subspecies have so strikingly different migration patterns and wintering grounds? Alerstam (1990) suggests that low competition, only with northerly distributed rough-legged buzzard, enables common buzzard to stay in Europe whereas severe potential competition with various buzzard species in Asia and Africa drives steppe buzzard to make a leapfrog migration wintering in Eastern and Southern Africa. Although Walls and Kenward (2020) doubted the role of migration behaviour differentiating the two subspecies, we believe that similarly to sister species pairs in raptors (Väli et al. 2018), timing, routes and destinations of migration may be important drivers of speciation. Furthermore, different migration patterns are reinforced (or vice versa) by different moulting habits (Herremans 2001, Zuberogoitia et al. 2005) and different wintering sites are ecologically connected with size differences as the larger common buzzard winters in cooler climates while the smaller steppe buzzard in warmer ones (Alerstam 1990). Therefore, we carefully examine migration patterns among buzzards always keeping the division of two subspecies in mind.

Knowledge Status

As one of the most obligate soaring flyers, migration of the Eurasian buzzard strongly depends on weather conditions and it prefers an anticyclonic situation with light winds, sun, and typical cumulus formations (Rudebeck 1950, Bildstein 2006, Newton 2008). Suitable conditions for soaring occur on land and the Eurasian buzzard avoids water crossings as much as possible (Bildstein 2006). To bypass wide water bodies large flocks concentrate on land-bridges and the species is often recorded in large numbers in migration observation stations (Table 1). As regards the common buzzard, the best known observation site is Falsterbo, in Southern Sweden, while very few cross the Strait of Gibraltar in Western Mediterranean (Onrubia 2016). On the other hand, hundreds of thousands of steppe buzzards are observed at watchsites east of the Mediterranean and Black Seas. In addition to direct observations, radar has been used to follow the migration there (e.g., Leshem and Yov-Tov 1996, Spaar and Bruderer 1997, Hilgerloh et al. 2011). Away from land bridges, the number of migrating Eurasian buzzard is difficult to quantify, being distributed across wider areas. We propose that this obstacle may be overcome in the future by using the potential of citizen scientists. Numerous observations deposited in open online databases have a potential to quantify migration and study its phenology (e.g., Väli 2015b) across large areas, at least in densely inhabited European countries. The two migratory buzzard subspecies are sometimes difficult to tell apart, particularly when passing high above in large flocks (Bildstein 2006). This may have caused identification mistakes leading to some erroneous conclusions. For example, the increased frequency of steppe buzzard in the Central Mediterranean region in recent decades can be partly attributed to improved field-identification skills and knowledge, together with better optical aids (Sammutt and Bonavia 2004). Moreover, given the extensive introgressive hybridisation the correct identification of the two subspecies may be challenging. In conclusion, the subspecies is not always noticed at migration watchsites and often not separated in publications (Table 1). Another important source for migration data are ring recoveries. Rather many research projects have used colour ringing (European Colour-ring Birding 2020), but we are not aware of any published results analysing migration of colour-ringed birds. However, conventional ornithological rings used by professional and amateur ornithologists in the

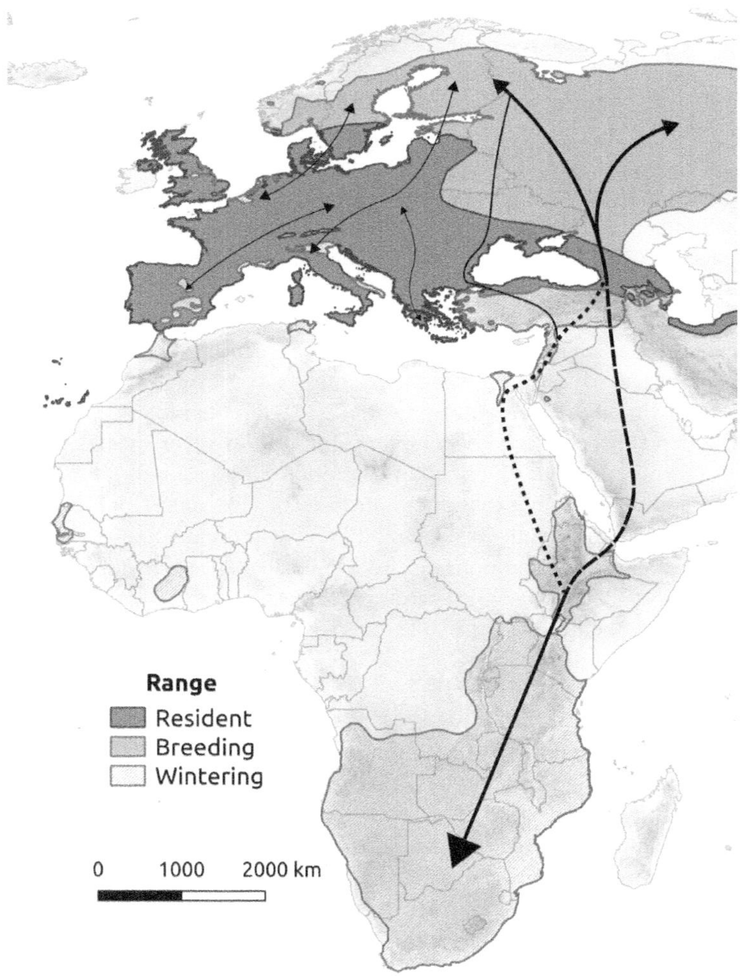

Figure 1. Schematic representation of the main migratory routes followed by the Eurasian buzzard. The dashed arrow, through the Arabian Peninsula, is followed only by steppe buzzard, almost exclusively during autumn, while the dotted arrow, through Suez, is followed mainly during spring.

course of monitoring programmes have provided large amount of information on migration. Many Eurasian buzzards are ringed as nestlings in breeding grounds and the massive ringing effort (e.g., more than 23 thousand only in Finland; Saurola et al. 2013) compensates low recovery rate, typical to ornithological rings. Ringing recovery data have been compiled in ringing atlases (e.g., Bønløkke et al. 2006, Saurola et al. 2013), databases (RRRCN 2020), general overviews on migration of bird species (e.g., Melde 1982, Shmeleva and Haritonov 2016) or detailed species-specific analyses (e.g., Saurola 1977, Väli and Vainu 2015). There are however, drawbacks in analyses of ring recoveries, because these originate mostly from Europe, i.e., on nominate subspecies, but less often from Africa on steppe buzzard. This might lead to the erroneous conclusion on abundance ratio of the two subspecies in breeding populations (see Saurola 2013). Only a small number of buzzards, all of the nominate subspecies, have been studied by means of telemetry. Pioneering work was carried out in Britain, where buzzards are sedentary. Walls and Kenward (e.g., 1995, 1998) have used radio-telemetry to evaluate their home ranges and dispersal. More recently, migration has been the subject of studies in Sweden (Strandberg et al. 2009a,b), Estonia (Väli et al. 2015, Väli 2017, Väli and Mirski unpubl. data) and Poland (Ptaki Silesiana 2020).

Post-breeding migration

Main routes

Highest numbers of common buzzard on migration in Europe have been registered at Falsterbo, located at the southern tip of Sweden. This is the concentration point of the flyway funnelling buzzards from the Scandinavian Peninsula, including birds from its northern part (Fig. 1; Rudebeck 1950, Alerstam 1990, Kjellén 1999). Also individuals from South-Western Finland join the Scandinavian flyway after crossing the Baltic Sea via Åland (Saurola et al. 2013). However, looking at the telemetry data of Swedish buzzards (Strandberg et al. 2009b), one may note that adults prefer to cross the Baltic sea close to Helsingborg (north of Falsterbo) where the distance over water is even over four times shorter. Probably juveniles are driven mainly by innate genetic mechanism, acting also in soaring raptors (Väli et al. 2018), directing them south-west while experienced adults have already navigational capabilities allowing them to use the optimal route. It means that overall numbers of birds migrating from Scandinavia are probably higher than those registered in Falsterbo. Exiting Scandinavia, buzzards settle in Denmark or head further to Germany, France, the Benelux countries or even Northern Spain (Alerstam 1990, Strandberg et al. 2009a). Although there is no clear subspecies divide detected in Sweden (Ulfstrand 1977, see above), northern birds leapfrog southern birds to winter in South-Western Europe or even in Western Africa (Alerstam 1990). Such medium and long-distance migration is more typical to steppe buzzards. Indeed, although majority of buzzards migrating in Falsterbo are nominate subspecies, few steppe buzzards are recorded too (Rudebeck 1950). Saurola (2013) suspects that easterly winds may also drive some Finnish steppe buzzards to use the western route when exiting the country. The same might happen in Sweden north of the Baltic Sea. On top of that, it is hard to predict movements of various level hybrids between the two subspecies. However, not many Eurasian buzzards these days pass the strait of Gibraltar. In the past, several thousands were counted (Bensusan et al. 2007), but nowadays only dozens occur annually (Onrubia 2016). Similarly, only very small numbers have been noted crossing the Sicily Channel and the Aegean sea (Agostini et al. 2005a, Lucia et al. 2011), the latter mainly leading to wintering in the island of Crete (Panuccio et al. 2019a). In addition to water bodies, mountains form an obstacle for migration too. Up to 20,000 individuals has been observed in Eastern France (LPO Haute-Savoie 2016) bypassing the Alps: few of these buzzards cross the Pyrenees (Filippi-Codaccioni et al. 2010). Ringing recoveries suggests that an important flow of buzzards cross the Alps in order to winter in Italy (Fig. 1; Spina and Volponi 2008, Saurola et al. 2013), however, still no watchsites where they concentrate during autumn have been found.

Another, somewhat smaller migration pathway circumvents the Baltic Sea on its southern coast (Polakowski et al. 2014). Several thousands of buzzards noted in autumn migration are coming from North-Eastern Europe, the Baltic States (Melde 1982) and Finland (Saurola et al. 2013), but probably also from North-Western Russia. Walls and Kenward (2020) suggested that common buzzards from Eastern Europe are likely to go around the east end of the Mediterranean Sea and cross into Turkey together with steppe buzzards. This might be true for Finnish and probably also Russian birds (Saurola 1977, Saurola et al. 2013, Shmeleva and Haritonov 2016), but ringing and GPS tracking of the breeders in the Baltic States clearly indicate their south-westward migratory direction leading to Central Europe, sometimes even further west and south (Melde 1982, Väli and Vainu 2015, Väli and Mirski unpubl. data). At least part of the breeding population in Central Europe is migratory too; migration is shorter but also directed towards south-west (Melde 1982). Various populations perform narrow-front migrations along this main direction (Alerstam 1990), but registered numbers are relatively low (e.g., Helbig and Laske 1989). However, even up to 580 individuals were noted in few hours during peak of migration in Western Poland (Czechowski 2004). Further east, a marked flyway has been recorded in Carpathians (Wilk et al. 2018). Up to almost twenty individuals were noted per day with flocks of maximum of 92 individuals, but mostly in numbers of one to five birds.

Birds breeding north-east of the Baltic Sea exhibit another migratory direction. Although buzzards in South-Western Finland follow general south-westerly migration route (194°; see above), many birds from rest of Finland, especially migrants departing early, follow a migration corridor to southeast towards the eastern coast of the Black Sea (176–178°; Saurola 1977, Saurola et al. 2013, Shmeleva and Haritonov 2016). A similar direction has been recorded for few birds breeding in Northern Sweden (Alerstam 1990, see also Kjellén 1999). However, birds from South-Central Finland, located on overlap of putative ranges of common and steppe buzzards, have intermediate mean direction (183°) and have been recovered in the Balkan Peninsula (Saurola 1977, Saurola et al. 2013). Individuals passing Dobruja (at Western Black Sea coast) were in great majority following a north-south axis, also indicating possible Finnish origin (Stanciu et al. 2017). Diversity of migration directions is even larger in Ukraine. Near Kiev, the dominant flight direction was south-westward (towards the Balkan Peninsula), significant portions flew also southward (towards the west coast of the Black Sea) and westward (towards Central Europe), while only a minority headed southeast (east coast of the Black Sea; Domashevsky and Demidenko 2009). Further east, near Kharkiv, most individuals migrated southward (i.e., towards the east coast of the Black Sea; Viter 2018). Only 270 km northwest, near Voronezh in South-Western Russia, as much as 1,042 buzzards were noted passing south and south-west in just few hours, but concentration was due to preceding 10 days of heavy rain (Kiselev 2005).

Many Eurasian buzzards concentrate on the flyway along the west coast of the Black Sea (Michev et al. 2011, Fülop et al. 2018) and leave Europe via Bosphorus (Fülop et al. 2014, Panuccio et al. 2017). Although the steppe buzzard dominates here, small numbers of the nominate form are recorded too. Notably, rather low numbers of buzzards are recorded in Iskanderun, at the eastern tip of the Mediterranean (Oppel et al. 2014). On the other hand, up to 300,000 steppe buzzards are recorded on the east coast of the Black Sea at the Batumi bottleneck, Georgia (Wehrmann et al. 2019). Further south, individuals from Bosphorus probably join the flyway coming from the east coast of the Black Sea. This flyway also collects birds from the eastern side of the Caspian Sea, including individuals from as far as 88°E (RRRCN 2020). Although some adult (but not 1st year) steppe buzzards may remain to winter in the Middle East (Saurola et al. 2013), the main flight now continues across the Arabian Peninsula until Yemen to cross Bab al Mandab Strait and enters Africa in Djibouti (Fig. 1; Yom-Tov 1984, Welch and Welch 1998, 1988, Shirihai et al. 2000). At Bab al Mandab, nearly 20,000 migrating Steppe Buzzards were recorded in 1985 and nearly 100,000 in 1987 (Welch and Welch 1988), while the flow through Israel and the Suez is much less important in this season (Shirihai and Christie 1992, Hilgerloh et al. 2011a). In Africa, some birds may stay wintering as close as in Ethiopia, but most birds probably continue their migration to southern regions of Africa (Melde 1982, Saurola 2013, Shmeleva and Haritonov 2016).

Timing

When migratory, the common buzzard is considered a rather late migrant in autumn. Swedish GPS-tracked birds started their migrations between late September and mid-October (Strandberg et al. 2009b). This fits well with the observations at Falsterbo where migration starts in September and peaks mainly in October (Kjellén 1992). In Estonia, east of the Baltic Sea, all 15 GPS-tracked buzzards started their migration somewhat later, during October (Väli et al. 2015, Ptaki Silesiana 2020, Väli and Mirski unpubl. data). The timing of completing migration was very variable both among Swedish and Estonian birds, even within a season, ranging from the end of September till the end of November (some being on the move even in December). However, observations in Falsterbo indicate that birds from Southern Scandinavia leave earlier than northern birds and adults earlier than juveniles (Kjellén 1992, 1999). In an observation station south of the Baltic coast the median of the migration was on 24th October and still a significant amount of individuals was passing in the first decade of November (Polakowski et al. 2014). This fits well with the data from Germany where the passage, but also migration of local birds, occurs mainly in October and early November and the median is on 23rd October (Helbig and Laske 1990, Holte et al. 2017). Carpathians are crossed mostly in the second and third week of October (Wilk et al. 2018).

In the North-Eastern and Eastern Europe, the migration of steppe buzzards occurs earlier. Migration peaks in early September in Finland (Saurola et al. 2013) and North-Western Russia (at the eastern coast of Lake Ladoga; Noskov 2016), at the end of September and the very beginning of October near Kiev in North-Central Ukraine (Domashevsky and Demidenko 2009) and mostly in September near Kharkiv in North-Eastern Ukraine (Viter 2018). In Dobruja, Eastern Romania, the highest numbers of Eurasian buzzards are recorded in late September and early October (median on 4th October; Fülop et al. 2018), but the migration peak of steppe buzzards occurs in the last days of September, while the nominate subspecies was observed later, in the first half of October (Stanciu et al. 2017). At Bourgas, Bulgaria, most of migrants are detected in October; the peak migration period stretched from 28th September to 3rd October, when on average 37% of the seasonal total occurred, with the overall peak date on 2nd October (Michev et al. 2011). At the Bosphorus, peak was on 30th September in 1972 (Beaman et al. 1975), and large number migrated between 22nd September and 10th October in 2008 (Fülop et al. 2014). Rather similar was timing in the other side of the Black Sea, in Batumi (Georgia), where the median of migration was on 29th September and 90% of buzzards migrated between 18th September and 13th October (Verhelst et al. 2011). At this bottleneck probably more eastern birds are migrating as well, for example from the Ural region of Russia where main autumn migration occurs in the middle of September (Karyakin 1998). Also in Northern Valleys of Israel 90% of birds migrate between 17th September and 21st October (peak counts between 25th September and 10th October; Alon et al. 2004). Entering Africa at Bab al Mandab Strait peaked 16–17 October in 1985 and 6–8 October in 1987 (Welch and Welch 1988). Finally, there are only few anecdotal observations on timing of autumn migration inside Africa. For example, in Rwanda, between lakes Victoria and Tanganjika, 3300 buzzards moved southward in October (Alerstam 1990). In the Western Cape Province, South Africa, adults arrived before first-year birds; peak populations of the former occurred from December to February, and of the latter from January onwards (Herremans 2001).

Speed

Autumn migration of common buzzards is rather slow, but varies remarkably between individuals, being probably affected by weather conditions (Walls and Kenward 2020). Shortest migration among Swedish buzzards lasted just 4 days (to Southern Sweden), the longest 64 days (to Belgium), and the average duration of migration was 19 days (Strandberg et al. 2009b). The mean distance covered was 702 km (range: 452–1449 km) and the mean daily distance covered was 84 km (range of individual means: 33–133 km per day; Strandberg et al. 2009b). Migration of twelve GPS-tracked buzzards from Estonia lasted 15–49 days (mean: 32 days) in 2019 (Väli and Mirski unpubl. data), but earlier a buzzard migrated from Estonia to Southern Germany in 67 days showing a mean daily flight distance of only 27 km (Väli et al. 2015). Another satellite tracked common buzzard from Estonia took around 5 weeks to cover a 1,160 km straight line to Czech Republic (Ptaki Silesiana 2020).

Pre-breeding Migration

Main routes

In general, spring routes are similar to those in autumn, although rarely buzzards follow exactly the same route on the way back (Väli et al. 2015, Ptaki Silesiana 2020, Väli and Mirski unpubl. data). The most significant difference is the outward route from Africa made by steppe buzzards. Only small numbers leave Africa in spring through Bab el Mandab, i.e., from Dijbouti to Yemen (Welch and Welch 1998). Instead, most buzzards continue the flight northward along the west coast of the Red Sea, and enter Asia from Israel, showing therefore a loop migration pattern (Fig. 1). For example, in Eastern Egypt, where only tens of Steppe Buzzards are counted in autumn, tens of thousands are seen in spring and even more than hundred thousand in North-Eastern Egypt, close to Suez (Hilgerloh 2009, Megalli and Hilgerloh 2013; Table 1). Further, spring flight follows the east coast of the Mediterranean and many are observed in observation stations in Israel (Table 1). Over 98% of buzzards caught and described in Israel are steppe buzzards and as many as 300,000 buzzards are noted there in spring (Yosef et al. 2002, Shirihai and Christie 1992). They continue their journey mainly east of the Black Sea, as in Batumi (Georgia) similar numbers to the autumn ones are observed also in spring (https://www.trektellen.nl/site/totals/1047/2020, accessed 1/7/2020). On the other hand, a very small share of this is recorded at the west of the Black Sea, in Dobruja (Stanciu et al. 2017). Further west, some tens of steppe buzzards are observed crossing the Mediterranean Sea (Corso 2001a, Panuccio et al. 2004, Gustin 2006). Interestingly, Melde (1982) reports two long-distance (ca 2200 km) north-eastward movements of putative steppe buzzards ringed in June in Southern Europe (Italy, Balkan Peninsula) and recorded next year (April and June) in Russia (Melde 1982). Common buzzards wintering in Southern France and Italy go back to their breeding areas in Central and Northern Europe migrating on a broad front, even if annual counts of up to 2,333 (Col de L'Escrinet, France; www.migraction.net) and 939 individuals (Cima Comer, Italian Alps; Leo et al. 2006) have been recorded. Finally, the passage of 187 and 282 birds was reported over the island of Antikythira (Southern Greece) in springs 2007 and 2008, respectively, probably comprising birds wintering in Crete (Lucia et al. 2011, Panuccio et al. 2019a). Considering the lower numbers recorded in autumn over Antikythira until mid-October (Table 1), it is likely that most of common buzzards wintering in Crete reach the island later in the season.

Timing

Steppe Buzzards leave their wintering grounds in South Africa in February, adults before young who stay longer there (Herremans et al. 2001). In March, thousands of buzzards moving northward have been recorded in Zambia and in Rwanda (Alerstam 1990). In North-Eastern Africa (Egypt), the migration is temporally distributed rather evenly over time between mid-March and late April with median in 3rd April (Hilgerloh et al. 2009, Megalli and Hilgerloh 2013). In Israel, most of migration occurs between 22nd March and 9th April, but two peaks are recognised there, higher in late March and early April, lower in the third week of April (Shirihai and Christie 1992, Leshem and Yov-Tov 1996). There, the estimated mean passage date is on 3rd April and the peak migration has been registered on 31st March (Leshem and Yov-Tov 1996). Yosef et al. (2002) estimated the median date for adults being 9th April and 26th April for juveniles, and Gorney and Yom-Tov (1994) as 6th and 28th April, respectively. In the Ural region of Russia, most birds arrive in the second half of April (Karyakin 1998).

At the Bosporus, Eurasian buzzards are among the first migrating raptor species with small numbers being observed already in late February, passage peaks in mid or late March and, secondly, in mid-April, but some migrants could be observed up to July (Üner et al. 2010, Arslangündoğdu et al. 2018). Migration of Eurasian buzzards has been observed in March (peak in mid-March) up to mid-April in Greece (Schindler et al. 2015). In Central Europe, migration of common buzzards may start in January, but usually in February peaking in March; although migration still is seen on the flyway in April, the numbers are 5–10 times smaller than in March (Melde 1982, Helbig and Laske 1989, Mirski 2008, Drapolicz 2015). These observational and ring recovery data are supported by the GPS-tracking; in 2020, 12 birds left wintering grounds between 13th February and 11th March (Väli and Mirski unpubl. data). Four Swedish GPS-tracked birds started their spring migration in the second half of March (16–27.03) and they arrived at breeding quarters after 6–10 days of flight without any stopover (Strandberg et al. 2009b). Migration speed of a male reached, on average, 148 km/day and was more than six times higher than that in autumn; females were slower moving on average 51–88 km per day, depending on individuals (Strandberg et al. 2009b). Similarly, spring migration of twelve GPS-tracked common buzzards to their breeding ground in Estonia lasted 3–19 days (mean: 8 days; Väli and Mirski unpubl. data). A male common buzzard started its migration from Southern Germany at 08th March and arrived on its breeding territory in Estonia at 24th March, but took two (one- and four-day) stopovers (Väli et al. 2015). One more Estonian buzzard left Czech Republic on 2nd March and reached its breeding site on 12th March with one stop-over lasting two days (Ptaki Silesiana 2020). In line with that, arrival in March, and sometimes in late February, has been observed in Estonia and South-Western Finland (Saurola et al. 2013, Väli 2015b, Väli and Mirski unpubl. data). However, steppe buzzards arrive to Finland only in mid-April (Saurola 2013). Migration peaks in mid-April also at the eastern coast of Lake Ladoga, North-western Russia (Noskov 2016).

Behavioural and Ecological Remarks

Migration patterns have been changing over time. Higher proportion of Eurasian buzzards remain sedentary nowadays (Bønløkke et al. 2006, Väli et al. 2014) and distances between breeding and wintering quarters have shortened during the 20th century in Europe (Martín et al. 2014). The change has been very obvious at the south-western migration bottleneck, the Strait of Gibraltar, where several thousand individuals were noticed in spring migration in the 1960's and 1970's, but numbers declined by five times after the break in the 1980's (Bensusan et al. 2007) and now the species is barely noticed on migration there (Onrubia 2016). Alerstam (1990) noticed that central point of wintering distribution of Swedish birds has moved northwards during the 20th century and connected this with general declining trend of Swedish buzzards (Kjellén and Roos 2000). Interestingly, continuous severe decline has been recorded during recent decades in Finland, occupied mainly by the Steppe Buzzard (Saurola et al. 2013, Meller et al. 2019). Moreover, in Finland declining steppe buzzards seem to be replaced by nominate subspecies, which is expanding according to field observations (Saurola et al. 2013). Could similar replacement of subspecies explain the controversial views on taxonomy over time in Sweden and Baltic States (c.f. Väli and Vainu 2015)? Replacement of (or changing ratio between) subspecies would explain also the more pronounced shortening of migration distance by breeders of Northern Europe (Martín et al. 2014), earlier breeding in Finland (Lehikoinen et al. 2009) and later autumn migration phenology recorded both at Falsterbo in the 1990's compared to 1940's and 1950's (Kjellén and Roos 2000) and Organbidexa, South-Western France (Filippi-Codaccioni et al. 2010). Expansion of or genetic introgression from the common buzzard (or change of migratory habits of the steppe buzzard) has been a possible explanation to advancement of spring migration and increased sedentarism in European Russia and Ukraine (Morozov 2007).

Juveniles tend to disperse various directions before migrating (Strandberg et al. 2009b), but also some adults may perform pre-migratory movements away from the main migratory direction. For example, a GPS-tagged female buzzard from Estonia moved 300 km NE to Karelia and thereafter migrated with fellow Estonian birds south-westward (despite general south-eastward passage in the region, see above). Interestingly, this female wintered in Serbia much further south than other Estonian birds (Väli and Mirski unpubl. data), and thus such migration pattern might be determined by genetic ancestry of steppe buzzard. A similar pre-migratory north-westward movement (347 km) has been detected by a ring recovery (Väli and Vainu 2015).

In Central Europe, where partial migration occurs, juveniles and immatures migrate more frequently than adults (Holte et al. 2017), but there seem to be no difference between migration distances of various age classes (Väli and Vainu 2015, Holte et al. 2017). Immature birds probably do not return to their exact natal site, although common buzzards, especially in partially migrating populations, are still rather close (Holte et al. 2017), compared to immature steppe buzzards that probably stay in their distant wintering grounds (Saurola et al. 2013).

The Eurasian buzzard is a model species in research of flight mechanics and physiology. For example, fat levels of steppe buzzards have been comparatively measured during wintering in South Africa and during migration in Israel suggesting the need of additional feeding en route (Gorney and Yom-Tov 1994). Steppe buzzards have been used in radar studies as one of model species in research of flight behaviour of raptors, and the influence of wind conditions (e.g., Spaar and Bruderer 1997, Malmiga et al. 2014). According to radar studies, Eurasian Buzzard compensates the elevation flying relatively high (434 m) the sea level, at intermediate altitude at sea level 300 m) and relatively low in upland (252 m; Spaar 1995, Malmiga et al. 2014). Only few steppe buzzards fly higher than 1,000 m above ground level in Israel (Spaar 1995), while this height is not reached in Sweden by common buzzards (Malmiga et al. 2014). The flight altitude of buzzards recorded at the Bosphorus during autumn migration using an optical range finder and counts revealed lower numbers passing during strong northerly (lateral) winds (Panuccio et al. 2017). The mean cross-coutry speed was 9.7 m/s in Israel; it was lower (7.7 m/s) with opposing winds and higher (11.4 m/s) with following winds (Spaar and Bruderer 1997). Comparatively, common buzzards fly slower in Europe, where mean cross-country speed has been 7.5 m/s in Switzerland (Schmid 1986) and 7.2 m/s in Sweden (Malmiga et al. 2014). This difference is probably due to different characteristics of landscape and wind and not due to subspecies identity (Spaar 1995).

What could be the next research questions regarding the migration of the Eurasian buzzard? First, telemetry studies on steppe buzzards are needed, because the migration strategy of this long-distance migrant is most probably different from that of short-distance migratory nominate form. Similarly to closely related, but hybridizing sister species (Väli et al. 2018), the two buzzard subspecies, exhibiting contrastingly different migration patterns, is a promising tool for advanced research of migration determinants. The second question relates migration to taxonomy. How large is the genetic differentiation between the subspecies, where are the actual (past and nowadays) borders of their ranges and how large is the genetic introgression? No answers have been obtained from genetic studies based on mitochondrial DNA and multilocus neutral genetic markers (e.g., Kruckenhauser et al. 2004, Jowers et al. 2019), but we hope that modern genomic tools will shed light on this question and tell us who are the buzzards wintering in Western and Northern Africa, and the Middle East. We hypothesize that such medium-distance migration may have resulted from interbreeding and exchange of genetic material between the two subspecies. Such patterns may result from combining genes responsible for long-distance migration with those determining south-western direction. After hybridisation over several generations, more diffuse pattern is generated. Obviously, morphological bimodality also disappears in the large hybrid zone with diffuse borders where subspecies are gradually merged to each other.

Table 1. Numbers of the Eurasian buzzard (annual means of multiyear counts) and their proportion among all migrating raptors in migration observations stations along the west-east axis. Share of the dominant subspecies is calculated if data is available in referred source.

Location	Longitude	Latitude	Years	Season	Period	Mean annual no. of raptors	Mean annual no. of Eurasian Buzzards	Proportion of the Eurasian Buzzard (%)	Subspecies considered	Reference
Strait of Gibraltar, Spain	–5°20’	36°06’	2009–2012	autumn	20.07–10.10	ca 240,000	86	< 0.1	Not separated	Onrubia 2016
Organbidexka, France	1°00’	43°02’	1981–2008	autumn	15.07–15.11	27,700	192	0.7	Not separated	Filippi-Codaccioni et al. 2010
Défilé de l’Écluse, France	5°53’	46°07’	2016	autumn	26.08–3.12	39,102	21,610	45.8	Not separated	LPO Haute-Savoie 2016
Bielefeld, Germany	8°27’	52°03’	1980–1985	autumn	10.08–21.11	791	344	43.5	Not separated	Helbig and Laske 1990
Falsterbo, Sweden	12°49’	55°23’	1986 –1995	autumn	1.08–20.11	33,585	10,850	32.3	Not separated	Kjellén 1998
Marettimo, Sicily, Italy	12°03’	37°58’	1997, 1998, 2000, 2002, 2007	autumn	26.08–14.09 15.09–30.09 03.10–18.10	4,765	30	< 0.1	*buteo*	Agostini et al. 2000, 2005a, Panuccio et al. 2011
Strait of Messina, Italy	15°39’	38°16’	2018–2019	autumn	10.08–20.10	22,955	71	0.3	*buteo*	Cento et al. 2018, 2019
Circeo promontory, Italy	13°03’	41°13’	2002	autumn	26.08–30.09	1,307	4	0.3	*buteo*	Agostini and Panuccio 2003
Buskett, Malta	14°23’	35°51’	1998–2002	autumn	15.08–31.10	3,105	8	0.3	84,6% *buteo*	Sammut and Bonavia 2004
Carpathians, Poland	18°43’–22°43’	49°03’–49°48’	2011–2013	autumn	14.08–19.11	2,354	1,509	64.1	Not separated	Wilk et al. 2018
Vistula Spit, Poland	19°31’	54°24’	2008–2011	autumn	15.08–15.11	10,965	3,170	28.9	Not separated	Polakowski et al. 2014
Antikythira, Greece	23°18’	35°52’	2007–2009	autumn	21.08–20.10	1,369	57	4.2	98.3% *buteo*	Lucia et al. 2011
Dardanelles, Turkey	26°44’	40°29’	2010	autumn	22.08–1.10	1,601	173	10.8	Not separated	Panuccio et al. 2017
Bourgas, Bulgaria	27°29’	42°33’	1979–2003	autumn	10.08–31.10	38,534	17,739	46.0	Mostly *vulpinus*	Michev et al. 2011

Table 1 contd. ...

...Table 1 contd.

Location	Longitude	Latitude	Years	Season	Period	Mean annual no. of raptors	Mean annual no. of Eurasian Buzzards	Proportion of the Eurasian Buzzard (%)	Subspecies considered	Reference
Măcin, Romania	28°12’	45°15’	2002–2007	autumn	16.08–21.10	11,297	6,254	55.4	Mostly *vulpinus*	Fülop et al. 2018
Dobruja, Romania	28°37’	44°07’	2012–2015	autumn	29.08–6.11	-	563	-	> 68% *vulpinus*	Stanciu et al. 2017
Bosphorus, Turkey	29°01’–29°06’	41°00’–41°11’	2008	autumn	22.09–10.10	141,844	74,055	52.2	Not separated	Fülop et al. 2014
Bosphorus, Turkey	29°08’	41°07’	2010	autumn	22.08–1.10	24,701	1,958	7.9	Not separated	Panuccio et al. 2017
Pogreby, Ukraine	30°34’	50°32’	2008	autumn	4.09–13.10	2,542	1,125	44.3	Not separated	Domashevsky and Demidenko 2008
Zeit bay, Egypt	33°31’	27°44’	2006	autumn	20.08–29.10	9,376	38	0.4	Only *vulpinus*	Hilgerloh et al 2011a
Cape Greco, Cyprus	34°01’–34°04’	34°57’–35°00’	2005	autumn	03.09–26.10	5,153	36	0.7	Not separated	Roth and Corso 2007
Eilat, Israel	34°55’	29°34’	1980,86,87	autumn	25.08–30.11	26,998	167	0.6	> 99% *vulpinus*	Shirihai and Christie 1992
Israel	34°55’–35°30’	29°34’–32°45’	1986–1989	autumn	01.08–30.11	510,354	767	0.2	Only *vulpinus*	Leshem and Yov-Tov 1996
Northern Valleys, Israel	35°00’–35°30’	32°25’–32°45’	1990–1999	autumn	10.08–15.10	450,995	1,306	0.3	>99% *vulpinus*	Alon et al. 2004
Iskenderun, Turkey	35°58’36°28’	36°30’–37°01’	2013	autumn	30.08–26.09	62,293	4,010	6.4	Not separated	Oppel et al. 2014
Batumi, Georgia	41°44’–41°47’	41°41’	2008–2009	autumn	21.08–13.10	829,025	257,829	31.1	Mainly *vulpinus*	Verhelst et al. 2011
Batumi, Georgia	41°44’–41°47’	41°41’	2012–2018	autumn	17.08–16.10	1.16 mln	296,030	25.5	Only *vulpinus*	Wehrmann et al. 2019
Bab al Mandab, Djibouti	43°09’–43°19’	12°28’–12°43’	1985	autumn	15.10–01.11	80,732	17,900	22.1	Only *vulpinus*	Welch and Welch 1988
Bab al Mandab, Djibouti	43°09’–43°19’	12°28’–12°43’	1987	autumn	03.10–09.11	246,478	98,339	39.9	Only *vulpinus*	Welch and Welch 1988

Strait of Gibraltar, Spain	–5°20'	36°06'	2009–2012	spring	1.02–10.06	ca 63 000	46	<0.1	single ind.	Onrubia 2015
Bielefeld, NW Germany	8°27'	52°03'	1979–1985	spring	28.02–25.05	116	45	38.8	Not separated	Helbig and Laske 1989
Silea, Italy	12°18'	45°39'	2006	spring	28.02–16.05	769	325	42.2	Not separated	Mezzavilla 2006
Strait of Messina, Italy	15°39'	38°16'	1996–2000	spring	01.04–27.05	26,172	79	0.3	72% *buteo*	Corso 2001a
			2006	spring	20.04–20.05	18,460	41	< 0.1	57.5% buteo	Gustin 2006
Ustica, Italy	13°12'	38°42'	2002	spring	22.03–20.05	3,938	9	0.2	67% buteo	Panuccio et al. 2004
Vistula Spit, Poland	19°31'	54°24'	2015	spring	1.03–5.06	9,691	6,752	69.7	Not separated	Drapolicz 2015
Zawady, Vistula, Poland	21°09'	52°09'	2007	spring	1.03–30.04	101	81	80.2	Not separated	Mirski 2008
Antikythira, Greece	23°18'	35°52'	2007–2008	spring	15.04–15.05	399	235	58.9	99.6% *buteo*	Lucia et al. 2011
Dadia, Greece	26°23'	41°13'	2003–2005	spring	01.03–31.07	495	238	48.1	Not separated	Schindler et al. 2015
Bosphorus, Turkey	29°04'	41°12'	2006	spring	18.03 - 31.05	46,862	16,185	34.5	Not separated	Üner et al. 2010
Anadolu, Bosphorus, turkey	29°05'	41°11'	2007	spring	02.03–03.06	38,313	23,954	62.5	Not separated	Arslangündoğdu et al. 2018
Batumi, Georgia	41°44'–41°47'	41°41'	2020	spring	2.03–25.5	> 700,000	338,419	> 40	Only *vulpinus*	https://www.trektellen.nl/site/totals/1047/2020
Ayn Sokhna, Egypt	32°21'	29°33'	2012	spring	01.03–02.05	154,276	116,560	75.6	Only *vulpinus*	Megalli and Hilgerloh 2013
Zait bay, Egypt	33°31'	27°44'	2007	spring	20.02–6.05	36,976	23,359	63.2	Only *vulpinus*	Hilgerloh 2009
Eilat	34°55'	29°34'	1977–1988	spring	2.02–31.05	790,000	326,556	41.3	Only *vulpinus*	Shirihai and Christie 1992
Israel	34°55'–35°30'	29°34'–32°45'	1986–89	spring	01.03–31.05	776,040	321,021	41.4	Only *vulpinus*	Leshem and Yov-Tov 1996a, 1996b

24

Lesser kestrel *Falco naumanni*

Maurizio Sarà,[1,*] *Javier Bustamante,*[2] *Jacopo G. Cecere*[3] and *Diego Rubolini*[4]

ABSTRACT

The lesser kestrel is a small colonial falcon with a broad breeding distribution range, extending from the Iberian Peninsula to eastern China. Most of the biological and ecological knowledge concerning this species, including migration, has been obtained from European populations. After the end of each breeding season, these populations move preferentially northward, showing pre-migratory movements (of 100–1,000 km) that are necessary to reach rich foraging grounds where they can fuel before commencing the true migration towards Africa. European lesser kestrels migrate on a broad front across the Mediterranean Sea. There is a strong connectivity between European breeding and African non-breeding grounds, with different populations being clearly segregated also during winter. Iberian lesser kestrels show a rather different migration phenology (i.e., earlier departures and shorter travels) compared to other European birds, likely due to the geographic conformation of continents and greater proximity to wintering areas. Tracking data indicate that departures span across two months, averaging ca. September 20th. After ca. 12 days of travel, individuals arrive in the Sahel where they stay 5–6 months. The spring migration routes are similar to autumn ones, being however less straight, with more pronounced westward detours and crossings in proximity of the Gibraltar and Sicily Straits. Spring departures are concentrated in the first two weeks of March. Spring travels usually last longer than autumn ones, i.e., 18–23 days, with more time spent on stopover. The pattern of migration of Middle East and Asian populations is mostly based on anecdotal information.

Introduction

The lesser kestrel (*Falco naumanni* Fleischer, 1818) is a small, sexually dimorphic (120–150 g males, 140–190 g females) and migratory falcon breeding in the Palaearctic (Ferguson-Lee and Christie 2001). Although some Chinese specimens have been earlier considered as a distinct subspecies,

[1] Dipartimento di Scienze Biologiche, Chimiche e Farmaceutiche, Università di Palermo, Via Archirafi 18 - 90123 Palermo, Italy.

[2] Department of Wetland Ecology, Estación Biológica de Doñana, CSIC. C/ Américo Vespucio 26 - 41092 Sevilla, Spain.

[3] Area Avifauna Migratrice, Istituto Superiore per la Protezione e la Ricerca Ambientale (ISPRA), via Ca' Fornacetta 9 - 40064 Ozzano Emilia (BO), Italy.

[4] Dipartimento di Scienze e Politiche Ambientali, Università degli Studi di Milano, via Celoria 26 - 20133 Milano, Italy.

* Corresponding author: maurizio.sara@unipa.it

F. naumanni pekinensis (Swinhoe 1870), the taxon is currently considered monotypic, and the *pekinensis* subspecies is put in synonymy with *F. n. naumanni* (Ferguson-Lee and Christie 2001, Dickinson and Remsen 2013). A recent morphological analysis (Corso et al. 2015, 2016) has questioned this taxonomic arrangement, proposing the resurrection of *F. n. pekinensis*, also considering the genetic distinctiveness of Asian and European populations (Wink et al. 2004, Bounas et al. 2018). The lesser kestrel's estimated area of occurrence is huge, spanning some 25 million km^2 (BirdLife International 2016c), although largely fragmented and distributed in Southern Europe, North Africa and Asia. The species' distribution is scattered from the Iberian Peninsula and Northern Morocco in the west, to South-Eastern Mongolia and North-Eastern China in the east of the range. Although some individuals remain in Europe during winter, mostly in southern areas of Spain, Italy and Turkey and their number is likely increasing (Negro et al. 1991, Tella and Forero 2000, Morganti et al. 2017), the bulk of populations behave as a typical long-distance migrant spending the non-reproductive season, which corresponds to the boreal winter, entirely in the Afrotropical region, south of the Sahara (Ferguson-Lee and Christie 2001).

The lesser kestrel was in the past one of the commonest birds of prey of Western Palaearctic and experienced a severe decline in the middle of the 20th century (Biber 1990), so that it was assessed as a 'Vulnerable' species in past IUCN Red Lists (1994–2008). Implementation of conservation actions along the lines suggested by International (Biber 1990, 1996, Iñigo and Barov 2010) and National Action Plans (e.g., Pilard et al. 2016, La Gioia et al. 2017) and largely carried out within the framework of several LIFE projects, mostly in the Iberian region, Italy and France, led to its status improvement. Since 2011, the species' world population has been considered to become stable or slightly increasing and consequently the category of threat was down-ranked to 'Least Concern' in the IUCN Red List (BirdLife International 2020).

The causes and extent of lesser kestrel decline have certainly differed across the regions of Western Palaearctic range, but can ultimately be brought back to the major topics of habitat loss and habitat degradation. Indeed, the lesser kestrel inhabits steppe-like habitats, natural or semi-natural grasslands, and extensively cultivated land, and feeds mostly upon small vertebrates (lizards, shrews, voles) during courtship and egg-laying, and upon large orthopterans and coleopterans during brooding of chicks (Rodríguez et al. 2006, Catry et al. 2012, Sarà et al. 2014, Di Maggio et al. 2018). Agricultural intensification and the associated land use changes have replaced grasslands and extensive dry cereal crops with perennial plantations (e.g., vineyards, olive orchards, etc.) in many European areas. Other changes negatively affecting the lesser kestrel have included the increasing cultivation of sunflower and maize that reduce the diversity and abundance of large insects, and/or taller cereal varieties that decrease the access to prey (Tella and Forero 2000, Bonal and Aparicio 2008, Catry et al. 2012, Catry et al. 2014). However, small predators like the lesser kestrel generally behave as income breeders (Dunn 2006), so that the outcome of their breeding season is determined by the overall ecological conditions in the breeding areas immediately before and during the start of reproduction (Podofillini et al. 2019). Together with the relatively high productivity (the species' mean breeding success ranged between 1.1 and 4.5 fledglings/pair in several studies reviewed in La Gioia et al. 2017), income breeding may allow rapid population fluctuations in response to improvement or degradation of environmental quality in the breeding grounds. As consequence, many populations in Italy, France and in the Iberian Peninsula have rapidly increased after the implementation of conservation measures (Iñigo and Barov 2010, La Gioia et al. 2017). Nonetheless, the latest census in Spain (Spanish Ornithological Society, unpublished) and Italy (e.g., Mascara et al. 2019) reveals a new phase of marked population decline. Indeed, the high sensitivity of lesser kestrel to climate and environment changes at small spatial and temporal scales (e.g., Morganti et al. 2019) makes conservation of this species a true challenge.

Like most *Falco* species, the lesser kestrel is a secondary-cavity nester, which lays its eggs directly on the cavity floor after scraping the substrate (Ferguson-Lees and Christie 2001). Nests are usually found under roof tiles and inside balconies and wall crevices in houses of villages, farms and often in large historic buildings (Serrano et al. 2001, 2003, Franco et al. 2005, Sarà 2010). Lesser kestrels frequently breed in urban habitats (Tella et al. 1996, Liven-Schulman et al. 2004, La Gioia et al. 2017) where colonies can be huge. In Southern Italy, the towns of Matera, Gravina in Puglia and Altamura host ca. 1,000 pairs each (La Gioia et al. 2017). Opportunistically, it can breed in peculiar sites like those consisting of piles of stones in the Crau Plain (Southern France) (Le Brun et al. 2001), or in the Scythian shell mounds in the Tuva Republic (Russia) (M. Sarà unpublished data).

In addition to land use changes, weather, especially rainfall, is the environmental factor mostly affecting the breeding success of the lesser kestrel (Rodríguez and Bustamante 2003), with hot and dry springs reducing the abundance of active colonies across years (Sarà 2010). The effect of rainfall extends beyond the breeding season, as the annual survival of French yearling kestrels depended on the severity of annual drought experienced in the African Sahel where they have spent the boreal winter (Mihoub et al. 2010). Furthermore, distribution models at the Italian scale showed that the amount of spring rainfall is the main determinant of climate suitability during breeding (Morganti et al. 2017).

One further remarkable aspect of this species' biological cycle is communal roosting in single trees and small wood patches, or on electric pylons and wires. Communal roosting may occur during breeding, in particular for males, but is more frequent during the post-fledging and pre-migratory period (Olea 2001, Olea et al. 2004, Franco et al. 2005, Sarà et al. 2014). From July to early October, soon after fledging, lesser kestrel families leave their colonies and carry on short-distance (usually within 100–1,000 km) pre-migratory movements that are not directed toward the usual overwintering grounds, and have been referred to as intermediate or intermittent migration (Berthold 1996). During such movements hundreds or even thousands of individuals may disperse in areas full of orthopteran prey to reach the optimal body conditions before post-breeding migration, and every night congregate at preferred roost-sites (Olea 2001, Olea et al. 2004, Sarà et al. 2014). Communal roosts are equally important in the non-breeding African range: for instance, in the Kaolack area (Senegal), thousands of kestrels congregate every year together with thousands of African swallow-tailed kites (*Chelictinia riocourii*) (Pilard et al. 2009), and similar concentrations occur in De Aar and other towns of South Africa's Karoo, where lesser kestrels mix with thousands of Amur falcons (*Falco amurensis*) (R. Visagie, personal communication).

Knowledge Status

Most data about the migration of the lesser kestrel have been traditionally collected at migration observation points (Zalles and Bildstein 2000), usually located on coastal uplands or other vantage points on islands and along the migration bottlenecks (e.g., Finlayson 1992, Giordano 1991, Shirihai and Christie 1992). This method, repeated during decades and with increasingly standardized protocols, has allowed to gather important long-term information on world population trends, mostly for soaring raptors (e.g., Alon et al. 2004, Martín et al. 2016, Panuccio et al. 2018a). Nonetheless, lesser kestrels, similarly to many other *Falco* species (e.g., Dixon et al. 2012, Mellone et al. 2011c), can rely much more on powered flight during migration, and are not so hesitant to cross extensive bodies of open waters, unlike soaring birds. So, they tend not to funnel at migration bottlenecks (i.e., Gibraltar, Sicily, Bosphorus) to cross the Mediterranean Sea, but rather to migrate on a broad front (Meyer et al. 2003, Sarà et al. 2019).

In addition, it is known that lesser kestrels can fly during night-time (Kopij 2002, Limiñana et al. 2012a, Gustin et al. 2014) and at high altitudes (Ferguson-Lee and Christie 2001). These traits, combined with the small size and the possibility of confusion with similar species (e.g., Eurasian kestrel *Falco tinnunculus*), resulted in low numbers of lesser kestrels being counted at observation points. For instance, only less than 0.5% (i.e., an average of < 500 individuals per year) of the Spanish population has been recorded annually at the Gibraltar Strait during the autumn migrations of 1999–2008 (Román et al. 2009). A similar underestimation has occurred during spring migration both at the Gibraltar Strait and at other bottlenecks, like the Messina Strait (Giordano et al. 2009). At the first site, only 248 individuals were counted in 2008 (Román et al. 2009); likewise, in the Messina Strait an average of 57 (range: 12–159) individuals were observed every spring (April–May) in the period 1991–2008 (Giordano et al. 2009). Even adding the quota of unidentified kestrels (annual range: 5.5%–29.8% of overall kestrel records in the cited references), the number remains irrelevant with respect to the estimates of the Italian population, increasing in the same period from ~ 3,000 to ~ 12,000 individuals (La Gioia et al. 2017). The same paucity of records occurred in other watch-sites on the Eastern flyway (Shirihai and Christie 1992, Alon et al. 2004, Panuccio et al. 2017). Nonetheless, the monitoring of seasonal migration at major bottlenecks and vantage points on islands and coasts of Southern Palaearctic has allowed the collection of some general information on migration phenology, like for instance the spreading of passages over all day-time (Román et al. 2009), and of basic knowledge about species' routes (Table 1).

Information on migration of Middle East and Asian lesser kestrel populations is much more anecdotal and fragmented (e.g., Curry-Lindahl 1981, Ferguson-Lee and Christie 2001, Moreau 1972). In autumn, birds belonging to these Oriental populations should migrate to the west and south-west and most of the individuals should arrive in Africa from the eastern edge of the Mediterranean Basin and continue to their final destination in Eastern and Southern Africa (Ferguson-Lee and Christie 2001). Together with many other species, they might be using either the 'Black Sea route' to reach Sinai and finally cross into Africa at the northern end of the Red Sea (Alon et al. 2004) as well as the more easterly 'Caspian/Arabian route' that would bring them to cross into Africa at the southern end of the Red Sea and to fly finally along the Rift Valley towards the south of the continent (Ferguson-Lee and Christie 2001, Moreau 1972). There, in Zimbabwe, Botswana and South Africa, they settle in one of the largest African areas of open dry-highlands and dry mix savannahs, where after their post-breeding migration all Palaearctic populations use to spend their non-breeding period. Other such areas have been found in the Sahel zone, from Senegal to Nigeria, and in East Africa, from Eritrea to Tanzania (Ferguson-Lee and Christie 2001, Pepler and Matin 2001).

Besides, according to older literature reviewed in Corso et al. (2016), the non-breeding grounds and migration areas of the supposed *F. n. pekinensis* should include India, Laos and Myanmar in Southern Asia; plus, possibly Eritrea, Ethiopia, Somalia, Kenya and Tanzania, in Eastern Africa. Actually, most of them do not seem to reach the Cape regions of South Africa leaving open the issue of the origin of this wintering population. The autumn monsoon could help these Asian lesser kestrels to use instead a third route to reach Africa (Pepler and Matin 2001). After leaving Mongolia and China they might pass first over Iran and India and then should cross the Indian Ocean directed to the Arabian Peninsula or more south to Ethiopia and Somalia (Ferguson-Lee and Christie 2001, Moreau 1972, Pepler and Matin 2001). The strongest, although indirect, arguments in favour of this latter route are the irregular species' observations in remote islands of the Indian Ocean like the Maldives and Seychelles (Ferguson-Lee and Christie 2001), and the strong interspecific association during migration and wintering with the Amur falcon (e.g., Auburn 1988), another small colonial species known to travel from Asian breeding grounds to Southern Africa, with a fast non-stop flight across the Indian Ocean (Meyburg et al. 2017b).

Several studies have investigated migratory movements of lesser kestrels breeding in Europe, using tracking devices, whereas very few data exist from Asian populations. A single geolocator (GLS) study just reported migration data of three birds from Central Mongolia (Millhouser 2016). The non-breeding areas of these three birds were found to extend over a broad area spanning between the Central African Republic, South Sudan and Kenya. During both post- and pre-breeding migration, they moved through Central Asian countries (Southern Kazakhstan, Turkmenistan), Iran and Saudi Arabia.

In Europe, migration tracks have been collected for lesser kestrels breeding in Spain (Rodríguez et al. 2009, Limiñana et al. 2012a), Portugal (Catry et al. 2010), France (Pilard et al. 2017), Italy (Sarà and Bondì 2016), Bulgaria (Gradev et al. 2016a) and Greece (Vlachopoulos et al. 2016), using GLS, Argos Platform Transmitter Terminals (hereafter Argos PTTs) and remote-downloading archival GPS tags (GPS-UHF devices). From these studies, Sarà et al. (2019) selected the migration data of individuals whose non-breeding area could be identified, added a new sample of migration tracks recorded by the same GLS, GPS-UHF and Argos PTTs devices, in order to study in a unique

Table 1. List of main European migration localities where the lesser kestrel has been recorded across years. Continental inland locations (e.g., mountain passes or valleys in the Alps, Apennines, etc.) and mix records of unidentified *F. naumanni/F. tinnunculus* have not been considered.

Watch-Site	Year	Range (N)	References
Autumn (post-breeding) passage			
Gibraltar Strait (SW Spain)	1972–1977	254–514	Bernis 1980
Gibraltar Strait (SW Spain)	1999–2008	100–600	Román et al. 2009
Balearic islands (SE Spain)	2008–2011	1–10	GOB 2009, 2011
Elba island (Tuscany, C Italy)	2006–2007	1–4	Paesani and Vanni 2006, Vanni and Paesani 2007
Messina Strait (Sicily, S Italy)	2015–2018	2–7	www.straitobservatory.com
Malta	1998–2002	22–58	Sammut and Bonavia 2004
Malta	2007	140–220	Raine 2007 2008 Raine and Vella 2009
Bosphorus Strait (Turkey)	2010	1	Panuccio et al. 2017
Antikythira island (S Greece)	2007–2009	1–4	Lucia et al. 2011
Northern Valleys (Israel)	1990–1999	4–34	Alon et al. 2004
Spring (pre-breeding) passage			
Gibraltar Strait (SW Spain)	2008	248	Román et al. 2009
Balearic islands (SE Spain)	2008–2011	1–10	GOB 2009 2011
Maritime Alps (S France)	2001–2004	0–1	Jardin 2005
S. Teresa Gallura (NE Sardinia, Italy)	2005	1	Premuda et al. 2007a
Mount Conero (Adriatic coast, C Italy)	2005–2010	2–12	Borioni and Baldoni 2005 2006, 2007 2008 2009 2010
Cape Otranto (Adriatic coast, S Italy)	2003–2006	123–208	Premuda et al. 2004, La Gioia 2009
Cape Gargano (Adriatic coast, S Italy)	2008	25	Pandolfi and Piccinetti 2008
Tremiti islands (Apulia, S Italy)	2008	1	Pandolfi and Piccinetti 2008
Messina Strait (Sicily, S Italy)	1991–2005	12–159	Giordano et al. 2009
Messina Strait (Sicily, S Italy)	2006–2010	72–153	Chiofalo et al. 2006, 2007, Gustin 2006 2008, 2009, Cutini et al. 2008, Ricciardi et al. 2009 2010
Panarea (Aeolian islands, S Italy)	2005–2009	11–87	Gustin 2005 2006, 2007, 2008, 2009
Cap Bon (N Tunisia)	2007	14	Premuda et al. 2007c
Marettimo (Egadi island, S Italy)	2007–2009	2–5	Gustin 2007 2008, 2009
Ustica island (Sicily, S Italy)	2002	30	Panuccio et al. 2004
Ustica island (Sicily, S Italy)	2006–2009	1–12	Gustin 2006 2008, 2009
Pantelleria island (Sicily, S Italy)	2006–2009	1–42	Gustin 2006, 2007, 2008 2009
Linosa island (Sicily, S Italy)	2006	20	Gustin 2006
Malta	2010	22	Raine and Vella 2010

framework, the phenology of migration and the connectivity between the breeding and non-breeding areas.

Overall, Sarà et al. (2019) reported data from birds trapped at 25 breeding colonies from the above-mentioned countries. The birds and their colonies were grouped in the three distinct regions, Iberia (Portugal, Spain and France), Italy (Southern Italy and Sicily) and Balkans (Bulgaria, Greece).

Birds from the small French population were considered as closely linked to the nearest Iberian core populations (Lelong and Riols 2009). Furthermore, the study combined a set of devices with different spatial and temporal resolution. After data homogenization (see Sarà et al. 2019 for further details), the overall sample of devices (i.e., including GLS) yielded 94 migration events from 87 individuals across most of the European breeding range, and is the most complete available study to report on the migration phenology and movements of lesser kestrels, as summarized in next paragraphs.

During the boreal winter, lesser kestrels distributed across a huge potential non-breeding area located in the arid and semi-arid regions of the Sahelian belt, extending more than 5 million km^2 between 20° 44' 24" N and 8° 57' 35" N, and 16° 44' 54" W and 20° 44' 34" E and including eight different African countries, all in the Sahel belt (except Burkina Faso, where it was not recorded before), in accordance with past literature (Elgood et al. 1964, Salvan 1967, Rodríguez et al. 2009, Limiñana et al. 2012a, Pilard et al. 2004, 2017).

Post-breeding Migration

At migratory watch-sites most of autumn observations were recorded at the Gibraltar Strait and Malta, whereas the Messina Strait seemed to be less used by lesser kestrels (Table 1). However, the advancement of count protocols (Agostini et al. 2015b) revealed, on the Calabrian side of the Messina Strait, a seasonal considerable passage of unidentified *naumanni*/*tinnunculus* kestrels (max. 3,142 in September 2013, Morabito et al. 2013) occurring mostly at sunset (Agostini et al. 2015b, Cento et al. 2018, 2019). Autumn migration has been recorded at the Gibraltar Strait from the very end of July to the half of October, peaking in the second half of September (Bernis 1980, Román et al. 2009), similarly in Greece it extended from early August to early October, peaking in mid-September with most birds flying over mainland Greece, Antikythira Island and Crete (Bounas et al. 2016).

According to the tracked individuals, the post-breeding migration of the European populations is initiated at the end of the boreal summer and in early autumn. It occurred on a broad front across the Mediterranean Sea, with only a minority of kestrels, even less than in spring, taking advantage of the narrowest stretches of sea (Fig. 1). Mean departures date from the post-breeding dispersal areas was September 20th (range: August 20th–October 24th) (Table 2). After an average journey of 12–13 days (range: 3–38 days in Table 3) spent in migration, the lesser kestrels arrived to their non-breeding grounds, on average October 1st (range: September 8th – November 9th). Males tended to leave the breeding areas on average a week later (September 19th) than females (September 26th), and therefore arrived a week later than females in their Sahel non-breeding areas, with these differences being statistically significant. Similar statistically significant regional differences occurred in the migration schedule, both in the dates of departure and in those of arrival, as Iberian lesser kestrels tended to depart earlier (on average September 18th) than Italian and Balkan conspecifics (on average September 22nd and 20th, respectively). As a consequence, the Iberian ones arrived earlier to Sahel areas, on average September 30th, while the Italian ones arrived on average on October 2nd and the Balkan ones on October 3rd (Table 2). During their post-breeding travel, migrating individuals covered about 370 km per day (range: 177–670 km/day in Table 3) totalling an average post-breeding migration distance of about 3,700 km (range: 2,545–4,634 km, Table 2). Regional track lengths differed, with the Iberian birds making the shortest and the Balkan the longest post-breeding journey. The track straightness in Table 2 is the ratio between the minimum migration distance and the real distance travelled, thus a track straightness value of 1 means that the bird covered exactly the minimum distance between the non-breeding and the breeding areas, hence without detours; while lower values mean a longer route and more detours with respect to the minimum distance. The average track straightness in Table 3 is 0.84, meaning that the post-breeding journey was on average some 16% longer than the minimum distance between the non-breeding area and the breeding colony. There was however some variability, as some lesser kestrels made quite direct travels to the non-breeding areas (track straightness about 0.90–0.96), while others performed consistent detours (track straightness = 0.60–0.70). Instead, the E-W deviation parameter quantifies the direction of detours from the shortest path (Table 2). There

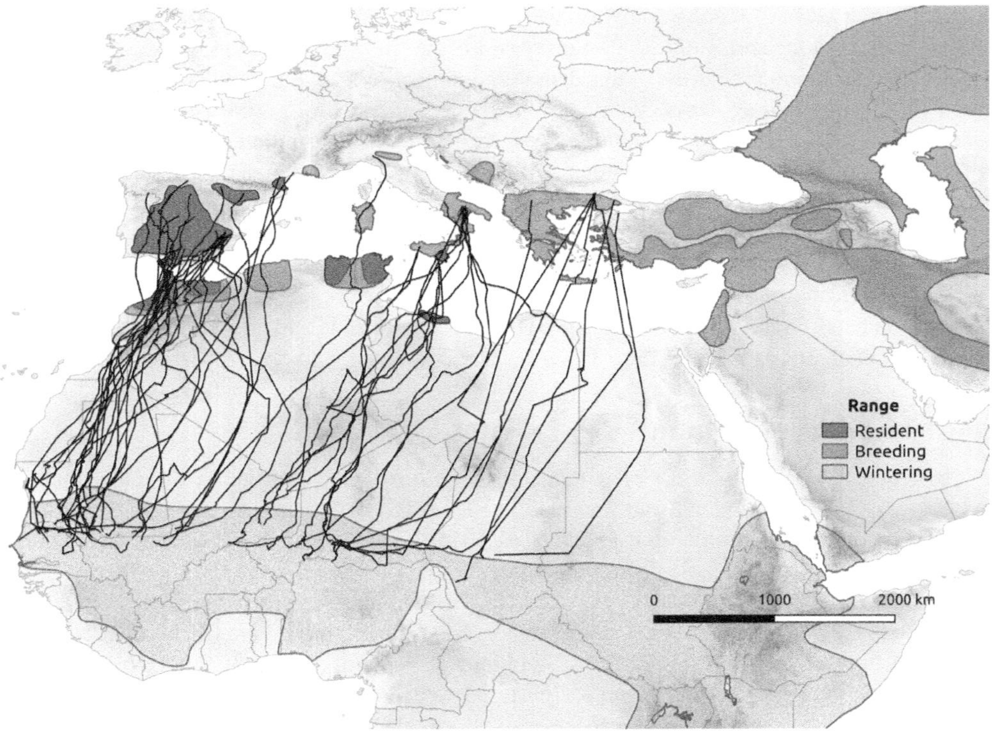

Figure 1. Migratory routes of lesser kestrels from the breeding European grounds to the non-breeding Sahel areas in Africa (modified from Sarà et al. 2019).

Table 2. Summary statistics of the migration phenology and speed of lesser kestrels from the three main European breeding regions during the post-breeding migration (modified from Sarà et al. 2019). Mean values ± standard deviation, in round brackets: minimum–maximum values and sample size referred to the number of migration events. The minimum migration distance is the orthodromic distance between breeding colony and non-breeding area. The track length is the sum of all distances between positions recorded on a migration track. The track straightness is the minimum migration distance divided per track length. The E-W deviation represents the deviation in km to the east (positive values) or west (negative values) of a track compared to the minimum migration distance line.

Post-breeding migration	**Iberia**	**Italy**	**Balkan**
Onset (day/month)	18/9 ± 14 (20/8 – 24/10; 44)	22/9 ± 10 (2/9 – 16/10; 36)	20/9 ± 9 (6/9 – 6/10; 12)
End (day/month)	30/9 ± 13 (8/9 – 9/11; 44)	2/10 ± 10 (18/9 – 25/10; 36)	3/10 ± 8 (17/9 – 14/10; 12)
Duration (days)	12 ± 8 (3 – 38; 44)	11 ± 4 (5 – 22; 36)	14 ± 7 (7 – 29; 12)
Speed (km/day)	373 ± 155 (177 – 670; 24)	318 ± 91 (206 – 558; 13)	425 ± 103 (266 – 535; 6)
Track length (km)	3,347 ± 386 (2,545 – 3,906; 25)	3,770 ± 547 (2,936 – 4,634; 13)	3,945 ± 307 (3,546 – 4,280; 7)
Track straightness	0.86 ± 0.09 (0.7 – 0.96; 25)	0.82 ± 0.1 (0.6 – 0.93; 13)	0.85 ± 0.03 (0.82 – 0.9; 7)
E-W deviation (km)	17 ± 182 (–302 – 401; 25)	175 ± 257 (–79 – 863; 13)	334 ± 153 (37 – 501; 7)

were no statistical regional or sex differences between track straightness and E-W deviations, but post-breeding migration tracks were significantly straighter and more shifted eastward than pre-breeding ones.

Patterns of post-breeding migration have been reported in Fig. 1, showing the overall routes of post-breeding migration to the non-breeding areas. There was a clear geographic separation in the

Table 3. Summary statistics of the migration phenology and speed of lesser kestrels from the three main European breeding regions during the pre-breeding migration (modified from Sarà et al. 2019). Mean values ± standard deviation, in round brackets: minimum–maximum values and sample size referred to the number of migration events. The minimum migration distance is the orthodromic distance between breeding colony and non-breeding area. The track length is the sum of all distances between positions recorded on a migration track. The track straightness is minimum migration distance divided per track length. The E-W deviation represents the deviation in km to the east (positive values) or west (negative values) of a track compared to the minimum migration distance line.

Pre-breeding migration	Iberia	Italy	Balkan
Onset (day/month)	2/3 ± 17 (5/2 – 19/4; 35)	16/3 ± 15 (3/2 – 24/4; 32)	11/3 ± 10 (26/2 – 25/3; 12)
End (day/month)	19/3 ± 21 (13/2 – 27/4; 34)	4/4 ± 18 (19/2 – 13/5; 31)	4/4 ± 13 (8/3 – 18/4; 9)
Duration (days)	18 ± 12 (3 – 50; 34)	20 ± 10 (8 – 44; 31)	23 ± 13 (5 – 41; 9)
Speed (km/day)	234 ± 114 (98 – 476; 20)	246 ± 123 (141 – 476; 9)	294 ± 286 (139 – 723; 4)
Track length (km)	3661 ± 506 (2854 – 4716; 20)	4203 ± 586 (3479 – 5219; 9)	4339 ± 696 (3616 – 5143; 4)
Track straightness	0.80 ± 0.09 (0.63 – 0.94; 20)	0.75 ± 0.09 (0.63 – 0.88; 9)	0.83 ± 0.10 (0.72 – 0.96; 4)
E-W deviation (km)	–134 ± 214 (–477 – 217; 20)	–210 ± 221 (–497 – 199; 9)	–59 ± 320 (–461 – 271; 4)
Permanence in non-breeding area (days)	153 ± 24 (101 – 201; 35)	166 ± 18 (123 – 210; 32)	160 ± 14 (141 – 181; 12)
Minimum migration distance (km)	2946 ± 350 (2107 – 3639; 45)	3098 ± 261 (2413 – 3570; 36)	3481 ± 236 (3045 – 3812; 13)

migratory routes travelled to reach the non-breeding areas of Sahel. Iberian kestrels passed over the Alboran Sea and at both sides (Atlantic and Mediterranean) of the Gibraltar Strait to land on Moroccan coasts. Some French kestrels passed over Ibiza, then crossed the sea to land in Algeria. Most post-breeding tracks of Iberian kestrels were quite straight to the non-breeding areas, and only a minority of birds, mostly of French origin, made small eastern detours to reach the Western Sahel (Fig. 1).

The Italian birds also crossed the sea on a broad front. Generally, both the south Italian and some Sicilian birds descended eastbound over the Ionian Sea or approached the eastern coasts of Sicily. However, other Sicilian birds started their post-breeding journey more westerly and from the locations of the summer roosts (see Sarà et al. 2014), passing over the Sicilian Channel. Remarkably, one south Italian bird, after spending the summer in North Italy, had started from there the post-breeding migration and crossed the Central Tyrrhenian Sea passing over Sardinia to land in Tunisia. All Italian individuals usually landed in Tunisia and Libya and most of them made a western detour in Libya and then did a fairly straight flight from north-east to south-west.

The Balkan population behaved similarly to lesser kestrels of the other two European regions, although more similarly to the Italian birds. They crossed the Mediterranean on a broad front, arrived to the African coasts in Libya and Egypt and maintained a fairly northeast-southwest trajectory to cross the Sahara Desert, until their final destination in Central Sahel areas. Indeed, some of them did large western detours, passing over Naxos (Aegean islands) and Crete before flying to African coasts. During the last days of migration, at about 15° N, the tracked individuals of all populations stopped their latitudinal descent, making east-west movements to reach their non-breeding quarters.

Pre-breeding Migration

At the watch-sites of the Gibraltar Strait (Table 1) the first migrants from Africa that directed to the Iberian Peninsula have been detected at in the end of January, and later with a bimodal peak of passage in mid-March and mid-April (Román et al. 2009). At the Messina Strait, usually observations started in early April, hence the peak passage in March could not be recorded, and only the mid-April peak was confirmed with few individuals detected also during May (Giordano et al. 2009). Based on ring recovery data, and visual observations, spring migration over Greece occurred in both western and eastern mainland, as well as through many Aegean islands, enduring from March till mid-May with a peak in mid-April (Bounas et al. 2016). Several dozens of lesser kestrels are observed every year at Cape Otranto (Table 1), but few of them cross the Adriatic Sea en route to the Balkans, and likely they mostly belong to the Southern Italy population (Premuda et al. 2004).

After a permanence in the Sahel areas of about five months and half, which was individually quite variable (Table 3), but without statistical differences attributable to sex or region of origin; the tracked lesser kestrels started their pre-breeding journey to the colonies for a new breeding season in Europe. The mean date of such spring departures from Sahel areas was March 9th (range: February 3rd–April 23rd), and the mean date of arrival to the colony was March 29th (range: February 14th – May 14th). Males and females did not show differences either in the dates of departure and in those of arrival to breeding areas. Contrariwise and alike to autumn, there were statistically significant differences related to the location of the breeding area: Iberian lesser kestrels tended to depart earlier (on average February 26th) than Italian and Balkan conspecifics (on average March 15th and 12th, respectively). As a consequence, the Iberian ones arrived earlier to breeding areas, on average March 16th, while the Italian ones arrived on average on April 4th and the Balkan ones on April 8th (Table 3). The average pre-breeding travel was about 20 days (range: 3–50 days, Table 3), during which the migrating individuals covered some 260 km per day (range: 98–723 km/day) totalling an average pre-breeding migration distance of about 4,000 km (range: 2,854–5,219 km). Alike the minimum migration distance, the regional track lengths differed with the Iberian making the shortest and the Balkan the longest pre-breeding travel. The average track straightness in Table 3 is about 0.80, revealing that the pre-breeding migration was on average 20% longer than the minimum distance between the non-breeding area and the breeding colony. Also at this stage, some lesser kestrels made a quite direct return journey to the breeding colony (track straightness about 0.95), while others made broad detours (track straightness = 0.63). There are no statistical regional and sex differences in track straightness and E-W deviation (Table 3). Patterns of pre-breeding migration have been reported in Fig. 2 and show the overall spring return routes to the breeding areas and the relative straightness and E-W deviation for each migratory event. There was a clear geographic separation either in the non-breeding areas and the migratory routes travelled to reach the breeding colonies in the main European peninsulas (i.e., migratory connectivity). Pre-breeding migration was not funnelled at the narrowest stretches of sea (i.e., Gibraltar Strait, Sicilian Strait, etc.), but occurred over a broad front.

From Western Sahel, Iberian lesser kestrels flied along the Atlantic coasts (one individual even over the Atlantic Ocean) to converge later to the Alboran Sea area where they crossed the Mediterranean, some of them also across Gibraltar. Instead, the Italian kestrels moved from their non-breeding areas in Western-Central Sahel, flied on broad front across the Sahara and converged later in Tunisia and Western Libya to cross on broad front the Central Mediterranean and reach Sicily and Southern continental Italy. Interestingly, several Italian birds reached their breeding colonies in Apulia (south eastern tip of Italy) from the north-west. They departed from Tunisia and crossed the Central Tyrrhenian Sea and the Apennines, thus making a much broader open sea crossing than would have been necessary. Only one bird reached Sicily, crossed the Messina Strait and migrated overland to Apulia doing the shorter and 'logical' path. The Bulgarian birds, the only representative of Balkan populations tracked so far with Argos PTTs, travelled straight from south Niger to Bosnia-Herzegovina and Serbia, flying across Southern Italy, and adjusted the route to the breeding colonies only when they were in the Balkans, at about 43° N. Instead, others individuals flied eastbound and adjusted the route very early, so doing the western detour at varying North-African latitudes (one in Libya, another in Egypt), and from the latter flied directly to their breeding colony (Fig. 2). One individual used Crete and Astipalaia islands, as spring stopover, before reaching its breeding colony.

Behavioural and Ecological Remarks

The minimum orthodromic distance of migration, representing the shortest distance on the curved surface of earth between every individual's breeding colony in Europe and the respective non-breeding

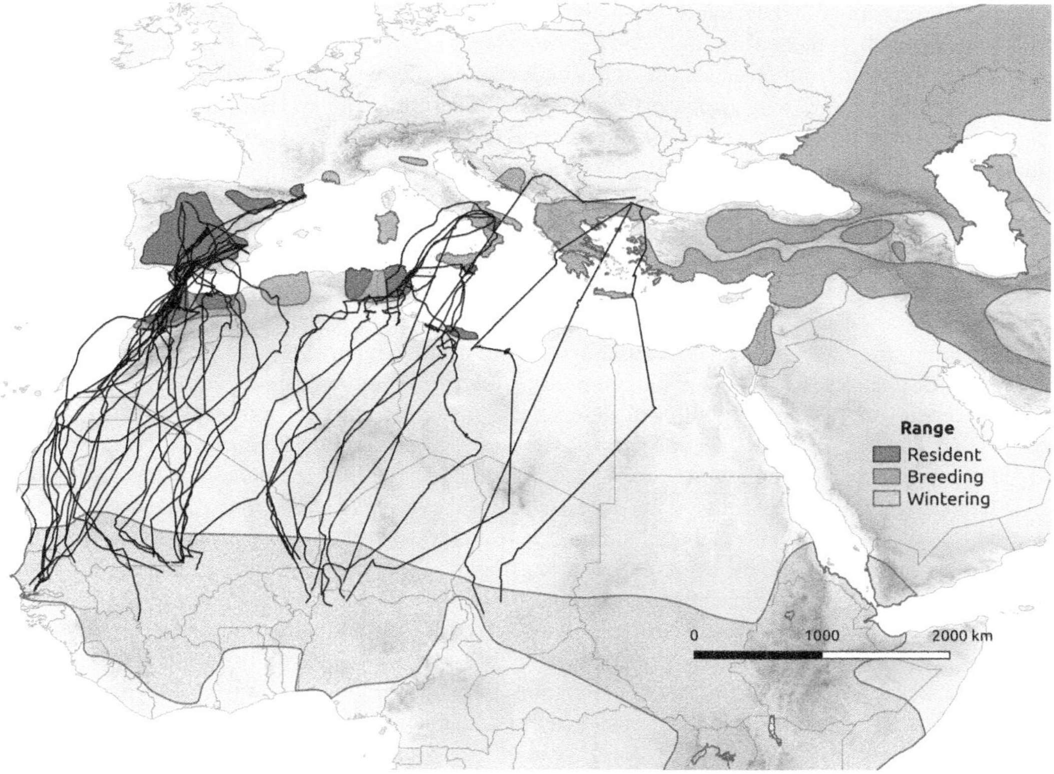

Figure 2. Migratory routes of lesser kestrels from the non-breeding Sahel areas in Africa to the breeding European grounds (modified from Sarà et al. 2019).

area in Sahel, was on average about 3,200 km (Table 3). Male and female kestrels covered the same migration distances, as statistical difference between sex was not found. Although the overall migratory behaviour of males and females was remarkably similar, some significant sex differences in migratory behaviour were found, as males started their post-breeding migration later than females (see § 4), and in addition, performed a significantly straighter migration compared to females. Contrariwise, the geographic region of origin mattered, because there was a significant regional difference, with Iberian kestrels making on average shorter travels than the Italian and those from the Balkans making the longest ones (Sarà et al. 2019). Such a significant regional difference leads to the most intriguing result of the Sarà et al. (2019) study; which is, that individuals from different breeding regions tended to migrate to distinct sectors of the Sahel (Fig. 1 and 2). The non-breeding areas of Iberian birds were mostly located in the Western Sahel, most of the Balkan breeding birds spent the non-breeding season in Central-Eastern Sahel (Niger, Nigeria and Chad), whereas the Italian lesser kestrels spread over a broad area of the Central Sahelian belt, from Eastern Mali (where they overlapped somewhat with Iberian birds) to Niger and Nigeria. Only a minority of individuals did not follow this general pattern. Such a spatial segregation of European populations in the Sahelian belt was attested by the strong statistical significance of parameters estimating the migratory connectivity. The longitude of the breeding sites and that of the non-breeding areas were positively correlated, and individuals from the same breeding region had non-breeding areas closer to each other than expected by chance (Sarà et al. 2019). Weak migratory connectivity is the most frequent pattern found in a suite of 98 populations of 45 long-distance migratory bird species (Finch et al. 2017). Therefore, the strong connectivity found in the lesser kestrel is a peculiar biological trait of this species, with remarkable implications on its conservation planning, and so far shared with few other migratory species in the Palaearctic. Geographical constraints on migration routes, mostly of Iberian kestrels, may have originated this pattern. The proximity of the Atlantic coast, implies that Iberian migrating kestrels cannot disperse over the Sahel as much as those from Italy and the Balkans. Besides, there exist geographical differences in resource abundance among Sahel sectors, the Western Sahel being much rainier and hence having greater primary productivity than the Central Sahel. Therefore, the Western Sahel might sustain a higher population density in a relatively narrower area compared to the Central Sahel, where most of the Italian and Balkan birds spend the non-breeding season (Sarà et al. 2019). The strong migratory connectivity observed among European populations pattern is further corroborated by the few birds tracked from Mongolia, which seem to occupy non-breeding areas located to the south-east of those of European individuals (Millhouser 2016).

There were intriguing differences in the seasonal migratory patterns of lesser kestrels. Post-breeding migration was noticeably faster and occurred in less time than pre-breeding migration, besides being straighter and with less detours (compare parameters in Tables 2 and 3). Marked detours especially during desert crossing are a common feature for raptors migrating across the Sahara and might happen as a response to both broad-scale seasonal wind patterns and local winds (Limiñana et al. 2013, Sarà et al. 2019). In European lesser kestrels detours occurred mostly during the pre-breeding migration, although with some noticeable cases during the post-breeding stage. They were very likely a consequence of initial wind drifts (in spring, severe westward crosswind over the Sahara; in autumn, tailwinds over the Sahara, and moderate eastward crosswinds over the Mediterranean), which were later corrected by overcompensation flights segments (Limiñana et al. 2013). Overall, more favourable winds during autumn, together with seasonal differences in stopover behaviour (which are discussed below), are the most likely reason of the seasonal differences in migration speed.

The seasonal variation in the stopover strategy of lesser kestrels is quite remarkable. Limiñana et al. (2012a) found that stopovers were much more frequent and longer during spring migration. An examination of 18 kestrels during spring and 24 ones during autumn migration from Spain, Italy and Bulgaria (M. Sarà, S. Bondì, J.G. Cecere, D. Rubolini, J. Bustamante, unpublished data), extended these findings. During spring migration all but one individuals have stopped on average 1–3 times, and stopovers lasted on average more than one week (range: 1–26 days). Contrariwise, only one third of the 24 individuals have stopped during their autumnal journey, each doing a single stopover, on average of about three days (range: 1–5 days).

Most of spring stopovers were in coastal agricultural areas, when the lesser kestrels reached North African coasts or just few hundred kilometres before. Most of these sites often coincided with circular irrigated fields created by pumping water from underground aquifers into the Sahara, in Algeria and Libya. They were likely performed to take advantage of local concentrations of prey, in order to get the best conditions for arrival to the breeding grounds and start reproduction (M. Sarà, S. Bondì, J.G. Cecere, D. Rubolini, J. Bustamante, unpublished data). Interestingly, one of the remainder spring stopover sites was along the Nile River and other two were in the Aegean islands (Naxos, Crete), right before reaching the breeding areas. Indeed, lesser kestrels used to fly over islands without stopping there, as most of Italian individuals did over Malta and Lampedusa; and Balkans did over the Ionian islands. Few individuals also did short pauses of few hours (hence not computed as stopovers), like the Iberian kestrel that has spent just two hours on Lanzarote (Canary Islands), before its long flight over the Atlantic Ocean, or the Italian kestrel that has remained 13 hours on Ischia island, in the Naples gulf, near the Italian coasts (M. Sarà, S. Bondì, J.G. Cecere, D. Rubolini, J. Bustamante, unpublished data). Moreover, tens of migrating lesser kestrels are observed foraging every spring, also for several days, in pseudo-steppe fields in Cape Otranto (Premuda et al. 2004). On the contrary, the few autumn stopovers were more scattered, being localized well within the desert, near dry river beds, like those located inside the Tassili N'Ajjer National Park, in Algeria, or at Wádí el Agial (Lybia), as well as in coastal agricultural areas.

Acknowledgments

Maurizio Sarà is indebted to Laura Zanca and Salvatore Bondì, Rosario Mascara, Michelangelo Morganti for the help in the field during years of study in Sicily. Ronelle Visagie provided photos and useful information about lesser kestrels in South Africa, being an exquisite host during the Maurizio Sarà and Laura Zanca stay in the Northern Cape Province. Figures 1–2 and Table 2–3 modified upon acceptance of Wiley. Diego Rubolini and Jacopo G. Cecere acknowledge financial support by the LIFE programme (LIFE11 NAT/IT/000068 and LIFE17 NAT/IT/000586). Javier Bustamante acknowledges funding writing by the "KESTRELS-MOVE" project (ref: CGL2016-79249-P) (AEI/FEDER, UE).

25

Common kestrel *Falco tinnunculus*

Michelangelo Morganti

ABSTRACT

The common kestrel (*Falco tinnunculus*) is the commonest raptor species in the whole of Eurasia, with different populations showing a variety of migration strategies and wide ranges of behavioural adaptations. Northern populations are mainly truly migratory, with long-distance sub-Saharan movements regularly performed by at least a fraction of the population. Central European birds behave as facultative migrants, with juveniles and females more prone to migrate than adults and males. Southern European populations show limited movements, mainly due to altitudinal migrations and breeding or natal dispersal. Dispersal can reach considerable distances, up to 500 km in northern populations. Common kestrel may migrate both by soaring and flapping, and has the ability to perform long sea-crossings. This is probably the main cause of a lack of huge concentrations of birds at migratory bottlenecks, where few tens to a few hundreds of individuals are observed. Observations, especially in the Central Mediterranean, seems more abundant in spring than in autumn but the overall numbers only represent a negligible proportion of the overall European population, suggesting that broad front migration is the norm for migratory kestrels. The lack of tracking data from migrating individuals appears as a striking gap in the current knowledge of the species, and it is advisable that individual-based movement data should be collected in the years to come.

Introduction

The common kestrel (*Falco tinnunculus* Linnaeus, 1758)—hereafter simply 'kestrel'—is a small falcon, (136–252 g male, female 154–314 g, wingspan 57–79 cm; Ferguson-Lees and Christie 2001) and probably has the widest breeding area among the ~ 40 species of the genus *Falco*. Coherently, it is considered as the most abundant diurnal raptor in Western Palearctic (Orta et al. 2020). The breeding areas of the kestrel, with its 11 subspecies (some of these considered as full species depending on the authors), distribute over the entire Palearctic and most of Africa, from 70° N on continental Europe towards tropical and sub-tropical populations down to include South Africa (Fig. 1, Orta et al. 2020). Main Atlantic archipelagos also hosting breeding populations of kestrels with the only exceptions of northern archipelagos and Iceland, where the species is considered only accidentally recorded (Orta et al. 2020). The kestrel is normally a solitary and cavity nester but easily adapts to artificial

CNR-IRSA Italian National Research Council – Water Research Institute, Via del Mulino 19, 20861 Brugherio (Monza-Brianza), Italy.
Email: michelangelo.morganti@irsa.cnr.it

nest-boxes and can occasionally breed in trees using old corvid nests (Ferguson-Lee and Christie 2001, Grattini 2008). It is rarely reported that kestrel sometimes aggregates in loose, small (3–10 pairs) colonies to breed both using corvid nests (see Ermolaev 2016 and references therein) or in rocky areas as in Central Greece (N. Tsiopelas—HOS/Birdlife pers. comm.). The kestrel can breed in almost every open habitat, from seacoasts to inland wetlands, including pure urban habitats and sub-desertic areas, only excluding too densely wooded areas which are probably unsuitable for the hunting technique of this raptor. It normally also breeds in high mountains range, frequently above 3,000 m a.s.l. in the Alps and up to 5,500 m a.s.l. in Himalaya (Ferguson-Lee and Christie 2001). The European population estimate ranges from 409,000 to 600,000 breeding pairs while the total world population is estimated in 4–6.5 millions of mature individuals (BirdLife International 2017a, 2020). The global conservation status of the species is overall classified as of 'Least Concern' by IUCN (2020) given the wide distribution of the species, despite the same IUCN report a recent slightly declining trend at a global level. For Europe, Birdlife International (2017a) classifies the species as a 'SPEC 3' (i.e., unfavourable European conservation status but population not concentrated in EU at a local scale), and scattered data shows how different populations recently had alternate trends, ranging from a clear increase as in Italy (Grattini 2008, Gustin et al. 2019) to steep declines as in United Kingdom (Petty et al. 2003, Clements 2008). Breeding phenology of the kestrel is centred around April–May (March–June) but dramatically differs among populations breeding at different latitudeswith a progressive delay of 6 days every 10° of latitude towards north (Carrillo and González-Dávila 2009). The kestrel predominantly behaves as an active predator, with a wide range of preys including mammals, reptiles, birds, bats and arthropods (i.e., Costantini and Dell'Omo 2020, Mikula et al. 2013, Orihuela-Torres et al. 2017, van Zyl 1994).

This chapter refers to the known migratory movements of the Western Palearctic kestrel populations, so belonging to the nominal race *F. t. tinnunculus*. As a general pattern, kestrels are not typical soarers and move over a wide front without concentrating at bottlenecks. Ringing recoveries (either direct or not) confirm the ability of the species for wide sea and desert crossing. Northern populations show the longest migrations, with most of the individuals performing trans-Saharan crossing, while Central and Southern European populations progressively lose in migration propensity, shaping an overall leap-frog migration pattern. Differential migration for age and sex classes has been described for several populations so that partial migration is likely to be the norm among middle-latitudes populations. During winter, and especially in open agricultural habitats, kestrels can reach

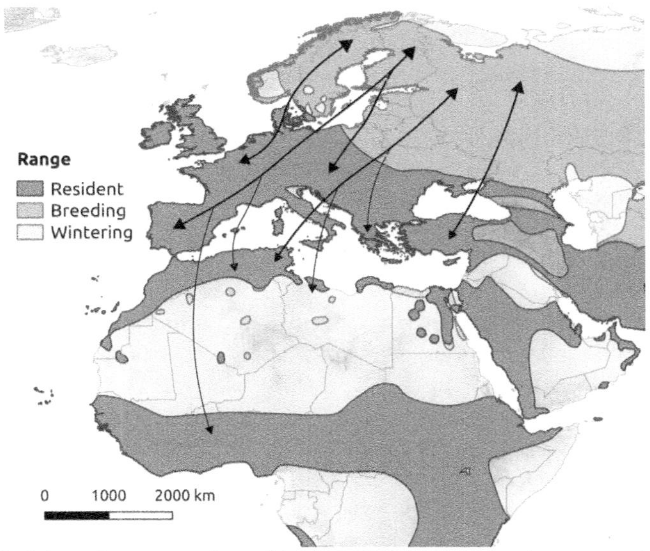

Figure 1. Schematic representations of the routes followed by migrating common kestrels.

considerably high densities (up to 6–8 birds/km^2, SEO/Birdlife 2012a), which can vary depending on prey availability (e.g., voles, Village 1990). Despite this propensity to concentrate in wintering areas, kestrels do not normally constitute communal roosts, but can occasionally join the roost of other species out of the breeding season (i.e., lesser kestrels, Dell'Omo G. and Grattini N. pers.obs.).

Knowledge Status

Migration patterns of the different populations of kestrel are inferable from the wide availability of ringing recoveries, which anyhow mainly concerns the Northern European populations. The high numbers of adults and chicks ringed at the nest allowed to discern among dispersal and migratory movements and to quantify the magnitude of these movements. All the national migration atlas published to the date are reporting valuable information on the species. Both published and informal reports from migration observatories include kestrel counts, however, the main problem in these reports is the potential confusion between the common and lesser kestrel (*Falco naumanni*) in southern observation sites. Discrimination among the two species is sometimes difficult, especially for birds observed by far and/or in mixed flocks of common and lesser kestrels. It is noteworthy the low number (a few hundreds in total) of kestrels counted at bottleneck migration sites every year if compared to the total European estimated population. This observation leads to the conclusion that, as a general rule, the kestrel migrates over a broad front, also undertaking long sea crossings. For example, very few kestrels are observed during both migratory passages at the most important bottlenecks sites of Eurasia, Batumi and the Strait of Gibraltar (Verhlest et al. 2011, batumiraptorcount.org accessed July 2020, Onrubia 2016). Local censuses in breeding and wintering periods allowed to estimate bird densities for different environments and at different latitudes and constituted the fundamental data to assess the resident behaviour at a population level. Overall, the almost complete lack of any type of individual tracking data on migration patterns stands out as a huge knowledge gap which should be filled out in the next years.

Migration Patterns

Migratory behaviour of the common kestrel well fits the classical definition of a partial migrant if assigned at a European level. Indeed, remarkable differences among populations breeding at different latitudes are observed. Northern European birds tend to migrate over medium and long distances (some of them being even trans-Saharan migrants), while Central and Southern European birds progressively lose in migration propensity (Saurola et al. 2013, Holte et al. 2016, Costantini and Dell'Omo 2020). The relationship between breeding latitude and migration tendency seems to promote a leap-frog pattern of migration, with northern populations overwintering at southern latitudes than those of Central Europe. Birds from Fennoscandia and Baltic regions are thought to be almost completely migratory, with very few birds wintering in these northern regions, mainly those originating from the south of the region (i.e., < 62° N, Saurola et al. 2013). The loss of migratory propensity in southern populations is clearly shown by the comparison of recoveries distances among birds belonging to Northern and Southern UK populations, with the latter showing significantly shorter migration distances (Snow 1968). A similar pattern is reported for birds ringed in Germany (Holte et al. 2016). Coherently with this, re-encounter analysis of birds ringed in Italy and Spain show that—after the removal of individuals captured during active migration across these countries—local birds only show movements of few tens of kilometres that possibly belongs to altitudinal migrations and/or natal dispersal (Spina and Volponi 2008; SEO/Birdlife 2012a, 2012b). Sedentariness of these populations seems confirmed by the case of 10 adult kestrels (6 females and 4 males) equipped with GPS/UHF during reproduction near Rome (Italy, 41° N) in 2019 and 2020. All these birds (movements data updated to July 2020) remained the whole year within 20 km from the nesting locality (Dell'Omo unpub. data). As a consequence of the above-mentioned movement patterns, entire Europe and north-Africa are classified as areas of year-round occurrence of the common kestrel, with the only exception of Russia and Fennoscandia and the Baltic regions where almost the whole population is considered truly migratory (del Hoyo 2020). Entire sub-Saharan Africa is also considered as an area of year-round occurrence, even if this includes both wintering birds from Northern Europe and local common kestrels belonging to other subspecies. Central- and Eastern Asian kestrels belong to two subspecies, one overwintering in India and South-Eastern Asia (*F. t. interstinctus*), and one mostly resident at southern latitudes of the regions (*F. t. objurgatus*, Orte et al. 2020). However, it must be stressed that movements of eastern populations/subspecies remain to the date poorly known due to the lack of ringing recoveries or telemetry-based studies. Differential migration (*sensu* Newton 2008) has been widely described for the common kestrel, with juveniles and females thought to be more prone to migration than adults and males, respectively: the median wintering location of adults lays significantly northern than those of juveniles both for birds ringed in Finland (Saurola et al. 2013) and in Germany (Holte et al. 2016). Additionally, adult males represent the biggest fraction of the common kestrels wintering in Southern Sweden (Kjellén 1994).

True long-distance migrations are observed in birds from most of the Northern European countries, with recoveries in the Sahelian belt of birds ringed in Germany, Czech Republic, Belgium, Finland and Sweden (Fig. 1; Adriansen et al. 1997, Saurola et al. 2013, Holte et al. 2016, Costantini and Dell'Omo 2020). It remains to be disclosed what is the proportion of long-distance migrants in the northern populations of the species since most of the recoveries come from Central and Southern Europe. Long-distance movements have also been registered towards the Atlantic islands (Canary, Madeira, Cape Verde), once again belonging to birds ringed in Central and Northern Europe (Holte et al. 2016, Saurola et al. 2013) but also for birds ringed in Spain and Italy, possibly captured in this region during migration (Spina and Volponi 2008, SEO/Birdlife 2012b). Although numbers of kestrels recorded at migratory bottlenecks are relatively low, some counts are worth to be mentioned, especially autumn ones. At Falsterbo (Sweden) in the 1986–1995 period a mean (± SD) of 427 ± 127 birds per autumn has been recorded, with a remarkably high percentage (72% in mean) of juveniles in the sample (Kjellén 1998). This unusually high proportion of juveniles suggests that adult birds tend to migrate over a wider front instead of concentrating at bottlenecks (Kjellén 1998). A much smaller number of kestrels are counted in autumn along the Baltic migratory corridor (i.e., Vistula observation point) with a mean (± SD) of 82 ± 67 birds per season in the 2008–2011 period, representing a significant proportion (~ 2.5%) of the Baltic population (Polakowksi et al. 2014). In Western France, a range of 264–1548 individuals has been observed in the decade 2010–2019 at Fort L'Ecluse bypassing the Alps towards south-west during autumn, while in the same period 123–383 individuals have been counted crossing the Pyrenees at Organbidexka (www.migraction.net, accessed 01/08/2020). Along the same flyway, a considerable flow is observed during spring at Col de l'Escrinet, Western France (range: 264–1023, www.migraction.net). Other noteworthy counts are reported in Central Italy (maximum of 241 in autumn at Capriglia, Premuda 2018) and at the continental shore of the Strait of Messina (75 during autumn 2018, Cento et al. 2018). Across the Central Mediterranean, 100–200 birds (mean ± SD = 148 ± 66) are counted in autumn in Malta (Sammut and Bonavia 2004, Sammut et al. 2013), with a peak of 282 in autumn 2010 (Bonavia et al. 2010). A significant flow of birds en route to Africa and crossing the Mediterranean Sea over a wide stretch is also observed at Mallorca (486 in autumn 2019, G.O.R.A. unpub. data).

The spring counts of common kestrel along the Central Mediterranean flyway are considerably higher (but note that autumn migration phenology of kestrel may be larger than the period normally covered by observation points), with regular counts of over 200 birds per season at watchsites on the Adriatic coast of Italy, e.g., 396 (Sonet et al. 2009) observed at Mount San Bartolo. In addition, migrating kestrels are regularly observed in northward migration on the Sicilian side of the Strait of Messina, with a mean (± SD) of 642 ± 179 birds observed in the 1996–2000 period (Corso 2001a). Scarce (< 40 individuals per site) but regular observations of migrating kestrels also come from spring observations on the Tyrrhenian islands (i.e., 39 in 2005 over Panarea, Sicily, Gustin 2005).

Both ringing recoveries and visual observations concur in clearly defining a NE-SW general direction of migratory movements, but it must be considered that very few birds have been ringed in Eastern European and Central Russian populations (Costantinti and Dell'Omo 2020), and that recovery probability is much reduced in these areas. Closest analyses of the ringing recoveries of German and Belgian kestrels disclosed that re-encounters occurred within 100 km from the natal or breeding sites almost lacking a clear directionality (so that can be attributed to dispersal), while south-eastern directions emerge for recoveries at longer distances (Adriansen et al. 1997, Saurola et al. 2013, Holte et al. 2016). However, natal and breeding dispersal (*sensu* Costantini and Dell'Omo 2020) of up to 300–500 km have been described for Finnish birds (Saurola et al. 2013).

As a consequence of the wide range of the species and its complex migration patterns, the migration phenology of the common kestrel results spanned over a wide period, with the bulk of the autumn movements apparently taking place during the second half of September (Polakowski et al. 2014).

Behavioural and Ecological Remarks

Common kestrels appear as a facultative, partial migrant in most of its distribution. The yearly association between dispersal distances and availability of food (i.e., longer dispersal in worst years, Adriansen et al. 1998) concurs to suggest high plasticity in the behavioural response of this species to local conditions. It is therefore not surprising that the common kestrel is considered to have progressively lose in migration propensity as a response of recent climate warming (Holte et al. 2016), and recent recovery data seems to propose that if in the 1990s the border segregating truly migratory from facultative migratory population was crossing Poland and Ukraine (Village 1990), nowadays this border should be moved significantly northward, at least until Denmark. Despite its main tendency to short or facultative migration, common kestrel shows great migration abilities and flying capacities as the long sea-crossing demonstrates. It would be a good subject species to study the trade-off between migration and residency strategies and to observe the development of migration strategies along the different life stages. Eventually, the recent miniaturization of the tracking device will certainly offer unprecedented possibilities for fine studies on the movement ecology of this species in the near future (Costantini and Dell'Omo 2020).

26

Red-footed falcon *Falco vespertinus*

Todd Katzner,[1,*] *Evgeny Bragin*[2,†] and *Tricia Miller*[3]

ABSTRACT

Red-footed falcons (*Falco vespertinus*) are a small, long-distance and obligate migrant falcon that breeds at the forest-steppe interface in Eurasia and winters in Southern Africa. Research carried out with geolocators and satellite transmitters show that during the southbound migration Central Asian birds migrate through the Caucasus and the Middle East, while those from Eastern Europe cross the Eastern Mediterranean. Once in Africa, they cross the Sahara, the Sahel, where they perform short stopovers, and the Congo rainforest or the Rift valley, heading towards the wintering grounds. Northbound migration involves stopovers in West Africa and long crossings over the Mediterranean Sea. Both visual observations and tracking studies show that at least some populations of the species perform a clockwise loop migration that involves extensive longitudinal movements and, for some sub-populations, crossing the Central Mediterranean only during spring. Pre-breeding migration has a longer duration than post-breeding migration, probably because stopovers are longer and wind assistance lower. Red-footed falcons also face substantial anthropogenic threats along migration, particularly the risk of shooting at migratory bottlenecks.

Introduction

The red-footed falcon (*Falco vespertinus* Linnaeus, 1766) is a small, long-distance and obligate-migrant falcon that breeds at the forest-steppe interface in Eurasia and winters in Southern Africa. The wings are long, narrow and pointed, and, when folded, they are of equal length to, or slightly protrude beyond, the edge of the tail. Red-footed falcons are striking and strongly sexually dimorphic in appearance. Adult males are uniform slate grey or greyish-blue, with bright orange cere, eye-ring, and legs, and reddish undertail feathers. Adult females have a rufous head, neck and belly, with a dark mask and malar stripe. The top of the body, wings and tail are gray with dark barring. Recent fledglings are dark-brownish with streaking on top and ocher-colored feathering of the chest and belly. They have a similar black mask and malar stripe as do the adult females.

The breeding distribution of red-footed falcons extends across Eurasia, from Central Europe to Northern Kazakhstan and Southern Russia and into extreme Northwestern Mongolia and China (BirdLife International 2018b). To the east of this distribution, in Mongolia, China, and south of

[1] U.S. Geological Survey, Boise, Idaho, USA.
[2] Faculty of Natural Science, Kostanay State Pedagogical, Kostanay, Kazakhstan.
[3] Conservation Science Global, Cape May, NJ, USA.
† Deceased
* Corresponding author: tkatzner@usgs.gov

Lake Baikal in Russia, this species is replaced by closely related Amur falcons (*F. amurensis*), once considered a subspecies of red-footed falcons. The northern border of the range crosses Southern Belarus and extends into Russia, through the middle of the Ural Range and into Western Siberia. Breeding habitats are generally forest-steppe landscapes, but in taiga of North-Eastern Russia and Western and Central Siberia they nest in river valleys with open meadows and swamps covered with sparse coniferous forest. During the non-breeding season, red-footed falcons are observed in large flocks in and around the Kalahari Desert of Southern Africa. They are regularly recorded in Northern South Africa, Namibia, Botswana, Zimbabwe, Southern Angola and Western Zambia, and occasionally seen in neighboring countries.

During migration, this species has been observed in many countries of Africa, Europe, and the Middle East. Although the bulk of the distribution of the red-footed falcon occurs within the former Soviet Union, there are only a few observations of the species on migration in ex-Soviet countries. In fact, this species is very rarely observed when it is outside of its breeding range in Siberia, Mongolia, or Central Asia (Boldbaatar et al. 2012, Isakov and Vorobev 1949, Pavlov and Molodovskyi 1962, Mitropolsky et al. 1987, Yasko 2017). Even at Chokpak Pass, where extensive migration counts have been conducted since ~1976, red-footed falcons have only been observed once in the past ~30 years (Korelov 1962, Kovshar and Berezovikov 1999, Gavrilov et al. 2002). Similarly, there are very few records from other areas west of Chokpak Pass that are hotspots of raptor migration—around the Aral Sea (Bostanzhoglo 1911, Zarudny 1916, Varshavsky 1957, Korelov 1962, Berezovsky 1983), along the western coast of the Caspian Sea (Butyev et al. 1983), and in Uzbekistan (Bogdanov 1882, Pleske 1888, Mitropolsky et al. 1987). In contrast, the vast majority of observations of autumn migration of red-footed falcons are from north of the Caspian and Black Seas (Katzner et al. 2016). Likewise, during spring migration there are a large number of observations from Mediterranean watch sites (Bounas et al. 2020). These records, together with tracking data, provide important insight into migration routes of this species.

Although there are reports of migration of red-footed falcons at many watch sites, published studies of nesting red-footed falcons are predominantly from two regions. In Central Asia (Kazakhstan), this species breeds in natural steppe settings, both in nests of corvids and in nest boxes. In contrast, in the Carpathian Basin of Hungary and the Balkans, where there is limited natural steppe habitat, they are associated with agricultural landscapes and nest boxes provided by humans. In both regions, red-footed falcons are gregarious and usually colonial breeders and foragers. Long-term data from Kazakhstan suggest little change in nesting phenology or productivity of the species over time (Bragin et al. 2017). These same data highlight the potential for an ecological trap from nest boxes, in which rates of both egg-laying and offspring loss are higher in nest boxes than in natural nests (Bragin et al. 2017).

Throughout their breeding range, red-footed falcons forage in open steppe habitats. Diet of red-footed falcons has been studied primarily during the breeding season. In this period, they predominantly eat insects, especially those caught on the wing, as well as small mammals and birds, and reptiles, especially lizards. In Africa, red-footed falcons are thought to feed predominately on insects and some authors have suggested that the large flocks of these falcons follow seasonally and regionally abundant insect populations (BirdLife International 2018b).

Populations of red-footed falcons, although large, appear to be declining throughout much of their range (BirdLife International 2018b). There has been an especially sharp decline in Central Europe and in northern parts of Eastern Europe. Likewise, in Siberia, red-footed falcons have disappeared from the regions near Lake Baikal (Popov 2000, Popov and Maleev 2008). The population in Northern Kazakhstan and in the Southern Ural region of Russia seems to be stable (E. Bragin, personal observations). Populations in Hungary, Italy, and Austria appear to be slightly increasing, in concert with extensive conservation efforts in these areas (Palatitz et al. 2009). At the end of 20th century the total population of red-footed falcon was estimated to be between 150,000 and 400,000 pairs (Ferguson-Lees and Christie 2001). Recent estimates suggest that the world population is stable and that most of those birds breed in Central Asia, with 30,000–64,000 breeding pairs in Europe (BirdLife International 2018b). In Central Europe the population is 2,500–3,000 pairs, which includes 1,000–1,100 pairs in Hungary and 1,300–1,600 pairs in Romania (Palatitz et al. 2009).

Knowledge Status

In comparison with other well studied long-distance migrants, there has been limited study of the migration of red-footed falcons. What is known of migration routes comes primarily from tracking studies of birds that were tagged in Kazakhstan (Katzner et al. 2016) and Central Europe (Palatitz et al. 2018). Likewise, there are also ring-return data in the Russian-language literature from birds marked in these areas (summarized in Katzner et al. 2016). There are records of the species from migration counts in and around the Mediterranean and the Black Sea, and in reports from Africa (see citations below). As the species is a broad front migrant, visual observations may not be as helpful to depict its migratory routes as they are for species that concentrate in bottlenecks, but a recent analysis of citizen science data helped to clarify the migration patterns in the Mediterranean area (Bounas et al. 2020). Key knowledge gaps for understanding the migration of the species are linked to migratory connectivity, routes of migration for untracked populations, and sources of mortality when on migration.

Post-breeding Migration

Routes of migration

Although migration data for this species are poor, it is clear that southbound migration by red-footed falcons is dramatically different from the northbound route. Tracking data from Central Asia (Katzner et al. 2016, EAB unpublished observations) suggest that autumn migration of red footed falcons first involves a westward flight that brings the birds north of the Caspian Sea and then either south, through the Caucasus or west again, north of the Black Sea and then south. From this point, birds either migrate through the Middle East or make a water crossing through the Eastern Mediterranean (Fig. 1). Once in Africa, they cross the Eastern Sahara and then the Sahel, heading to the wintering grounds. It is not clear what proportion of birds follows the East African Rift Valley (as suggested by geolocator data) or travel west of there, through the Congo Basin (as suggested by satellite telemetry data).

Satellite tracking data from birds breeding in the Carpathian Basin are, to a large degree, similar to those from Kazakhstan, except that Carpathian birds start west of the Black Sea (Palatitz et al. 2018). The majority of these birds appear to make a water crossing in the Eastern Mediterranean and their migration through Africa may occur to the west of their Central Asian kin. Telemetry data suggest that these birds slow their travel in the Sahel, possibly to recover from their long water and desert crossings (Palatitz et al. 2018; see also Eleonora's falcons for a similar strategy: Mellone et al. 2013b, Chapter 28).

Despite the existence of these tracking data, there are almost no data on migratory connectivity of these populations. However, as the wintering range is much smaller than the breeding range, mixing among different populations is likely to occur (see also Kassara et al. 2017).

Data from watch sites are largely consistent with the tracking data, although they also suggest gaps in knowledge. There are no data on observations of red-footed falcons to the east of the Ural Mountains. To the west of the Urals, there are observations of small flocks of falcons along the Ural River and its tributaries (Berezovikov et al. 2000). Farther west, flocks of > 200 individuals are observed in the North Caucasus (in the Stavropol region), the Eastern Azov and Black Seas regions (Ilyukh and Zabolotnyi 1999, Malovichko et al. 2003, Tilba 2010, Abuladze et al. 2003, Komarov 2010, Abuladze 2013), and in Bulgaria (Iankov et al. 2007). Evidence from telemetry and observational data suggests that some of these birds cross portions of the Eastern Black Sea (Barabashin 2010,

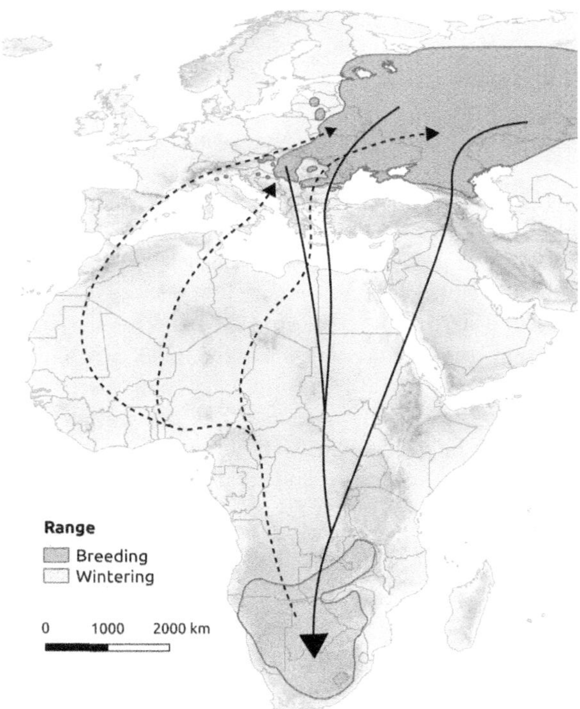

Figure 1. Schematic representation of the routes followed during autumn (solid lines) and during spring (dashed lines) by red-footed falcons.

Dinkevich 2012, Stanciu et al. 2017). There are also records of thousands of birds roosting communally in south Ukraine, Hungary, and Romania (Strautman 1963, Petrovich and Redinov 2008, Palatitz et al. 2009). Observations have been reported of large concentrations of birds passing through Georgia, at Batumi and other sites (Verhelst et al. 2011, Abuladze 2006, 2013). Finally, there are observations of birds migrating through the Dardanelles (Panuccio et al. 2017) and along the Mediterranean coast of Turkey (Oppel et al. 2014).

Literature reports suggest that Central Asia may have once hosted an alternative migration route along the Altai and Tien-Shan mountains and through Iran to the west-south-west. Along this route there are historical observations of migrant red-footed falcons from Kazakhstan, at south Balkhash, near Almaty and the Kopa River valley, at Chokpak, near Tashkent, and at Iskanderkul Lake, in Afghanistan (Korelov 1962, Ivanov 1969, Mitropolsky et al. 1987). Records from the middle of the 20th century in Iran were from Gilan province in the north-west, Hamadan province in the west-central part of the country, and from Qazvin, near Tehran (Roselaar and Mansour 2009).

Timing of migration

Departure of five red-footed falcons tracked from breeding grounds in Northern Kazakhstan occurred at the end of August or in early September (Katzner et al. 2016), while 22 individuals tagged with satellite transmitters in Eastern Europe initiated migration between 20 September and 10 October (Palatitz et al. 2018). Duration of migration for Kazakhstan birds was about 30 days, with arrival in early October, and, for Carpathian birds, sometimes only two weeks, with arrival between the 5th of October and the 20th of November.

Observational data largely corroborate these telemetry data. In Siberia, red-footed falcons disappear from breeding areas in the last 10 days of August (Rogacheva 1988, Rogacheva et al.

1988). In Western Siberia and in Kazakhstan, migration is observed from about mid-August until mid-September (Kuchin 1976, Danilov 1976, Chernishov et al. 1983). Farther west, in the Southern Ural region, falcons disappear at the start of September (Zakharov 2006). In the Volga-Ural watershed migration begins at the end of August and continues through September or early October (Shevchenko et al. 1978, Berezovikov et al. 2000, Kovshar 2007). In the Don River basin, the last red-footed falcons of the year are observed in mid-October (Barabash-Nikiforov and Semago 1962). Finally, the peak of red-footed falcon migration on the Western Black Sea coast of Bulgaria is between 11 September and 14 October, with the greatest numbers in the late part of September. At this site, 15,895 individuals were observed in the 2017, including 10,517 in one day, the 27th of September (Michev et al. 2011). This is the highest count known for a single watchsite. During September–October, hundreds of birds cross the Dardanelles and the Bosphorus (Fülöp et al. 2014, Panuccio et al. 2017). A similar phenology has been observed by Bounas et al. (2020).

Huge flocks of red-footed falcons may be encountered on the plains of the Russian North Caucasus, in September and early October (Verhelst et al. 2011). Likewise, in the western part of the region near the Black Sea, falcons are observed as late as mid-November (Khokhlov and Ilyukh 1998, Ilyukh and Zabolotnyi 1999, Komarov 2010, Tilba 2010). In Southern Ukraine, just to the west of this part of Russia, migration takes place almost at the same time, although mass migration is observed during the second half of September (Zubarovsky 1977, Domashevsky 2006).

Around the Mediterranean basin, there are regular observations of red-footed falcon migration at a number of watch sites. In late September and early August, > 40 falcons may be observed within a day at migration sites in the south of Cyprus (Roth 2008). These birds are some of those making water crossings in the Eastern Mediterranean. In Israel, yearly variation is great, with a record count of 11,371 individuals in 1997, with numbers peaking on the 28th of September (Shirihai et al. 2000).

Pre-breeding Migration

Routes of migration

Routes of migration of red-footed falcons are generally only known from animal tracking studies and from a few ring recoveries. The birds are known to winter in Southern Africa, especially in Namibia and Botswana, although they are occasionally also reported in neighboring countries. In March, before migration initiates, a great portion of the global population come together in Central Angola where they are observed in roosting flocks (Palatitz et al. 2018).

Data from both geolocator studies (Katzner et al. 2016) and satellite tracking (Palatitz et al. 2018) suggest that regardless of wintering grounds, northbound migration of red-footed falcons occurs in Western Africa, through Angola, western parts of the Democratic Republic of Congo, and through central and western countries of the Sahara Desert. Before the Sahara crossing, at around 10° N latitude, red-footed falcons may spend two to six weeks in West Africa, presumably replenishing energy stores (Palatitz et al. 2018). Heading north from Africa, birds make a water crossing over the Mediterranean Sea (Bounas et al. 2020; Fig. 1). Telemetry data suggest crossing occurs on a broad front, with some birds reaching Europe at Gibraltar, and the majority crossing to Sicily or other parts of Italy directly from mainland Africa or via Malta (Palatitz et al. 2018).

Observational data suggest large numbers of birds crossing from North Africa to Sicily (Galea and Massa 1985), and then traveling north into mainland Italy. At the Strait of Messina in Southern Italy (between Sicily and mainland Italy), counts of 135–1,012 red-footed falcons have been recorded (Corso 2001). Crossings also regularly occur at the Tyrrhenian Sea, with up to 294 individuals observed from the island of Ustica (Gustin 2008). These yearly differences among watch sites located roughly at the same latitude suggest that the migratory corridor is broad and may shift greatly every year, possibly according to local weather conditions. Once on mainland Italy, where just a few tens of pairs breed, the majority of red-footed falcons cross the Adriatic Sea *en route* to the Balkans, with counts of up to 341 individuals in the Conero promontory and 725 in San Bartolo Regional Park (Borioni and Baldoni 2007, Premuda et al. 2008, P.R. Monte San Bartolo 2015). The Sardinia-Corsica corridor is also used, with 204 individuals observed in the 2005 (Premuda et al. 2007a) In Spain, the species is much less frequently observed than in Italy (Barriocanal et al. 2004), but in some years the main flyway may shift westward so greatly that hundreds of individuals are recorded (Prieta 2015). During spring, contrary to what takes place in autumn, the species is not frequently observed in Israel (max. 44 in 1985; Shirihai and Christie 1992, Shirihai et al. 2000).

Strong westward winds may be very critical during spring migration, negatively affecting survival rates. This is evidenced by both observations of falcons in the Canary Islands in spring and death of tagged individuals (Palatitz et al. 2018). Both these observations occurred as a result of apparently unfavourable wind conditions for the Sahara crossing.

Once in Eastern Europe, falcons fly north and east, towards nesting grounds. Those birds that breed in Ukraine, Russia, or Kazakhstan appear to travel north of the Black Sea. This route is longer than a direct route from wintering grounds, but presumably it provides more time in steppe habitats and, thus, more refueling opportunities for these long-distance migrants (Katzner et al. 2016).

Timing of migration

The best data on departure date of spring migration come from tracking studies of birds tagged in Central Asia and Central Europe. Tracked birds left wintering grounds in late March (Katzner et al. 2016) or early April (Palatitz et al. 2018). For Kazakhstan birds, arrival was in early May, suggesting a trip of about 60 days. Atypically, this trek is longer than is autumn migration, and it appears lengthened by stops, possibly for refueling, along the way (Katzner et al. 2016). The typical journey of an Eastern European bird lasts about one month. Duration of travel for a tagged female from Crimea was 34 and 28 days in two subsequent years, with 11 and 8 days, respectively, needed for the Sahara crossing (Palatitz et al. 2018).

There are substantially more data on dates of passage of red-footed falcons from raptor migration sites and from other observations in the literature, especially from the Mediterranean region. Northbound migration of red-footed falcons is generally from southwest to northeast and timing of observations tracks this geography. Records from the Central Mediterranean region (Tunisia, Malta, Italy) suggest that movements of falcons peak in early May (Bounas et al. 2020).

Within the breeding range, spring migration and arrival of red-footed falcons occurs from mid-April to the second half of May, with stragglers returning in June. In Slovakia, the earliest observations of red-footed falcons are from mid-late April, with the bulk of passage in May (Noga et al. 2017). Similarly, on the coast of the Black Sea, the peak of migration occurs in the middle of April, with the last migrants observed in mid-May (Abuladze 2006, 2013). The first nesting falcons arrive in Southern Moldova and Ukraine around mid-April (Averin et al. 1971, Zubarovsky 1977). Arrival in Northern Ukraine is later, at the end of April and peaking in early May (Zubarovsky 1977, Golovushkin 1992, Domashevsky 2006, Knish et al. 2008). Farther west, around Krasnodar in North Caucasus region, migration of red-footed falcons continues from mid-April to the beginning of May (Dinkevich and Lastovetsky 1997, Tilba 2010). Timing of migration is roughly similar at the Azov Sea (Ochapovsky 1967, Emtyl and Lokhman 1994) and in the delta of the Volga River (Krivonosov 1963, Bondarev and Reutsky 1989). In the forest-steppe zone in European Russia, the Southern Urals and in Northern Kazakhstan, red-footed falcons arrive in late April or early May (Puzanov 1955, Artemiev et al. 1977, Sotnikov 1999, Bragin 1989). Finally, in Western Siberia, red-footed falcons arrive at the end of April and early May (Danilov 1976, Kuchin 1976) and farther east in Siberia between 13 and 21 May (Rogacheva 1988, Rogacheva et al. 2008). These arrival dates in the literature are largely consistent with arrival dates as estimated from animal tracking of birds from the Carpathian Basin and Northern Kazakhstan.

Behavioural and Ecological Remarks

Our review of the literature on red-footed falcon migration points to several patterns that are unique to the species or that may have important implications for its conservation. For example, there are mixed reports as to the degree to which red-footed falcons are flocking migrants. Some have suggested that these birds fly in flocks that may number many hundreds or thousands (Ferguson-Lees and Christie 2001, Barriocanal et al. 2004, Iankov et al. 2007). Elsewhere, especially in the Russian language literature, there are suggestions that, when together, individuals fly in small groups of 3–50 individuals (Knish et al. 2008, Ochapovsky 1967, Emtyl and Lokhman 1994, Dinkevich and Lastovetskyi 1997). It may be that both these reports are correct, and that flocking behaviour is region-specific, with larger flocks in some regions and smaller flocks in other places, and also weather-dependent. Indeed, at the Dardanelles, red-footed falcons were recorded mainly with high levels of air humidity and lower air pressure, probably because they concentrated there to avoid crossing the Eastern Mediterranean during unfavourable weather conditions (Panuccio et al. 2017). It has also been suggested that red-footed falcons migrate in large flocks at very high altitudes and that eastern populations migrate to the west of the Black Sea where they join up with Western European breeders (Ferguson-Lees and Christie 2001).

Migration by red-footed falcons generally follows an unusual clockwise loop pattern. Southbound traveling falcons from Central Asia and Eastern Europe appear to migrate west before turning south and traveling through Eastern Africa. Their return trip takes them north through Western Africa and across the Mediterranean Sea, before turning east to return to nesting grounds (Bounas et al. 2020; Fig. 1). This detour in both directions takes them north of the Caspian Sea and, often, north of the Black Sea. Such behaviour lengthens migration distances but may allow these birds to migrate through suitable foraging habitat for a greater proportion of their overall migration route and to skip potentially food-poor Central Asian deserts. On northbound migration, there appears to be evidence, from both satellite and geolocator tracking, of a long refueling period in West Africa. After this, falcons cross the massive Sahara Desert and then face a substantial water crossing. It makes sense that this would be an important time for building up fat stores to support these dangerous crossings. This information suggests that unlike pure soaring migrants, red-footed falcons need to replenish their energy stores during migration. This concept also is supported by observations from Southern Italy during spring migration, when is not uncommon to observe individuals stopping over in the same area for several days (Premuda et al. 2004, Mellone unpub. data). A recent analysis highlighted the importance of mosaics of cropland and natural vegetation as stopover habitat in the Mediterranean basin (Bounas et al. 2020).

Overall, both visual observations and tracking studies show that at least some populations of red-footed falcons perform a clockwise loop migration that involves crossing the Central Mediterranean only during spring (Bounas et al. 2020). It is not clear if this pattern is shaped by different wind regimes or distribution of suitable stopover areas. Palatitz et al. (2018), suggest that winds are more favourable in the western part of the Sahara during spring, thus promoting the loop migration. Moreover, even if the spring passage is more concentrated than the autumn one in the Mediterranean basin (Bounas et al. 2020), tracking data show that the pre-breeding migration lasts longer than post-breeding one. This is in contrast with the rule for other migratory birds, and possibly occurs because the birds have longer stopovers and less wind assistance. An important priority for both ecological understanding and conservation management is therefore to study the drivers of seasonal variation in migratory behaviour of this species.

Available data suggests that migration and overwintering appear to be times of high mortality of red-footed falcons. In geolocator studies, although 20 units were deployed, only 2 were recovered (Katzner et al. 2016). Among birds tracked by satellite telemetry, 17 of 27 stopped sending data, either because of bird death or transmitter malfunction, during winter (Palatitz et al. 2018). Furthermore, both these studies show that the migration route of these falcons takes them through some of the areas of greatest shooting pressure in Eurasia. These include the Black Sea Coast (van Maanen et al. 2001, Sándor and Anthony 2018), the Central and Eastern Mediterranean, and Malta (Panuccio 2005).

Red-footed falcons are of conservation concern due to their moderately rapid population declines (BirdLife International 2018b). The limited data on causes of death for this species imply that mortality on migration may be an important contributor to ongoing population declines of this species (BirdLife International 2018b). That said, the general lack of knowledge about the species suggests that more intensive population monitoring and study may contribute to effective conservation for this species.

27

Amur falcon *Falco amurensis*

Ugo Mellone[1,2]

ABSTRACT

The Amur falcon (*Falco amurensis*) is a small raptor breeding in Eastern Asia and wintering in Southern Africa. The 14,000 km journey among these two ranges includes a long stopover in NE India, where probably the whole world population concentrates for several weeks in November. After this fattening period, Amur falcons embark into the crossing of the Indian Ocean, which is aided by tailwinds and involves at least two-three days flying no-stop. Amur falcons reach Southern Africa after travelling 1.5–3 months, and during winter congregate in large roosts. Spring migration begins in March and, after a significant stopover in Eastern Africa, Amur falcons again cross the Indian Ocean, once more with the aid of tailwinds, but using a slightly different route in comparison with the autumn one. Once in Asia, they reach their breeding grounds, avoiding the Himalaya crossing as they did in autumn. Such strategy, finely matched with resource availability and global wind patterns, makes the Amur falcon's journey one of the most remarkable ones of migratory organisms.

Introduction

The Amur falcon (*Falco amurensis* Radde, 1863) is a sexually dimorphic small falcon (weight: 120–160 g, wingspan 75 cm) breeding in temperate areas of Eastern Asia (Russia Mongolia, China) and wintering in Southern Africa (Bildstein 2006). This mainly insectivorous species may constitutes loose colonies with nests spaced ca. 200 m from each other, even if it is not strictly colonial as the closely related red-footed falcon (*Falco vespertinus*), probably also because it breeds re-using nests of the non-colonial Eurasian magpie (*Pica pica*; Burner et al. 2019). The species trend is considered stable, and therefore it is classified at "Least concern", with an estimated population of 1,000,000 individuals (Ferguson-Lees and Christie 2001). Counts carried out at 32 winter roost in South Africa yielded a total 111,291 individuals in 2009 (Symes and Woodborne 2010), a number which is way far lower than the global estimates. However, many winter roosts may remain undetected, also in other African countries, and until a few years ago more than 120,000 individuals (part of a much bigger flow) were estimated to be trapped and killed in the post-breeding stopover areas in India (Dalvi et al. 2013). As other falcon species, Amur falcons may fly more independently of thermal updrafts, in comparison with soaring migrants, and therefore they migrate on a wide front. However, the concentrations in intermediate goal areas make it different from typical broad front migrants. The Amur

[1] MEDRAPTORS (Mediterranean Raptor Migration Network), Rome, Italy.
[2] Vertebrates Zoology Research Group, University of Alicante, Alicante, Spain.
Email: u.mellone@gmail.com.

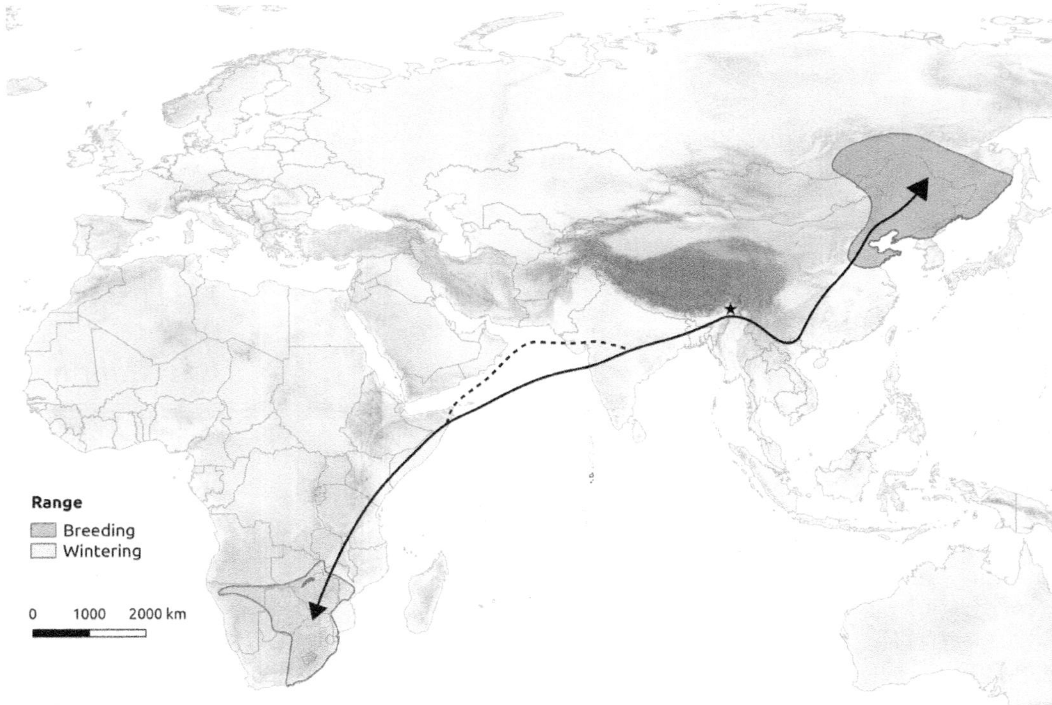

Figure 1. Schematic representation of the migration routes followed by Amur falcons (see references in the text for the sources). The dashed line represents the average sea-crossing track followed during spring. The star marks the location of the stopover area used during autumn in India.

falcon is mostly known for its trans-continental and trans-equatorial migratory route, which ranks it among the raptors performing the longest and most hazardous journeys worldwide (Bildstein 2006).

Knowledge Status

The migratory route of the Amur falcon has been already hypothesized since some decades on the basis of its seasonal distribution and on sporadic observations of flocks carried out especially in India (Ali and Ripley 1968, Clement and Holman 2001). Only recently some individuals have been followed by means of satellite telemetry (Argos transmitters), but comprehensive analyses of their travel patterns are not yet available. An adult female has been tagged in its nesting area in 2009, but its signal has been lost during the crossing of the Indian Ocean (Dixon et al. 2011). In early 2010, 10 adults have been tagged at a winter roost in South Africa, and provided valuable data, especially thanks to a female followed during four years (Meyburg et al. 2017b). Finally, 12 individuals, mainly adults, have been tagged since 2013 in the renowned stopover area in Nagaland (India; Kumar 2015, https://www.satellitetracking.eu).

Migration Patterns

On the basis of the available sources, it is possible to draw a general description of the routes followed by adult Amur falcons during both pre- and post-breeding migrations. They start the journey in the second half of September, flying to Southern China and then turning towards the west in order to reach the intermediate goal areas located in North-East India and Bangladesh. In this way they extend significantly the travel distance, but avoid the crossing of the Tibet highlands and of Himalaya (Dixon et al. 2011). In the above mentioned areas, especially in the region of Nagaland (India), Amur falcons perform a stopover that may last from two to five weeks, building stunning concentrations of tens of thousands of individuals that hunt on recently emerged swarming termites. This fattening period is mandatory to perform the crossing of the Indian Ocean. Aided by the tailwinds of the winter monsoon (Clement and Holman 2001), Amur falcons undertake the crossing on a broad front, leaving Asia generally from the region of Maharashtra (India), in the second half of November, and landing in Somalia after 2,400–3,150 km over water, the longest sea crossing performed on a regular basis by any raptor worldwide (Bildstein 2006, Meyburg et al. 2017b). This crossing involves a no-stop flight of at least two-three days. Kumar (2015) reports that an individual (Naga) travelled no-stop from NE India to Africa, covering a distance of 5,600 km in 5 days and 10 hours. The species is absent or rare vagrant on islands in the Indian Ocean, confirming this flight strategy (Clement and Holman 2001). Once in Africa, they head south, towards the wintering areas in Southern Africa, generally without performing further stopovers. Overall, such journey may last 1.5 to 3 months, for a total distance of almost 15,000 km (Meyburg et al. 2017b, https://www.satellitetracking.eu).

The pre-breeding migration largely retraces the post-breeding one, even if there are some differences, probably triggered by the crossing of the Indian Ocean, and the seasonal pattern of prevailing winds. For example, after departing in March, Amur falcons concentrate on a narrow front in coastal areas of Kenya and especially Somalia, showing slow paced and non-directional movements which may last up to three weeks, mainly during the first half of April (Meyburg et al. 2017b, https://www.satellitetracking.eu). During this period they probably build up the fat stores necessary for the ocean crossing. Future studies should clarify if concentrations similar to those occurring in India in November take place also in Eastern Africa during spring, and if they involve as well individual fidelity to intermediate goal areas. In this scenario, the observations of flocks of up to 250 individuals in Southern Somalia (Clement and Holman 2001) are very interesting and fits perfectly in the pattern shown by satellite tracked individuals (Meyburg et al. 2017b, https://www.satellitetracking.eu).

During spring, the over water portion of the route lies more in the north, in comparison to the autumn one, showing a curved, almost detoured, shape. Such differences probably occur since Amur falcons take advantage of the predictable tailwinds of the summer monsoon. Once in Asia, falcons fly directly to the breeding areas, even if stopovers in SE Asia lasting some days have been recorded. Arrival dates span from May until mid-June (Meyburg et al. 2017b, https://www.satellitetracking.eu).

Behavioural and Ecological Remarks

The Amur falcon journey shows many features that may be identified as textbook examples for a wide array of behaviours that make bird migration so fascinating for scientists and mankind in general. Some of them are the length of the journey, the massive foraging concentrations matching the phenology of the termites, and the no-stop flights, which are shaped by large-scale wind patterns. Especially the latter phenomenon may make them vulnerable to climate change, while massive concentrations put at risk the entire population in a restricted area. Indeed, thousands of falcons were killed until a few years ago taking advantage of these gatherings but, at least according to the latest available reports, the capture of Amur falcons in Nagaland does not take place anymore. Thanks to the effort of conservationists, nowadays the town of Pangti is even home of a festival celebrating the gathering of the falcons (Weidensaul 2018).

28

Eleonora's falcon *Falco eleonorae*

Ugo Mellone[1,2]

ABSTRACT

The Eleonora's falcon (*Falco eleonorae*) is colonial species showing a delayed breeding season, taking place in summer/autumn. It is a trans-equatorial migrant breeding in the Mediterranean ecoregion and wintering mainly in Madagascar. Therefore, twice a year it migrates across the African continent on a broad front, showing also long no-stop flights involving nocturnal migration, especially over ecological barriers like the desert and the sea. The post-breeding journey of juveniles may even entail a double duration in comparison with adults, as these inexperienced individuals slow down their migration speed, especially in the Sahel, in order to replenish their energy stores. The spring routes lies more in the east than autumn ones, shaping thus a loop migration pattern, promoted mainly by differences in the seasonal distribution of trophic resources, with Eleonora's falcons stopping-over for some days in Ethiopia/Somalia. This species shows stunning orientation capabilities, since, after a complex route, juveniles are able to reach alone their restricted wintering grounds in the island of Madagascar.

Introduction

The Eleonora's falcon (*Falco eleonorae* Gené, 1839) belongs to that group of falcons whose size is intermediate to small ones, like kestrels, and bigger ones, such as the peregrine. The species is polymorphic, with the pale morph being commoner than the dark one, and body mass ranges between 360 and 490 grams (Gschweng et al. 2008) and has an average wingspan of 1.2 m (Bruderer and Boldt 2001). It is a long-distance migrant, an endemic colonial breeder of the Mediterranean ecoregion (Walter 1979): all the colonies are located in Mediterranean islands, apart from a few on the Atlantic coast of Morocco and in the Canary islands. This species has a unique adaptation among Western Palearctic birds: it shows a delayed breeding season, starting in late July, so that eggs hatch at the end of August, coinciding with the migration season of passerines crossing the Mediterranean Sea, en route to sub-Saharan Africa. Breeding on islands, the Eleonora's falcons perform foraging trips on the open sea, where the preys have nowhere to hide and hunting is more successful (Rosen et al. 1999, Mellone et al. 2012c, Xirouchakis and Panuccio 2019, Gangoso et al. 2020). During the rest of the year the species feeds mainly on insects (Ristow 2004, Gschweng et al. 2008, Mellone et al. 2013b). Another special characteristics of this species is that it shows one of the smaller wintering

[1] MEDRAPTORS (Mediterranean Raptor Migration Network), Rome, Italy.
[2] Vertebrates Zoology Research Group, University of Alicante, Alicante, Spain.
Email: u.mellone@gmail.com

range known among Palearctic migrants, with the great majority of individuals, especially the adults, concentrating in North-East Madagascar (Kassara et al. 2017). There are some winter records from mainland East Africa (Gschweng et al. 2008, Mellone et al. 2013b), but none from the breeding region. The Eleonora's falcon probably takes advantage of good thermal conditions during most of their journeys over land (Rosen and Hedenstrom 2002). However, its high aspect ratio (relatively long and pointy wings; Agostini et al. 2015a) allows energy efficient powered flight, enabling it to also travel under poor soaring conditions: the Eleonora's falcon performs very long sea-crossings (up to 1,500 km, Mellone et al. 2011c), with the longest recorded continuous flight entailing 3,530 km covered in 85 h (Hadjikyriakou et al. 2020). Also, contrary to obligate soaring migrants, Eleonora's falcons do not rely only on fat reserves during migration, and therefore have to perform a fly-and-forage migration strategy, even remaining in a given area for some days (Mellone et al. 2013b, Hadjikyriakou et al. 2020). There is no evidence that Eleonora's falcons migrate in flocks, and local concentrations may be related to foraging individuals on stopover.

Knowledge Status

Being a broad front migrant, rarely observed in migration bottlenecks, all our knowledge on the migration strategies of the Eleonora's falcon has been gathered in the last 10–15 years through satellite telemetry studies. On the basis of sporadic observations and anecdotes it has been speculated for several decades that in order to reach the wintering grounds in Madagascar the species uses a coastal route,

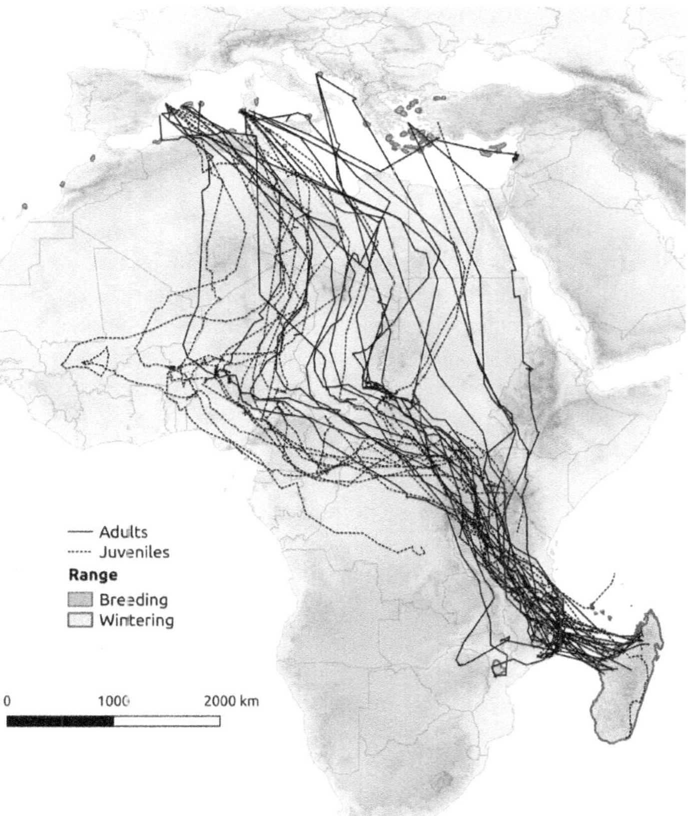

Figure 1. Post-breeding migratory routes followed by Eleonora's falcons tagged in Spain, Sardinia, Croatia and Greece (modified from Gschweng et al. 2008, Kassara et al. 2012, Mellone et al. 2013b). Greek data have been collected by the Hellenic Ornithological Society, funded by the A.G. Leventis Foundation through the project "Survey and Conservation of Seabirds in Greece".

circumnavigating the African continent (Stresemann 1954). However, the ring recovery in Mali of an individual belonging to the Canarian population suggested that actually the falcons were crossing inland Africa, using a much shorter route (Delgado and Quilis 1990). This has been confirmed by the first satellite telemetry studies (Gschweng et al. 2008, Lopez-Lopez et al. 2009). Nowadays, tracking studies across the whole breeding range of the species (Morocco, Spain, Italy, Croatia, Greece, Cyprus) confirm this: all falcons head to Madagascar attempting to follow the shortest possible route (Fig. 1; Gschweng et al. 2008, Kassara et al. 2012, Mellone et al. 2013b, Hadjikyriakou et al. 2020), without avoiding any ecological barrier (Lopez-Lopez et al. 2010). Probably, the coastal-route hypothesis has been believed and cited so widely because the species was known mainly for its coastal/marine habits that, however, only characterize the species during the breeding season. This led to a bias in the perception of the natural history of this species, which for the rest of the year is mainly an insectivorous tropical bird. The available studies on the migration behaviour have been based all on Argos satellite telemetry, apart the last one, from Cyprus, which also involved GPS devices: 12 individuals have been tagged in Balearic and Columbretes islands (Spain; Mellone et al. 2013b), 13 in Sardinia (Italy, Gschweng et al. 2008), two in Croatia (Mellone et al. 2013b), four in Aegean islands (Greece, Kassara et al. 2012) and 12 in Cyprus (Hadjikyriakou et al. 2020).

Post-breeding Migration

Eleonora's falcons begin to migrate during the second half of October, apparently without age differences in their departure schedules (Gschweng et al. 2008, Kassara et al. 2012, Mellone et al. 2013b, Hadjikyriakou et al. 2020).

After crossing the Mediterranean Sea, they continue over the Sahara desert, on a broad front and without following special topographical features. This is the region where they are more affected by wind conditions, with adults being able to cover more than 700 km in one day in case of good tailwind conditions (> 5 m/sec). Falcons also tend to drift with strong sidewinds in this region, which explains the S-shaped migration route typical of the species (Fig. 1; Mellone et al. 2015b), more evident in western populations than in eastern ones (compare with Hadjikyriakou et al. 2020). The temporal resolution of Argos data prevented a detailed analyses of orientation behaviour like those that have been conducted for GPS tracked soaring raptors (e.g., Klaassen et al. 2011), but apparently Eleonora's falcon overcompensate (especially adults) as much as they can until 20° N of latitude, being then drifted towards the South-West (especially juveniles) just before entering the Sahel. In this region there are no significant differences in the regional speed, straightness and number of stopover days between adult and juvenile falcons. Moreover, all those parameters promote overall higher migration speeds in comparison to the remaining regions. The Sahara desert thus acts an ecological barrier where restricted foraging opportunities and rapidly changing weather conditions can even result in the birds' deaths (Strandberg et al. 2009e). Consequently, both adults and juveniles try to accomplish its crossing as soon as possible (Mellone et al. 2013b). The observed variability in travel rates likely depends on weather conditions. This is also the region where birds travel most hours per day, even after sunset. The rate of nocturnal migration is significantly higher over the desert than in the other mainland regions and they never stop during the day (Lopez-Lopez et al. 2010, see also Hadjikyriakou et al. 2020).

Below 15° N, in the Sahel, even if adults slow down their migration progress, the drop in travel rates among juveniles is so strong that age differences become very evident and significant: 200 vs 100 km/day (ad. vs juv; average regional speed), 1.5 vs 6.4 (number of stopover days), 0.9 vs 0.6 (straightness; Mellone et al. 2013b). A similar pattern has been observed also for juveniles from Sardinia and Greece, stopping in the same regions and landscape type (Acacia savannah) for approximately two weeks (Gschweng et al. 2008, Kassara et al. 2012). Juveniles stopped as soon as the habitat became suitable for foraging, just south of the edge of the Sahara: the Sahel hosts benign habitats where an insectivorous bird such as the Eleonora's falcon can find large amounts of food in the right season (Zwarts et al. 2009). The difference between age classes might be explained by juveniles being less skilled in hunting, resulting in less efficient fat deposition before starting migration. Similar age-dependent patterns have also been observed in other species of raptors (Restani 2000, Ueta and Higuchi 2002, Strandberg and Alerstam 2007). The Sahel belt is known to be an important stopover area for many trans-equatorial migrants, some even stopping for several weeks (Pearson and Lack 1992, Tøttrup et al. 2012a) to exploit the favourable foraging conditions occurring after the end of the summer rains. Therefore, it is possible to argue that, the Sahel in particular, is a vast, rather "diffuse" stopover area, where especially juveniles Eleonora's falcons use a fly-and-forage strategy. GPS tracking showed that falcons fly slower over areas with higher greenness, and that nocturnal travelling segments are quicker than daily ones, thus confirming the fly-and-forage strategy (Hadjikyriakou et al. 2020). South of the Sahel, the Eleonora's falcons have to cross regions where the main habitats are equatorial forests and savannahs: here, age differences are still very evident, but both age classes migrate quicker than in the Sahel (Mellone et al. 2013b). Apparently, the equatorial forest does not act as an ecological barrier for the species (Lopez-Lopez et al. 2010). The last critical stage of migration is the Mozambique Channel: during autumn migration falcons try to cross it in the narrowest stretch, or at least avoiding the widest part. Even if this species is well equipped for long sea crossings, existing data show that this is a risky stage of the journey: among Sardinian birds, an adult delayed the crossing until March, a juvenile accomplished it only at the second attempt (one month after the first), and another juvenile decided to stop in Comoros (Gschweng et al. 2008) and then sent the last signals North of Madagascar, probably from a boat (Gschweng pers. comm.). A Greek juvenile delayed the crossing for 15 days, remaining close to the coast of Mozambique (Kassara et al. 2012). Among Spanish birds, a subadult bird sent the last data just in the middle of the Channel, probably from a boat where it subsequently died (Mellone et al. 2013b). Migration is almost over after birds reach Madagascar, given that the majority of falcons especially adults then quickly reach the North-East regions where they will spend the winter. Migration duration varies greatly among the studied populations, perhaps mainly because, given the low sample sizes, individual differences strongly influence population-averages. Generally speaking, adults need 3–4 weeks to reach the wintering grounds, while juveniles normally need more or less twice this time, as a result of the higher number of stopover days and, also, the lower straightness.

Pre-breeding Migration

The strategies used by Eleonora's falcons for return migration are much less known because of lower sample sizes (due to lifespan of the transmitters and mortality during autumn migration and winter). Adults leave Madagascar around mid-April and generally cross the Mozambique Channel in a northward direction, somewhat perpendicular to the routes via which they arrived to Madagascar during autumn migration. Before entering Kenya, around latitude 5° S, all tracked falcons shift their course towards the North-East, heading towards Ethiopia and Somalia, where all return migrants performed some stopover days (Fig. 1). All tracked birds across the different populations show this behaviour: for example two adults, from Croatia and Sardinia stopped for one week, and a Spanish immature for 27 days.

A Sardinian adult, after stopping in Ethiopia, remained also for 29 days in West Africa (Nigeria). Finally, although anecdotal, it is interesting to report that three simultaneously tracked individuals from Spain, Croatia and Greece used basically the same area in Ethiopia during the same days in April 2010, being just some tens of km apart (Mellone et al. 2013b, Kassara et al. 2012), or even in the same spot: the spatial and temporal resolution of Argos data does not allow to confirm this with pinpoint precision. These regions experience rain peaks just during spring ("belg" season, Glantz 1987; see also Pearson and Lack 1992), that could boost insect abundance (Janzen and Schoener 1968, Moreau 1972, Trierweiler et al. 2012). The high inter-year variation in rainfall patterns (Glantz 1987), as well as individual conditions, could explain the spatial and temporal variation in the occurrence of such stopover areas (see also Tøttrup et al. 2012b). Such stopovers may be very important to prepare the falcons to accomplish the Sahara crossing. Concerning the last leg of the journey, sample sizes are

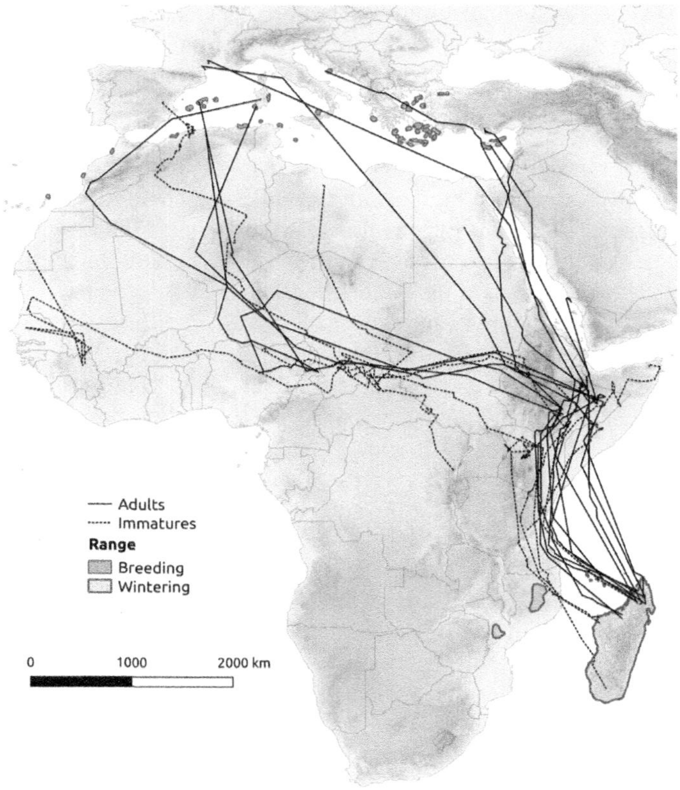

Figure 2. Pre-breeding migratory routes followed by Eleonora's falcons tagged in Spain, Sardinia, Croatia and Greece (modified from Gschweng et al. 2008, Kassara et al. 2012, Mellone et al. 2013b). Greek data have been collected by the Hellenic Ornithological Society, funded by the A.G. Leventis Foundation through the project "Survey and Conservation of Seabirds in Greece".

very small, but show that Croatian, Greek and Cyprus birds fly between the Nile valley and the Red sea, while the remaining tracked individuals cross the Central Sahara (Kassara et al. 2012, Mellone et al. 2013b, Hadjikyriakou et al. 2020). Probably the route choice is very flexible and depends a lot upon wind conditions, especially among those individuals crossing the Central Sahara. It seems that Cyprus birds follow the eastern shore of the Red Sea (Arabian Peninsula) in order to avoid the headwinds they would experience in Africa (Hadjikyriakou et al. 2020). The spring journey of adults is completed, normally, within one three-four weeks, while for juveniles it may take up to two months. No firm conclusions can be drawn on seasonal differences in migration duration (Gschweng et al. 2008, Kassara et al. 2012, Mellone et al. 2013b, Hadjikyriakou et al. 2020).

There is a time lapse of two months between the end of the spring migration and the beginning of the breeding season. As during this period food resources are very scarce around the colonies, Eleonora's falcons perform inland movements, sometimes hundreds of kilometres away from the respective breeding colonies, visiting several habitats, from forests to arable lands, probably taking advantage of high densities of insects. Then, adults established their colonies around mid-July, while non-breeding immatures continue to wander until the beginning of the autumn migration (Mellone et al. 2013c). Observations of this species away from the breeding colonies, sometimes even in raptor migration bottlenecks, are sometimes believed to be related to migrating individuals, but actually they are mostly wandering immatures or adults in their pre-breeding movements.

Behavioural and Ecological Remarks

Repeated journeys of Eleonora's falcons are available for a limited sample of individuals, but, especially individuals showing the longest migration route, e.g., those breeding in the Western Mediterranean, show a high flexibility, mainly during the crossing of ecological barriers such as the Sahara desert and the Indian ocean. The same individual may cross the Sahara desert using routes separated by more than 2,800 km, despite showing high wintering site fidelity (Mellone et al. 2012b) and after using stopover areas in Ethiopia with only ca. 200 km in between (Mellone et al. 2013b). Such variation may be related to wind conditions. The effect of weather conditions on route flexibility has been highlighted using repeated tracks of falcons crossing the Mozambique channel in spring (Mellone et al. 2011c): despite the fact that falcons were migrating individually, they started the sea crossing within a time window of 14 h and therefore experienced similar weather conditions. In 2009, three individuals used all the same direct route towards Somalia, involving a no-stop sea-crossing of more than 1,500 km. During the following year (2010), two of these individuals used a very different route in comparison to the 2009 one, flying both on a shorter sea-crossing to Tanzania, similar to the one followed in the majority of the available tracked birds (Gschweng et al. 2008, Kassara et al. 2012, Mellone et al. 2013b, Hadjikyriakou et al. 2020). In 2009, a low pressure area was present just in the area crossed by the falcons in 2010. Bad weather conditions likely caused the falcons to fly directly to Somalia in 2009, thereby avoiding more risky flights through a low pressure system with unpredictable weather conditions (Mellone et al. 2011c).

Despite considerable flexibility in regional route choice depending on weather conditions, large-scale seasonal differences are not related only to local weather conditions and seem to be well defined across all the studied populations, with falcons migrating more in the East during spring and showing therefore a clear loop migration pattern. Among Spanish birds, these differences are significant between latitude 10° N and 10° S, which is the area where the majority of stopovers take place. Therefore, falcons deliberately choose to deviate towards suitable foraging grounds in order to replenish their energy stores in between two major ecological barriers (the ocean and the desert). Such loop migration pattern seems to be widespread among other trans-equatorial insectivorous migrants stopping-over in East Africa, as well for example the red-backed shrike; Tøttrup et al. 2012a). According to Pearson and Lack (1992), the differential use of this region according to the season may be explained by climatic patterns that drive seasonal habitat availability, since in April Ethiopia and Somalia are green, while much of the Sahelian belt is very dry, with the opposite occurring in early autumn, just when juvenile Eleonora's falcons slow down their migration speed in the Sahel (see also Hadjikyriakou et al. 2020).

Eleonora's falcons exhibit very high orientation abilities, as despite the long and complex migration routes and substantial variation in individual route choice across consecutive years, they always return to the same restricted wintering area. Especially the performance of juveniles, which are able to reach the island of Madagascar independently of adults, is impressive. Indeed, it is hard to explain this feat assuming that juvenile birds rely on a simple clock-and-compass strategy (Thorup et al. 2010). We can only speculate about the mechanisms falcons use to navigate between their seasonal destinations. The main cues used by falcons may be magnetic inclination and geographic landmarks (Gschweng et al. 2008), but the fact that juveniles may migrate also by night and are able to compensate for wind drift suggests that another mechanism may be involved, for example infrasounds: especially when crossing the Mozambique Channel for the first time, naïve individuals may detect infrasounds emitted by a major landmass such as Madagascar, in order to find the right flight direction (Gschweng et al. 2008). It must be also stressed that close to the Equator magnetic inclination is 0 and therefore its detection is useless for orientation. It is interesting to note that, during autumn migration, between 2° N and 6° N, some juveniles deviated from their normal south-east bearing, heading more to the north in some cases for up to ca. 500 km (Gschweng et al. 2008, Mellone et al. 2013b): were they experiencing orientation issues when approaching the Equator? Some astonishing data obtained from GPS tracked adults breeding on the Canary islands show that around the magnetic equator and far from mountainous areas (that may be used as landmarks) Eleonora's falcons may fly up to 5,500 m above ground level, at sunrise, probably in order to "calibrate their map with visual cues on compass information obtained from polarized light and the rising sun" (Bouten et al. 2015). GPS tracking including altitudinal measurements with pressure sensors and behavioural measurements using accelerometers is very promising and will improve a lot our knowledge on the migratory behaviour of the species, especially concerning their orientation behaviour and their fly-and-forage strategy.

29

Sooty falcon *Falco concolor*

Ugo Mellone[1,2]

ABSTRACT

The sooty falcon (*Falco concolor*) is a medium sized falcon breeding mainly in the Middle East, and wintering in Madagascar. It is a facultative colonial species, showing a delayed breeding phenology, similar to the Eleonora's falcon (*Falco eleonorae*), as both species match their breeding season with the post-breeding migration of passerines. The few data available on its migratory strategies have been got through satellite tracked birds tagged in the eastern portion of the Arabian Peninsula. They show that the species migrate on a broad front, reaching Eastern Africa and then flying to the South in order to cross to Madagascar from Mozambique, in a journey lasting a couple of weeks at least. Juveniles are slower than adults as they follow a less direct route and perform stopovers in Sahel countries. The few data available for the pre-breeding migration suggest the existence of staging areas inland in Ethiopia.

Introduction

The sooty falcon (*Falco concolor*, Temminck 1825) is a medium sized falcon with an average body mass of 250 g and a wingspan of 97 cm (Bruderer and Boldt 2001) showing a patchy distribution, spanning from the eastern regions of the Sahara desert (Lybia) until Pakistan. Its breeding strongholds are located especially around the Red Sea, and the wintering range on the south-eastern coast of Africa and in Madagascar (Ferguson-Lees and Christie 2001). From the conservation point of view, the species is considered "Vulnerable", and a world population of 1,400–2,000 pairs has been estimated (BirdLife International 2020), breeding both inland in desert areas and on small islands, singly or in colonies including tens of pairs. The average brood size for a given population may range from 1.8 to 2.5 chicks (see McGrady et al. 2017 and references therein). As the sooty falcon shows a delayed breeding season, in order to feed on migrating passerines, its ecology is considered very similar to the Eleonora's falcon (*Falco eleonorae*), as the two species are also very close from the phylogenetic point of view (Wink and Ristow 2000). During the boreal winter, in Madagascar, sooty falcons may be observed feeding on insects, sometimes in groups of up to 19 individuals, also taking advantage of artificial lights (Zefania 2001). The likely decreasing trend of the species is probably related also

[1] MEDRAPTORS (Mediterranean Raptor Migration Network), Rome, Italy.
[2] Vertebrates Zoology Research Group, University of Alicante, Alicante, Spain.
Email: u.mellone@gmail.com

to human disturbance in the breeding sites (McGrady et al. 2010, 2019). Given the morphological similarity with the Eleonora's falcon, the sooty falcon is expected to migrate in a similar way, travelling on a broad front, independently of thermal currents availability and therefore using flapping flight to cross large water bodies, and probably extending the travel also during the night.

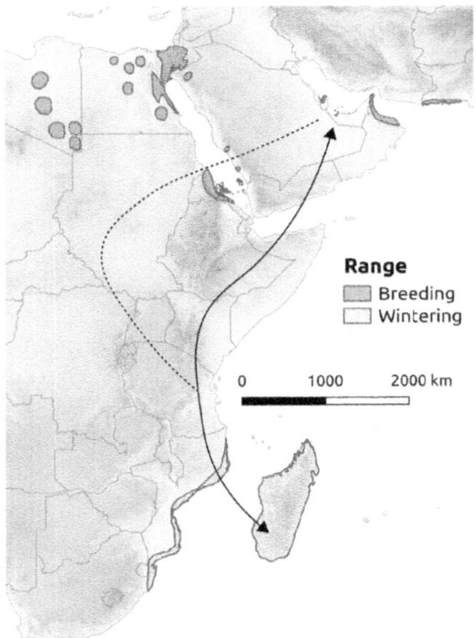

Figure 1. Schematic representation of the migration routes followed by sooty falcons breeding in the Eastern Arabian Peninsula. The dashed arrow represents the route followed mainly by juveniles during the post-breeding migration (Gschweng 2010, McGrady et al. 2010, 2016, Javed et al. 2012, McGrady and Al Fazari 2013).

Knowledge Status

The sooty falcon has been the subject of few studies focused on its breeding biology, and even less on the migration strategies. The only existing paper on the subject describes the autumn journey of an adult tagged with a satellite transmitter in the United Arab Emirates (Javed et al. 2012). Other 12 individuals, both adults and juveniles, have been tagged in Oman, and even one during winter in Madagascar (Gschweng 2010, McGrady et al. 2010, 2016, McGrady and Al Fazari 2013). However, their migration patterns have not been subjected to detailed analyses, so far.

Migration Patterns

The adult tracked by Javed et al. (2012), began its post-breeding migration probably at night, on the 31th October, and took 13 days to reach Madagascar, after a journey lasting at least 5,656 km. It crossed the Arabian Peninsula, and then Ethiopia, Kenya and Tanzania, finally crossing from Mozambique en route to Southern Madagascar. The authors of this study highlight the use of some stopover areas, but probably these stationary data were related to a night break and not to a true stopover aimed at foraging (Javed et al. 2012). Instead, the tracking of juveniles of Oman highlighted that their journey is less direct than adult ones: while it seems that the latter reach Africa crossing the Gulf of Aden, juveniles have a more northerly component in their migration direction, and cross mainly the Red Sea (Gschweng 2010, McGrady et al. 2010, McGrady and Al Fazari 2013; Fig. 1). They depart in early November and, once in Africa, they perform non-directed movements in concrete areas, which may last 2–3 weeks, especially in Ethiopia, South Sudan and Chad, but also on coastal areas of Eritrea. Such behaviour reminds the one recorded in juvenile Eleonora's falcons in the same period (Mellone et al. 2013b, see Chapter 28). After two abrupt changes of direction, first towards the south and then towards south-east, sooty falcons reach Madagascar. A juvenile tracked for the entire journey reached its wintering area in Central Madagascar on the 18th December (Gschweng 2010).

Concerning spring migration, less data are available. Once in mainland Africa, an individual tagged in the previous year in Oman as a chick, stopped for several weeks in Ethiopia, before reaching its natal area in late July. In Ethiopia, roughly the same staging area was used also by the adult tagged in Madagascar, which was found to belong to the population of the Dahlak islands in the Red Sea. This individual moved among the breeding area and the inland one for a couple of times before settling in the islands (in mid-July; McGrady and Al Fazari 2013), remarkably in a similar manner to adult Eleonora's falcons during their pre-breeding movements (Mellone et al. 2013c).

Behavioural and Ecological Remarks

Preliminary data on the sooty falcon distribution during winter in Madagascar (Zefania 2001, Javed et al. 2012, Gschweng 2015) suggest that the species is more likely to be present also on the western, dry, side of the island, unlike the Eleonora's falcon, which mostly concentrates in the humid forest located in the North-East (Kassara et al. 2017). Such pattern makes easier the detectability of the sooty falcon and indeed, they resulted ten times more abundant than the Eleonora's falcon in a sample area of Western Madagascar (del Hoyo et al. 1994). On the basis of this proportion, and using the breeding population estimates of the Eleonora's falcons (Walter 1979), an estimation of 40,000 breeding pairs was extrapolated, which nowadays has proven to be unfortunately much exaggerated (BirdLife International 2020).

The behaviour recorded for the sooty falcon tagged in Madagascar and belonging to the population of the Dahlak islands, which spent the pre-breeding season mostly inland, is very similar to the one recorded in Eleonora's falcons breeding in small Mediterranean islands: before the beginning of the breeding season, they may spend several weeks in staging areas rich of insect preys, and located even at hundreds of km from the breeding areas, where foraging conditions are unfavourable before the beginning of passerine migration (Mellone et al. 2013c).

30

Merlin *Falco columbarius*

Nicolantonio Agostini

ABSTRACT

The merlin (*Falco columbarius*) is a Holarctic species with a large range and population size. In the Western Palearctic most populations are migratory, breeding in Northern Europe and with wintering grounds widely distributed in Central and Southern Europe, in the Middle East and, to a small extent, in NW Africa. This species migrates on a broad front, mostly using the flapping flight and undertaking long water crossings. As a result, very few individuals are observed at raptor migration watchsites. Substantial concentrations occur in Southern Scandinavia during both post and pre-breeding movements peaking late September and late April-mid May. During post-breeding migration, adult females migrate earlier than juveniles which in turn migrate earlier than adult males. Conversely, during pre-breeding movements, males precede females. Ringing recoveries data of Scandinavian birds showed that they winter mostly in Western Europe, with some reaching the Mediterranean region. Adults migrate further than first-year birds. Among birds ringed in the British isles, a minority were recovered in continental Europe, mostly in France. Icelandic birds may reach mainland Europe, undertaking extremely long water crossings.

Introduction

The merlin (*Falco columbarius* Linnaeus, 1758) is a small falcon with an average body mass of 206 g (male: 125–234 g; female: 164–300 g) and a wing span ranging between 52 and 64 cm. It has relatively short-wings if compared with other falcons. It is a Holarctic species with nine subspecies, three breeding in Nearctic (*F. c. columbarius, F. c. suckleyi* and *F. c. richardsoni*), two in the Western Palearctic (*F. c. subaesalon* and *F. c. aesalon*) and four in the Eastern Palearctic (*F. c. insignis, F.c. pacificus, F. c. lymani* and *F. c. pallidus*), with a world population of 500,000–2,000,000 mature individuals. Because of its large range and population size its status has been classified as "Least Concern" (BirdLife International 2020). This species, such as other raptors, has declined in 1950s and 1960s mostly due to DDT poisoning, with a recovery in 1970s (Nygård 1999). This chapter refers to the movements of the two subspecies breeding in the Western Palearctic, from Iceland, Faroes (*F. c. subaesalon*), Britain and Ireland, Fennoscandia, Baltic states and Northern Russia (*F. c. aesalon*). The European population has been estimated in 32,000–51,600 pairs (BirdLife International 2017a). Most populations are migratory, but partial migrant on Iceland and sedentary especially in Britain

MEDRAPTORS (Mediterranean Raptor Migration Network), Roma, Italy.
Email:nicolantonioagostini@gmail.com

and Ireland, where they can show altitudinal movements (Forman 1999, Ferguson-Lees and Christie 2001). Wintering grounds are widely distributed in Europe, Middle East and locally in NW Africa. Two subspecies breeding in the Eastern Palearctic, *F. c. insignis* and *F. c. pallidus*, are seen in winter in the Middle East (Ferguson-Lees and Christie 2001).

The merlin breeds in heather moorland (especially in Britain and Ireland), on tundra, around the fringes of forest, Arctic scrubland and prairie, but it is also found in urban areas. The nest is built on the ground, but also on rocks or in trees, sometimes in colonies. The breeding season lasts from late April to August, beginning about one month later in north of range than in south, and the clutches consist of 3–6 eggs. It mostly feeds on small birds caught in flight and often low to the ground. It also hunt insects and small mammals, but also larger birds. During the winter it needs open grounds to hunts although it is also seen in urban areas (Ferguson-Lees and Christie 2001, Boal and Dykstra 2018). The merlin moves on a broad front during migration (Fig. 1), flying singly or in small flocks (< 10 individuals). It mostly uses the flapping flight, undertaking long water crossings (> 100 km) and hunting at stopover sites (Kerlinger 1989, Aborn 1994, Bildstein 2005). As a result, with rare exception, very few are observed at raptor migration watchsites (Zalles and Bildstein 2000). During the wintering season, it usually forms communal roosts (Dickson 1973, Grattini et al. 2006)

Knowledge Status

Migration patterns of this species have been studied in the Baltic region, where a substantial concentration occurs during both post- and pre-breeding migration. In particular, remarkable counts have been reported along the Vistula Spit (Poland; Polakowsky et al. 2014) and at Falsterbo (Sweden), where the post-breeding migration in relation to sex and age classes has been studied (Kjellén 1992). In addition, ringing recoveries data provided further information on this secretive species.

Migration Patterns

As mentioned above, during post-breeding movements a substantial concentration has been recorded at some watchsites located in the Baltic region. At Falsterbo, in Southern Sweden, on average 236 birds were counted per season in the period 1973–2019 (maximum 585 in 2011; https://www.falsterbofagelstation.se/). There, mostly juveniles are observed. As regards migration phenology, adult females migrate earlier than juveniles, which peak at the end of September, and juveniles earlier than adult males. Males and females combined migrate earlier than juveniles (Kjellén 1992). Other significant passages were reported at Stevns Klint (Denmark; maximum 130 in 1988), Kongelunden (Denmark; maximum 138 in 1982) and on the Vistula Spit (Poland), along the southern coast of the Baltic Sea (maximum 151 in 2011; Zalles and Bildstein 2000, Polakowsky et al. 2014). Through the Western Pyrenees, up to 100 per season have been reported in cumulating counts made at three watchsites, Organbidexska, Lindux and Lizarrieta, peaking in mid October (De Juana and Garcia 2015). Finally, very few (< 5) are observed at the Strait of Gibraltar en route to Morocco (Onrubia 2016), where the species is a regular but very scarce winter visitor (Thevenot et al. 2003).

During pre-breeding movements remarkable counts were reported in Denmark at Skagen (N Jutland), where on average 241 merlins were observed per season peaking late April-mid May in the period 1972–1998 (maximum 538 in 1991), and at Rørvig (maximum 139 in 1993; Zalles and Bildstein 2000). Unlike post-breeding migration, males migrate earlier than females in this period (Newton et al. 1978).

Ringing recoveries data mostly concern birds ringed in Northern Europe as nestlings. In particular thousands were ringed in Norway, Sweden, Finland, Great Britain and Ireland. Birds ringed in Northern Norway were recovered mostly in Central Europe (e.g., in Italy), while those ringed in Central-Southern Norway have their wintering quarters in Western Europe (Nygård 1999). Non-juvenile Finnish birds move further than Swedish ones as also than Finnish juveniles: in particular, their median position is located 735 km further, in a south-west direction, in comparison to the first-

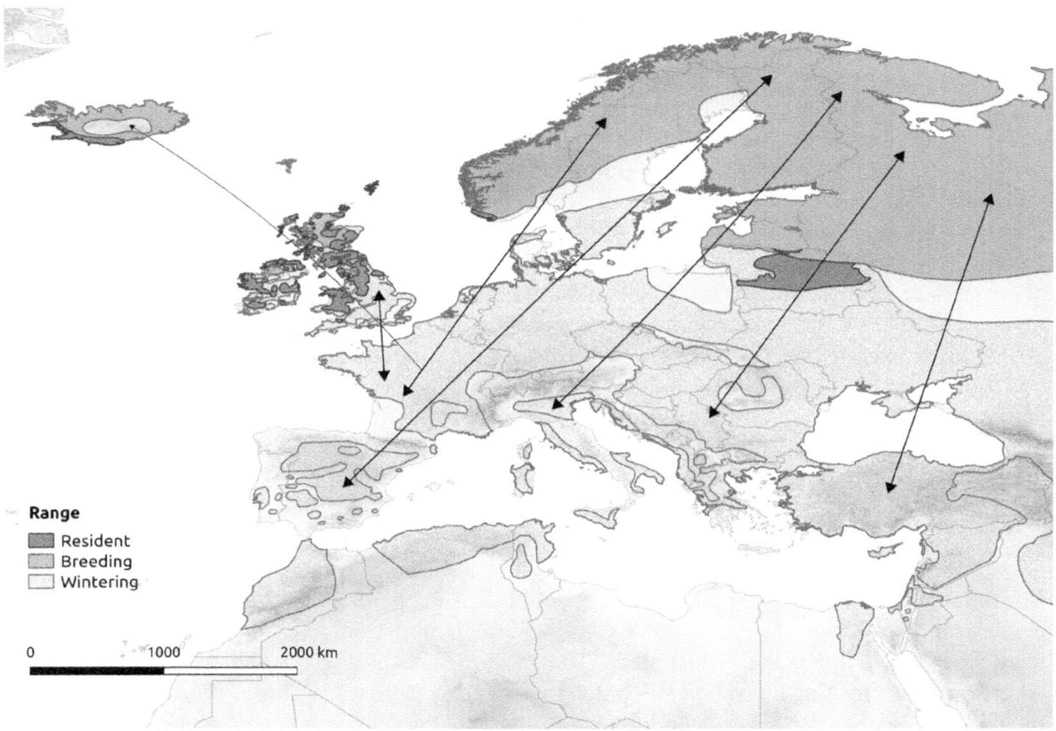

Figure 1. Schematic representation of the migration routes followed by merlins.

year birds from the same country. The longest distance covered by a Finnish birds was 4,159 km, recovered in Southern Spain (Saurola et al. 2013). Among birds ringed in Great Britain and Ireland, a minority were recovered in continental Europe, mostly in France. Conversely, 71% of first-year birds and 82% of older birds spend the winter within 100 km of their natal/breeding area. Merlins belonging to the Icelandic subspecies *subaesalon* were recorded in Great Britain, Ireland and Spain (Williamson 1954, Heavisides 1987, Sunyer and Viñuela 1990), highlighting that the species is able to cross large stretches of sea. The winter records of Asian subspecies such as *insignis* and *pallidus* in Egypt and the Middle East (Shirihai et al. 2000) show that they can perform long distance movements.

Behavioural and Ecological Remarks

As mentioned above, among merlins aged during post-reproductive movements at Falsterbo, juveniles largely outnumber adults. The constant prevalence of juveniles, reported in a ten-year study (1986–1995), suggested fairly low annual differences in the production of young during the study period. On the other hand, the stable number of birds seen each season in that period could be also due to the fact that this species is less affected by weather conditions during migration, not being dependent on thermal currents (Kjellén 1998). Anyway, although the merlin mostly hunts small birds, a study made in Scandinavia showed that its breeding success is more or less affected by vole densities being generally better in rodent years (Hagen 1969).

31

Eurasian hobby *Falco subbuteo*

Giuseppe Bogliani

ABSTRACT

The Eurasian hobby (*Falco subbuteo*) is a small falcon breeding in the Western Palearctic and wintering in Africa, mainly south of the Equator. This fly-and-forage migrant performs a trans-equatorial journey and travels on a broad front, without avoiding large sea crossings. Consequently, visual counts from traditional watchsites are scarce and not helpful to draw the migration routes. On the other hand, satellite telemetry studies revealed the routes followed by some individuals from Sweden and Germany. These birds cover ca. 9,000 km in 40–60 days, using a wide corridor, but concentrating significantly in order to cross the equatorial rainforest in its narrower region. Further research is needed to compare spring vs. autumn migration, also tracking birds from more eastern areas.

Introduction

The Eurasian hobby (*Falco subbuteo*, Linnaeus 1758) is a small, slender falcon, with a wingspan of 82–92 cm, and a sexual dimorphism fairly accentuated as regards size (male: 131–232 g, female: 141–340 g; Snow and Perrins 1998). During the breeding season, the main habitats of the species are open woodlands and plantations, often but not exclusively near wetlands and more generally open areas interspersed with woods, in which it nests mainly using old corvid nests built on taller trees. Recently, it has also started to use corvid nests built on artificial structures, such as electric pylons. In winter, it frequents savannah and open areas south of the Equator. Outside the nesting period, it feeds mainly on large insects that are caught in flight. During nesting, small and medium-sized birds are the main prey (Sergio and Bogliani 1999). A peculiar characteristic of this species is the nesting phenology. In Central and Southern Europe, pairs occupy the territory around May, but do not start egg laying until mid-June. Therefore, the young remain in the nest until mid-August (Chapman 1999). The Eurasian hobby is, therefore, intermediate between the other small falcons that nest in the same area and breed much earlier, and the Eleonora's falcon (*Falco eleonorae*), which nests six weeks later along sea coasts. Probably, the late nesting of the hobby allows it to prey birds most available in mid-summer, both the newly fledged passerines and swifts and the first migrants passing across its nesting areas and directed to the wintering grounds. The hobby is a summer visitor to the western Palearctic and overwinters in the southern third of the African continent, where the conventional

Department of Earth and Environmental Sciences, University of Pavia, Via Adolfo Ferrata 9, I-27100 Pavia–Italy.
Email: giuseppe.bogliani@unipv.it

observations gave it as associated with the rainy fronts that generate large temporary abundances of flying insects, especially termites (summarized in Sergio et al. 2001). The most recent estimates reported a world population of 500,000–999,999 breeding individuals (Status: Least Concern; BirdLife International 2020). The breeding population in Europe is between 92,100 and 147,000 pairs with overall stable populations. The westernmost fraction of the Asian nesting population, east of the Urals, most likely overwinters in Africa and partly passes through the territory where European populations also pass, but probably an unknown fraction reaches Africa via the Indian Ocean. The most eastern Asian populations of the subspecies *Falco s. subbuteo* probably overwinter in Southeast Asia and the Indian subcontinent. The hobby moves on a broad front during migration, mostly using the flapping flight and undertaking long water crossings. As a result, with rare exception, very few are observed at raptor migration watchsites (Zalles and Bildstein 2000).

Knowledge Status

The knowledge on the migration of the species was due, until a few years ago, almost exclusively to direct observations. Ringing did not provide much information and of the 5,720 individuals ringed until 2002 in 10 European countries, only 9 of the 203 recovered were returned from Africa south of the Sahara (Chapman 2020), with further two until 2011 (Meyburg et al. 2011). Data on passage are

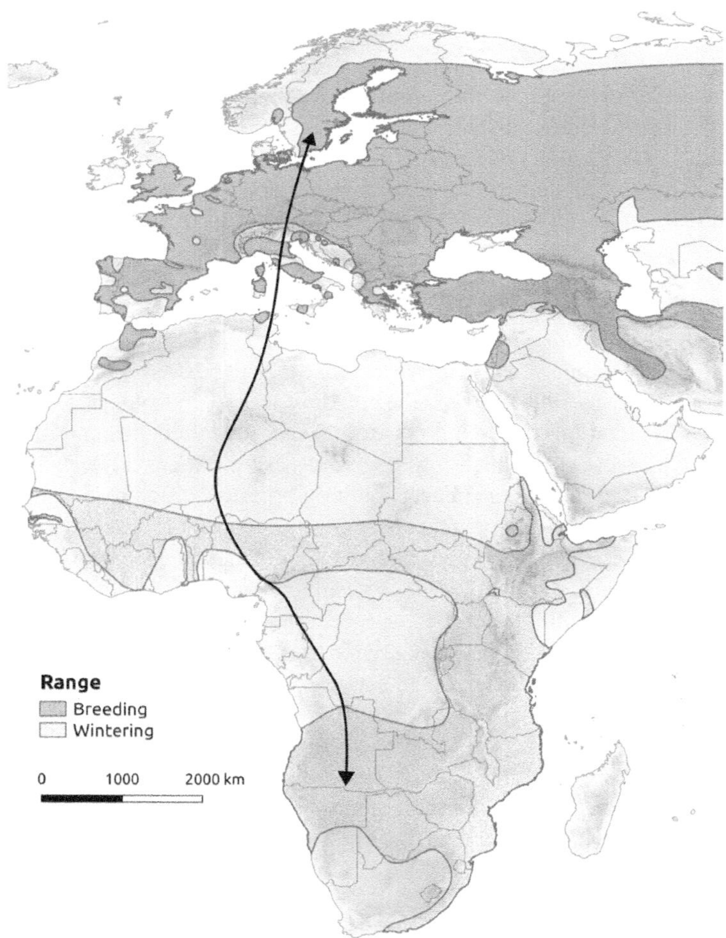

Figure 1. Schematic representation of the route followed by satellite tracked Eurasian hobbies (Strandberg et al. 2009b, Meyburg et al. 2011).

available for a number of observation sites across Europe, North Africa and the near East, which are mostly devoted to count soaring birds concentrating around land bridges to cross the sea. However, the migration across the Mediterranean Sea and the Sahara takes place on a wide front and the animals do not concentrate at bottlenecks, unlike what happens for several species of soaring raptors. The general picture that emerges is of a scarcity of individual hobbies passing through all the sites, if compared with the numbers of soaring birds of prey. For instance, in the Turkish Straits, during autumn 2010, Panuccio et al. (2017) counted just 19 migrating Eurasian hobbies out of 24,701 raptors of 25 species crossing the Bosphorus and 33 Eurasian hobbies out of 1,601 raptors of 20 species in the Dardanelles area. This scarcity must not be associated with a low number of hobby individuals present in the nesting areas, but with the fact that this species migrates on a broad front and therefore it is not easy to count with visual methods. Moreover, with favourable weather, hobbies fly at relatively high altitudes and therefore are out of sight. Nonetheless, visual observations had made it possible to gather very interesting information on some aspects of migration, such as changes in first arrival dates and differential timing of autumn migration between age group (Eddowes 2011, Kjellen 1992). Panuccio et al. (2017) suggest that migrating falcons, including the Eurasian hobby, of the eastern part of Europe tend to migrate following an innate NE-SW axis and that unlike soaring species, falcons cross the Mediterranean on a broad front (see above).

As regards more precise information of the routes followed by migrating Eurasian hobbies, insights derive from studies carried out by two groups of researchers using tracking devices applied to 6 individual falcons (4 adults and 2 juv, the latter both died during Sahara crossing) by Strandberg et al. (2009b, 2009c, 2009e) and to one by Meyburg et al. (2011). A further 16 individuals were tagged using 5g PTTs by B-U. Meyburg and C. Meyburg between 2008 and 2012, but data are not yet available (https://www.researchgate.net/project/Tracking-the-intercontinental-migrations-of-Eurasian-Hobbies-Falco-subbuteo-using-5-g-satellite-transmitters; consulted 10th July 2020). Unfortunately, we only have tracking data from a few individuals and from a homogeneous nesting area of Northern Europe. It would be necessary to tag more individuals, especially from the more eastern areas, to study which routes are actually followed.

Post-breeding Migration

At Falsterbo (Sweden), an average of 55 individuals per year has been observed between 1973 and 2019. Here, adults pass in late August and juveniles in the first half of September, with median day for juveniles being 14 days later than adults (Kjellen 1992).

The satellite tracking study of Strandberg et al. (2009b, 2009c) concerns four adults equipped with PTT transmitters in Southern Sweden and followed to just south of the Equator (all individuals) or to wintering areas (three individuals). Although coming from a restricted area, the four individuals followed different paths during the crossing of Europe and the Mediterranean, spreading over a broad front and crossing the sea also during night (Fig. 1; Strandberg et al. 2009b, Sammut et al. 2013). In particular, three crossed the Mediterranean Sea through Italy and Sicily to Libya, while the other one undertook a non-stop flight of 740 km, mostly over the sea, from France to Algeria, which lasted ca. 27 hours. The low counts reported from Mediterranean islands fit well in this scenario, as the highest ones are still not very relevant: in Malta, between 2009 and 2012, Sammut et al. (2013) counted 759 hobbies (max 229 per year) out of a total of 19,632 birds of prey. In the autumn of 2005, in Cyprus, Roth and Corso (2007), counted 152 hobbies out of a total of 5,153 observed birds of prey. During the European step of the journey, hobbies spend flying 7.9–10.5 hours per day, showing lower flight speeds especially during afternoons, which are probably related to a fly-and-forage strategy (Strandberg et al. 2009c).

Once in Africa, the Eurasian hobbies cross the Sahara, where all, although following parallel paths, changed direction by heading more or less towards the south-west, probably in response to prevailing easterly winds (Fig. 1; Strandberg et al. 2009b). Through this ecological barrier, hourly speeds are higher and the number of travel hours increases (11.7–14.6), even if true nocturnal migration was not recorded (Strandberg et al. 2009c). This translates into higher daily distances, reaching almost 400 km/day, the double than in the other regions, allowing hobbies to cross the desert in 4–4.5 days (Meyburg et al. 2011). Up to 10° N, the routes of the four individuals were up to 2,000 km away. When they passed the parallel of 10° N, the most western three began to correct the route towards south-east, then converged in a restricted area of about 70 km wide near the Equator. The fourth individual apparently continued on its route but also passed through the same restricted area. Interestingly, the tracking data of Meyburg et al. (2011), confirm the importance of the same narrow corridor, used four times, during both autumn and spring journeys, by a single adult individual tagged in Germany. The explanation may be the fact that at that point the equatorial forest belt is narrower, because a wedge of open areas consisting of cultivated areas, wet savannah and degraded woodlands enters the Equatorial rainforest from the north of Kinshasa. The stretch of forest in this section is only 450–600 km wide while elsewhere it is at least twice as wide. The passage in such a restricted area of all five individuals considered during the two researches suggests that the rainforest constitutes an ecological barrier for the hobby, which therefore tries to cross in the shortest time possible. The reasons why the rainforest is a barrier might be different and concurrent. On the one hand, the hobby would find few preys above the tree canopies and would not be able to hunt under them. Another reason is that during the day the forest is frequently affected by very heavy rains, which would however make the search for food inefficient. Moreover, in this region, flight segments beginning before sunrise or extending into the night were recorded, too (Strandberg et al. 2009c). It seems strange that a species capable of crossing the Mediterranean Sea and the Sahara desert apparently without problems experiences difficulties in overcoming the rainforest belt; however, the available tracking data seems unambiguous in this respect. Beyond this area, the flight tracks began to diverge, before stopping the long migratory flight and settle in the wintering area.

The journey of the Swedish birds lasted on average 61 days of which 39 were travel days (daily distance > 50 km). The mean distance travelled was 9,223 km (Strandberg et al. 2009c).

Pre-breeding Migration

For the spring migration, only data from a single tagged adult female, followed for two years by Meyburg et al. (2011) are available. The bird left the same wintering area both years and successfully reached the nesting site. During both spring migrations, it passed through the narrowest part of the equatorial forest, in the stretch already used during the previous autumn migrations by itself and by the individuals followed two years earlier by Strandberg et al. (2009c). However, after passing the Equator, it followed different routes and kept different migration paces. In 2009 the departure had taken place in the second half of March and the journey to the nesting site in Germany, reached at the end of April, had lasted 35 days, after covering 7,640 km, with a speed of 218 km/day. During the Sahara crossing, 1,184 km were covered in just three days. In 2010 the female had left the wintering area on March 31, following the same itinerary as the previous year only until after passing the Equatorial rainforest, after which she bent decidedly west-north-west to Western Mali, then headed north and passed the Strait of Gibraltar on 30 April. Crossing the Sahara, she covered 1,243 km on 28th–29th April; this average speed of 621 km/day is the fastest by this falcon on all the recorded migration routes. She later kept migration rates different from day to day. The latest fix came from a location 470 km from the nesting site. Up to this point she had covered 8,678 km in 38 days (228 km/day). Interestingly, these data show that travel duration is lower than the one recorded during autumn. As during autumn migration, low numbers of hobbies are recorded at migration watchsites: for example, a simultaneous study has been carried out on the Strait of Messina and on the promontory of the Conero from 20 April to 30 May 2000 (Gustin et al. 2001) with 180 and 153 individuals counted at each location, respectively.

Behavioural Remarks

Eddowes (2011) observed a conspicuous anticipation in the first arrival date of the hobby in an area of UK since the middle '70s. The rate, days per year, was –1.13, ($N = 37$, $p < 0.001$, $R = 0.71$), the highest among the considered species, including several passerines. The author attributed this anticipation to the fact that the species has become increasingly abundant in the study area and this would have increased the probability of early observation; he did not attribute the advancement of first arrival dates to a response to climate change.

Meyburg et al. (2011) suggested that Eurasian hobbies follow a loop migration strategy, as they use more westerly routes during spring migration, especially when crossing the Sahara. This finding is not surprising, as such pattern has been recorded in several other trans-Saharan migratory raptors. However, this conclusion is based just on an individual tracked for two years, and more tracks are needed to draw firmer conclusions.

32

Saker falcon *Falco cherrug*

*Mátyás Prommer** and *János Bagyura*

ABSTRACT

The saker falcon (*Falco cherrug*) is a large falcon species inhabiting open grassland and agricultural areas of Eurasia. Sakers are specialized hunters on ground-dwelling mammals, but they also hunt birds. Similarly to other falcons, sakers are superb flyers and they can cover large distances in migration or roaming. First year birds may disperse far from the fledging place in the post-fledging period. More directed movements towards concrete wintering areas start in the October–December period and regardless their actual location all 1st cy saker falcon keeps an almost identical south-westerly bearing resulting in parallel migratory routes. Individual migration distance, however, varies greatly, with females showing a tendency to cover longer distances. Adults usually remain in the territory in Central Europe all year around and do not migrate except for a few individuals. Overall, saker falcons can be considered partial migrants, with a significant proportion of juveniles performing short and medium to long-distance movements characterized by a strong directionality. The journeys may be repeated for a few years, showing fidelity to wintering sites. On the other hand, it seems that very few breeding adults migrate.

Introduction

The saker falcon (*Falco cherrug*, Gray 1834), hereafter "saker", is the second largest falcon species in the world with an average body mass of 0.90 kg (male: 0.66–0.90 kg; female: 0.90–1.30 kg) and a wing span of 1.16 m (Clark 1999), which inhabits the steppe zone and adjacent zones—forested steppes, semi-deserts, mountain steppes, etc.—of Eurasia. It is an iconic species of the steppes and steppe-like habitats and is specialized on hunting on ground-dwelling small mammals. The distribution range of the species is very broad, stretching from Central Europe to East China. The southernmost populations can be found in the Tibetan Plateau and Kashmir (Sangha et al. 2014), while the northernmost populations inhabit the forested steppes in Russia. Due to the impact of human activities, the once more or less continuous range became heavily fragmented in the last century, and the species has disappeared from vast areas in the former distribution range especially in East Europe (Ukraine and the European part of Russia) and West Asia (Kazakhstan). The most recent estimates reported a world population of 12,200–29,800 breeding individuals (Status: Endangered; BirdLife

MME/BirdLife Hungary, Költő u. 21., 1121 BUDAPEST, Hungary.
* Corresponding author: prommer.matyas@mme.hu

International 2020). Only less than 10% of the estimated global population lives in Europe (Kovács et al. 2014). Recent DNA research suggests significant genetic distance between the European and Central Asian saker falcons (Gábor Sramkó, *pers. comm.*).

The only relatively stable population lives in the Central European countries—most pairs in Hungary, but also in Austria, Slovakia, Romania, Serbia, Croatia, and the Czech Republic (Prommer et al. 2012). This population was estimated to 252–278 pairs in 2018 (Bagyura et al. 2017). The East European—Ukraine and the Crimea—population was estimated 285–312 in 2012 (Milobog 2012), however, based on the trend of regularly monitored pairs, the saker population in Ukraine has probably gone through a steep decline since then (Yuriy Milobog *pers. comm.*).

Both the European and Asian populations are partially migratory, although European sakers seem to overwinter in the breeding area at a higher percentage. Migrating individuals usually do not congregate at bottlenecks, but they migrate on broad front, crossing large water bodies through flapping flight. They cover large daily distances and they may migrate also during the night. Indeed, observations at the major watchsites located along the range crossed by satellite- tracked individuals are extremely scarce (Shirihai et al. 2000, Corso 2001a, Lucia et al. 2011, Verhelst et al. 2011, Oppel et al. 2014).

Knowledge Status

Migration pattern of European saker falcons has been revealed in the past decades by regular ringing programmes, as well as satellite-tracking in Central and East Europe. Most research has been done in Hungary, as it has the largest population of the species in Central Europe. Systematic ringing started in 1980, when the national saker conservation programme was launched; only a few individuals were ringed in the previous decades. Overall, in the period 1951–2019, in total 4,362 saker falcons were ringed, out of which 4,231 were pulli. A total of 221 ring recoveries are registered by 2019 and 211 of them are linked to birds ringed in Hungary: 151 recovery data are from within the country and 60 are from abroad.

Long-term ringing has been in place since the 1980s also in Slovakia and a few juvenile individuals were satellite-tracked. In the Czech Republic, chicks of the less than a dozen pair are regularly ringed. In the two countries 312 sakers were ringed until 2002 including the period when they formed one country, Czechoslovakia (Schröpfer 2008). Between 1999 and 2009 in the Czech Republic 151 Sakers were ringed (Beran et al. 2012), while 240 saker nestlings were ringed in Slovakia between 2007–2010.

In addition to the ringing effort, 127 saker falcons were satellite-tracked (Fig. 1) between 2007 and 2018 in the frame of European Union funded projects: LIFE06 NAT/HU/000096, LIFE09 NAT/HU/000384, LIFE13 NAT/HU/000183 in Hungary (94 individuals), Slovakia (8) and Romania (13); three projects were supported by the Rufford Foundation in Ukraine (10); and private projects in Austria (1), the Czech Republic (1) and Serbia (2).

In Austria there is no ringing scheme or tracking programmes for the species, but a few captive-bred and wild-hacked juvenile saker falcons were satellite tracked (Gamauf and Dosedel 2012). Similarly, captive-bred and wild-hacked juvenile saker falcons were satellite-tracked in Bulgaria in the framework of a reintroduction project (Dixon et al. 2020).

Table 1. Distribution of saker falcons tagged in Europe between 2007 and 2018 by age and sex.

	Female	**Male**	**Total**
Juvenile/Immature	44	35	71
Adult	9	39	48
Total	45	74	127

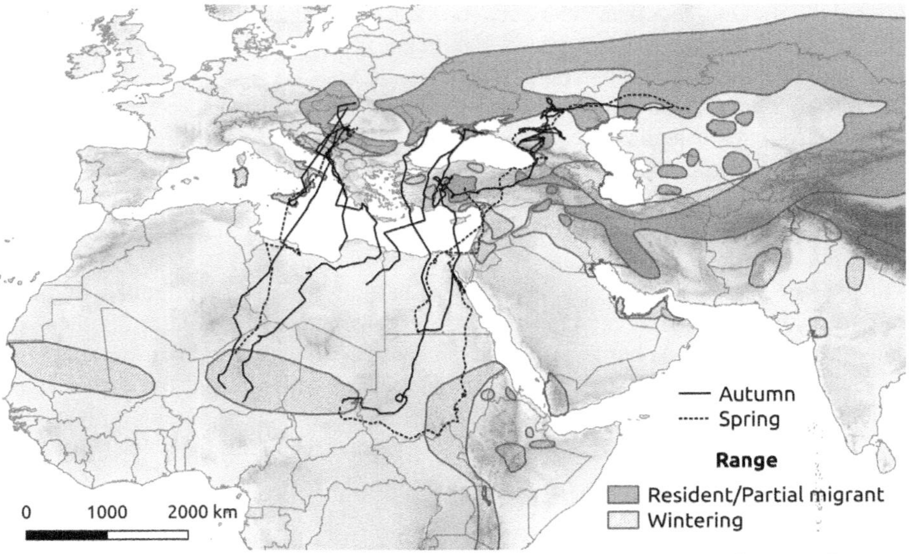

Figure 1. Sub-sample of the satellite tracking data belonging to saker falcons tagged in Eastern Europe.

Post-breeding Movements

Most recoveries occur in the first six life years of sakers and recovery distances decrease according to their age. The furthest recoveries are linked to the first two years suggesting that juvenile and immature saker falcons are more likely to move over long distances, while adults are more sedentary (Fig. 2).

Ring recoveries of Hungarian sakers suggest that autumn movements have mainly a south–south-westerly direction. There are fifteen recoveries from North-Africa and the Middle East—twelve from Libya and one each from Saudi-Arabia, Egypt and Mauritania. All but three of those recoveries belong to 1st cy juveniles trapped in the region in the autumn. In one case, there was a 2nd cy saker trapped

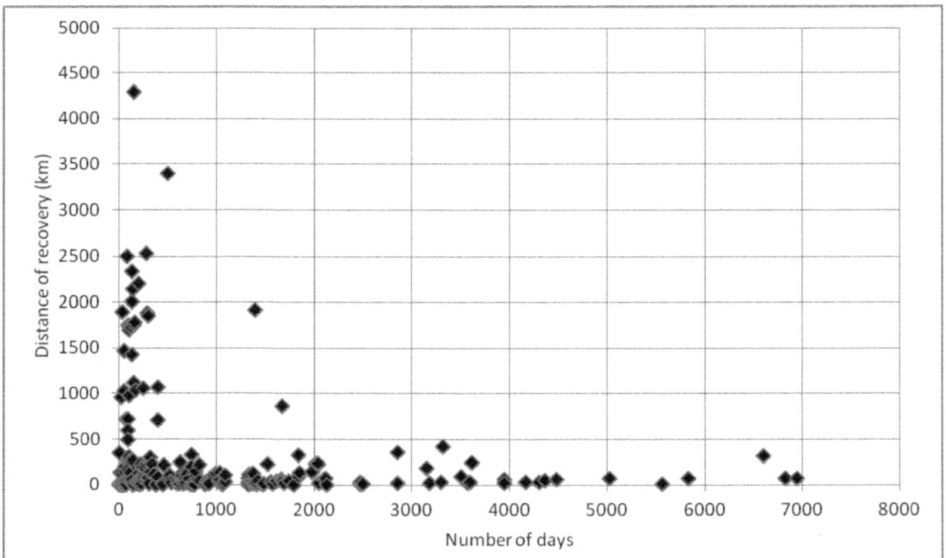

Figure 2. Distance of recovery of saker falcons ringed as chicks in Hungary in relation to elapsed time (number of days) between ringing and recovery.

in October. Recoveries proved that European sakers are still targeted by falcon trapping in North Africa and in the Middle East. Rough estimates suggest that about 0.5–5% of the annual cohort of the European population is being trapped in North Africa.

Satellite-tracking provided a more detailed insight of saker falcons' movements. Most of the fledged juveniles established 1–3 temporary settlement areas (TSA) in the post-fledging dispersal period before moving to proper wintering areas. They regularly made longer exploratory flights from their TSAs sometimes covering considerable distances and then returning to the TSAs. The direction and distance of post-fledging dispersal varied greatly. Most tagged juvenile sakers remained in the Pannonian Basin, however some of them dispersed as far as Spain, the Baltic states or Kazakhstan. None of the birds moved directly towards south in the post-fledging dispersal period. Based on the movement patterns the distinction between dispersal and migration is very straightforward because of that.

Most sakers start to move towards their wintering areas in October, but departure date can be as late as December. All juveniles started a south-westerly movement at some point in the autumn regardless their actual position. Thus, juveniles remaining in the Pannonian basin, as well as individuals dispersed to Kazakhstan or Spain start to migrate to the south-southwestern direction showing parallel movements (Prommer et al. 2012). Satellite-tracking data showed that 16.1% of juvenile saker falcons remained in the Central European breeding range without displacing significantly, while a 35.4% moved short distances (< 300 km). About one-third of the annual cohort (32.2%) left the breeding range showing medium distance movements (300 km < 1,200 km, e.g., leaving sakers' breeding range, but staying in Europe), remaining in the European part of the Mediterranean basin—mostly in Italy (especially in Sicily) or a few in Greece. It is only less than one-fifth (16.1%) of the annual Central European cohort that shows real long-distance movements. Those individuals cross the Mediterranean Sea and the Sahara, and overwinter in the Sahel. Satellite-tracking confirmed that only very few breeding adults migrate, but none of them crossed the Mediterranean Sea. Migrating immatures and adults returned to the same wintering sites in the subsequent years.

Overall, the movement pattern shown by saker falcons resembles the one occurring in species such as the griffon vulture (Chapter 6) and the eastern imperial eagle (Chapter 12), partial migrants, in which a significant proportion of juveniles, even if not the majority, perform long-distance movements characterized by a strong directionality. The journeys may be repeated for a few years, until individuals become mainly sedentary.

Established adults are mainly residents, although there are exceptions (Prommer et al. 2012). In severe winters, when hunting in the home territory is difficult or impossible, adults may leave the eyrie for shorter or longer—from a few days to one or two months—period. In those cases, the birds usually move to 100–200 km away from their home area. The wintering site may even be north from the nesting ground.

Migratory sakers departing from the Pannonian basin cross the West Balkan and reach the Adriatic coastline. Juveniles usually keep the bearing of 190°–200° (Prommer et al. 2012), however, they may detour when encountering large bodies of water. Some individuals cross the water immediately, others follow the coastline as long as they can. Once they reach the southernmost tip of the mainland or the last island, they cross the sea. Duration of autumn migration may range between 1–2 to 35–40 days, depending on the distance and the age of the bird. As the wintering site is the same after the first successful winter, older birds take advantage of their experience and migrate more directly, thus faster to the wintering site, while 1st cy birds spend more time choosing an appropriate wintering site. Flight speed of migrating sakers, regardless the sex and whether it is autumn or spring migration, is remarkably constant: they fly roughly about 40 km/h over land and about 50 km/h above the sea on average, with average daily distances of approximately 260 km. Among migrating juveniles, females tend to migrate over longer distances than males.

Pre-breeding Movements

There is very little information on spring migration of far travelling sakers despite the relatively large number of tracked individuals, as most sakers reaching North Africa were trapped or killed (Prommer et al. 2012).

According to the limited data, spring migration of 2nd cy sakers are remarkably similar: birds leave the wintering area between late March and early April regardless whether they spend the winter in the Pannonian Basin or in the Sahel. The birds wintering in the Pannonian Basin leave the wintering ground a few days earlier than birds migrating long distances. Spring migration of 2nd cy sakers is slower than the autumn migration of the previous year. In spring, immature saker falcons return to the fledging area and/or the last (pre-migration) autumn TSA and then they 'overshoot': they continue to move north-northeast before starting summer roaming without any specific bearing. The few birds that were successfully tracked from fledging to breeding showed annually "closing loops" around the fledging area.

Spring migration starts earlier as the sakers get older. The same individuals that as immatures were starting in late March/early April, as adults may start spring migration as early as the end of February.

Behavioural and Ecological Remarks

The earliest indication of saker falcons' possible long-distance migration is dated back to the 19th century's Hungary, when in the early 1840s, a silk pad with "Arabic or Persian script" was found attached to a leg of a breeding adult female saker falcon shot near Adony, about 40 km southwest of Budapest (Csörgey 1897). Falconry has been a long tradition in the Middle East and North Africa and juvenile saker falcons were preferred hunting birds. In traditional Arab falconry falcons were caught in autumn, trained, and used throughout the winter and they were released in spring. That time Arab falconers did not have the possibility to keep their falcons in the extreme hot summer. Satellite tracking provided us with many insights on interesting individual stories about saker falcons. For example, the following ones, all related to juvenile females: one crossed the Mediterranean Sea between Albania and Libya (approximately 1,100 kms) in less than 24 hours, flying also over Italy without landing. That is longest non-stop flight of a saker falcon recorded so far. Another one crossed the Adriatic Sea to Italy from Croatia, but she turned around almost immediately, crossed the sea again and returned to Hungary.

In 2009, one dispersed to North Spain, then crossed the Strait of Gibraltar and followed the Atlantic coast as far as a peninsula located in the NW corner of Mauritania, and perished near the town of Nouadhibou. The carcass was found. Exactly thirty years before this recovery, another saker was found drowned in the bay just 5 km away (Dejonghe 1980). According to local rangers, the peninsula is a natural trap for migrating birds of prey as it lacks food and fresh water. Rangers collect and save a number of dehydrated and starving raptors every year.

33

Peregrine falcon *Falco peregrinus*

Giacomo Dell'Omo

ABSTRACT

Eurasian peregrines (*Falco peregrinus*) belong to four of the nineteen subspecies of the World. Three of these that breed in Northern Europe and Russia are characterized by true migration movements whereas populations of the *F. p. brookei* occurring in Southern Europe are mostly sedentary. Information on the migration of the species gathered by direct observations at watchpoints and by ringing recoveries is scarce. The few tracking studies have involved birds breeding in the northernmost regions of Russia and have shown straight migration routes of several thousand km towards the south, encompassing a large area which spans from the Iberian Peninsula to Southern China. Some individuals belonging to the *calidus* subspecies may even reach South Africa. Peregrines in the Mediterranean region do not seem to migrate, and short-range movements are related to the dispersal phase of juveniles.

Introduction

The peregrine falcon (*Falco peregrinus* Tunstall, 1771) is a large falcon inhabiting all regions of the Earth except Antarctica (White 1994). Nineteen subspecies have been recognized with four living in Eurasia: *F. p. peregrinus, calidus, japonensis,* and *brookei* (White 1994). Its wingspan ranges between 79 and 114 cm, and the weight from 330 to 1500 g, with females up to 50% heavier than males (Ferguson-Lees and Christie 2001). The species is considered of Least Concern (BirdLife International 2020). Peregrines prefer open habitats where they can hunt for birds which represent their main prey. Peregrines do not build their nest but use cavities on cliffs and other elevated structures from which they have a good sight of the surroundings. They usually lay four eggs that incubate for about 37 days and chicks are ready to fledge in about five weeks and become independent from their parents at around eight-ten weeks.

Peregrine in Latin is "*Peregrinus*", the wanderer. The name has been assigned to the species supposedly because these large falcons appeared in winter in Central and Southern European countries likely coming from northern regions several thousands of km away. This is in general true for those populations breeding at northern latitudes, however, peregrine falcon populations vary largely in their propensity to migrate and some do not migrate at all. Although this species is very charismatic, individuals are elusive, ringing activities limited, and recoveries rare, and the various subspecies are difficult to recognize in the field. The technology nowadays would allow the deployment of smaller

Ornis italica, Rome, Italy.
Email: giacomo.dellomo@gmail.com.

and more efficient tags, however, only few tracking studies are available so far, showing that mostly the populations in the Russian tundra are those performing true migratory movements.

Knowledge Status

The knowledge on the migratory movements of peregrines has been gathered so far by observations at watchpoints, by ringing recoveries, and tracking studies. Being a species in which individuals migrate singly, the visible passage at the watchpoints used for monitoring other raptor species is less conspicuous and therefore, numbers detected are typically low. As a matter of fact, it is not a typical soaring species (although it can soar) as its flight sustained by strong flapping does not need thermal uplifts. In addition, when detected, it is difficult to distinguish peregrines performing a true migration from local birds. Populations of Southern and Central Europe are mainly resident or showing short distance movements (Spina and Volponi 2008). As far as the *F. p. brookei*, the only published study in Central Mediterranean on the natal dispersal of this subspecies suggests that birds remain in the vicinity of their natal site and do not perform any migratory displacement (Bondì et al. 2018). Data gathered on the dispersal of urban peregrines from Rome (18 tracked birds during 2017–2019) confirm these findings, with four birds followed for up to three years after they fledge remaining in the vicinity of the natal area (G. Dell'Omo, unpublished data). These two studies point for a lack of true migratory behaviour in *F. p. brookei*, at least in the populations breeding in Central Mediterranean. As an additional consideration, it should be taken into account that there is no clear separation between *F. p. peregrinus* and *F. p. brookei* in Europe as genetic differentiation in both mtDNA and microsatellites loci is almost absent and it is not possible to distinguish geographical groups according to taxonomic designation at the subspecies level (Mengoni et al. 2018). Therefore, the lack of a migratory behaviour might also be present in *F. p. peregrinus* breeding in other areas of Central and Southern Europe. Peregrines migrating over long distances are mainly those breeding at northernmost latitudes, e.g., the *calidus* subspecies (Corso 2001b, Grussu and Paddeu 2015); they move from Siberia to wintering grounds located from the Mediterranean basin until Southern Africa, showing therefore a leap-frog migration pattern (Jenkins and Stephenson 1997, Meyburg et al. 2017c).

Peregrines are easier to ring as chicks at nest. For example, more than 4,000 have been tagged in Finland (Saurola et al. 2013). Of them, 276 have been recovered, with some providing information on migration behaviour. Tracking studies on peregrines are very few, and in Europe less than in North America. There are at least three main reasons for this. First the species has been considered endangered for more than four decades, during and soon after the "DDT period"; birds were rare and the scientific interest was focused on the physiological effects of the pesticides, especially on the reproductive success, such as on improvement of captive breeding for supporting the recovery of the populations. Second, the species is elusive and adult birds are difficult to trap. Third, tags for the size of the peregrine were too large and only in the last two decades the technology has produced suitable tracker devices. As a result, there are only few tracking studies on the species in Eurasia and mostly are based on females, the sex with the larger size. The first was pioneered by Ganusevich et al. (2004) which in 1994 succeeded to tag with satellite-received transmitters (PTT) four females from the Kola Peninsula (Russia). One year later Meyburg et al. (2017c) were able to track with PTT one female that migrated from Saudi Arabia to South Africa. Finally, in the Yamal peninsula (Russian Arctic), 12 individuals have been tagged in 2009 and 2010 (Sokolov et al. 2014, Sokolov et al. 2018).

Migration Patterns

As stated above, in the Western Palearctic, only the populations from Northern Eurasia perform long-distance migrations (Fig. 1). Ringing recoveries of Finnish birds show that their wintering areas span from Denmark to Northern Africa, being concentrated in France and Spain, with juvenile males being those displacing more to the South and adult females showing the opposite pattern (Saurola et al. 2013, see also Ollé et al. 2016). On the broad front of migration shown by these individuals, some

Figure 1. Schematic representation of the routes followed by long-distance migrating peregrine falcons.

of the most important watchsites of the species are located: Falsterbo (Sweden), where an average of 92 birds has been observed in the period 2010–2019 (www.falsterbofagelstation.se), and the Southern Baltic coast of Poland where a range of 69–170 has been observed in the period 2008–2011 (Polakowski et al. 2014). Ganusevich et al. (2004) got data regarding the complete autumn journeys of two adult females from the Kola peninsula (Russia): they started to migrate in September, in agreement with the phenology observed in Poland (median data: 4th October; Polakowski et al. 2014), reaching their wintering grounds in France and Spain three weeks later. Sokolov et al. (2018) tracked 12 birds (mainly adult females) from the Yamal peninsula in Siberia, some across repeated journeys and showing high site fidelity to the wintering areas. This is the most comprehensive tracking study on the species carried out in the Western Palearctic. Their wintering areas spanned from Portugal to Iran and from Russia to South Sudan. The post-breeding migration phenology was similar to the one of birds from the Kola Peninsula, while in spring the departure was after mid-April. Average travel rates of 230 km/day have been recorded taking into account both migration seasons, including also stopover days, with extremely long daily flights exceeding 1,000 km. From West to East, starting from Scandinavia until the Russian far East there is a transition of the breeding populations from *F. p. peregrinus* to *F. p. calidus* and to *F. p. japonensis*; The areas of three subspecies, however, are not precisely defined as there seem to be a clinal variation of the peregrine characters (Dixon et al 2012). Peregrines tagged in the Eastern Taymir and in the Lena delta spent the winter from Pakistan to South-East Asia, mostly in two distinct clusters, clearly segregated from those from the Yamal Peninsula, showing therefore an interesting pattern of strong connectivity. Actually, only one of the birds tracked in the above mentioned studies went to Africa and none reached South Africa, therefore that tracked by Meyburg et al. (2017c) remains the only one showing such long route: it involves a *calidus* bird trapped by local falconers in Saudi Arabia, which, after release, covered 6,346 km reaching South Africa in three weeks.

Behavioural and Ecological Remarks

Peregrine falcons are solitary during migration and fly at low altitude. Studies on North American birds revealed that they fly in general at altitudes less than 100 m above ground (Kerlinger 1989).

Peregrines also soar during migration and glide between thermals. However, they can also withstand extended periods of flapping flight. Using a variety of radio-telemetric techniques and selecting periods when the flight path was known to be approximately straight, Cochran and Applegate (1986) could measure precisely the speed of gliding and flapping flight of six individuals (three males and three females) trapped in Illinois (USA) during migration. The horizontal gliding airspeed (measured on one individual) was about 64 km/h while the average flapping speed was 48.8 km/h. The authors suggested that, although flight speeds for peregrines are in general emphasized to show the maximal capabilities of the species, the speed they measured, although apparently low, were in the ranges providing good aerodynamic efficiency. The actual flight speed, however, cannot be considered as the speed of migration. This latter is calculated over a period of days and is much lower. For example, the speed measured during the migration of Siberian peregrines over the Central Siberian Plateau, the South Siberian Mountains, the Steppe and the Gobi Desert and the Himalaya, ranged between 9.4 and 15.5 km/h, the Himalaya being the geographical barrier crossed with faster speed (Dixon et al. 2017). A recent analysis of the spring migratory paths of peregrines from North America and Russia related the movements of some tracked individuals with the progression of the snow cover and compared their timing with that of the snowy owl (*Bubo scandiacus*), an irruptive migrant, and the rough-legged buzzard (*Buteo lagopus*), with an intermediate migratory pattern (Curk et al. 2020). The three species used snow presence as a cue to adjust their arrival at the breeding grounds, however, contrary to the others, peregrines stayed behind the temporal peak in snowmelt and migrated only when the starting area was free of snow and occupied areas with no more than 10% snow cover.

References

Aborn, D.A. 1994. Correlation Between Raptor and Songbird Numbers at a Migratory Stopover Site. Wilson Bulletin. 106: 150–154.

Abuladze, A.V. 1996. Ecology of the imperial eagle *Aquila heliaca* in Georgia. pp. 447–457. *In*: Meyburg, B.-U. and R. Chancellor [eds.]. Eagle Studies. World Working Group on Birds of Prey and Owls. Berlin, Germany.

Abuladze, A. and J. Shergalin. 1998. The Egyptian Vulture *Neophron percnopterus* in the former USSR. pp. 183–195. *In*: Chancellor, R.D., B.-U. Meyburg and J.J. Ferrero [eds.]. Holarctic Birds of Prey: Proceedings of an International Conference. Asociación para la Defensa de la Naturaleza y los Recursos de Extremadura, Badajoz.

Abuladze, A.V. 1999. On the flight of birds of prey in Georgia in the autumn of 1997. pp. 13–16. *In*: Proceedings of the 3rd Conference on Birds of Prey of Eastern Europe and Northern Asia. Part 2 (Kislovodsk, September 15–18, 1998). Stavropol, Russia (in Russian).

Abuladze, A.V. and G.V. Edisherashvili. 2003. Flight of birds of prey in Georgia in the spring and autumn of 1998, 113–117. *In*: Galushin, V.M. [ed.]. Proceedings of Conference on Birds of Prey. Penza, Russia (in Russian).

Abuladze, A.V., V.E. Eligulashvili, G.G. Rostiashvili and G.V. Edisherashvili. 2003. About the status of some rare birds of prey in Georgia. pp. 106–109. *In*: Galushin, V.M. [ed.]. Proceedings of the IV Conference on Birds of prey of Northern Eurasia. Penza, Russia. Publisher (in Russian).

Abuladze, A.V. 2006. Birds of prey of Georgia. PhD Dissertation, Institution of Zoology of Georgia, Tbilisi, Georgia. (in Russian).

Abuladze, A.V. 2008. On the winter records of the greater spotted eagle in Georgia. pp. 5–8. *In:* Research and Conservation of the Greater and the Lesser Spotted Eagle in Northern Eurasia. Ivanovskii Gosudarstvennyi Universitet, Ivanovo.

Abuladze, A. 2013. Birds of Prey of Georgia: Materials towards a Fauna of Georgia Issue VI, Ilia State University, Tbilisi, Georgia.

Adami, I., S. Cutini, A. Giordano, O.F. Larsen, D. Ridente, D. Ricciardi and R. Vella. 2019. 36° Campo internazionale per lo studio e la protezione dei rapaci e delle cicogne in migrazione sullo Stretto di Messina. Infomigrans. 43: 2–4.

Adriansen, F., N. Verwimp and A.A. Dhondt. 1997. Are Belgian Kestrels *Falco tinnunculs* migratory: an analysis of ringing recoveries. Ringing & Migration. 18: 91–101.

Adriansen, F., N. Verwinp and A.A. Dhondt. 1998. Between-cohort variation in dispersal distance in the European Kestrel *Falco tinnunculus* as shown by ringing recoveries. Ardea. 2: 147–152.

Aebischer, A. 2010a. Distribution and recent population changes of the Red Kite in the Western Palaearctic: results of a recent comprehensive inquiry. pp. 12–14. *In*: Proceedings of the International Red Kite symposium 2009. Monbéliard, France.

Aebischer, A. 2010b. Migration and wintering of Swiss Red Kites assessed by satellite telemetry. pp. 48–51. *In*: Proceeedings of the International Red Kite symposium 2009. Monbéliard, France.

Agafonov, A.V., D.S. Rezinko, A.A. Rozhkov and N.M. Semenov. 1957. Towards the ecology of the Steppe Eagle. Bulletin MOIP, Dep. Biology. 62: 33–41 (in Russian).

Agostini, N. 1992. Spring migration of honey buzzards (*Pernis apivorus*) at the Straits of Messina in relation to atmospheric conditions. Journal of Raptor Research. 26: 93–96.

Agostini, N. 2001. Spring migration in relation to sex and age of Marsh Harriers *Circus aeruginosus* over a central Mediterranean island. Ardeola. 48: 71–73.

Agostini, N. 2002. La migrazione dei rapaci in Italia. pp. 157–182. *In:* Brichetti, P. and A. Gariboldi [eds.]. Manuale Pratico di Ornitologia 3. Edagricole-Il Sole 24 Ore, Bologna.

Agostini, N. 2004. Additional observations of age-dependent migration behaviour in western honey buzzards *Pernis apivorus*. Journal of Avian Biology. 35: 469–470.

Agostini, N. 2005. Are earlier estimates of Accipitriformes crossing the Channel of Sicily (Central Mediterranean) during spring migration accurate? Journal of Raptor Research. 39: 184–186.

Agostini, N. and A. Duchi. 1994. Water-crossing behaviour of Black Kites (*Milvus migrans*) during migration. Bird Behaviour. 10: 45–48.

Agostini, N. and D. Logozzo. 1995. Autumn migration of honey buzzards in southern Italy. Journal of Raptor Research. 29: 275–277.

Agostini, N. and D. Logozzo 1997. Autumn migration of Accipitriformes through Italy en route to Africa. Avocetta. 21: 174–179.

Agostini, N. and G. Malara. 1997. Entità delle popolazioni di alcune specie di Accipitriformi migranti, in primavera, sul Mediterraneo centrale. Rivista italaliana di Ornitologia. 66: 174–176.

Agostini, N. and D. Logozzo. 1998. Primi dati sulla migrazione primaverile dei rapaci Accipitriformi sull'isola di Marettimo (Egadi). Rivista Italiana di Ornitologia. 68: 153–157.

Agostini, N. and D. Logozzo. 2000. Migration and wintering distribution of the Marsh Harrier (*Circus aeruginosus*) in southern Italy. Buteo. 11: 19–24.

Agostini, N. and M. Panuccio. 2003. How do Accipitriformes behave during autumn migration at the Circeo promontory? Rivista Italiana di Ornitologia. 73: 165–167.

Agostini, N. and M. Panuccio. 2005. Analysis of the spatial migration patterns of adult Honey Buzzards (*Pernis apivorus*) during spring and autumn in the Central Mediterranean. The Ring. 27: 215–220.

Agostini, N. and U. Mellone. 2007. Migration strategies of Oriental Honey-buzzards *Pernis ptilorhyncus* breeding in Japan. Forktail. 23: 182–183.

Agostini, N. and M. Panuccio. 2008. Are migration pathways of European Honey Buzzards *Pernis apivorus* passing over Cyprus age-related? Sandgrouse. 30: 97–99.

Agostini, N. and U. Mellone. 2008. Does migration flyway of short-toed snake-eagles breeding in central Italy reflect the colonization history? Journal of Raptor Research. 42: 158–159.

Agostini, N. and M. Panuccio. 2010. Western Marsh Harrier (*Circus aeruginosus*) migration through the Mediterranean Sea: a review. Journal of Raptor Research. 44: 136–142.

Agostini, N., D. Logozzo and C. Coleiro. 1999. The orientation/navigation hypothesis: an indirect evidence in migrating Honey Buzzards. Rivista Italiana di Ornitologia. 69: 153–159.

Agostini, N., D. Logozzo and M. Panuccio. 2000. The island of Marettimo (Italy), important bird area for the autumn migration of raptors. Avocetta. 24: 95–100.

Agostini, N., C. Coleiro and M. Panuccio. 2003. Autumn migration of Marsh Harriers across the central Mediterranean in 2002. The Ring. 25: 47–52.

Agostini, N., C. Coleiro and M. Panuccio. 2004b. Analysis of the autumn migration of juvenile Honey Buzzards (*Pernis apivorus*) across the central Mediterranean. Journal of Raptor Research. 38: 283–286.

Agostini, N., M. Panuccio and C. Pasquaretta. 2015a. Morphology, flight performance, and water crossing tendencies of Afro-Palearctic raptors during migration. Current Zoology. 61: 951–958.

Agostini, N., A. Scuderi, G. Chiatante, G. Bogliani and M. Panuccio. 2015b. Factors affecting the visible southbound migration of raptors approaching a water surface. Italian Journal of Zoology. 82: 186–193.

Agostini, N., M. Gustin and M. Panuccio. 2016a. Short-toed Snake-Eagles *Circaetus gallicus* (Gmelin, 1788) (Aves: Accipitridae) approaching a water barrier show reverse direction of migration. Italian Journal of Zoology. 83: 543–548.

Agostini, N., L. Baghino, M. Panuccio and G. Premuda. 2002b. A conservative strategy in migrating Short-toed Eagles (*Circaetus gallicus*). Ardeola. 49: 287–291.

Agostini, N., P. Amato, A. Provenza and M. Panuccio. 2005a. Do Common Buzzards *Buteo buteo* migrate across the Channel of Sicily during autumn? Avocetta. 29: 19.

Agostini, N., M. Gustin, J. von Hardenberg and M. Panuccio. 2016b. Wind patterns affect migration flyways and flock size of a soaring bird over sea. Avian Biology Research. 9: 159–166.

Agostini, N., M. Gustin, M. Cento and M. Panuccio. 2017a. Analisi della migrazione primaverile del falco di palude *Circus aeruginosus* attraverso il Mediterraneo Centrale. *In:* Fasano, S.G. and D. Rubolini [eds.]. Riassunti del XIX Convegno Italiano di Ornitologia. Tichodroma. 6: 26.

Agostini, N., L. Baghino, W.M.G. Vansteelant and M. Panuccio. 2017c. Age-related timing of Short-toed Snake Eagle *Circaetus gallicus* migration along detoured and direct flyways. Bird Study. 64: 37–44.

Agostini, N., M. Prommer, M. Vaczi and M. Panuccio. 2019. Repeated large scale loop migrations of an adult European Honey Buzzard. Avocetta. 43: 13–21.

Agostini, N., G. Chiatante, G. Dell'Omo and M. Panuccio. 2021. Differential autumn migration between sex and age groups in the Western Marsh Harrier: a longitudinal pattern analysis. Ethology Ecology & Evolution. 33: 73–82.

Agostini, N., G. Malara, F. Neri, D. Mollicone and S. Melotto. 1994. Flight strategies of Honey Buzzards during spring migration across the central Mediterranean. Avocetta. 18: 73–76.

Agostini, N., L. Baghino, C. Coleiro, F. Corbi and G. Premuda. 2002a. Circuitous autumn migration in the Short-toed Eagle (*Circaetus gallicus*). Journal of Raptor Research. 36: 111–114.

Agostini, N., L. Baghino, M. Panuccio, G. Premuda and A. Provenza. 2004a. The autumn migration strategies of juvenile and adult Short-toed Eagles (*Circaetus gallicus*) in the central Mediterranean. Avocetta. 28: 37–40.

Agostini, N., M. Panuccio, A. Pastorino, N. Sapir and G. Dell'Omo. 2017b. The migration of the Western Marsh Harrier to the African wintering quarters along the Central Mediterranean flyway: a 5-year study. Avian Research. 8: 24.

Agostini, N., M. Panuccio, U. Mellone, G. Lucia, S. Wilson and J. Ashton-Booth. 2007. Do migration counts reflect population trends? A case study of the honey buzzard *Pernis apivorus*. Ardeola. 54: 339–344.

Agostini, N., G. Premuda, U. Mellone, M. Panuccio, D. Logozzo, E. Bassi and L. Cocchi. 2004c. Crossing the sea en route to Africa: autumn migration of some Accipitriformes over two central Mediterranean islands. The Ring. 26: 71–78.

Agostini, N., G. Premuda, U. Mellone, M. Panuccio, D. Logozzo, E. Bassi and L. Cocchi. 2005b. Influence of wind and geography on orientation behavior of adult honey buzzards *Pernis apivorus* during migration over water. Acta Ornithologica. 40: 71–74.

Agostini, N., G. Premuda, L. Cocchi, R. Molajoli, C. Cardelli, M. Gustin and L. Baghino. 2006. Spring migration of Western Honey-buzzards (*Pernis apivorus*) along the Sardinia-Corsica corridor (Central Mediterranean). Journal of Raptor Research. 40: 244–246.

Agostini, N., G. Lucia, U. Mellone, M. Panuccio, J. von Hardenberg, A. Evangelidis and T. Kominos. 2012. Loop migration of adult European Honey Buzzards *Pernis apivorus* through the central-eastern Mediterranean. Italian Journal of Zoology. 79: 280–286.

Agostini, N., M. Panuccio, G. Lucia, C. Liuzzi, P. Amato, A. Provenza, M. Gustin and U. Mellone. 2009. Evidence for age-dependent migration strategies in the short-toed eagle. British Birds. 102: 506–508.

Agudo, R., C. Rico, F. Hiraldo and J.A. Donázar. 2011. Evidence of connectivity between continental and differentiated insular populations in a highly mobile species. Diversity and Distributions. 17: 1–12.

Alarcón, P.A.E. and S.A. Lambertucci. 2018. A three-decade review of telemetry studies on vultures and condors. Movement Ecology. 6: 13.

Alerstam, T. 1987. Radar observations of the stoop of the Peregrine Falcon *Falco peregrinus* and Goshawk *Accipiter gentilis*. Ibis. 129: 267–273.

Alerstam, T. 1990. Bird Migration, Cambridge University Press, Cambridge, UK.

Alerstam, T. and A. Hedenström. 1998. The development of bird migration theory. Journal of Avian Biology. 29: 343–369.

Alerstam, T. and Å. Lindström. 1990. Optimal bird migration: the relative importance of time, energy and safety. pp. 331–351. *In*: Gwinner, E. [ed.]. Bird migration: physiology and ecophysiology. Springer, Berlin.

Alerstam, T., A. Hedenström and S. Åkesson. 2003. Long-distance migration: evolution and determinants. Oikos. 103: 247–260.

Alerstam, T., M. Hake and N. Kjellén. 2006. Temporal and spatial patterns of repeated migratory journeys by ospreys. Animal Behaviour. 71: 555–566.

Alexeyenko, M.N. and I.V. Fefelov. 2018. Autumn migration of Greater Spotted Eagle in the South Baikal migratory pass. pp. 26–33. *In:* Research and Conservation of the Greater and the Lesser Spotted Eagle in Northern Eurasia. Ivanovskii Gosudarstvennyi Universitet, Ivanovo.

Ali, S. and S.D. Ripley. 1968. Handbook of the Birds of India and Pakistan, vol. 1, Oxford University Press.

Alivizatos, H., D. Papandropoulos and S. Zogaris. 2004. Winter diet of the Greater Spotted Eagle (*Aquila clanga*) in the Amvrakikos wetlands, Greece. Journal of Raptor Research. 38: 371–374.

Alléon, A. and J. Vian. 1869. Les migrations des oiseaux de proie sur le Bosphore de Constantinople. Revue et Magasin de Zoologie. 21. 258–273, 305–315, 342–348, 369–374, 401–409.

Alléon, A. and J. Vian 1870. Les migrations des oiseaux de proie sur le Bosphore de Constantinople; nouvelles observations. Revue et Magasin de Zoologie. 22: 81–86, 129–138, 161–165.

Alon, D. and H. Shirihai. 1991. A survey of wintering raptors in Israel. 1987/88. pp. 46–52. *In*: Yekutiel, D. [eds.]. Raptors in Israel: Passage and Wintering Populations. Eilat, Israel: International Birdwatching Centre.

Alon, D., B. Granit, J. Shamoun-Baranes, Y. Leshem, G.M. Kirwan and H. Shirihai. 2004. Soaring-bird migration over northern Israel in autumn. British Birds. 97: 160–182.

Alonso, J.A. 1984. Sur les quartiers d'hiver des vautours fauves migrateurs. Alauda. 52: 308–309.

AlSheikhly, O.F. and A.J. AlAzawi. 2019. Migration pattern and wintering population of the Eurasian marsh harrier (*Circus aeruginosus*) in the Central Marshes, a wetland of international importance in southern Iraq. Raptor Journal. 13: 127–137.

Altenburg, W., I. Bruinenberg-Rinsma, P. Wildschut and M. Zijlstra. 1987. Colonisation of a new area by the Marsh Harrier. Ardea. 75: 213–220.

Altenburg, W., S. Daan, J. Starkensberg and M. Ziljstra. 1982. Polygamy in the marsh harrier, *Circus aerunginosus*: individual variations in hunting performance and number of mates. Behaviour. 79: 272–312.

Amartuvshin, P. and S. Gombobaatar. 2012. The assessment of high risk utility lines and conservation of globally threatened pole nesting steppe raptors in Mongolia. Ornis Mongolica. 1: 2–12.

Amosov, P.N. 2011. Steppe Eagle *Aquila nipalensis* in the vicinity of Lake Baskunchak. Russian Ornithological Journal. 20: 1937–1940.

Anadon, J.D., J.A. Sanchez-Zapata, M. Carrete, J.A. Donazar and F. Hiraldo. 2010. Large-scale human effects on an arid African raptor community. Animal Conservation. 13: 495–504.

Andevski, J. [ed.]. 2013. Vulture conservation in the Balkan peninsula and the adjacent regions. 10 years of vulture research and conservation. Vulture Conservation Foundation, Skopje, 39

Andrews, J., M. Beaman, P. Fisher, T. Hereward, M. Heubeck, M. Morton, R. Porter and P. Round. 1977. A new raptor migration route through N.E. Turkey. Bulletin of the Ornithological Society of Turkey. 14: 2–5.

Angelov, I., I. Hashim and S. Oppel. 2013b. Persistent electrocution mortality of Egyptian Vultures *Neophron percnopterus* over 28 years in East Africa. Bird Conservation International. 23: 1–6.

Angelov, I., B. Abdu, N. Terziev and S. Zelleke. 2013a. Possible sighting of the Indian subspecies of the Egyptian Vulture *Neophron percnopterus ginginianus* in Africa. Vulture News. 64: 44–49.

Angelov, I., T. Yotsova, M. Sarrouf and J. McGrady. 2013c. Large increase of the Egyptian vulture *Neophron percnopterus* population on Masirah island, Oman. Sandgrouse. 34: 140–152.

Antipov, S.M. and V.P. Shubenkin. 1984. About some rare and endangered bird species in the Kaplan-Kyrskii Reserve. pp. 46–47. *In:* Elkin, K.F. [ed.]. Study and Protection of Protected Objects. Kaynar Press, Alma-Ata, USSR (in Russian).

AquilaSystem. 2019. Eastern Imperial Eagle Russia. https://gps.aquila-it.pl/en/migration-maps/eastern-imperial-eagle-russia [accessed 15 September 2019].

Arkumarev, V., V. Dobrev, Y.D. Abebe, G. Popgeorgiev and S.C. Nikolov. 2014. Congregations of wintering Egyptian Vultures *Neophron percnopterus* in Afar, Ethiopia: present status and implications for conservation. Ostrich. 85: 139–145.

Arkumarev, V., D. Dobrev and A. Stamenov. 2019. First record of Eurasian Griffon Vulture *Gyps fulvus* from the Balkans migrating to South Sudan revealed by GPS tracking. Scopus. 39: 27–35.

Arrondo, E., M. Moleon, A. Cortés-Avizanda, J. Jimenez, P. Beja, J.A. Sánchez-Zapata and J.A. Donazar. 2018. Invisible barriers: differential sanitary regulations constrain vulture movements across country borders. Biological Conservation. 219: 46–52.

Arroyo, B., A. Amar, F. Leckie, G.M. Buchanan, J.D. Wilson and S. Redpath. 2009. Hunting habitat selection by hen harriers on moorland: implications for conservation management. Biological Conservervation. 142: 586–596.

Arroyo, B.E., J.T. García and V. Bretagnolle. 2002. Conservation of the Montagu's harrier (*Circus pygargus*) in agricultural areas. Animal Conservation. 5: 283–290.

Arslangündoğdu, Z., L. Smith, Ü. Yardim, P.J. Vanmarcke and M. Payne. 2018. Soaring bird migration research at the Bosphorus Strait, Turkey. Applied Ecology and Environmental Research. 16: 7953–7968.

Artemiev, Y.T., A.S. Ayupov, T.V. Vodolazhskaya, E.M. Vorontsov, V.I. Garanin, P.K. Gorshkov, N.D. Grigoriev, Y.E. Egorov, P.V. Efremov, V.S. Zaletayev, R.A. Zatsepina, T.M. Kulaeva, I.D. Nazarova, B.V. Nekrasov, P.V. Pleskiy, V.A. Popov, Y.K. Popov and S.G. Priklonsky. 1977. Birds of the Volga-Kama region, Non-passerines, Moscow, Russia, Science Publisher (in Russian).

Ash, J. and J. Atkins. 2009. Birds of Ethiopia and Eritrea: an Atlas of Distribution, Christopher Helm, London.

Attar, W. 2005. National Report on Hunting. Syrian Arab Republic. Building capacity for sustainable hunting of migratory birds in Mediterranean third countries, Life project.

Auburn, J. 1988. Do some African Lesser Kestrels migrate from the Far East across the Indian Ocean? Gabar. 3: 5–7.

Augiron, S., B. Gangloff, S. Brodier, F. Chevreux, J.-F. Blanc, P. Pilard, A. Coly, A. Sonko, A. Schlaich, V. Bretagnolle and A. Villers. 2015. Winter spatial distribution of threatened acridivorous avian predators: Implications for their conservation in a changing landscape. Journal of Arid Environments. 113: 145–153.

Averin,Y.V., I.M. Ganya and G.A. Uspensky. 1971. Birds of Moldavia. vol. 2, Shtinitsa, Chisinau, Moldova (in Russian).

Azmanis, P.N., R. Tsiakiris and H. Alivizatos. 2009. Alimentación suplementaria para buitres en Grecia, Chipre y la península Balcánica. Munibe. 29: 352–373.

Babbington, J. and O. Campbell. 2016. Recent status and occurrence of Crested Honey Buzzards *Pernis ptilorhynchus* in the Arabian Peninsula, with emphasis on Saudi Arabia and the United Arab Emirates. Sandgrouse. 38: 12–22.

Babushkin, M., A. Kuznetsov and M. del Mar Delgado. 2019. Autumn migratory patterns of North West Russian Ospreys *Pandion haliaetus*. Ardeola. 66: 119–128.

Baghino, L., G. Premuda, M. Gustin, A. Corso, U. Mellone and C. Cardelli. 2007. Exceptional wintering and spring migration of the booted eagle Hieraaetus pennatus in Italy in 2004 and 2005. Avocetta. 31: 47–52.

Baghino, L. 2016. La migrazione autunnale nel Parco del Beigua: il sito di Arenzano (GE). Infomigrans. 38: 6.

Baghino, L. Premuda, G. and L. Giraudo. 2012. Nuove analisi sulla migrazione post-riproduttiva del biancone *Circaetus gallicus* nell'Italia nord-occidentale. Avocetta. 36: 107–111.

Bagyura, J., M. Prommer, T. Cserkész, M. Váczi and P. Tóth. 2017. A Kerecsensólyom (Falco cherrug) állományváltozásának okai az elmúlt 120 évben, különös tekintettel a 2007–2018 közötti időszakra. (Reasons behind the population changes of the Saker Falcon (*Falco cherrug*) in Hungary in the past 120 years, in particular with regard to the period between 2007 and 2018. Heliaca. MME/Birdlife Hungary, Budapest.

Bahat, O. 1995. Physiological adaptations and foraging ecology of an obligatory carrion eater—the griffon vulture (*Gyps fulvus*). PhD thesis, Tel Aviv University.

Bahat, O. 1997. Long-Range movements of Griffon Vultures from Israel. Torgos. 28: 19–24.

Bahat, O. 2001. Conservation of threatened raptor populations in Israel. pp. 177–189. *In:* Leshem, Y., A. Froneman, P. Mundy and H. Shamir [eds.]. Wings over Africa. Proceedings of the International Seminar on Bird Migration: Research, Conservation, Education and Flight Safety, Tel Aviv Univ.

Bahat, O., A. Kaplan, Y. Baidatz, L. Kaplan, I. Choshniak and D. Houston. 1993. Radio-tracking a juvenile Griffon Vulture (*Gyps fulvus*) in the north of Israel. Israel Journal of Zoology. 39: 47–77.

Bahat, O., O. Hatzoffe, A. Kaplan and B. Woodley. 2001. Foraging range and movements of Griffon vultures (*Gyps fulvus*) in Israel, as determined by satellite tracking. pp. 11–12. *In:* Proceedings of the 4th Eurasian Congress on raptors, Doñana Biological Station, Raptor Research Foundation, Seville.

Bai, M.L. and D. Schmidt. 2012. Differential migration by age and sex in central European Ospreys *Pandion haliaetus*. Journal of Ornithology. 153: 75–84.

Bairlein, F., J. Dierschke, V. Dierschke, V. Salewski, O. Geiter, K. Huppop, U. Koppen and W. Fiedler. 2014. Atlas des Vogelzugs: Ringfunde deutscher Brut- und Gastvogel, Aula-Verlag, Wiebelsheim.

Bakaloudis, D.E. 2009. Implications for conservation of foraging sites selected by Short-toed Eagles (*Circaetus gallicus*) in Greece. Ornis Fennica. 86: 89–96.

Bakaloudis, D.E. and C.G. Vlachos. 2011. Feeding habits and provisioning rate of breeding short-toed eagles *Circaetus gallicus* in northeastern Greece. Journal of Biological Research. 16: 166–177.

Bakaloudis, D.E., C.G. Vlachos and G.J. Holloway. 2005. Nest spacing and breeding performance in Short-toed Eagle *Circaetus gallicus* in northeast Greece. Bird Study. 52: 330–338.

Bakka, S.V. and N.Y. Kiseleva. 2010. The Imperial Eagle in the Nizhniy Novgorod District, Russia. Raptors Conservation. 20: 84–88 (in Russian).

Bakken, V., O. Runde and E. Tjørve. 2003. Norsk ringmerkningsatlas, vol. 1. Stavanger Museum, Stavanger.

Balaban, S. and E. Yamaç. 2018. Breeding success and diet of the Egyptian vulture (*Neophron percnopterus* L.) in Middle and Upper Sakarya Region, Turkey. North-Western Journal of Zoology. 14: 91–95.

Balachandran, S., T. Katti and R. Manakadan. 2018. Indian Bird Migration Atlas, Bombay Natural History Society, Oxford University Press.

Balfour, E. and J.C. Cadbury. 1979. Polygyny, spacing and sex ratio among Hen Harriers *Circus cyaneus* in Orkney, Scotland. Ornis Scandinavica. 10: 133–141.

Barabashin, T.O. 2010. Migration of birds of prey through the Black and Azov Seas - hopeless attempts? pp. 200–204. *In:* Birds of Prey and their Protection. No. 20. City, publisher (in Russian).

Barabash-Nikiforov, I.I. and L.L. Semago. 1962. Birds of the southeast of the Chernozem center, Publisher, Voronezh, Russia (in Russian).

Bari, F., E.U. Rehman, M. Kabir and A. Shakeel. 2020. An extension to the known wintering range of the steppe eagle *Aquila nipalensis* in the Poonch and Jhelum valleys, Azad Jammu and Kashmir, Pakistan. Ardeola. 67: 111–118.

Barriocanal, C., D. Burgas, C. Salvà and J. Camps. 2004. Unusual concentration of the red-footed falcon *Falco vespertinus* in the spring migration in NE Spain. Scientia Gerundensis. 27: 31–34.

Batbayar, N. and H. Lee. 2017. Steppe eagle migration from Mongolia to India. pp. 117–127. *In*: Bird Migration Across the Himalayas: Wetland Functioning Amidst Mountains and Glaciers. Cambridge University Press, Cambridge, UK.

Beaman, M. 1973. Bosphorus migration, autumn 1971. Bulletin of the Ornithological Society of Turkey. 8: 3–4.

Beaman, M. and C. Galea. 1974. Visible migration of raptors over the Maltese Islands. Ibis. 116: 419–431.

Beaman, M., R.F. Porter and A. Vittery. 1975. Turkey Bird Report 1970–1973, Ornithological Society of Turkey, London.

Becciu, P., M. Panuccio, C. Catoni, G. Dell'Omo and N. Sapir. 2018. Contrasting effects of tailwind and asymmetrical response to crosswinds in soaring migrants. Behavioral Ecology and Sociobiology. 72: 28.

Bednarz, J. C., D. Jr. Klem, L.J. Goodrich and S.E. Senner. 1990. Migration counts of raptors at Hawk Mountain, Pennsylvania, as indicators of population trends, 1934–1986. The Auk. 107: 96–109.

Bekmansurov, R., M.V. Babushkin and I.V. Kayakin. 2018. Study of territorial connections in eagles by color ringing methods within the program of the Russian Raptor Research and Conservation Network. Raptors Conservation Supplement. 1: 52–53 (in Russian).

Belik, V.P. 1998. Towards the avifauna of the South Pre-Urals. pp. 13–16. *In:* Ryabitsev, V.K. [ed.]. Materials on the Distribution of Birds in the Urals, the Pre-Urals and Western Siberia. Publishing House Yekaterinburg Publisher, Yekaterinburg, Russia (in Russian).

Belik, V.P. 1999. Some elements of the ethology and ecology of the Imperial Eagle in Eastern Europe. pp. 105–121. *In:* Belik, V.P. [ed.]. Royal Eagle: Distribution, State of Populations and Prospects for the Protection of the Imperial Eagle (*Aquila heliaca*) in Russia. Series: Rare Bird Species, V.1. Russian Bird Conservation Union, Moscow, Russia (in Russian).

Belik, V.P. 2000. Birds of Don valley steppes, Rostov State Pedagogical University Press, Rostov-on-Don.

Belik, V.P. 2004. Dynamics of the Caspian Steppe Eagle population and assessment of limiting factors. Strepet. 2: 116–133.

Belik, V.P. and V.M. Galushin. 1999. The population structure of the Imperial eagle in Northern Eurasia. pp. 129–139. *In:* Belik, V.P. [ed.]. Royal Eagle: Distribution, State of Populations and Prospects for the Protection of the Imperial Eagle (*Aquila heliaca*) in Russia. Series: Rare Bird Species, V.1. Russian Bird Conservation Union, Moscow, Russia (in Russian).

Belik, V.P. and V.V. Vetrov. 1998. Distribution and abundance of the greater spotted eagle in the steppe part of the Don basin. pp. 7–8. *In:* III Conf. for Birds of Prey in Eastern Europe and Northern Asia, Kislovodsk.

Bensusan, K.J., E.F.J. Garcia and J.E. Cortes. 2007. Trends in abundance of migrating raptors at Gibraltar in spring. Ardea. 95: 83–90.

Beran, V., V. Škorpíková, M. Valášek and D. Horal. 2012. The breeding population of Saker Falcon (*Falco cherrug*) in the Czech Republic between 1999–2010. Aquila. 119: 21–30.

Berezovikov, N.N., A.A. Vinokurov and O.V. Belialov. 2008. Birds of Mountain Valley of Central and Northern Tien Shan. Russian Ornithological Journal. 17: 67–93.

Berezovikov, N.N., A.V. Kovalenko, V.V. Khrokov and F.F. Karpov. 1992. Steppe Eagle in the Utva-Ilek interfluve. pp. 27–30. *In*: Rare species of Plants and Animals of the Orenburg Region. Information Materials. Orenburg, Russia (in Russian).

Berezovikov, N.N., V.V. Khrovkov, F.F. Karpov and A.V. Kovalenko. 2000. Birds of the Utva-Ilek interfluve. Falconiformes, Strigiformes. Russian Ornithological Journal, Express Edition. 122: 3–12 (in Russian).

Berezovsky, V.G. 1983. Flight of birds of prey on the eastern coast of the Aral Sea in 1978–1979. pp. 7–9. *In*: Galushin, V.M., S.A. Polozov and A.V. Davygora [eds.]. Ecology of Birds of Prey. Materials of the 1st All-Union Conference on the Ecology and Protection of Birds of Prey. Nauka Press, Moscow, Russia (in Russian).

Bermejo, A. and J. De La Puente. 2016. Programa Migra. In: SEO/BirdLife. Programas de seguimiento y grupos de trabajo de SEO/BirdLife 2015. pp. 30–33. SEO/BirdLife. Madrid.

Bernis F. 1975. Migración de Falconiformes y *Ciconia* spp. por Gibraltar. Parte II, Análisis descriptivo del verano-otoño 1972. Ardeola. 21: 489–580.

Bernis, F. 1980. La migración de las aves en el estrecho de Gibraltar. Vol. 1: Aves planeadoras. pp. 451–462. Cátedra de Zoología de Vertebrados, Facultad de Biología, Universidad Complutense de Madrid, Madrid, Spain.

Bernis, F. 1983. Migration of the common griffon vulture in the western Palearctic. pp. 185–196. *In:* Wilbur, S.R. and J.A. Jackson [eds.]. Vulture biology and management. University of California, Berkeley.

Berthold, P. 1996. Control of Bird Migration, Chapman & Hall, London, UK.

Berthold, P., J. Griesinger, E. Nowak and U. Querner. 1991. Satellite tracking of a migrating Griffon Vulture in Spain. Satelliten-telemetrie eines gansegeiers (*Gyps fulvus*) in Spanien. Journal für Ornithologie. 132: 327–329.

Biber, J.-P. 1990. Action plan for the conservation of western Lesser Kestrel *Falco naumanni* populations. Cambridge, U.K.: International Council for Bird Preservation. Study Report no. 41: 1–26.

Biber, J.-P. 1996. International action plan for the Lesser Kestrel *Falco naumanni*. pp. 191–203. *In:* Heredia, B., L. Rose and M. Painter [eds.]. Globally Threatened Birds in Europe: Action Plans. Council of Europe and BirdLife International, Strasbourg.

Bijlsma, R.G. 1983. The migration of raptors near Suez, Egypt, Autumn 1981. Sandgrouse. 5: 19–44.

Bijlsma, R.G. 1987. Bottleneck areas for migratory birds in the Mediterranean region. An assessment of the problems and recommendations for actions. ICBP Study Report n. 18, Cambridge, England.

Bildstein, K.L. 1998. Long-term counts of migrating raptors: a role for volunteers in wildlife research. The Journal of Wildlife Management. 62: 435–445.

Bildstein, K.L. 2006. Migrating Raptors of the World: their Ecology and Conservation. Cornell University Press, Ithaca, NY.

Bildstein, K.L. 2018. Raptor migration. *In*: Sarasola, J., J. Grande and J. Negro [eds.]. Birds of Prey. Springer, Cham.

Bildstein, K.L., M.J. Bechard, C. Farmer and L. Newcomb. 2009. Narrow sea crossings present major obstacles to migrating Griffon Vultures *Gyps fulvus*. Ibis. 151: 382–391.

Bird, D.M. and K.L. Bildstein [eds.]. 2007. Raptor Research and Management Techniques. Hancock House Publishers, Surrey, Canada.

BirdLife International. 2001. Threatened Birds of Asia: the BirdLife InternationalRed Data Book. Birdlife International, Cambridge, UK.

BirdLife International. 2004. Birds in Europe: population estimates, trends and conservation status. BirdLife International (BirdLife Conservation series No. 12), Cambridge UK.

BirdLife International. 2015. European Red List of Birds. Office for Official Publications of the European Communities, Luxembourg.

BirdLife International. 2016a. *Pernis ptilorhynchus*. The IUCN Red List of Threatened Species 2016: e.T22694995A93483912., vol. 2019pp.

BirdLife International. 2016b. *Hieraaetus pennatus*. The IUCN Red List of Threatened Species 2016: e.T22696092A93543946.

BirdLife International. 2016c. *Falco naumanni*. The IUCN Red List of Threatened Species 2016: e.T22696357A87325202. http://dx.doi.org/10.2305/IUCN.UK.2016-3.RLTS.T22696357A87325202.en. Accessed November 2018.

BirdLife International. 2017a. European birds of conservation concern: populations, trends and national responsibilities Cambridge, UK: BirdLife International.

BirdLife International. 2017b. *Clanga clanga* (amended version of 2016 assessment). The IUCN Red List of Threatened Species 2017: e.T22696027A110443604. http://dx.doi.org/10.2305/IUCN.UK.2017-1.RLTS.T22696027A110443604.en.

BirdLife International. 2017c. *Aquila nipalensis* (amended version of 2016 assessment). The IUCN Red List of Threatened Species 2017: e.T22696038A118576408. http://dx.doi.org/10.2305/IUCN.UK.2017-3.RLTS.T22696038A118576408.en. Downloaded on 22 October 2019.

BirdLife International. 2017d. *Milvus milvus* (amended version of 2016 assessment). The IUCN Red List of Threatened Species 2017: e.T22695072A110921280. http://dx.doi.org/10.2305/IUCN.UK.2017-1.RLTS.T22695072A110921280.en. Downloaded on 22 August 2018.

BirdLife International. 2017e. *Buteo buteo* (amended version of 2016 assessment). The IUCN Red List of Threatened Species 2017: e.T61695117A119279994. http://dx.doi.org/10.2305/IUCN.UK.2017-3.RLTS.T61695117A119279994.en. Accessed in 20.12.2019.

BirdLife International. 2018a. *Circus macrourus*. The IUCN Red List of Threatened Species 2018: e.T22695396A132304131. http://dx.doi.org/10.2305/IUCN.UK.2018-2.RLTS.T22695396A132304131.en. Downloaded on 03 June 2019.

BirdLife International. 2018b. Species factsheet: *Falco vespertinus*. Downloaded from http://www.birdlife.org on 02 July 2017.

BirdLife International. 2019a. *Buteo rufinus* (amended version of 2017 assessment). The IUCN Red List of Threatened Species 2019: e.T22736562A155442127. https://dx.doi.org/10.2305/IUCN.UK.2019-3.RLTS.T22736562A155442127.en. Downloaded on 07 June 2020.

BirdLife International. 2019b Species factsheet: *Pandion haliaetus*. Downloaded from http://www.birdlife.org on 20/05/2019.

BirdLife International. 2019c. Species factsheet: *Gyps fulvus*. Downloaded from http://www.birdlife.org on 01/01/2019

BirdLife International 2019d. Species factsheet: *Aquila nipalensis*. Downloaded from http://www.birdlife.org on 22 October 2019.

BirdLife International. 2019e. Species factsheet: *Aquila heliaca*. Downloaded from http://www.birdlife.org 4 August 2019.

BirdLife International. 2020. IUCN Red List for birds. Downloaded from http://www.birdlife.org on 11/07/2020.

Biricik, M. and R. Karakaş. 2011. Black Kites (*Milvus migrans*) winter in southeastern Anatolia, Turkey. Journal of Raptor Research. 45: 370–373.

Blanc, J.F., A. Sternalski and V. Bretagnolle. 2013. Plumage variability in Marsh Harriers. British Birds. 106: 145–158.

Blanco, G. 1994. Seasonal abundance of Black Kites associated with the rubbish dump of Madrid, Spain. Journal of Raptor Research. 28: 242–245.

Blanco, G. 1997. Role of refuse as food for migrant, floater and breeding Black Kites (*Milvus migrans*). Journal of Raptor Research. 31: 71–76.

Blanco, G., J.A. Lemus, B. Arroyo, F. Martínez and M. Garcia. 2009. The dilema of extensive livestock and the myth of muladares: the implications of pharmaceutical contamination and its impact on the health of avian scavengers. Munibe. 29: 434–451.

Blanco, J.C., F. Hiraldo and B. Heredia. 1990. Variations in the diet and foraging behaviour of a wintering red kite (*Milvus milvus*) population in response to changes in food availability. Ardeola. 37: 267–278.

Boal, C.W. and C.R. Dykstra [eds.]. 2018. Urban raptors: ecology and conservation of birds of prey in cities, Island Press, Washington, DC.

Bogdanov, M.N. 1882. Essays of nature of the Khiva oasis and the Kyzyl-Kum desert. *In:* Trotskii. V.N. [ed.]. Description of the Khiva campaign in 1873. Tashkent. 12 (in Russian).

Bold, A. and S. Boldbaatar. 1990. Steppe Eagle *Aquila nipalensis*. *In*. Proceedings of the National University of Mongolia 9: 103–121.

Boldbaatar S., S.A. Bukreev and B.M. Zvonov. 2012. Falcons of Mongolia. pp. 123–130. *In:* Gavrilyuk, M.N. [ed.]. Birds of Prey in the Dynamic Environment of the Third Millennium: State And Prospects. Proceedings of the VI International Conference on Falconiformes and Owls of Northern Eurasia. Krivoy Rog, Russia. Publisher (in Russian).

Bonal, R. and J.M. Aparicio. 2008. Evidence of prey depletion around lesser kestrel *Falco naumanni* colonies and its short term negative consequences. Journal of Avian Biology. 39: 189–197.

Bonavia, E. 2019. Buskett Raptor Observatory report autunno 2019. Infomigrans. 44: 4.

Bonavia, E. and R. Vella. 2016. La migrazione post riproduttiva a Malta nel 2016. Infomigrans. 38: 2.

Bonavia, E., R. Galea and R. Vella. 2010. La migrazione post-riproduttiva dei rapaci a Malta 2010. Infomigrans. 26: 2.

Bondarev, D.V. and R.D. Reutsky. 1989. Nesting of the Red-footed Falcon in the Volga Delta. pp. 56–58. *In:* Breeding biology of birds. Publisher, City, Country, Permian (in Russian).

Bondì, S., E. Guzzo, R. Mascara and M. Sarà. 2018. Onset of natal dispersal in Peregrine Falcon from Mediterranean islands (Italy). Ornis Hungarica. 26: 201–221.

Bønløkke, J., J.J. Madsen, K. Thorup, K.T. Pedersen, M. Bjerrum and C. Rahbek. 2006. Dansk Trækfugleatlas: the Danish bird migration atlas. Rhodos, Humlebæk.

Borioni, M. and M.R. Baldoni. 2005. La migrazione primaverile sul Monte Conero. Infomigrans. 15: 8.

Borioni, M. and M.R. Baldoni. 2007. La migrazione primaverile sul Monte Conero. Infomigrans. 19: 7.

Borioni, M. and M.R. Baldoni. 2008. La migrazione primaverile sul Monte Conero. Infomigrans. 21: 8.

Borioni, M. and M.R. Baldoni. 2009. La migrazione primaverile sul Monte Conero. Infomigrans. 23: 8.

Borioni, M. and M.R. Baldoni. 2010. La migrazione primaverile sul Monte Conero. Infomigrans. 25: 6.

Borodin, O.V. 2003. Breeding group of imperial eagles in the conditions of modern management in the Ulyanovsk region. pp. 156–158. *In:* Galushin, V.M. [ed.]. Materials of the IV Conference on Birds of Prey of Northern Eurasia, Rostov State Pedagogical University, Penza, Russia (in Russian).

Borodin, O.V. and S.L. Smirnova. 2006. Monitoring of a local cluster of breeding imperial eagles in the south of the Middle Volga. pp. 86–87. *In:* Kurochkin, E.N. [ed.], Ornithological Studies in Northern Eurasia. Abstracts of the XII International Ornithological Conference, Stavropol State University, Stavropol, Russia (in Russian).

Bosch, J. and B.-U. Meyburg. 2012. The Lesser Spotted Eagle *Aquila pomarina* in Catalonia (Spain)—Breeding attempt and migration. Vogelwelt. 133: 89–97.

Bostanzhoglo, V.N. 1911. Ornithological fauna of the Aral-Caspian steppes. Publisher. Moscow, Russia (in Russian).

Botha, A., J. Andewski, C. Bowden, M. Gudka, R. Safford and N.P. Williams. 2017. Multi-species action plan to conserve African-Eurasian vultures Technical Publication, UNEP/Raptors MOU Coordination Unit, Abu Dhabipp.

Bougain, C. 2016. Identification of important migration concentration areas of Egyptian vultures *Neophron percnopterus* from the Balkan population tracked by satellite telemetry. M.S. Thesis, University of Strasbourg, Strasbourg, France.

Bounas, A., M. Panuccio, A. Evangelidis, K. Sotiropoulos and C. Barboutis. 2016. The migration of the Lesser Kestrel *Falco naumanni* in Eastern Europe: a ringing recovery and direct observation approach. Acrocephalus. 37: 49–56.

Bounas, A., D. Vavylis, L. Sidiropoulos, E. Kret, J.M. Rodriguez and V. Saravia. 2015. Your face looks familiar: individual identification using facial markings and plumage patterns for the monitoring of the endangered Egyptian Vulture (*Neophron percnopterus*), pp. 10. 13th International Congress on the Zoogeography, Ecology and Evolution of Eastern Mediterranean, Heraclion, Crete.

Bounas, A., D. Tsaparis, R. Efrat, G. Gradev, M. Gustin, K. Mikulic, A. Rodriguez, M. Sarà, G., Kotoulas and K. Sotiropoulos. 2018. Genetic structure of a patchily distributed philopatric migrant: implications for management and conservation. Biological Journal of Linnean Society. 124: 633–644.

Bounas A., M. Solanou, M. Panuccio, S. Barišić, T. Bino, K. Erciyas-Yavuz, P. Iankov, C. Ieronymidou and C. Barboutis. 2020. Mining citizen science data to explore stopover sites and spatiotemporal variation in migration patterns of the red-footed falcon. Current Zoology. 66: 467–475.

Bouten, W., L. Gangoso, A. Dokter, J. Shamoun-Baranes, D. Viana and J. Figuerola. 2015. Dawn ascents to astonishing altitudes during migration of Eleonora's Falcons (*Falco eleonorae*). Oral presentation, EOU 10th Conference, Badajoz, Spain, 24–28 August 2015.

Bragin, E.A. 1980. The phenology of nesting of rare birds of prey in the Naurzum Reserve. *In:* Kibalchich, O.A. [ed.]. Seasonal Rhythm of Rare and Endangered Species of Plants and Animals, Proceedings of Conference. Moscow Branch of Geographical Society of the USSR, Moscow, Russia (in Russian).

Bragin, E.A. 1987. The Eagles, Kainar Press, Alma-Ata, Kazakhstan (in Russian).

Bragin, E.A. 1989. Biology of birds of prey of pine forests of the Kustanay Steppe. Ph.D. Dissertation, Kazakhstan Academy of Sciences, Almaty, Kazakhstan (in Russian).

Bragin, E.A. 1983. The nature and timing of the flight of birds of prey in the Naurzum Reserve. pp. 97–101. *In:* Kovshar, A.F. and E.I. Gavrilov [eds.]. Bird Migrations in Asia, vol. 8. Nauka Press, Alma-Ata, USSR (in Russian).

Bragin, E.A., A.E. Bragin and T.E. Katzner. 2017. Demographic consequences of nest box use for Red-footed Falcons *Falco vespertinus* in Central Asia. Ibis. 159: 841–853.

Brazil, M. and D. Nurney. 2009. Birds of East Asia: China, Taiwan, Korea, Japan, and Russia, Princeton University Press, Princeton,NJ.

Brichetti, P. and G. Fracasso. 2018. Birds of Italy. vol. 1, *Anatidae-Alcidae*, Edizioni Belvedere, Latina, Italy.

Brochet, A.L., W. Vav Den Bossche, V.R. Jones, H. Arnardottir, D. Damoc, M. Demko, G. Driessens, K. Flensted, M. Gerber, M. Ghasabyan, D. Gradinarov, J. Hansen, M. Horvath, M. Karlonas, J. Krogulec, T. Kuzmenko, L. Lachman, T. Lehtiniemi, P. Lorge, U. Lotberg, J. Lusby, G. Ottens, J.Y. Paquet, A. Rukhaia, M. Schmidt, P. Shimmings, A. Stipnieks, E. Sultanov, Z. Vermouzek, A. Vintchevski, V. Volke, G. Willi and S.H.M. Butchartm. 2017. Illegal killing and taking of birds in Europe outside the Mediterranean: assessing the scope and scale of a complex issue. Bird Conservation International. 1–31.

Brown, A. and P. Grice. 2005. Birds in England, London, T. & A.D. Poyser.

Brown, L.H., E.K. Urban and K. Newman. 1982. The Birds of Africa, vol. 1, Academic Press, London and New York.

Bruderer, B. and A. Boldt. 2001. Flight characteristics of birds: I. Radar measurements of speeds. Ibis. 143: 178–204.

Bruderer, B, S. Blitzblau and D. Peter. 1994. Migration and flight behaviour of Honey Buzzards *Pernis apivorus* in southern Israel observed by radar. Ardea. 82: 111–122.

Bruderer, B., F. Liechti, M. Kestenholz, D. Peter, R. Spaar, H. Stark and T. Steuri. 1997. Vogelzugstudien mit Zielfolgeradar im Süden Israels. Ornithologische Beobachter. 97: 21–44.

Bruun, B. 1985. Raptor migration in the Red Sea area. ICBP Technical Publication No. 5: 251–255.

Buechley, E.R. and Ç.H. Şekercioğlu. 2016. The avian scavenger crisis: Looming extinctions, trophic cascades, and loss of critical ecosystem functions. Biological Conservation. 198: 220–228.

Buechley, E.R., M.J. McGrady, E. Çoban and Ç.H. Şekercioğlu. 2018a. Satellite tracking a wide-ranging endangered vulture species to target conservation actions in the Middle East and East Africa. Biodiversity and Conservation. 27: 2293–2310.

Buechley, E.R., S. Oppel, W.S. Beatty, S.C. Nikolov, V. Dobrev, V. Arkumarev, V. Saravia, C. Bougain, A. Bounas, E. Kret, T. Skartsi, L. Aktay, K. Aghababyan, E. Frehner and Ç.H. Şekercioğlu. 2018b. Identifying critical migratory bottlenecks and high-use areas for an endangered migratory soaring bird across three continents. Journal of Avian Biology. 49: e01629.

Bukreev, S.A. 1998. On the passage of the Steppe Eagle in the Western Kopetdag. pp. 19–20. *In:* Proceedings of the 3rd Conference on Birds of Prey of Eastern Europe and Northern Asia. Part 1. (Kislovodsk, September 15–18, 1998). Stavropol, Russia (in Russian).

Burfield, I.J. 2008. The conservation status and trends of raptors and owls in Europe. Ambio. 37: 401–407.

Burner, R., S. Gombobaatar, P. van Els, L. Burner, D. Usukhjargal and M. Bayasgalantselmeg. 2019. Nesting ecology of solitary-nesting Amur Falcons (*Falco amurensis*) in central Mongolia. Journal of Field Ornithology. 90: 266–276.

Butyev, V.T., A.V. Mikheev and A.O. Shubin. 1983. Seasonal flight of birds of prey along the western coast of the Caspian Sea, 11–13. *In:* Ecology of Birds of Prey. Materials of the 1st All-Union Conference on the Ecology and Protection of Birds of Prey. Moscow, Russia. Publisher (in Russian).

Butyev, V.T., A.V. Mikheev, A.B. Kostin, E.A. Koblik and E.A. Lebedeva. 1989. Notes on rare bird species of the Caucasus coast of the Caspian. pp. 137–152. *In:* Khokhlov, A.N. [ed.]. Ornithological Resources of the North Caucasus. Stavropol Regional Council of the All-Russian Society for Nature Protection, Stavropol, Russia (in Russian).

Cadahía, L., P. López-López, V. Urios and J.J. Negro. 2010. Satellite telemetry reveals individual variation in juvenile Bonelli's eagle dispersal areas. European Journal of Wildlife Research. 56: 923–930.

Cameron, R.A.D., L. Cornwallis, M.J.L. Percival and A.R.E. Sinclair. 1967. The migration of raptors and storks through the Near East in autumn. Ibis. 109: 489–501.

Campora, M. and G. Cattaneo. 2005. Ageing and sexing Short-toed Eagles. British Birds. 98: 370–376.

Can, O. 2001. Studies on soaring bird migration at the Belen Pass and Hatay Province. M.S. Thesis, Middle East Technical University, Ankara.

Candler, G.L. and P. Kennedy. 1995. Flight strategies of migrating Osprey: fasting vs. foraging. Journal of Raptor Research. 29: 85–92.

Capotosti, S., M. Scacco, L. Nelli, G. Dell'Omo and M. Panuccio. 2019. Hypatia-trackRadar: A software for animal tracking using marine surveillance radars. Ecological Informatics. 53: 100972.

Careau, V., J.F. Therrien, P. Porras, D. Thomas and K.L. Bildstein. 2006. Soaring and gliding flight of migrating broad-winged hawks: behavior in the nearctic and neotropics compared. The Wilson Journal of Ornithology. 118: 471–477.

Carrete, M., J.A. Sánchez-Zapata, J.R. Benítez, M. Lobón and J.A. Donázar. 2009. Large scale risk-assessment of wind-farms on population viability of a globally endangered long-lived raptor. Biological Conservation. 142: 2954–2961.

Carrillo, J. and E. González-Dávila. 2009. Latitudinal variation in breeding parameters of the Common Kestrel *Falco tinnunculus*. Ardeola. 56: 215–228.

Castelijns, H. and W. Castelijns. 2008. Het overwinteren van de Bruine Kiekendief *Circus aeruginosus* in Zeeland. Limosa. 81: 41–49.

Catchot, S. 2017. Illa de L'Aire - Estudi de la Migració postnupcial. Memòria any 2017.

Catry, I., A.M. Franco and F. Moreira. 2014. Easy but ephemeral food: exploring the trade-offs of agricultural practices in the foraging decisions of Lesser Kestrels on farmland. Bird Study. 61: 447–456.

Catry, I., T. Amano, A.M.A. Franco and W.J. Sutherland. 2012. Influence of spatial and temporal dynamics of agricultural practices on the lesser kestrel. Journal of Applied Ecology. 49: 99–108.

Catry, I., M.P. Dias, T. Catry, V. Afanasyev, J. Fox, A.M.A. Franco and W.J. Sutherland. 2010. Individual variation in migratory movements and winter behaviour of Iberian Lesser Kestrels *Falco naumanni* revealed by geolocators. Ibis. 153: 154–164.

Ceballos, O. and J. Donázar. 1990. Roost-tree characteristics, food habits and seasonal abundance of roosting Egyptian Vultures in Northern Spain. Journal of Raptor Research. 24: 19–25.

Cecere, J.G., M. Panuccio, A. Ghiurghi, F. Urbano, S. Imperio, C. Celada and P. Lopez-Lopez. 2018. Snake species richness predicts breeding distribution of short-toed snake eagle in central Italy. Ethology, Ecology & Evolution. 30: 178–186.

Cento, M., N. Agostini, G. Dell'Omo and M. Panuccio. 2018. Strait of Messina bird observatory report Autunno 2018. Infomigrans. 42: 2–3.

Cento, M., N. Agostini, M. Gustin and G. Dell'Omo. 2019. Osservatorio dello Stretto, rapporto autunno 2019. Infomigrans. 44: 2–3.

Cepák, J., P. Klvana, J. Skopek, L. Schropfer, M. Jelinek, D. Horak, J. Formanek and J. Zarybnicky [eds.]. 2008. Atlas migrace ptaku Ceske republiky a Slovenska, Aventinum, Praha.

Chapman, A. 1999. The Hobby, Arlequin Press, Chelmsford.

Chapman, A. 2002. Eurasian hobby. pp. 253–255. *In*: Wernham, C.V., M.P. Toms, J.H. Marchant, J.A. Clark, G.M. Siriwardena and S.R. Baillie [eds.]. 2002. The Migration Atlas: Movements of the Birds of Britain and Ireland, Poyser, London.

Chavko, J., L. Deutschová, Š. Danko, J. Mihók. B. Landsfeld, J. Pavelka, L. Šnírer, S. Harvancík, A. Dubravský, L. Prešinský and R. Galaš. 2014. Status of the eastern imperial eagle population in Slovakia between 1977 and 2013. Slovak Raptor Journal. 8: 9–15.

Chernishov, V.M., T.K. Blinova and V.N. Blinov. 1983. Comparative characteristics of seasonal migrations of birds of prey in the south of Baraba and in the Upper Ob region. pp. 47–49. *In:* Ecology of Birds of Prey. Materials of the 1st All-Union Conference on the Ecology and Protection of Birds of Prey. Moscow, Russia. Publisher. (in Russian).

Chevallier, D., F. Jiguet, T. Nore, F. Baillon and P. Cavallin. 2010. Satellite tracking of a Booted Eagle (*Aquila pennata*) during migration. Ringing & Migration. 25: 62–64.

Chiofalo, G., S. Cutini, A. Giordano and D. Ricciardi. 2006. La migrazione primaverile sullo Stretto di Messina - Monti Peloritani (ME). Infomigrans. 16: 2.

Chiofalo, G., S. Cutini, D. Ricciardi, A. Spampinato and A. Giordano. 2007. La migrazione sullo Stretto di Messina - Monti Peloritani 2007. Infomigrans. 19: 4.

Christensen, S. and U. Sorensen. 1989. A review of the Migration and wintering of *Aquila pomarina* and *Aquila nipalensis orientalis*. pp. 139–150. *In:* Meyburg, B.-U. and R.D. Chancellor [eds.]. Raptors in the Modern World, Berlin, London, Paris, World Working Group on Birds of Prey.

CIBIO (Centro de Investigação em Biodiversidade e Recursos Genéticos). 2011. Reintroduction of the Osprey (*Pandion haliaetus*) in Portugal. Annual report 2011.

Clark, W.S. 1999. A Field Guide to the Raptors of Europe, the Middle East and North Africa, Oxford, Oxford University Press.

Clark, W.S. 2005. Steppe Eagle *Aquila nipalensis* is monotypic. Bulletin of the British Ornithologists' Club. 125: 149–153.

Clark, W.S. and E. Gorney. 1987. Oil contamination of raptors migrating along the Red Sea. Environmental Pollution. 46: 307–313.

Clark, W.S. and P. Christy. 2006. First record of Oriental Honey Buzzard Pernis ptilorhynchus for Gabon and sub-Saharan Africa. Bulletin of the African Bird Club. 13: 207–210.

Clark, W.S. and R. Davies. 2018. African Raptors, Helm, Bloomsbury, London.

Clarke, R. 1995. The Marsh Harrier, Hamlyn species guides, London.

Clarke, R. and D. Watson. 1990. The Hen Harrier *Circus cyaneus* winter roost survey in Britain and Ireland. Bird Study. 37: 84–100.

Clarke, R., A. Bourgonje and H. Castelijins. 1993. Food niches of sympatric Marsh Harriers *Circus aeruginosus* and Hen Harriers *C. cyaneus* on the Dutch coast in winter. Ibis. 135: 424–431.

Clarke, R., V. Prakash, W.S. Clark, N. Ramesh and D. Scott. 1998. World record count of roosting harriers *Circus* in Blackbuck National Park, Velavadar, Gujarat, north-west India. Forktail. 14: 70–71.

Clement, P. and D. Holman. 2001. Passage records of Amur Falcon Falcoamurensis from SE Asia and southern Africa including first records fromEthiopia. Bulletin of British Ornithologists' Club. 121: 222–230.

Clements, R. 2008. The Common Kestrel population in Britain. British Birds. 101: 228–234.

Cochran, W.W. and R.D. Applegate. 1986. Speed of flapping flight of Merlins and Peregrine Falcons. Condor. 88: 397–398.

Coleiro, C. 1999. Large flock of Short-toed Eagles *Circaetus gallicus* in late autumn. Il Merill. 29: 28.

Collmait, J. and J. Croxall. 1967. Spring migration at the Bosphorus. Ibis. 109: 359–372.

Copete, J.L., J.A. Lorenzo, E. Amengual, D. Bigas, P. Fernández, D. López-Velasco, G. Rodríguez and M. García-Tarrasón. 2015. Observaciones de aves raras en España, 2012 y 2013. Ardeola. 62: 453–508.

Cormier, J.-P. and F. Baillon. 1991. Concentration de Busards cendrés *Circus pygargus* (L.) dans la région de M'Bour (Sénégal) durant l'hiver 1988–1989: utilisation du milieu et régime alimentaire. Alauda. 59: 163–168.

Corso, A. 2001a. Raptor migration across the Strait of Messina, southern Italy. British Birds. 94: 196–202.

Corso, A. 2001b. First data on the migration of the Siberian peregrine *Falco peregrinus calidus* at the Strait of Messina and comments on its status in Italy. Avocetta. 25: 196.

Corso, A. 2005. Avifauna di Sicilia, L'EPOS, Palermo. Italy.

Corso, A. 2012. The status of second-calendar-year Honey-buzzards in Europe. British Birds. 105: 484–485.

Corso, A. and C. Cardelli. 2004. The migration of Pallid Harrier across the central Mediterranean. British Birds. 97: 238–246.

Corso, A., L. Starnini, M. Viganò and J.J.F.J. Jansen. 2015. A quantitative morphological geographical study from a widely distributed raptor: the Lesser Kestrel *Falco naumanni* Fleischer, 1818 (Falconiformes Falconidae). Biodiversity Journal. 6: 285–296.

Corso, A., M. Viganò, J.J.F.J. Jansen and L. Starnini. 2016. Geographical plumage variation in Lesser Kestrel. Dutch Birding. 38: 271–292.

Corso, A., O. Janni, H. Larsson, M. Viganò and M. Gustin. 2009. First data on migration of raptors at the Pelagie Islands, Sicilian Channel. pp. 216–218. *In:* Brunelli, M., C. Battisti, F. Bulgarini, J.C. Cecere, F. Fraticelli, M. Gustin, S. Sarrocco and A. Sorace [eds.]. Atti XV Convegno Italiano di ornitologia, Sabaudia, Italy.

Cortes-Avizanda, A., P. Almaraz, M. Carrete, J.A. Sanchez-Zapata, A. Delgado, F. Hiraldo and J.A. Donazar. 2011. Spatial heterogeneity in resource distribution promotes facultative sociality in two Trans-Saharan migratory birds. PLoS One. 6: e21016.

Costantini, D. and G. Dell'Omo. 2020. The Kestrel: ecology, behaviour and conservation of an open-land predator, Cambridge University Press, Cambridge, UK.

Cramp, S. and K.E.L. Simmons. 1980. Handbook of the birds of Europe, the Middle East and North Africa: The birds of the Western Palearctic, vol. 2, Oxford University Press, Oxford.

Cristol, D.A., B.B. Mitchell and C. Carbone. 1999. Differential migration revisited: latitudinal segregation by age and sex class. pp. 33–88. *In:* Nolan, V. [ed.]. Current Ornithology 15, Academic/Plenum Publisher, New York.

Csörgey, T. 1897. Falco sacer, Brisson 1760 – Kerecsensólyom. Aquila. 1–3: 105–139.

Csörgo, T., Z. Karcza, G. Halmos, G. Magyar, J. Gyurácz, T. Szép, A. Bankovics, A. Schmidt and E. Schmidt. 2009. Magyar madárvonulási atlasz. Kossuth Kiadó, Budapest.

Curk, T., I. Pokrovsky, N. Lecomte, T. Aarvak, K. Burnham, A. Dietz, A. Franke, G. Gauthier, K.O. Jacobsen, J. Kidd, S.B. Lewis, I.J. Øien, R. Solheim, K. Wiebe, M. Wikelski, J.F. Therrien and K. Safi. 2020. Arctic avian predators synchronise their spring migration with the northern progression of snowmelt. Scientific Reports. 10: 7220.

Curry-Lindahl, K. 1981. Bird Migration in Africa: movements between six continents, Academic Press, London.

Cuthbert, R., R.E. Green, S. Ranade, S. Saravanan, D.J. Pain, V. Prakash and A.A. Cunningham. 2006. Rapid population declines of Egyptian vulture (*Neophron percnopterus*) and red-headed vulture (*Sarcogyps calvus*) in India. Animal Conservation. 9: 349–354.

Cutini, S., G. Chiofalo, J.P. Fyott, L. Vanni and D. Ricciardi. 2008. Stretto di Messina: 25° anno di attività 2 aprile - 27 maggio 2008. Infomigrans. 21: 2–3.

Czechowski, P. 2004. Przeloty i zimowanie ptaków szponiastych Falconiformes w Dolinie Odry koło Zielonej Góry. Ptaki Śląska. 15: 89–103.

Dalvi, S., R. Sreenivasan and T. Price. 2013 Exploitation in Northeast India. Science. 339(6117): 270.

Dan, A., B. Granit, Z. Labinger and Y. Perlman. 2005. Population dynamics of wintering Greater Spotted Eagles *Aquila clanga* in the Hula Valley. pp. 77–79. *In:* Proceedings of the international meeting on spotted eagles (*Aquila clanga*, *A. pomarina* and *A. hastata*) – Research and conservation. Biebrza National park, Osowiec–Poznan–Berlin.

Danilov, O.N. 1976. Birds of prey and owls of Baraba and northern Kulunda. Science Publisher, Novosibirsk, Russia (in Russian).

Danko, S., B.-U. Meyburg, T. Belka and D. Karaska. 1996. Individuelle Kennzeichnung von Schreiadlern *Aquila pomarina*: methoden, bisherige Erfahrungen und Ergebnisse. pp. 209–243. *In:* Meyburg, B.-U. and R.D. Chancellor. [eds.]. Eagle Studies. Berlin, London and Paris: World Working Group on Birds of Prey.

Danko, S., A. Ranner, J. Lontkowski, I. Sandor, J. Sirek and I. Gorban. 2013. Ocurrence of the griffon vulture and the black vulture in Central Europe in the 21st century. Slovak Raptor Journal. 7: 1–16.

De Frutos, Á. and P.P. Olea. 2008. Importance of the premigratory areas for the conservation of lesser kestrel: space use and habitat selection during the post-fledging period. Animal Conservation. 11: 224–233.

de Juana, E. and E. Garcia. 2015. The Birds of the Iberian Peninsula, Bloomsbury, London.

De la Cruz, A., G.M. Arroyo, A. Onrubia, L. Barrios and A.R. Muñoz. 2011a. Migración primaveral de aves planeadoras en el Estrecho de Gibraltar. Diseño de un programa de seguimiento a largo plazo. Migres. 2: 79–88.

De la Cruz, A., A. Onrubia, B. Pérez, C. Torralvo, G.M. Arroyo, J. Elorriaga, J. Ramírez, M.González and R. Benjumea. 2011b. Seguimiento de la migración de las aves en el estrecho de Gibraltar: resultados del Programa Migres 2009. Migres. 2: 65–78.

De Lucas, M., M. Ferrer, M.J. Bechard and A.R. Muñoz. 2011. Griffon Vulture mortality at wind farms in southern Spain: distribution of facilities and active migration measures. Biological Conservation. 147: 184–189.

De Roder, F.E. 1989. The migration of raptors south of Annapurna, Nepal, autumn 1985. Forktail. 4: 9–17.

De Sante, D.F. 1983. Vagrants: when orientation or navigation goes wrong. Point Reyes Bird Observatory Newsl. 61: 12–16.

DeCandido, R. 2013. Autumn 2012 Raptor Migration in Nepal. Technical Report. https://doi.org/10.13140/RG.2.1.2697.2001.

DeCandido, R. and C. Nualsri. 2009. Timing and abundance of Grey-faced Buzzards *Butastur indicus* and other raptors on northbound migration in southern Thailand, spring 2007–2008. Forkail. 25: 90–95.

DeCandido, R., D. Allen and K. Bildstein. 2006a. Spring migration of Oriental Honey-buzzards and other raptors at Tanjung Tuan, Malaysia, 2000–2001. Forktail. 22: 156.

DeCandido, R., R.O. Bierregaard, Jr., M.S. Martell and K.L. Bildstein. 2006b. Evidence of nocturnal migration by osprey (*Pandion haliaetus*) in North America and western Europe. Journal of Raptor Research. 40: 156–158.

DeCandido, R., D. Allen, R. Yosef and K. Bildstein. 2004a. A comparison of spring migration phenology of bee-eaters and Oriental Honey-buzzards *Pernis ptilorhyncus* at Tanjung Tuan, Malaysia, 2000–01. Ardea. 92: 169–174.

DeCandido, R., C. Nualsri, D. Allen and K.L. Bildstein. 2004b. Autumn 2003 raptor migration at Chumphon, Thailand: a globally significant raptor migration watch site. Forktail. 20: 49–54.

DeCandido, R.O., S.U. Gurung, T.U. Subedi and D.E. Allen. 2013. The east–west migration of Steppe Eagle *Aquila nipalensis* and other raptors in Nepal and India. BirdingASIA. 19: 18–25.

DeCandido, R., C. Kasorndorkbua, C. Nualsri, C. Chinuparawat and D. Allen. 2008. Raptor migration in Thailand. BirdingASIA. 10: 16–22.

DeCandido, R., M. Siponen, H. Smit, A. Pierce and D. Allen. 2015. Flight identification and migration pattern of the Oriental Honey Buzzard *Pernis ptilorhynchus orientalis* in southern Thailand, 2007–2014. BirdingASIA. 23: 27–33.

Dejonghe, J.-F. 1980. Analyse de la migration prénuptiale des rapaces et des cigognes au Cap Bon (Tunisie). L'Oiseau et la Revue Française d'Ornithologie, 50: 137–138.

del Hoyo, J. [ed.]. 2020. Handbooks of the birds of the World – Alive. Lynx Editions, Barcelona, Spain, www.hbw.com.

del Hoyo, J., A. Elliott and J. Sargatal [eds.]. 1994. Handbook of the birds of the world, Vol. 2, New World vultures to guineafoul, Lynx Edicions, Barcelona, Spain.

del Hoyo, J., N.J. Collar, D.A. Christie, A. Elliott and L.D.C. Fishpool. 2014. HBW and BirdLife International Illustrated Checklist of the Birds of the World, Lynx Edicions, BirdLife International, Barcelona, Spain and Cambridge, UK.

Del Moral, J.C. and B. Molina [eds.]. 2018. El buitre leonado en España, población reproductora en 2018 y método de censo, SEO/BirdLife, Madrid.

Delgado, G. and V. Quilis. 1990. New data concerning the migration of Eleonora's Falcon *Falco eleonorae*. Ringing & Migration. 11: 111–112.

Dementiev, G.P. and N.A. Gladkov. 1951. Birds of the Soviet Union, vol. 1, Sovetskaya Nauka, Moscow.

Dementiev, G.P., N.A. Gladkov, Y.A. Isakov, N.N. Kartashev, S.V. Kirikov, A.V. Mikheev and E.S. Ptushenko. 1951. Birds of the Soviet Union. Israel Program for Scientific Translations, Jerusalem, Israel.

Demerdzhiev, D., M. Horváth, A. Kovács, S. Stoychev and I. Karyakin. 2011. Status and population trend of the eastern imperial eagle (*Aquila heliaca*) in Europe in the period 2000–2010. Acta Zoologica Bulgarica Suppl. 3: 5–14.

Demerdzhiev, D., S. Stoychev, D. Dobrev, S. Spasov and N. Terziev. 2014. Conservation measures undertaken to improve the population status of eastern imperial eagle (*Aquila heliaca*) in Bulgaria. Slovak Raptor Journal. 8: 27–39.

Den Besten, J.W. 2004. Migration of Steppe Eagles *Aquila nipalensis* and other raptors along the Himalayas past Dharamsala, India, in autumn 2001 and spring 2002. Forktail. 20: 9–13.

Dendaletche, C. [ed.]. 1988. Grands Rapace setCorvidés des Montagnes d'Europe. Acta Biologica Montana 8 (Special vol.).

deNobel, W., F. deRoder, C.L.E. Marteijn, L.P. Meiniger, J.J. Stuart, F. Schepers and R. Van Westrienen. BirdsinNE-Greece, spring 1987. pp. 109–263. *In:* Meininger, P.L. [ed.]. 1990. Birds of the Wetlands in North-east Greece, spring 1987WIWO-report, vol 20. WIWO, Zeist, Germany.

Desmet, E. and W.M.G. Vansteelant. 2014. Recordaantallen Wespendieven over de Lage Landen in augustus-september 2013. Natuur.oriolus. 80: 83–90.

Di Maggio, R., D. Campobello and M. Sarà. 2018. Lesser kestrel diet and agricultural intensification in the Mediterranean: An unexpected win-win solution? Journal of Nature Conservation. 45: 122–130.

Di Vittorio, M., P. López-López, G. Cortone and L. Luiselli. 2017. The diet of the Egyptian vulture (*Neophron percnopterus*) in Sicily: temporal variation and conservation impications. Vie et Milieu-Life and Environment. 67: 1–8.

Di Vittorio, M., S. Henriquet, E. Kobierzycki, L. Luiselli, E.M. Hema, L. Murabito, G. Rannisi and P. López-López. 2016b. Dispersal of Egyptian Vultures *Neophron percnopterus*: the first case of long-distance relocation of an individual from France to Sicily. Ringing & Migration. 31: 111–114.

Di Vittorio, M., A. Barbera, E. Di Trapani, F.P. Faraone, A. Ciaccio, N. Sciagura, D. D'Amico, G. Giacalone, M. Zafarana, S. Grenci and A. Sarto. 2016a. Wintering of Egyptian Vultures (*Neophron percnopterus*) in Sicily: new data. Arxius de Miscel·lània Zoològica. 14: 114–116.

Di Vittorio, M., E.M. Hema, D. Dendi, G.C. Akani, G. Cortone, P. López-López, N. Amadi, G. Hoinsoudé Ségniagbeto, C. Battisti and I. Luiselli. 2018. The conservation status of West African vultures: an updated review and a strategy for conservation. Vie et milieu-Life and Environment. 68: 33–43.

Díaz-Ruiz, J. and F. Cebollada-Baratas. 2011. Monitoring and conservation of Booted Eagle (*Aquila pennata*) in the sierra de Guadarrama (Central Spain). pp. 93–98. *In:* Zuberogoitia, I. and J.E. Martínez [eds.]. Ecology and Conservation of European Forest-dwelling Raptors. Diputación Foral de Bizkaia. Bilbao.

Dickinson, E.C. and J.V. Jr. Remsen. 2013. The Howard and Moore Complete Checklist of the Birds of the World, 4th. edition, vol. 1, Aves Press Limited, Eastbourne.

Dickson, R.C. 1973. A Merlin roost in Wigtownshire. Scottish Birds. 7: 288–292.

Dierschke, J., V. Dierschke, K. Huppop, O. Huppop and K.F. Jachmann. 2011. Die Vogelwelt der Insel Helgoland, OAG Helgoland, Helgoland.

Dinkevich, M.A. 2012. Meetings of birds of prey over the water area of the Azov and Black Seas in the spring time (according to the data of ship observations). pp. 479–485. *In:* Gavrilyuk, M.N. [ed.]. Birds of Prey in the Dynamic Environment of the third Millennium: State and Prospects. Proceedings of the VI International Conference on Falconiformes and Owls of Northern Eurasia. Krivoy Rog, Russia. Publisher(in Russian).

Dinkevich, M.A. and V.E. Lastovetsky. 1997. Birds of prey and owls of the city of Krasnodar and adjacent territories. pp. 45–47. *In:* Topical issues of ecology and nature conservation of the Caucasus ecosystems. Stavropol, Russia. Publisher (in Russian).

Dinnik, N.Y. 1881. Ornithological observations in the vicinity of Stavropol, winter and spring of 1880. Nature and Hunting April: 68–71 (in Russian).

Dixon, A., N. Batbayar and G. Purev-Ochir. 2011. Autumn migration of an Amur Falcon *Falco amurensis* from Mongolia to the Indian Ocean tracked by satellite. Forktail. 27: 81–84.

Dixon, A., A. Sokolov and V. Sokolov. 2012. The subspecies and migration of breeding Peregrines in northern Eurasia. Falco. 39: 4–9.

Dixon, A., G. Purev-Ochir, B. Galtbalt and N. Batbayar. 2013. The use of power lines by breeding raptors and corvids in Mongolia: nest-site characteristics and management using artificial nests. Journal of Raptor Research. 47: 282–291.

Dixon, A., L. Rahman, A. Sokolov and V. Sokolov. 2017. Peregrine falcons crossing the 'Roof of the World'. pp. 128–142. *In*: H. Prins and T. Namgail [eds.]. Bird Migration across the Himalayas: Wetland Functioning amidst Mountains and Glaciers (pp. 128–142), Cambridge, Cambridge University Press.

Dixon, A., R. MaMing, A. Gunga, G. Purev-Ochir and N. Batbayar. 2013. The problem of raptor electrocution in Asia: case studies from Mongolia and China. Bird Conservation International. 23: 520–529.

Dixon, A., D. Ragyov, D. Izquierdo, D. Weeks, D., Md. Lutfor Rahman and I. Klisurov. 2020. Movement and Survival of Captive-Bred Saker Falcons *Falco cherrug* Released by Wild Hacking: Implications for Reintroduction Management. Acta Ornithologica. 54: 157–170.

Dobler, G., R. Schneider and A. Schweis. 1991. Die Invasion des Rauhfugbussards (*Buteo lagopus*) in Baden-Wiirttemberg im Winter 1986/87. Vogelwarte. 36: 1–18.

Dobrev, V., Z. Boev, V. Arkumarev, D. Dobrev, E. Kret, V. Saravia, A. Bounas, D. Vavylis, S.C. Nikolov and S. Oppel. 2016. Diet is not related to productivity but to territory occupancy in a declining population of Egyptian Vultures *Neophron percnopterus*. Bird Conservation International. 26: 273–285.

Domashevsky, S.V. 2005. Towards ecology of the Greater and Lesser Spotted Eagles in the north of Ukraine. Berkut. 14: 180–188.

Domashevsky, S.V. 2006. Materials on ecology of small falcons in the north of Ukraine. Strepet. 4: 42–58 (in Russian).

Domashevsky, S.V. and Yu. A. Demidenko. 2009. Migration of raptors in the vicinity of Kyiv in autumn 2008. https://raptors.org.ua/en/2009/04/migration-of-raptors-in-the-vicinity-of-kyiv-in-autumn-2008/ Accessed in 31.03.2020.

Dombrovski, V. 2010. The diet of the greater spotted eagle (*Aquila clanga*) in Belarusian Polesie. Slovak Raptor Journal. 4: 23–36.

Dombrovski, V., U. Sellis and A. Godino. 2011. The programs of colour ringing and wing-tagging of the Greater Spotted Eagle (*Aquila clanga*) in Belarus. pp. 15–16. *In:* Abstracts of the International Conference "The Birds of prey and owls of Caucasus", Tbilisi.

Dombrovski, V., Ü Väli, U. Sellisa and V.A. Fenchuk. 2018. Migration and Wintering of Greater Spotted Eagles of Belarus in 2017–2018: the first results of GSM-GPS tracking, pp. 144–149. *In:* Proceedings of the Conference "Actual problems of protection of fauna in Belarus and neighboring regions", Minsk.

Dombrovski, V.C. 2003. Use of breeding area by immature Spotted Eagles *Aquila pomarina* and *Aquila clanga* in Belarus: different strategies. pp. 575–580. *In:* Chancellor, R.D. and B.-U. Meyburg [eds.]. 2003. Raptors Worldwide. WWGBP, Budapest.

Dombrovski, V.C. 2005. Hybridation entre Aigles criard Aquila clanga et pomarin A. pomarina en Biélorussie: conséquence taxonomique. Nos Oiseaux. 52: 27–30.

Dombrovski, V.C. 2006. Morphological characteristics and diagnostic features of the Greater Spotted (*Aquila clanga*), Lesser Spotted (*A. pomarina*) eagles, and their hybrids. Ornithologia. 33: 42–52.

Dombrovski, V.C. 2013. Results of the monitoring of eagle numbers in Belarus. Raptors Conservation. 27: 92–101.

Dombrovski, V.C. 2016. Hybridization of lesser and greater spotted eagles as a consequence of anthropogenic transformation of their natural habitats. pp. 246–251. *In:* Belik, V.P. [ed.]. Birds of Prey in the Northern Eurasia: Problem and Adaptations in Current Environment. Proceedings of the VII International Conference on Birds of Prey and Owls of Northern Eurasia, Sochi, Sochi.

Dombrovski, V.C. 2019. Timing, Diet and Parental Care in a Spotted Eagle Nest in Chernobyl Exclusion Zone (Belarus) in 2018 as Revealed by a Camera Trap. Raptors Conservation. 38: 203–212.

Dombrovski, V.C. and V.V. Ivanovski. 2005. New data on numbers and distribution of birds of prey breeding in Belarus. Acta Zoologica Lituanica. 15: 218–227.

Dombrovski, V.C. and P.S. Panchenko. 2009. On the question about breeding of the Greater Spotted Eagle in South-West Ukraine. Berkut. 18: 41–44.

Donázar, J.A. 1993. Los Buitres Ibéricos. Biología y Conservación. J.M. Reyero Editor, Madrid.

Donazar, J.A. and A. Margalida. 2009. European vultures on a knife-edge: the consequences of changes in animal husbandry and sanitary legislation. Munibe. 29: 526–551.

Donázar, J.A. and O. Ceballos. 1989. Growth rates of nestling Egyptian vultures. Ardea. 77: 2.

Donázar, J.A., A. Cortés-Avizanda and M. Carrete. 2010. Dietary shifts in two vultures after the demise of supplementary feeding stations: consequences of the EU sanitary legislation. European Journal of Wildlife Research. 56: 613–621.

Donazar, J.A., A. Margalida, M. Carrete and J.A. Sánchez-Zapata. 2009. Too sanitary for vultures. Science. 326: 664.

Donázar, J.A., C.J. Palacios, L. Gangoso, O. Ceballos, M.J. González and F. Hiraldo. 2002. Conservation status and limiting factors in the endangered population of Egyptian vulture (*Neophron percnopterus*) in the Canary Islands. Biological Conservation. 107: 89–97.

Drapolicz. 2015. Sprawozdanie z realizacji zadania „Nowe aspekty Mierzei Wiślanej jako jednego z najważniejszych korytarzy ekologicznych w kraju" realizowanego w ramach Umowy nr WFOŚ /D/756/70/2015. Report.

Dravecký, M., U. Sellis, U. Bergmanis, V. Dombrovski, J. Lontkowski, G. Maciorowski, B. Maderič, B.-U. Meyburg, T. Mizera, M. Stój, R. Treinys and J. Wójciak. 2008. Colour ringing of the Spotted Eagles (*Aquila pomarina*, *Aquila clanga* and their hybrids) in Europe – a review. Slovak Raptor Journal. 2: 37–52.

Drost, W. A. 1938. Über den Einfluss von Verfrachtungen zur Herbstzugzeit auf den Sperber *Accipiter nisus* (L.). Proceedings of the International Ornithological Congress. 9: 502–509.

Dunn, P. 2006. Breeding Rate and Reproductive Performance. pp. 69–88. *In*: Møller, A.P., W. Fiedler P. Berthold [eds.]. Bird and Climate Change, Academic Press Elsevier.

Dupuy, A. 1969. Catalogue ornithologique du Sahara Algerien. Oiseau et R.F.O. 39: 140–160.

Duriez, O., G. Peron, D. Gremillet, A. Sforzi and F. Monti. 2018. Migrating ospreys use thermal uplift over the open sea. Biology Letters. 14: 20180687.

Duriez, O., A. Kato, C. Tromp, G. Dell'Omo, A.L. Vyssotski, F., Sarrazin, F. and Y. Ropert-Coudert. 2014. How cheap is soaring flight in raptors? A preliminary investigation in freely-flying vultures. PLoS One. 9: e84887.

Dzhamirzoev, G.S. and S.A. Bukreev. 2009. Steppe Eagle—*Aquila rapax* [*sic*] (Temminck, 1828). pp. 434–435. *In:* Abdurakhmanov, G.M. [ed.]. Red Book of the Republic of Dagestan. Makhachkala, Dagestan (in Russian).

Eagle Conservation Committee. 2014. Greater Spotted Eagles – satellite tracking. http://en.orlikgrubodzioby.org.pl/artykul/satellite-tracking.

Eddowes, M.J. 2011. Longer term trends in arrival timing of long distance migrants: the influence of abundance and population change. Ringing & Migration. 26: 56–63.

Eisaguirre, J.M., T.L. Booms, C.P. Barger, C.L. McIntyre, S.B. Lewis and G.A. Breed. 2018. Local meteorological conditions reroute a migration. Proceedings of the Royal Society B. 285: 20181779.

El Khamlichi, R. 2017. Censo prenupcial 2017 de buitres y otras rapaces en migración por el área sur del estrecho de Gibraltar, Jbel Moussa-Marruecos. Go South. 14: 219–229.

Elaev, E.N. 1997. On the phenology of bird migration in the Barguzinsky depression (north-eastern Baikal region). pp. 82–88. *In:* Dorzhiev, C.Z. [ed.]. Status and Problems of the Protection of Natural Complexes of the North-Eastern Baikal region. Proceedings of the Dzherginski Reserve, vol. 2, Buryat State University, Ulan-Ude, Russia (in Russian).

Elaev, E.N. 2008. Spotted Eagle in the south of eastern Siberia. pp. 96–98. *In*: Research and Conservation of the Greater and the Lesser Spotted Eagle in Northern Eurasia. Ivanovskii Gosudarstvennyi Universitet, Ivanovo.

Elgood, J.H., R.E. Sharland and P. Ward. 1964. Palearctic migrants in Nigeria. Ibis. 108: 98.

Ellis, D.H., S.L. Moon and J.W. Robinson. 2001. Annual movements of a Steppe Eagle (*Aquila nipalensis*) summering in Mongolia and wintering in Tibet. Bombay Natural History Society. 98: 335–340.

Elorriaga, J., I. Zuberogoitia, I. Castillo, A. Azkona, S. Hidalgo, L. Astorkia, F. Ruiz-Moneo and A. Iraeta. 2009. First documented case of long-distance dispersal in the Egyptian Vulture (*Neophron percnopterus*). Journal of Raptor Research. 43: 142–145.

Elósegui, J. 1989. Vautour fauve (*Gyps fulvus*), Gypaete barbu (*Gypaetus barbatus*), Percnoptère d'Egypte (*Neophron percnopterus*): Synthèse Bibliographique et Recherches. Acta Biologica Montana 3. Centre de Biologie des Ecosystèmes d'Altitude, Pau.

Elosegui, J. and R. Elosegui. 1977. Desplazamientos de buitres comunes pirenaicos. Munibe. 29: 97–104.

Emtyl, M.Kh. and Y.V. Lokhman. 1994. To the spring ornithofauna of the Kalininsk swampy lowland. pp. 170–0177. *In*: Topical Issues of Ecology and Nature Conservation of Steppe Ecosystems and Adjacent Territories. Part 1. Publisher, Krasnodar, Russia (in Russian).

Eng, T.C., L.K. Chye and L.K. Hong. 2014. Bedong, Kedah–amajor raptor migration site in peninsular Malaysia. Journal of Wildlife and Parks. 28: 51–62.

Ermolaev, A.I. 2016. Common kestrel (*Falco tinnunculus*, Falconiformes, Falconidae) in colonial settlements of rook (*Corvus frugilegus*, Passeriformes, Corvidae) in steppe ecosystems of the Manych valley. Biology Bulletin. 43: 870–875.

Espín, S., J. Terraube, B. Arroyo, P. Camarero, R. Mateo, R. Limiñana, X. Vázquez Pumariño, A. Pinilla, J.T. García and F. Mougeot. 2018. Blood concentrations of *p,p'*-DDE and PCBs in harriers breeding in Spain and Kazakhstan. Science of the Total Environment. 624: 1287–1297.

Esteller, V. and M. Tirado. 2011. Migración postnupcial de rapaces por el PN del Desierto de las Palmas (Castellón). 2003–2010, 85–100. Actes del 1r Congrés sobre Fauna Castellonenca.

Etheridge, B. and R.W. Summers. 2006. Movements of British hen harriers *Circus cyaneus* outside the breeding season. Ringing & Migration. 23: 6–14.

European colour-ring Birding. 2020. Common Buzzard. https://www.cr-birding.org/taxonomy/term/2243. Last accessed in 31.03.2020.

Evangelidis, A. 2017. Technical Report (Action D1): Monitoring of Egyptian Vulture migration on Antikythira Island. Annual Report 2017, under action D1 of the LIFE project "Egyptian Vulture New LIFE" (LIFE16 NAT/BG/000874), Hellenic Ornithological Society, Athens, Greece.

Evans, P.R. and G.W. Lathbury. 1973. Raptor migration across the Straits of Gibraltar. Ibis. 115: 572–585.

Falsterbo Fågelstation. 2020. https://www.falsterbofagelstation.se/index_e.html. Accessed 13/04/2020.

Faveyts, W., M. Valkenburg and B. Granit. 2011. Crested Honey Buzzard: identification, western occurrence and hybridisation with European Honey Buzzard. Dutch Birding. 33: 149–162.

Fefelov, I.V., M.N. Alekseenko and V.Y. Malysheva. 2004. The abundance and behavior of *Falconiformes* during the autumn migration on South Baikal. Bulletin of the Buryat University. 2: 61–85 (in Russian).

Fenech, N. 1992. Fatal flight: the Maltese obsession with killing birds, Quiller Press, London.

Ferguson-Lees, J. and D. Christie. 2001. Raptors of the World, Helm, London.

Filippi-Codaccioni, O., J. Moussus, J. Urcun and F. Jiguet. 2010. Advanced departure dates in long-distance migratory raptors. Journal of Ornithology. 151: 687–694.

Finch, T., S.J. Butler, A.M.A. Franco and W. Cresswell. 2017. Low migratory connectivity is common in long-distance migrant birds. Journal of Animal Ecology. 86: 662–673.

Finlayson, C. 1992. Birds of the Strait of Gibraltar, T. & A.D. Poyser, London.

Fisher, P.R., S.F. Newton, M.A. Tatwany and C.R. Goldspink. 2001. The status and breeding biology of the Osprey *Pandion haliaetus* in the Middle East. Vogelwelt. 122: 191–204.

Forsman, D. 1994. Field identification of Crested Honey Buzzard. Birding World. 7: 396–403.

Forsman, D. 1999. The Raptors of Europe and the Middle East: A handbook of field identification, T. & A.D. Poyser, London.

Forsman, D. 2016. Flight identification of raptors of Europe, North Africa and the Middle East, Christopher Helm, London.

Franco, A.M.A., J.T Marquesand and W.J. Sutherland. 2005. Is nest-site availability limiting lesser kestrel populations? A multiple scale approach. Ibis. 147: 657–666.

Fransson, T. and J. Pettersson. 2001. Svensk ringmärkningsatlas. Naturhistoriska riksmusset and Sveriges Ornitologiska Förening.

Fraser, P.A. and M.J. Rogers. 2002. Report on scarce migrant birds in Britain in 2000. British Birds. 95: 606–630.

Friedemann, G., Y. Yom-Tov, U. Motro and Y. Leshem. 2011. Shift in nesting ground of the long-legged buzzard (*Buteo rufinus*) in Judea, Israel—an effect of habitat change. Biological Conservation. 144: 402–406.

Friedemann, G., Y. Leshem, L. Kerem, A. Bar-Massada and I. Izhaki. 2017. Nest-site characteristics, breeding success and competitive interactions between two recently sympatric apex predators. Ibis. 159: 812–827.

Friedemann, G., Y. Leshem, K.M. McClain, G. Bohrer, A. Bar-Massada and I. Izhaki. 2020. Poleward non-breeding migration of a breeding population: Challenging the traditional perspective of avian migration. Journal of Avian Biology. 51: e02362.

Friedemann, G., Y. Leshem, L. Kerem, B. Shacham, A. Bar-Massada, K.M. McClain, G. Bohrer and I. Izhaki. 2016. Multidimensional differentiation in foraging resource use during breeding of two sympatric top predators. Scientific Reports. 6: 35031.

Frost, R. 1994. Observations of autumn raptor migration at Akrotiri salt lake, Cyprus 19 September–9 October 1992. Royal Air Force Ornithological Society Journal. 23: 65–73.

Frumkin, R., B. Pinshow and S. Kleinhaus. 1995. A review of bird migration over Israel. Journal of Ornithology. 136: 127–147.

Fulco, E., J. Angelini, G. Ceccolini, L. De Lisio, D. De Rosa, A. De Sanctis, M. Giannotti, G. Giglio, M. Grussu, A. Minganti, M. Panella, M. Sarà, A. Sigismondi, S. Urso and M. Visceglia. 2017. Il Nibbio reale *Milvus milvus* svernante in Italia, sintesi di cinque anni di monitoraggio. Alula. 24: 53–61.

Fülöp, A., I. Kovács, E. Baltag, S.J. Daróczi, A.S. Dehelean, L.A. Dehelean, R.B. Kis, I.S. Komáromi, H. Latková, T. Miholcsa, A. Nagy, S.Z. Ölvedi, T. Papp, A.K. Sándor, R. Zeitz and M.A. Kelemen. 2014. Autumn migration of soaring birds at Bosporus: validating a new survey station design. Bird Study. 61: 264–269.

Fülöp, A., S.J. Daróczi, A.S. Dehelean, L.A. Dehelean, Z. Domahidi, A. Dósa, G. Gyékény, Z. Hegyeli, R.B. Kis, I.S. Komáromi, I. Kovács, T. Miholcsa, A.A. Nagy, A. Nagy, S.Z. Ölvedi, T. Papp, L.G. Pârâu, A.K. Sándor, T. Sos and R. Zeitz. 2018. Autumn passage of soaring birds over Dobrogea (Romania): a migration corridor in Southeast Europe. Ardea. 106: 61–77.

Galarza, A. and I. Zuberogoitia. 2012. Osprey restoration project in the Urdaibai Biosphere Reserve (Basque Country). Urdaibai Bird Center/County Council of Biscay. Bilbao.

Galarza, A., M. del Arco, J. Elorriaga, E. Unamuno, J. Arizaga and I. Zuberogoitia. 2017. First Evidence Provided by Satellite Telemetry of Nocturnal Flight Overland by an Osprey (*Pandion haliaetus*). Journal of Raptor Research. 51: 184–186.

Galea, C. and B. Massa. Notes on the raptor migration across the central Mediterranean. pp. 257–261. *In:* Newton, I. and R.D. Chancellor [eds.]. 1985. Conservation Studies on Raptors. ICBP Technical Publication No. 5. International Council for Bird Preservation, Cambridge.

Galligan, T.H., T. Amano, V.M. Pranash, M. Kulkarni, R. Shringarpure, N. Prakash, S. Ranade, R.E. Green and R.J. Cuthbert. 2014. Have population declines in Egyptian Vulture and Red-headed Vulture in India slowed since the 2006 ban on veterinary diclofenac? Bird Conservation International. 24: 272–281.

Galushin, V.M. 1962. The Greater Spotted Eagle in the valley of Oka river and its impact on the density of some birds. Uchenye Zapiski Moskovskogo Gosudorstvennogo Pedagogicheskogo Instituta. 186: 115–151.

Galushin, V.M. 2006. Numbers and population trends in birds of prey nesting in Russia. Birds of Prey and Owls in Zoos and Breeding Stations. 15: 22–30 (in Russian).

Gamauf, A. and R. Dosedel. 2012. Satellite telemetry of Saker Falcons (*Falco cherrug*) in Austria: juvenile dispersal at the westernmost distribution limit of the species. Aquila. 119: 111–134.

Ganesh, T., P. Kanniah, H. Harrier and A. Pradesh. 2000. Roost counts of harriers *Circus* spanning seven winters in Andhra Pradesh, India. Bird Study. 16: 1–3.

Gangoso, L., D.S. Viana, A.M. Dokter, J. Shamoun-Baranes, J. Figuerola, S.A. Barbosa and W. Bouten. 2020. Cascading effects of climate variability on the breeding success of an edge population of an apex predator. Journal of Animal Ecology. 89: 2631–2643.

Ganusevich, S.A., T.L. Maechtle, W.S. Seegar, M.A. Yates, M.J. McGrady, M. Fuller, L. Schuek, J. Dayton and C.J. Henny. 2004. Autumn migration and wintering areas of Peregrine Falcons *Falco peregrinus* nesting on the Kola Peninsula, northern Russia. Ibis. 146: 291–297.

Garcia, E.F. and K.J. Bensusan. 2006. Northbound migrant raptors in June and July at the Strait of Gibraltar. British Birds. 99: 569–575.

García, J.T. and B.E. Arroyo. 1998. Migratory movements of western European Montagu's Harrier *Circus pygargus*: a review. Bird Study. 45: 188–194.

García, J.T., J. Viñuela and C. Sunyer. 1998. Geographic variation of the winter diet of the Red Kite *Milvus milvus* in the Iberian Peninsula. Ibis. 140: 302–309.

García-Dios, I.S. 2016. Aguililla calzada – *Hieraaetus pennatus*. *In:* Salvador, A. and M.B. Morales [eds.]. Enciclopedia Virtual de los Vertebrados Españoles. Museo Nacional de Ciencias Naturales, Madrid.

García-Dios, I.S. 2017. El águila calzada. Tundra ediciones.

García-Ripollés, C., P. López-López and V. Urios. 2010. First description of migration and wintering of adult Egyptian Vultures *Neophron percnopterus* tracked by GPS satellite telemetry. Bird Study. 57: 261–265.

García-Ripollés, C., P. López-López and V. Urios. 2011. Ranging behaviour of non-breeding Eurasian griffon vultures *Gyps fulvus*: a GPS-telemetry study. Acta Ornithologica. 46: 127–134.

Garrido, J.R. and C.G. Sarasa. 1998. Imortancia del vertadero de Los Barrios (Campo de Gibralpar, Cadiz) en la ruta migratoria occidental del Milano Negro *Milvus migrans* y el Buitre Leonado *Gyps fulvus*. Almoraima. 19: 217–224.

Garrido, J.R., C.G. Sarasa and M. Fernandez-Cruz. 2002. Rubbish dumps as key habitats for migration and wintering in the Griffon Vulture (*Gyps fulvus*) at a migratory bottleneck: implications for conservation. pp. 143–151. *In*: Yosef, R., M.L. Miller and D. Pepler [eds.]. Raptors in the New Millennium. Eilat, Israel: The International Birding and Research Center in Eilat.

Gatter, W. 2000. Vogelzug und Vogelbestande in Mitteleuropa. Aula-Verlag, Wiesbaden.

Gauthreaux, S.A. 1985. Differential migration of raptors: The importance of age and sex. pp. 9–106. *In:* Harwood, M. [ed.]. Proceedings of the Fourth Hawk Migration Conference, Rochester, New York.

Gavrilov, E.I. and A.P. Gistsov. 1985. Seasonal bird flights in the foothills of the Western Tien Shan, Nauka Press, Alma-Ata, Kazakhstan (in Russian).

Gavrilov, E.I., A.E. Gavrilov, A.V. Kovalenko, A.N. Dykhanbaev and K.A. Sarsekova. 2002. Expedition at the Chokpak Station. Kazakhstan Ornithological Bulletin volume: 18–19 (in Russian).

Genero, F. 2009. Estado actual y conservación de las aves carroñeras en Italia. Munibe. 29: 178–199.

Genero, F., M. Franchini, Y. Fanin and S. Filacorda. 2020. Spatial ecology of non-breeding Eurasian Griffon Vultures in relation to natural and artificial food availability. Bird Study. 67: 53–70.

Gensbøl, B. 2008. Collins Birds of Prey, Collins, London.

Germi, F. 2005. Raptor migration in east Bali, Indonesia: observations from a bottleneck watch site. Forktail. 21: 93.

Gessaman, J.A. 1979. Premigratory fat in the American Kestrel. Wilson Bulletin. 91: 625–626.

Gil Velasco, M., M. Rouco, J. Ferrer, M. García-Tarrasón, F.J. García-Vargas, A. Gutierrez, R. Hevia, F. López, D. López-Velasco, A. Ollé, G. Rodríguez, J. Sagardía and J.A. Salazar. 2017. Observación de aves raras en España, 2014. Ardeola. 64: 161–235.

Gil, J.A., J.L. Lagares and M. Alcántara. 2009. Seguimiento radio-telemétrico de Buitre Leonado (*Gyps fulvus*) en el Sistema Ibérico Oriental (Aragón-España). Teruel. 92: 137–164.

Giordano, A. 1991. The migration of Birds of prey and Storks in the Straits of Messina. Birds of prey Bulletin. 4: 239–250.

Giordano, A., D. Ricciardi, G. Candiano, S. Celesti and A. Irrora. 1998. Anti-poaching on the Straits of Messina: results after 15 years of activities. pp. 623–630. *In:* Meyburg, B.-U., R.D. Chancellor and J.J. Ferrero [eds.]. Holartic Birds of Prey. ADENEX-WWGBPO, Berlin, Germany.

Giordano, A., I. Adami, A. Battaglia, S. Cutini, D.E. Ricciardi and R. Vella. 2016. 33° campo internazionale per lo studio e la protezione dei rapaci e delle cicogne in migrazione sui monti Peloritani, Stretto di Messina (ME). Infomigrans. 37: 2.

Giordano, A., D. Ricciardi, G. Chiofalo, J. Ramirez, M. Guerrero, R. Vella, J.P. Fiott, L. Vanni, E. Macumelli and S. Cutini. 2009. The spring migration of the Lesser Kestrel *Falco naumanni* on the Straits of Messina, data from 1991 to 2008. pp. 155–158. *In*: Pilard, P. [ed.]. Actas del VII Congreso Internacional sobre el Cernícalo Primilla. LPO Mission Rapaces. Imprimerie Lagarde, France.

Giraudo, L. and R. Toffoli. 2003. La migrazione postnuziale del Nibbio bruno *Milvus migrans* attraverso le Alpi Marittime. Avocetta. 27: 60–62.

Glantz, M. 1987. Drought and Hunger in Africa, Cambridge University Press.
Global Raptor Information Network. 2019. Species account: Griffon Vulture *Gyps fulvus*. Downloaded from http://www.globalraptors.org on 1 Jan. 2019.
GOB (Grup Balear d'Ornitologia i Defensa de la Naturalesa). 2009. Annex II: Estatus de l'Avifauna Balear. Anuari Ornitolgic de les Balears. 23: 255–265.
GOB (Grup Balear d'Ornitologia i Defensa de la Naturalesa). 2011. Annex II: Estatus de l'Avifauna Balear. Anuari Ornitolgic de les Balears. 26: 269–280.
Golovushkin, M.I. 1992. Phenology of the spring arrival of birds in the vicinity of Kiev. pp. 242–249. *In*: Seasonal Migration of Birds on the Territory of Ukraine. Kiev, Ukraine, NaukovaDumka (in Russian).
Gombobaatar, S. and E.M. Monks [eds.]. 2011. Regional Red List Series Vol 7. Birds. Zoological Society of London, National University of Mongolia and Mongolian Ornithological Society, Ulaanbaatar, Mongolia.
Gombobaatar, S., R. Yosef, B. Odkhuu and D. Sumiya. 2012. Breeding ecology of the Steppe Eagle (*Aquila nipalensis*) in Mongolia. Ornis Mongolica. 1: 13–19.
Gonzàlez, J.L. 1991. El Aguilucho lagunero *Circus aeruginosus* (L., 1758) en Espana. Situacion, biologia de la reproduccion, alimentacion y conservacion. ICONA, Madrid, Spain.
González, L.M., J. Oria, A. Margalida, R. Sánchez Mateos, L. Prada, J. Caldera, A. Aranda and J.I. Molina. 2006. Effective natal dispersal and age of maturity in the threatened Spanish Imperial Eagle Aquila adalberti: conservation implications. Bird Study. 53: 285–293.
Goodwin, D. 1949. Notes on the migration of birds of prey over Suez. Ibis. 91: 59–63.
Gorman, G. 1998. The status of Levant Sparrowhawk in Europe. Alula. 4: 144–147.
Gorney, E. and Y. Yom-Tov. 1994. Fat, hydration condition, and moult of Steppe Buzzards *Buteo buteo vulpinus* on spring migration. Ibis. 136: 185–192.
Gorney, E., W.S. Clark and Y. Yom-Tov. 1999. A test of the condition-bias hypothesis yields different results for two species of Sparrowhawks (*Accipiter*). Wilson Bulletin. 111: 181–187.
Goroshko, O.A. 2012. Steppe Eagle *Aquila rapax* [*sic*] (Temminck, 1828). pp. 99–100. *In*: Red Book of the Trans-Baikal Territory. Animals. Novosibirsk, Russia.
Gradev, G., S. Marin and P. Zhelev. 2016a. Satellite tracking of Lesser Kestrel (*Falco naumanni*) from a recovered breeding population in Bulgaria. International Lesser Kestrel Expert Workshop, Plovdiv, Bulgaria, 4-8.10.2016. http://greenbalkans.org/birdsofprey/lesserkestrellife/en/International_Expert_Workshop_-c150.
Gradev, G., S. Marin, P. Zhelev and J. Antolín. 2016b. Recovering the Lesser kestrel (*Falco naumanni*) as a breeder in Bulgaria. pp. 136–144. *In*: Proceedings First National Conference of Reintroduction of Conservation-reliant Species, Sofia 2015. University Press, Sofia.
Grande, J.M., D. Serrano, G. Tavecchia, M. Carrete, O. Ceballos, R. Díaz-Delgado, J.L. Tella and J.A. Donázar. 2009. Survival in a long-lived territorial migrant: effects of life-history traits and ecological conditions in wintering and breeding areas. Oikos. 118: 580–590.
Grattini, N. 2008. Distribuzione, consistenza ed espansione territoriale del Gheppio *Falco tinnunculus* nidificante in provincia di Mantova. Alula. 15: 189–194.
Grattini, N., D. Longhi and F. Novelli. 2006. Indagine su alcuni roost invernali di smeriglio in provincia di Mantova. Avocetta 30: 73–75.
Griesinger, J. 1998. Juvenile dispersion and migration among Griffon Vultures *Gyps fulvus* in Spain. pp. 613–621. *In*: Chancellor, R.D., B.-U. Meyburg and J.J. Ferrero [eds.]. Holarctic Birds of Prey. ADENEX-WWGBP.
Grimmet, R., C. Inskipp and T. Inskipp. 1998. Birds of the Indian Subcontinent, Christopher Helm, London.
Grisesinger, J. 1996. Autumn migration of Griffon Vultures in Spain. *In:* Muntaner, J. and J. Mayol [eds.]. Biologia y Conservación de las Rapaces Mediterráneas. Monografía 4. SEO, Madrid.
Grote, H. 1930. Wanderungen und Winterquartiere der paläarktischen Zugvögel in Afrika. Mitteilungen aus dem Museum für Naturkunde in Berlin. 16: 1–116.
Grussu, M. and R. Paddeu. 2015. Russian-ringed Tundra Peregrine Falcon in Sardinia, Italy, in November 2011 and winter movements. Dutch Birding. 37: 80–85.
Gschweng, M. 2010. http://Sootyfalconoman.blogspot.co.at/ [Accessed 8 July 2020].
Gschweng, M., E.K.V. Kalko, U. Querner, W. Fiedler and P. Berthold. 2008. All across Africa: highly individual migration routes of Eleonora's falcon. Proceedings of the Royal Society B. 275: 2887–2896.
Gurung, S., H.J. Smit, R. DeCandido and T.R. Subedi. 2014. Raptor Migration Summary. Nepal – Autumn 2013. Technical Report.
Gurung, S., S. Baniya and T.R. Subedi. 2019. East-West Migration of Endangered Steppe Eagle and other raptors in Thoolakharka watch site, Nepal: Migration count summary- autumn 2019 Technical Report 1/2019, Himalayan Nature and Nepalese Ornithological Union, Kathmandu.
Gurung, S., H.J. Smit, V. Zembal, R. DeCandido and T.R. Subedi. 2013. Raptor Migration Summary. Nepal – Autumn 2012. Technical Report.
Gustin, M. 2005. Progetto Rapaci Migratori. Infomigrans. 15: 10–11.
Gustin, M. 2006. Progetto Rapaci Migratori 2006 Primavera. Infomigrans. 16: 3–4.
Gustin, M. 2007. Progetto Rapaci Migratori Primavera 2007. Infomigrans. 19: 5–6.
Gustin, M. 2008. Progetto Rapaci Migratori LIPU. Anno 2008. Infomigrans. 21: 4.
Gustin, M. 2009. Progetto Rapaci Migratori LIPU. Anno 2008. Infomigrans. 23: 6.

Gustin, M. 2012. Progetto rapaci migratori LIPU primavera 2012. Infomigrans. 29: 4–5.

Gustin M. and A. Sorace. 2004. Is the Conero promontory, central Italy, an important bridge for migrant raptors entering eastern Europe in spring? British Birds. 97: 403–406.

Gustin, M. and T. Pizzari. 1998. Migratory pattern in the genus *Circus*: sex and age differential migration in Italy. Ornis Svecica. 8: 23–26.

Gustin, M., M. Brambilla and C. Celada. 2019. Conoscerli, proteggerli. Guida allo stato di Conservazione degli uccelli in Italia, Lipu/Birdlife, Parma, Italy.

Gustin, M., A. Ferrarini, G. Giglio, S.C. Pellegrino and A. Frassanito. 2014. First evidence of widespread nocturnal activity of Lesser Kestrel (*Falco naumanni*) in Southern Italy. Ornis Fennica. 91: 256–260.

Gustin, M., A. Corso, M. Sorace, A. Giordano, C. Cardelli and D. Ricciardi. 2001. Confronto tra lo stretto di Messina ed il promontorio del Conero delle osservazioni di Lodolaio, *Falco subbuteo*, durante la migrazione primaverile 2000. Avocetta. 25: 56.

Gutiérrez, R., J.A. Lorenzo, G. Gorospe, P. Gutiérrez, D. López-Velasco, J. Martí-Aledo, S. Sales and C. Vidal. 2011. Observación de aves raras en España, 2010. Ardeola. 59: 353–411.

Gutiérrez, R., J.A. Lorenzo, J. Elorriaga, G. Gorospe, D. López-Velasco, J. Martí-Aledo, G. Rodríguez and S. Sales. 2013. Observaciones de aves raras en España 2011. Ardeola. 60: 437–506.

Hadjikyriakou, T.G. and A.N.G. Kirschel. 2017. Migration of the easternmost population of Eleonora's falcon *Falco eleonorae*. Poster presentation. BOU 2017 Annual Conference University of Warwick, UK, 28–30 March 2017.

Hadjikyriakou, T.G., E.C. Nwankwo, M.Z. Virani and A.N.G. Kirschel. 2020. Habitat availability influences migration speed, refueling patterns and seasonal flyways of a fly-and forage migrant. Movement Ecology. 8: 10.

Hagemeijer, W.J. and M.J. Blair. 1997. The EBCC atlas of European breeding birds, Poyser, London, 479.

Hagen, Y. 1969. Norwegian studies on the reproduction of birds of prey and owls in relation to micro-rodent population fluctuations. Fauna. 22: 73–126.

Hake, M., N. Kjellén and T. Alerstam. 2001. Satellite tracking of Swedish Ospreys *Pandion haliaetus*: Autumn migration routes and orientation. Journal of Avian Biology. 32: 47–56.

Hake, M., N. Kjellén and T. Alerstam. 2003. Age dependent migration strategy in honey buzzards *Pernis apivorus* tracked by satellite. Oikos. 103: 385–396.

Handrinos, G. and T. Akriotis. 1997. The birds of Greece, Helm, London, U.K.

Hardey, J., C. Humphrey, C. Wernham, H. Riley, B. Etheridge and D. Thompson. 2013. Raptors, a field guide for surveys and monitoring, TSO, Edinburgh, U.K

Harel, R., N. Horvitz and R. Nathan. 2016b. Adult vultures outperform juveniles in challenging thermal soaring conditions. Scientific Reports. 6: 27865.

Harel, R., O. Spiegel, W.M. Getz and R. Nathan. 2017. Social foraging and individual consistency in following behaviour: testing the information Centre hypothesis in free-ranging vultures. Proceedings of the Royal Society B-Biol. Sci. 284: 20162654.

Harel, R., O. Duriez, O. Spiegel, J. Fluhr, N. Horvitz, W.M. Getz, W. Bouten, F. Sarrazin, O. Hatzofe and R. Nathan. 2016a. Decision-making by a soaring bird: time, energy and risk considerations at different spatio-temporal scales. Philosophical Transactions Royal Society B. 371: 20150397.

Heavisides, A. 1987. British and Irish Merlin recoveries. 1911–1984. Ringing & Migration. 8: 29–41.

Hedenström, A. 1993. Migration by soaring or flapping flight in birds: The relative importance of energy cost and speed. Philosophical Transactions Royal Society B. 342: 353–361.

Hein, C. and M. Kisling. 1991. Der Frühjahrszug von Greifvögeln und Störchen auf der mittleren Zugschiene. Naturschutzjugend und Naturschtzbund Deutschland, Darmstadt, Germany.

Heiss, M. 2013. The importance of Besh Barmag bottleneck (Azerbaijan) for Eurasian migrant birds. Acta Ornithologica. 48: 151–164.

Heiss, M. and K. Gauger. 2011. Coastal bird migration at the Caspian shore of the Azerbaijan Republic in October 2007. Podoces. 6: 59–71.

Helbig, A.J. and D.R. Wells. 1990. Steppe Eagles in peninsular Malaysia in 1987. Dutch Bird. 12: 77–79.

Helbig, A.J. and V. Laske. 1989. Broadfront Raptor Migration in Interior NW Germany. pp. 109–114. *In:* Meyburg, B.-U. and R.D. Chancellor [eds.]. Raptors in the Modern World. WWGBP, Berlin, London and Paris.

Henry, M. 2018. The changing status of the Pallid harrier in western Europe. British Birds. 111: 744–760.

Heredia, B. 1996. International Action Plan for the Imperial Eagle (*Aquila heliaca*), BirdLife International/MME, Budapest, Hungary.

Herremans, M. 2001. Age-related differences in abundance and habitat occupation of Steppe Buzzards, *Buteo buteo vulpinus*, in the Western Cape Province, South Africa. Ostrich. 72: 105–108.

Hidalgo, S., J. Zabala, I. Zuberogoitia, A. Azkona and I. Castillo. 2005. Food of the Egyptian vulture (*Neophron percnopterus*) in Biscay. Buteo. 14: 23–29.

Higuchi, H. 2010. Satellite tracking the migration of birds in eastern Asia. British Birds. 103: 284.

Higuchi, H., H.-J. Shiu, H. Nakamura, A. Uematsu, K. Kuno, M. Saeki, M. Hotta, K.-I. Tokita, E. Moriya, E. Morishita and M. Tamura. 2005. Migration of Honey-buzzards *Pernis apivorus* based on satellite tracking. Ornithological Science. 4: 109–115.

Hilgerloh, G. 2009. The desert at Zait Bay, Egypt: a bird migration bottleneck of global importance. Bird Conservation International. 19: 338–352.

Hilgerloh, G., A. Michalik and B. Raddatz. 2011a. Autumn migration of soaring birds through the Gebel El Zeit Important Bird Area (IBA), Egypt, threatened by wind farm projects. Bird Conservation International. 21: 365–375.

Hilgerloh, G., G. Pegram and A. Schreiber. 2011b. The influence of wind conditions and topography on soaring migrants on the western side of the southern gulf of Suez, Egypt. Sandgrouse. 33: 139–148.

Hilgerloh, G., J. Weinbecker and I. Weiss. 2009. The timing of spring passage of soaring birds at Zait bay, Egypt. Sandgrouse. 31: 26.

Holte, D., U. Koppen and A. Schmitz-Ornés. 2016. Partial migration in a central European raptor species: an analysis of ring re-encounter data of common kestrels *Falco tinnunculus*. Acta Ornithologica. 51: 39–54.

Holte, D., U. Köppen and A. Schmitz-Ornés. 2017. A comparison of migratory strategies of partial migratory raptors from Germany. Journal of Ornithology. 158: 579–592.

Horton, T.W., R.O. Bierregaard, P. Zawar-Reza, R.N. Holdaway and P. Sagar. 2014. Juvenile osprey navigation during trans-oceanic migration. PLoS One. 9: e114557.

Horváth, M. and A. Kovács. 2009. Imperial eagle (*Aquila heliaca* Savigny, 1809). pp. 231–234. *In:* Csörgő, T., Z. Karcza, G. Halmos, G. Magyar, J. Gyurácz, T. Szép, A. Bankovics, A. Schmidt and E. Schmidt [eds.]. Hungarian Bird Migration Atlas. Kossuth Kiadó, Budapest, Hungary (in Hungarian with English summary).

Horváth, M., T. Szitta, J. Bagyura, I. Fatér, G. Firmánszky and C. Moskát. 2014. Simultaneous effect of habitat and age on reproductive success of Imperial Eagles (*Aquila heliaca*) in Hungary. Ornis Hungarica. 22: 57–68.

Horváth, M., B. Solti, I. Fatér, T. Juhász, L. Haraszthy, T. Szitta, Z. Ballók and S. Pásztory-Kovács. 2018b. Temporal changes in the diet composition of the Eastern Imperial Eagle (*Aquila heliaca*) in Hungary. Ornis Hungarica. 26: 1–26.

Horváth, M., T. Juhász, I. Fatér, M. Schmidt, M. Dvorak, B. Wendelin, G. Wichmann, A. Gamauf, M. McGrady, R. Raab, P. Spakovszky, S. Stoychev, D. Demerdzhiev, S. Spasov, D. Dobrev, B.-U. Meyburg, D. Horal, I. Literak, J. Chavko, T. Veselovský, Z. Guziová, N. Javakhishvili, B. Tatar, C. Özcan, E. Lisichanets, H. Zsolt and M. Prommer. 2018a. Juvenile dispersal movements of eastern imperial eagles in the resident populations. pp. 132–134. *In:* Abstracts of the II International Scientific and Practical Conference "Eagles of Palearctic: Study and Conservation". Raptors Conservation Suppl. 1. Sibecocenter, Novosibirsk, Russia.

Horvitz, N., N. Sapir, F. Liechti, R. Avissar, I. Mahrer and R. Nathan. 2014. The gliding speed of migrating birds: slow and safe or fast and risky? Ecology Letters. 17: 670–679.

Howes, C., C.T. Symes and P. Byholm. 2019. Highly skewed sex and age ratios of European Honey Buzzard in southern Africa. Journal of Ornithology. 161: 539–548.

Iankov, P., D. Georgiev and B. Ivanov. 2017. Large scale migration of Red-footed Falcon *Falco vespertinus* over Kaliakra (NE Bulgaria). Acrocephalus. 28: 37–38.

Iankov, P., I. Mateeva and H. Sierdsema. 2019. Migration of the Lesser Spotted Eagle *Clanga pomarina* Brehm, 1831 (*Aves: Accipitridae*) in Bulgaria. Acta Zoologica Bulgarica. 14: 91–103.

Ilyukh, M.P. and A.N. Khokhlov. 1999. The Imperial Eagle on Stavropol. pp. 87. *In:* Belik, V.P. [ed.]. Royal Eagle: Distribution, State of Populations and Prospects for the Protection of the Imperial Eagle (*Aquila heliaca*) in Russia. Series: Rare Bird Species, V.1. Russian Bird Conservation Union, Moscow, Russia (in Russian).

Ilyukh, M.P. and N.L. Zabolotniy. 1999. Birds of prey and owls of the Kuban. pp. 59–65. *In:* Proceedings of III Conference on birds of prey of Eastern Europe and Northern Asia, Part 2. Kislovodsk, Russia, Publisher (in Russian).

Iñigo, A. and B. Barov. 2010. Action plan for the lesser kestrel *Falco naumanni* in the European Union, SEO-BirdLife and BirdLife International for the European Commission. 55 pp.

Irisov, E.A. and N.L. Irisova. 1984. Rare birds of Altai, Altai Book Publishing House, Barnaul, Russia.

Isakov, Y.A. and K.A. Vorobev. 1940. Review of wintering and migration of birds on the Southern Caspian Sea. pp. 5–160. *In:* Makarov, V.N. [ed.]. Proceedings of the All-Union Ornithological Reserve Gassan-Kuli. Issue. 1. The General Directorate for Nature Reserves of the People's Commissars of the RSFSR, Moscow, Russia (in Russian).

Isenmann, P. and A. Moali. 2005. Persistance de l'Aigle pomarin *Aquila pomarina* en Algérie. Alauda. 73: 461–462.

IUCN. 2016. The IUCN Red List of Threatened Species. Version 2016-3. Available at: www.iucnredlist.org. (Accessed: 07 December 2016).

IUCN. 2020. The IUCN Red List of Threatened Species. Version 2020-1. Available at: https://www.iucnredlist.org.

Ivanov, A.I. 1969. Birds of the Pamiro-Alai, Science Publisher, Leningrad, Russia (in Russian).

Ivanovski, V.V. 2012. Raptorial birds of Belarusian Lakeland. Vitebsk State University, Vitebsk.

Janzen, D.H. and T.W. Schoener. 1968. Differences in insect abundance and diversity between wetter and drier sites during a tropical dry season. Ecology. 49: 96–110.

Jardin, F. 2005. La migrazione autunnale sulla Costa Azzurra: Fort de la Revère, Eze. Infomigrans. 15: 2–3.

Javed, S., D.C. Douglas, S. Khan, J.N. Shah and A.A. Al Hammadi. 2012. First description of autumn migration of Sooty Falcon *Falco concolor* from the United Arab Emirates to Madagascar using satellite telemetry. Bird Conservation International. 22: 106–119.

Javed, S., S. Khan, J. Shah, A. Ahmed, A.A. Hammadi and E.A. Hammadi. 2014. Satellite tracking of a young Steppe Eagle from the United Arab Emirates during two spring and autumn migrations. Ostrich. 85: 131–138.

Jenkins, A.R. and A. Stephenson. 1997. From Arctic tundras to African dunes. Siberian Peregrines in southern Africa. Africa-Birds & Birding. 2: 62–68.

Jetz, W., G.H. Thomas, J.B. Joy, K. Hartmann and A.O. Mooers. 2012. The global diversity of birds in space and time. Nature. 491: 444–448.

Jiguet, F., D. Chevallier, F. Baillon, J. Ventroux and P. Cavallin. 2011. Sub-Saharan staging areas of a first-summer Short-toed Snake Eagle *Circaetus gallicus*. Bird Study. 59: 102–107.

Jonzén, N. and J. Pettersson. 1999. Autumn migration of raptors on Capri. Avocetta. 23: 65–72.

Jowers, M.J., S. Sánchez-Ramírez, S. Lopes, I. Karyakin, V. Dombrovski, A. Qninba, T. Valkenburg, N. Onofre, N. Ferrand, P. Beja, L. Palma and R. Godinho. 2019. Unravelling population processes over the Late Pleistocene driving contemporary genetic divergence in Palearctic Buzzards. Molecular Phylogenetics and Evolution. 134: 269–281.

Jubete, F. and F. Mougeot. 2019. Primera cita de reproducíon confirmada del aguilucho papialbo en España. Quercus. 402: 12–15.

Juhant, M.A. and K.L. Bildstein. 2017. Raptor migration across and around the Himalayas. pp. 98–116. *In:* Prins, H.H.T., H.H.T. and T. Namgail [eds.]. Bird Migration across the Himalayas, Cambridge University Press.

Juhász, T., T. Zalai, T. Hadarics. 2018. Occurence of Greater Spotted Eagle in Hungary between 1992 and 2016. Heliaca. 16: 99–100.

Kapelj, S., I. Budinski, M. Zec, I. Katanović, I. Šarić and I. Šoštarić. 2017. The first results of GPS tracking of the Short-toed Eagle Circaetus gallicus in Croatia. Larus. 52: 65–68.

Karlsson, L. and P.-G. Bentz. 2004. Wings over Falsterbo. Falsterbo Bird Observatory.

Karyakin, I.V. 1998. Raptors of the Ural Region. Birds of Prey (*Falconiformes*), Owls (*Strigiformes*). CPI SOZH Ural/CoES, Perm, Russia (in Russian).

Karyakin, I.V. 1999. On the ecology of the Imperial Eagle in the Ural region. pp. 96–104. *In*: Belik, V.P. [ed.]. Royal Eagle: Distribution, State of Populations and Prospects for the Protection of the Imperial Eagle (*Aquila heliaca*) in Russia. Series: Rare Bird Species, V.1. Russian Bird Conservation Union, Moscow, Russia (in Russian).

Karyakin, I.V. 2008a. Ecology of the greater spotted eagle in Volga-Ural region. pp. 119–137. *In*: Research and Conservation of the Greater and the Lesser Spotted Eagle in Northern Eurasia. Ivanovskii Gosudarstvennyi Universitet, Ivanovo.

Karyakin, I.V. 2008b. Ecology of the Greater Spotted Eagle in Western Siberia. pp. 153–164. *In*: Research and Conservation of the Greater and the Lesser Spotted Eagle in Northern Eurasia.

Karyakin, I.V. 2013. Review of current status of Steppe Eagle in the World and in the Russia. Raptors Conservation. 26: 22–43.

Karyakin, I.V. 2015. The Steppe Eagle global conservation status was raised. Raptors Conservation. 30: 21–30.

Karyakin, I.V. 2018. Status of the Steppe Eagle in the world: "white spots" in distribution, population numbers, ecology and threats. Raptors Conservation Suppl. 1: 81–84.

Karyakin, I.V., A.V. Kovalenko and A.N. Barashkova. 2013. Monitoring of the Steppe Eagle populations in the trans-border zone of Russia and Kazakhstan in 2012. Raptors Conservation. 26: 61–83.

Karyakin, I.V., R. Bekmansurov and E. Nikolenko. 2014. Imperial Eagle in Russia and Kazakhstan: monitoring results 2008–2013. Slovak Raptor Journal. 8: 66–67.

Karyakin, I.V., A.I. Matsyna and E.G. Nikolenko. 2016a. Current state of the Steppe Eagle population in Kalmykia, Russia. Raptors Conservation. 33: 61–89.

Karyakin, I.V., L.S. Zinevich, D.M. Schepetov and S.Y. Sorokina. 2016b. Population structure of the Steppe Eagle range and preliminary data on the population genetic diversity and status of subspecies. Raptors Conservation. 32: 67–88.

Karyakin, I.V., R.H. Bekmansurov, E.G. Nikolenko and G.S. Dzhamirzoev. 2018a. Monitoring results from the breeding group of Egyptian Vulture at Narat-Tyube Ridge and GPS/GSM tracking data from juvenile Egyptian Vultures (Republic of Dagestan, Russia). Raptors Conservation. 36: 108–135.

Karyakin, I.V., E.G. Nikolenko, E.P. Shnayder, M. Horváth, M. Prommer, T. Juhász, A.S. Pazhenkov and L.S. Zinevich. 2018b. Direction, nature and timing of migration of the Imperial Eagles from the Volga-Ural region and Russian Altai (Russia) on data of the GSM/GPS and Argos/GPS-telemetry, 140–143. Proceeding of Conference, Raptors Conservation, Suppl. 1 (in Russian).

Karyakin, I.V., E.G. Nikolenko, E.P. Shnayder, L.S. Zinevich, G.I. Pulikova, K. Bartoszuk, M. Horváth, T. Juhász, K. Aghababyan and N.G. Andreyenkova. 2019. Results of the GPS/GSM-Tracking of juvenile Steppe Eagles from Russia and Kazakhstan. Raptors Conservation. 39: 71–227.

Kasprzykowski, Z. and P. Cieśluk. 2011. Rough-legged Buzzards *Buteo lagopus* wintering in central eastern Poland: population structure by age and sex, and the effect of weather conditions. Ornis Fennica. 88: 98–103.

Kassara, C., J. Fric, M. Gschweng and S. Sfenthourakis. 2012 Complementing the puzzle of Eleonora's Falcon (*Falco eleonorae*) migration: new evidence from an eastern colony in the Aegean Sea. Journal of Ornithology. 153: 839–848.

Kassara, C., L. Gangoso, U. Mellone, G. Piasevoli, T.G. Hadjikyriakou, N. Tsiopelas, S. Giokas, P. López-López, V. Urios, J. Figuerola, R. Silva, W. Bouten, A.N.G. Kirschel, M.Z. Virani, W. Fiedler, P. Berthold and M. Gschweng. 2017. Current and future suitability of wintering grounds for a long-distance migratory raptor. Scientific Reports. 7: 8798.

Katzenberger, J., E. Tabur, B. Şen, S. İsfendiyaroglu, I.L. Erkol and S. Oppel. 2019. No short-term effect of closing a rubbish dump on reproductive parameters of an Egyptian Vulture population in Turkey. Bird Conservation International. 29: 71–82.

Katzner, T.E. 2003. Ecology and behavior of four coexisting eagle species at Naurzum Zapovednik, Kazakhstan. Ph.D. dissertation, Arizona State University, Tempe, AZ, USA.

Katzner, T.E., E.A. Bragin, A.E. Bragin, M. McGrady, T.A. Miller and K.L. Bildstein. 2016. Unusual clockwise loop migration lengthens travel distances and increasespotential risks for a central Asian, long-distance, trans-equatorial migrant, the Red-footed Falcon. Bird Study. 63: 406–412.

Keijmel, M., J. Babbington, P. Roberts, M. McGrady and B.-U. Meyburg. 2020. World's largest gathering of Steppe Eagles *Aquila nipalensis* discovered in central Saudi Arabia. Sandgrouse. 42: 60–68.

Keller, V., S. Herrando, P. Voříšek, M. Franch, M. Kipson, P. Milanesi, D. Martí, M. Anton, A. Klvaňová, M.V. Kalyakin, H.-G. Bauer and R.P.B. Foppen. 2020. European Breeding Bird Atlas 2: Distribution, Abundance and Change, European Bird Census Council & Lynx Edicions, Barcelona.

Kemp, A. 2000. Concentration of non-breeding Lesser Spotted Eagles *Aquila pomarina* at abundant food: a breeding colony of Red-billed Quelea *Quelea quelea* in the Kruger National Park, South Africa. Acta Ornithoecologica. 4: 325–329.

Kennedy, A.S. and R.M. Marsh. 2016. First record of Crested Honey Buzzard *Pernis ptilorhynchus* for Kenya and East Africa. Bulletin of the British Ornithologists' Club. 136: 145–146.

Kerlinger, P. 1989. Flight strategies of migrating hawks, Univ. Chicago Press, Chicago, USA.

Kern, P., G. Lopez and G. Joannes. 2011. Migrazione post riproduttiva a Bellet (06 – Francia). Infomigrans. 28: 4.

Khaleghizadeh, A., D.A. Scott, M. Tohidifar, S. Babak Musavi, M. Ghasemi, M.E. Sehhatisabet, A. Ashoori, A. Khani, P. Bakhtiari, H. Amini, C. Roselaar, R. Ayé, M. Ullman, B. Nezami and F. Eskandari. 2011. Rare birds in Iran in 1980–2010. Podoces. 6: 1–48.

Khokhlov, A.N. 1995. The current state of the fauna of the Falconiformes of the Stavropol Territory and Karachay-Cherkessia. pp. 25–94. *In:* Polivanov, V.M. [ed.]. Birds of Prey and Owls of the North Caucasus. The Proceedings of Teberdinskii State Reserve, Vol. 14, Stavropol Book Publisher, Russia (in Russian).

Khokhlov, A.N. and M.P. Ilyukh. 1998. New information on birds of prey in the Stavropol Region. pp. 119–123. *In:* Proceedings of III Conference on birds of prey of Eastern Europe and Northern Asia, Part 2. Kislovodsk, Russia, Publisher (in Russian).

Khosravifard, S., V. Venus, A.K. Skidmore, W. Bouten, A.R. Muñoz and A.G. Toxopeus. 2018. Identification of Griffon Vulture's flight Types using high-resolution tracking data. International Journal of Environmental Research. 12: 313–325.

Kikuchi, D.M., T. Nakahara, W. Kitamura and N.M. Yamaguchi. 2019. Estimating potential costs of cumulative barrier effects on migrating raptors: a case study using global positioning system tracking in Japan. pp. 51–65. *In:* Bispo, R., J. Bernardino, H. Coelho, J. Lino Costa [eds.]. Wind Energy and Wildlife Impacts : Balancing Energy Sustainability with Wildlife Conservation. Springer International Publishing, Cham.

Kirwan, G.M., K. Boyla, P. Castell, B. Demirci, M. Ozen, H. Welch and T. Marlow. 2015. The birds of Turkey, Christopher Helm, London.

Kiselev, O.G. 2005. Observation for autumn migration of Buzzards near Voronezh. Raptors Conservation. 2: 60–61.

Kitowsky, I. 2009. Social learning of hunting skills in juvenile marsh harriers *Circus aeruginosus*. Journal of Ethology. 27: 327–332.

Kjellén, N. 1992. Differential timing of autumn migration between sex and age groups in raptors at Falsterbo, Sweden. Ornis Scandinavica. 23: 420–434.

Kjellén, N. 1994. Differences in age and sex ratio among migrating and wintering raptors in southern Sweden. The Auk. 111: 274–284.

Kjellén, N. 1997. Importance of a bird migration hot spot: proportion of the Swedish population of various raptors seen on autumn migration at Falsterbo 1986–1995 and population changes reflected by the migration counts. Ornis Svecica. 7: 21–34.

Kjellén, N. 1998. Annual variation in numbers, age and sex ratios among migrating raptors at Falsterbo, Sweden from 1986–1995. Journal of Ornithology. 139: 157–171.

Kjellén, N. 1999. Different migration strategies among Swedish Common Buzzards *Buteo buteo* revealed by the proportion of white birds. Ornis Svecica. 9: 11–18.

Kjellen, N. 2004. Migration counts. pp. 98–101. *In:* Karlsson, L. [ed.]. Wings over Falsterbo. Anser, Supplement 50. Lund.

Kjellén, N. 2017. Migration Counts at Falsterbo, SW Sweden. www.falsterbofagelstation.se

Kjellén, N. 2018. Migration Counts at Falsterbo, SW Sweden. www.falsterbofagelstation.se.

Kjellen, N. and G. Roos. 2000. Population trends in Swedish raptors demonstrated by migration counts at Falsterbo, Sweden. 1947–1997. Bird Study. 47: 195–211.

Kjellén, N., M. Hake and T. Alerstam. 2001. Timing and speed of migration in male, female and juvenile ospreys *Pandion haliaetus* between Sweden and Africa as revealed by field observations, radar and satellite tracking. Journal of Avian Biology. 32: 57–67.

Kjellén, N., M. Hake and T. Alerstam. 2001. Timing and speed of migration in male, female and juvenile Ospreys *Pandion haliaetus* between Sweden and Africa as revealed by field observations, radar and satellite tracking. Journal of Avian Biology. 32: 286–286.

Klaassen, R.H.G., R. Strandberg, M. Hake and T. Alerstam. 2008. Flexibility in daily travel routines causes regional variation in bird migration speed. Behavioral Ecology and Sociobiology. 62: 1427–1432.

Klaassen, R.H.G., A.E. Schlaich, W. Bouten and B.J. Koks. 2017. Migrating Montagu's harriers frequently interrupt daily flights in both Europe and Africa. Journal of Avian Biology. 48: 180–190.

Klaassen, R.H.G., M. Hake, R. Strandberg and T. Alerstam. 2011. Geographical and temporal flexibility in the response to crosswinds by migrating raptors. Proceedings of the Royal Society B. 278: 1339–1346.

Klaassen, R.H.G., R. Strandberg, M. Hake, P. Olofsson, A.P. Tøttrup and T. Alerstam. 2010. Loop migration in adult marsh harriers *Circus aeruginosus*, as revealed by satellite telemetry. Journal of Avian Biology. 41: 200–207.

Klaassen, R.H.G., M. Hake, R. Strandberg, B.J. Koks, C. Trierweiler, K.M. Exo, F. Bairlein and T. Alerstam. 2014. When and where does mortality occur in migratory birds? Direct evidence from long-term satellite tracking of raptors. Journal of Animal Ecology. 83: 176–184.

Klimov, A.S. 1986. Brief reports on the Steppe Eagle. pp. 127. *In*: Dolgushin, I.A. [ed.]. Rare animals of Kazakhstan. Nauka. Alma-Ata, Kazakhstan (in Russian).

Knish, N.P., A.I. Stativa, I.A. Bugaev, V.M. Savostyan and Y.V. Kuksa. 2008. Spring migration of Falconiformes in the forest-steppe of the north-eastern Ukraine. pp. 162–167. *In:* Gavrilyuk, M.N. [ed.]. New Research on Birds of Prey and Owls. Materials of the III International scientific conference "Birds of prey of Ukraine". Kriviy Rig, Ukraine, Publisher (in Russian).

Kokko, H. 1999. Competition for early arrival in migratory birds. Journal of Animal Ecology. 68: 940–950.

Koks, B.J., C. Trierweiler, E.G. Visser, C. Dijkstra and J. Komdeur. 2007. Do voles make agricultural habitat attractive to Montagu's Harrier *Circus pygargus*? Ibis. 149: 575–586.

Kolesnikov, A.D. 1960. Raptorial birds of the forest of Dnepropetrovsk Region. Nauchnye Zapiski Dnepropetrovskogo. Gosudarstvennogo Universiteta. 62: 29–34.

Komarov, Y.E. 2010. TheRed-footedfalconinNorthOssetia. Strepet 8: 99–100 (in Russian).

Komarov, Y.E. and A.D. Lipkovich. 1999. Imperial Eagle. Red Book Data of the Republic of Northern Ossetia. Alania, Vladikavkaz, Russia (in Russian).

Kopij, G. 2002. Food of the Lesser Kestrels *Falco naumanni* in its winter quarters in South Africa. Journal of Raptor Research. 36: 148–152.

Korelov, M.N. 1962. The order of birds of prey—*Falconiformes*. pp. 488–707. *In:* Dolgushin, I.A. [ed.]. Birds of Kazakhstan, vol. 2, Nauka Press, Alma-Ata, Kazakhstan (in Russian).

Korepov, M.V., V.V. Kovalev, M.M. Erokhina, S.G. Adamov, D.A. Korepova, S.A. Strukov, E.S. Turok and E.I. Zuzina. 2019. Migrations, wintering and summering sites of juvenile Imperial Eagles from the Volga region, Russia. Raptors Conservation. 38: 83–92 (in Russian).

Korovin, V.A. 1983. To the ecology of the Imperial Eagle on the northern border of the range. Protection of birds of prey. pp. 123–124. *In:* Galushin, V.M., S.A. Polozov and A.V. Davygora [eds.]. Materials of the 1st Meeting on the Ecology and Conservation of Birds of Prey. Nauka Press, Moscow, Russia (in Russian).

Kotkaklubi and 5D Vision. 2019. Birdmap. birdmap.5division.ee.

Kovács, A., N.P. Williams and C.A. Galbraith. 2014. Saker Falcon Falco cherrug Global Action Plan (SakerGAP), including a management and monitoring system, to conserve the species. Raptors MOU Technical Publication No. 2. CMS Technical Series No. 31. Coordinating Unit - CMS Raptors MOU, Abu Dhabi, United Arab Emirates.

Kovalenko, A.V., E.I. Gavrilov and A.E. Gavrilov. 2002. Brief reports on the Steppe Eagle. pp. 62. *In:* Belyalov, O.V. and V.A. Kovshar [compilers]. Kazakhstan Ornithological Bulletin, Issue 1, Tethys, Almaty, Kazakhstan.

Kovshar, A.F. 1966. The birds of Talas Alatau. Proceedings of Reserves of Kazakhstan. Issue 3, vol. 1. *In:* Dolgushin, I.A. [ed.]. Proceedings of National Reserve Aksu-Dzhabagly. Kainar Press, Alma-Ata, USSR (in Russian).

Kovshar, A.F. and N.N. Berezovikov. 1999. Materials on the phenology of spring migration of birds in the Kopa River valley (south-eastern Kazakhstan). Selevinia 1998–1999: 170–175 (in Russian).

Kovshar, A.F., A.S. Levin an O.V. Belyalov. 2004. Birds of the Betpak-Dala desert. pp. 85–125. *In*: A. Meldebekov [ed.]. Proceedings of the Institute of Zoology. Vol. 48. Ornithology. Almaty, Kazakhstan. (in Russian).

Kovshar, V.A. 2007. Observations over birds in the interfluve of Utva and Ilek rivers in September 2007. Selevinia 2007: 174–175 (in Russian).

Kralj, J., S. Barišić, D. Ćiković and V. Tutiš [eds.]. 2013. Atlas selidbe ptica Hrvatske. Croatian Bird Migration Atlas, Croatian Academy of Sciences and Arts, Zagreb.

Krasnoshtanova, M.N. 2001. Autumn migration of Falconides on Southern Baikal in 1995–1998, 2000s. pp. 117–118. *In:* Bardunov, L.V. [ed.]. Modern Problems of Baikal Science. Irkutsk State University Publisher, Irkutsk, Russia (in Russian).

Kretzmann, M.B., N. Capote, B. Gautschi, J.A. Godoy, J.A. Donázar and J.J. Negro. 2003. Genetically distinct island populations of the Egyptian vulture (*Neophron percnopterus*). Conservation Genetics. 4: 697–706.

Krivonosov, G.A. 1963. Essays on ecology and geographical distribution of birds of prey of the Volga delta. pp. 245–293. *In*: Fauna and Ecology of Birds in the Delta of the Volga and the coasts of the Caspian Sea. Proceedings of the Astrakhan Reserve. Issue. VIII. Publisher, Astrakhan, Russia (in Russian).

Krivosheev, V.G. 1960. Materials on bird migration in northern Kyzylkum. pp. 55–57. *In:* Dementyev, G.P. [ed.]. Animal Migrations, vol. 2. Nauka Press, Moscow, Russia (in Russian).

Kruckenhauser, L., E. Haring, W. Pinsker, M.J. Riesing, H. Winkler, M. Wink and A. Gamauf. 2004. Genetic vs. morphological differentiation of Old World buzzards (genus *Buteo*, Accipitridae). Zoologica Scripta. 33: 197–211.

Krumenacker, T. 2013. The migration of Lesser Spotted Eagle *Aquila pomarina*, European Honey Buzzard *Pernis apivorus*, Levant Sparrowhawk *Accipiter brevipes* and White Stok *Ciconia ciconia* over Israel—a balance over 30 years of counts. Studia i Materialy CEPL w Rogowie R. 15. Zeszyt. 36: 247–267.

Kuchin, A.P. 1976. Birds of the Altai. Publisher, Barnaul, Russia (in Russian).

Kumar, N., U. Guptac, Y.V. Jhala, Q. Qureshi, A.G. Gosler and F. Sergio 2020. GPS-telemetry unveils the regular high-elevation crossing of the Himalayas by a migratory raptor: implications for definition of a "Central Asian Flyway". Scientific Reports. 10: 15988.

Kumar, R.S. 2015. Tracking the Incredible Journey of the Amur Falcon. Conservation India. https://www.conservationindia.org/articles/tracking-the-incredible-journey-of-the-amur-falcon [accessed 8/7/2020].

La Gioia, G. 2009. La migrazione primaverile dei rapaci a Capo d'Otranto, Edizioni Publigrafic Trepuzzi, Lecce.

La Gioia, G., L. Melega and L. Fornasari. 2017. Piano d'Azione Nazionale per il grillaio (*Falco naumanni*). Quaderni Conservazione Natura, 41, MATTM - ISPRA, Roma.

Lanovenko, E.N. and B.B. Abdunazarov. 1983. On the influence of power lines on the number of birds of prey in Uzbekistan. pp. 29–30. *In:* Galushin, V.M., S.A. Polozov and A.V. Davygora [eds.]. Ecology of Birds of Prey. Materials of the 1st Meeting on the Ecology and Conservation of Birds of Prey. Nauka Press, Moscow, Russia (in Russian).

Lathbury, G. 1969. Migración de aves de presa (Accipitriformes) a través de Gibraltar. Ardeola. 14: 19–95.

Ławicki, Ł. and A.B. van den Berg. 2016. WP reports. Dutch Bird. 38(5): 322–336.

Le Brun, L., M. Lepley and P. Pilard. 2001. El Cernícalo primilla (*Falco naumanni*) en Francia: historia, estado y conservación. pp. 305–314. *In:* Toledano Garcés, J.F. and M. Corroto Matesanz [eds.]. Biologia y Conservación del Cernícalo primilla: actas IV Cogreso Nacional sobre el Cernícalo primilla. Consejería Medio Ambiente Y Grefa.

Lehikoinen, A. 2009. Climate forcing on avian life history. Dissertation, University of Helsinki, Helsinki, Finland.

Lehikoinen, A., P. Saurola, P. Byholm, A. Linden and J. Valkama. 2010. Life history events of the Eurasian sparrowhawk *Accipiter nisus* in a changing climate. Journal of Avian Biology. 41: 627–636.

Lelong, V. and C. Riols. 2009. Suivi des dortoirs postnuptiaux du Faucon crécerellette dans le sud de la France de 2003 a 2008. pp. 162–167. *In:* Pilard, P. [ed.]. Actes du VII Congrès International sur le Faucon crécerellette. Imprimerie Lagarde, Breuillet, France.

Leo, R., A. Pasqua and A. Micheli. 2006. La migrazione primaverile dei rapaci diurni nel Parco Alto Garda Bresciano (BS). Infomigrans. 17: 11.

Leshem, Y. 1987. Wings over Israel. BBC Wildlife. 5: 31–34.

Leshem, Y. 1989. Following Raptor Migration from the Ground, Motorized Glider and Radar at a Junction of three Continents. *In*: Meyburg, B.-U. and R.D. Chancellor [eds.]. Raptors in the Modern World: 43–52.

Leshem, Y. and O. Bahat. 1999. Flying with the Birds, Yedioth Ahronoth & Chemed Books, Tel Aviv.

Leshem, Y. and Y. Yom-Tov. 1996a. The magnitude and timing of migration by soaring raptors, pelicans and storks over Israel. Ibis. 138: 188–203.

Leshem, Y. and Y. Yom-Tov. 1996b. The use of thermals by soaring migrants. Ibis. 138: 667–674.

Leshem, Y. and Y. Yom-Tov. 1998. Routes of migrating soaring birds. Ibis. 140: 41–52.

Levin, A.S. and G.A. Kurkin. 2013. The scope of death of eagles on power lines in western Kazakhstan. Raptors Conservation 27: 240–244 (in Russian).

Liberatori, F. and V. Penteriani. 2001. A long-term analysis of the declining population of the Egyptian vulture in the Italian peninsula: distribution, habitat preference, productivity and conservation implications. Biological Conservation. 101: 381–389.

Lieury, N., M. Gallardo, C. Ponchon, A. Besnard and A. Millon. 2015. Relative contribution of local demography and immigration in the recovery of a geographically-isolated population of the endangered Egyptian vulture. Biological Conservation. 191: 349–356.

Limiñana, R., A. Soutullo and V. Urios. 2007. Autumn migration of Montagu's harriers *Circus pygargus* tracked by satellite telemetry. Journal of Ornithology. 148: 517–523.

Limiñana, R., A. Soutullo, P. López-López and V. Urios. 2008. Pre-migratory movements of adult Montagu's Harriers *Circus pygargus*. Ardea. 96: 81–90.

Limiñana, R., M. Romero, U. Mellone and V. Urios. 2012a. Mapping the migratory routes and wintering areas of Lesser Kestrels *Falco naumanni*: new insights from satellite telemetry. Ibis. 154: 389–399.

Limiñana, R., A. Soutullo, B. Arroyo and V. Urios. 2012b. Protected areas do not fulfil the wintering habitat needs of the trans-Saharan migratory Montagu's harrier. Biological Conservation. 145: 62–69.

Limiñana, R., A. Soutullo, V. Urios and A. Reig-Ferrer. 2012c. Migration and wintering areas of adult Montagu's Harriers (*Circus pygargus*) breeding in Spain. Journal of Ornithology. 153: 85–93.

Limiñana, R., M. Romero, U. Mellone and V. Urios. 2013. Is there a different response to winds during migration between soaring and flapping raptors? An example with the Montagu's harrier and the lesser kestrel. Behavioral Ecology and Sociobiology. 67: 823–835.

Limiñana R., B.E. Arroyo, J. Terraube, M. McGrady and F. Mougeot. 2015. Using satellite telemetry and environmental niche modelling to inform conservation targets for a long-distance migratory raptor in its wintering grounds. Oryx. 49: 329–337.

Literák, I., D. Horal, H. Alivizatos and H. Matušík. 2017. Common wintering of Black Kites (*Milvus migrans migrans*) in Greece, and new data on their wintering elsewhere in Europe. Slovak Raptor Journal. 11: 91–102.

Literák, I., R. Raab, S.Vyhnal, P. Spakovszky and J. Steindl. 2018. Occurrence of Red Kites *Milvus milvus* in Serbia based on birds tracked by telemetry devices. Acrocephalus. 39: 27–32.

Liven-Schulman, I., Y. Leshem, D. Alon and Y. Yom-Tov. 2004. Causes of population declines of the Lesser Kestrel *Falco naumanni* in Israel. Ibis. 146: 145–152.

Lõhmus, A. and Ü. Väli. 2005. Habitat use by the Vulnerable greater spotted eagle *Aquila clanga* interbreeding with the lesser spotted eagle *Aquila pomarina* in Estonia. Oryx. 39: 170–177.

Londei, T. 1991. Winter record of the Booted Eagle, *Hieraaetus pennatus*, and the Short-toed Eagle, *Circaetus gallicus*, in Crete. Rivista Italiana Ornitologia. 61: 128–129.

Lontkowski, J. and G. Maciorowski. 2010. Identification of juvenile Greater Spotted Eagle, Lesser Spotted Eagle and hybrids. Dutch Birding. 32: 384–397.

Lopez-Iborra, G., R. Limiñana, D. Pavon and J. Martinez-Perez. 2011. Modelling the distribution of short-toed eagle (*Circaetus gallicus*) in semi-arid Mediterranean landscapes: identifying important explanatory variables and their implications for its conservation. European Journal of Wildlife Research. 57: 83–93.

López-López, P. 2016. Individual-based tracking systems in ornithology: welcome to the era of big data. Ardeola. 63: 103–136.

López-López, P., R. Limiñana and V. Urios. 2009. Autumn migration of Eleonora's falcon *Falco eleonorae* tracked by satellite elemetry. Zoological Studies. 48: 485–491.

López-López, P., C. García-Ripollés and V. Urios. 2014. Individual repeatability in timing and spatial flexibility of migration routes of trans-Saharan migratory raptors. Current Zoology. 60: 642–652.

López-López, P., R. Limiñana, U. Mellone and V. Urios. 2010. From the Mediterranean Sea to Madagascar. Are there ecological barriers for the long-distance migrant Eleonora's falcon? Landscape Ecology. 25: 803–813.

López-López, P., J. Benavent-Corai, C. García-Ripollés and V. Urios. 2013. Scavengers on the move: behavioural changes in foraging search patterns during the annual cycle. PLoS ONE. 8: e54352.

López-López, P., J. De la Puente, U. Mellone, A. Bermejo and V. Urios. 2016. Spatial ecology and habitat use of adult Booted Eagles (*Aquila pennata*) during the breeding season: implications for conservation. Journal of Ornithology. 157: 981–993.

LPO Haute-Savoie. 2016. Suivi de la migration postnuptiale au Défilé de l'Écluse http://files.biolovision.net/haute-savoie.lpo.fr/userfiles/RapportDfildelEcluse2016.pdf. [accessed 8/5/2020].

Lucia, G., N. Agostini, M. Panuccio, U. Mellone, G. Chiatante, D. Tarini and A. Evangelidis. 2011. Raptor migration at Antikythira, in southern Greece. British Birds. 104: 266–270.

Lukopulos, D. 1940. Modern Greek Mythology: Animals—Plants, Society for the publicity of useful books, Athens.

Maciorowski, G. 2012. Tagging, autumn migration and wintering of the Biebrza *Aquila clanga* and *Aquila pomarina* population. pp. 8. *In:* Conservation of the Greater Spotted Eagle. Proceedings of the International Workshop, Goniądz, Poland.

Maciorowski, G. and P. Mirski. 2014. Habitat alteration enables hybridisation between Lesser Spotted and Greater Spotted Eagles in north-east Poland. Bird Conservation International. 24: 152–161.

Maciorowski, G., J. Lontkowski and T. Mizera. 2014. The spotted eagle–vanishing bird of the marshes. UNIGRAF, Poznań.

Maciorowski, G., P. Mirski and Ü. Väli. 2015. Hybridisation Dynamics between the Greater Spotted Eagles *Aquila clanga* and Lesser Spotted Eagles *Aquila pomarina* in the Biebrza River Valley (NE Poland). Acta Ornithologica. 50: 33–41.

Maciorowski, G., A. Galanaki, T. Kominos, M. Dretakis and P. Mirski. 2019. The importance of wetlands for the Greater Spotted Eagle *Clanga clanga* wintering in the Mediterranean Basin. Bird Conservation International. 129: 115–123.

Maciorowski, G., J. Kosicki, M. Polakowski, M. Urbanska, P. Zduniak and P. Tryjanowski. 2019. Autumn migration of immature Red Kites *Milvus milvus* from a central European population. Acta Ornithologica. 54: 45–50.

Mackrill, T.R. 2017. Migratory behaviour and ecology of a trans-saharan migrant raptor, the osprey *Pandion haliaetus*. PhD Thesis, University of Leicester.

Malmiga, G., C. Nilsson, J. Bäckman and T. Alerstam. 2014. Interspecific comparison of the flight performance between sparrowhawks and common buzzards migrating at the Falsterbo peninsula: A radar study. Current Zoology. 60: 670–679.

Malovichko, L.V., V.N. Moseikin, E.V. Moseikin and V.N. Fedosov. 2003. About formation of migratory accumulations of Falconiformes in the Eastern part of North Caucasus. pp. 220–225. *In:* Galushin, V.M. [ed.]. Materials of the IV Conference on Birds of Prey of Northern Eurasia, Penza, Russia. Publisher (in Russian).

MaMing, R. and X. Zhao. 2013. Distribution patterns and ecology of the Steppe Eagle in China. Raptors Conservation. 27: 172–179.

Mammen, U. and A. Stubbe. 1995. Alterseinschatzung und brutbeginn des Rotmilans *Milvus milvus*. pp. 91–98. *In*: Mueller, F. [ed.]. Vogel und Umvelt, Berlin, Germany.

Margalida, A. and J. Boudet. 2003. Dynamics and temporal variation in age structure at a communal roost of Egyptian vultures (*Neophron percnopterus*) in northeastern Spain. Journal of Raptor Research. 37: 252–256.

Margalida, A., J.R. Benitez, J.A. Sanchez-Zapata, E. Ávila, R. Arenas and J.A. Donázar. 2012. Long-term relationship between diet breadth and breeding success in a declining population of Egyptian Vultures *Neophron percnopterus*. Ibis. 154: 184–188.

Marques, A.T., C.D. Santos, F. Hanssen, A.R. Muñoz, A. Onrubia, M. Wikelski, F. Moreira, J.M. Palmeirim and J.P. Silva. 2020. Wind turbines cause functional habitat loss for migratory soaring birds. Journal of Animal Ecology. 89: 93–103.

Martell, M.S., Englund, J.V. and Tordoff, H.B. 2002. An urban Osprey population established by translocation. Journal of Raptor Research. 36: 91–96.

Martín, B., A. Onrubia and M.A. Ferrer. 2014. Effects of climate change on the migratory behavior of the Common Buzzard *Buteo buteo*. Climate Research. 60: 187–197.

Martín, B., A. Onrubia, A. de la Cruz and M. Ferrer. 2016. Trends of autumn counts at Iberian migration bottlenecks as a tool for monitoring continental populations of soaring birds in Europe. Biodiversity Conservation. 25: 295–309.

Martín, B., C. Perez-Bacalu, A. Onrubia, M. De Lucas and M. Ferrer. 2018. Impact of wind farms on soaring bird populations at a migratory bottleneck. European Journal of Wildlife Research. 64: 33.

Martín, B., C.A. Torralvo, G. Elias, J. Tomás, A. Onrubia and M. Ferrer. 2019. Are Western European ospreys (*Pandion haliaetus*) shortening their migration distances? Evidence from trends of the wintering population in the Iberian Peninsula. European Journal of Wildlife Research. 65: 72.

Martínez, J.E. and J.A. Sánchez-Zapata. 1999. Wintering of Booted Eagles (*Hieraaetus pennatus*) and Short-toed Eagles (*Circaetus gallicus*) in Spain. Ardeola. 46: 93–96.

Mascara, R., A. Nardo and M. Sarà. 2019. Nuovi dati sulle principali specie di uccelli steppico-cerealicole della Piana di Gela (Sicilia). Uccelli D'Italia. 44: 43–53.

Mateos-Rodríguez, M. and F. Liechti. 2012. How do diurnal long-distance migrants select flight altitude in relation to wind? Behavioral Ecology. 23: 403–409.

Mateo-Tomás, P. and P.P. Olea. 2010. Diagnosing the causes of territory abandonment by the Endangered Egyptian vulture *Neophron percnopterus*: the importance of traditional pastoralism and regional conservation. Oryx. 44: 424–433.

Mateo-Tomás, P., P.P. Olea and I. Fombellida. 2010. Status of the Endangered Egyptian vulture *Neophron percnopterus* in the Cantabrian Mountains, Spain, and assessment of threats. Oryx. 44: 434–440.

Matsyna, A.I., E.L. Matsyna, M.A. Korolkov, V.E. Badmaev and V.B. Badmaev. 2012. Assessment of the scale of annual bird deaths due to electric shock on 6–10 kV overhead power lines in Kalmykia, Russia. Raptors Conservation. 24: 186–201.

Mauve, L. 1938. Der Zug der Großvögel über den Bosporus. Journal für Ornithologie. 86: 261–301.

McClain, K.M. 2015. Environmental drivers of migration in two Israeli raptor species. M.Sc.Thesis. The Ohio State University.

McGrady, M.J. and A. Gavashelishvili. 2006. Tracking vultures from the Caucasus into Iran. Podoces. 1: 21–26.

McGrady, M.J. and W.A. Al Fazari. 2013. http://sootyfalcon.blogspot.com/2013/[Accessed 8 July 2020].

McGrady, M.J., M. Gschweng and W.A. Al-Fazari. 2010. Report on fieldwork to study the status and distribution of breeding sooty falcons (*Falco concolor*) on the northern islands of Oman – 2010. Natural Research, Ltd.

McGrady, M.J., H.A. Rayaleh, A.M. Dara and E. Abdillahi. 2014. Migration of raptors across the Bab el Mandeb Straits during 2–10 March 2013. Bulletin of the African Bird Club. 21: 65–72.

McGrady, M.J., W.A. Al Fazari, M.H. Al Jahdhami and M.K. Oli. 2016. Survival of Sooty Falcons (*Falco concolor*) in Oman. Journal of Ornithology. 157: 427–437.

McGrady, M.J., W.A. Al-Fazari, M.H. Al-Jahdhami, M.A.C. Nicoll and M.K. Oli. 2017. Sooty Falcon *Falco concolor* reproduction and population dynamics on the islands in the Sea of Oman. Ibis. 159: 828–840.

McGrady, M.J., D.L. Karelus, H.A. Rayaleh, M. Sarrouf Willson, B.U. Meyburg, M.K. Oli and K.L. Bildstein. 2018. Home ranges and movements of Egyptian Vultures *Neophron percnopterus* in relation to rubbish dumps in Oman and the Horn of Africa. Bird Study. 65: 544–556.

McGrady, M.J., W.A. Al Fazari, M.H. Al Jahdhami, M. Fisher, A. Kwarteng, H. Walters and M.K. Oli. 2019. Island accessibility and distance from beach influence nesting success of Sooty Falcons *Falco concolor* in Oman. Ibis. 161: 162–171.

McKnight, A., A.J. Allyn, D.C. Duffy and D.B. Irons. 2013. 'Stepping stone' pattern in Pacific Arctic tern migration reveals the importance of upwelling areas. Marine Ecology Proggress Series. 491: 253–364.

Mebs, T. and D. Schmid. 2006. Die Greifvögel Europas, Nordafrikas und Vorderasiens, Franckh-Kosmos, Stuttgart.

Medzhidov, R.A., V.M. Muzaev and V.B. Badmaev. 2011. About the state of the Steppe Eagle population in Kalmykia. Steppe Bulletin. 32: 33–37 (in Russian).

Megalli, M. and G. Hilgerloh. 2013. The soaring bird spring migration bottleneck at Ayn Sokhna, northern gulf of Suez, Egypt. Sandgrouse. 35: 28–35.

Melde, M. 1971. Der Mäusebussard. A. Ziemsen Verlag, Wittenberg Lutherstadt.

Melde, M. 1982. Buzzard *Buteo buteo* L. pp. 74–92. *In:* Mihelsons, H.A. and J.A. Viksne [eds.]. Migrations of Birds of Eastern Europe and Northern Asia. Falconiformes – Gruiformes. Nauka, Moscow.

Meller, K., H. Björklund, P. Saurola and J. Valkama. 2019. Kuuma kesä suosi haukkoja – myyräkato masensi pöllöja. Linnut Vuosikirja. 2018: 80–95.

Mellone, U. 2017. Estrategias de migración. pp. 43–58. *In:* Urios, V., A. Bermejo, J. Vidal-Mateo and J. De la Puente [eds.]. Migración y ecología espacial de la población española de águila calzada. Monografía n. 2 del programa Migra. SEO/BirdLife. Madrid.

Mellone, U. 2020. Sea crossing as a major determinant for the evolution of migratory strategies in soaring birds. Journal of Animal Ecology. 89: 1298–1301.

Mellone, U., N. Agostini, M. Panuccio and M. Gustin. 2007. La migrazione primaverile dei rapaci Accipitriformi attraverso le isole Eolie: primi dati. pp. 67. *In:* Barbagli, F. [ed.]. Proceedings XIV Italian Ornithological Congress, Trieste.

Mellone, U., R. Limiñana, E. Mallia and V. Urios. 2011a. Extremely detoured migration in an inexperienced bird: interplay of transport costs and social interactions. Journal of Avian Biology. 42: 468–472.

Mellone, U., P. López-López, R. Limiñana and V. Urios. 2011c. Weather conditions promote route flexibility during open ocean crossing in a long-distance migratory raptor. International Journal of Biometeorology. 55: 463–468.

Mellone, U., P. López-López, R. Limiñana and V. Urios. 2012b. Wintering habitats of Eleonora's Falcons *Falco eleonorae* in Madagascar. Bird Study. 59: 29–36.

Mellone U., P. López-López, R. Limiñana and V. Urios. 2013c. Summer pre-breeding movements of Eleonora's Falcons *Falco eleonorae* revealed by satellite telemetry: implications for conservation. Bird Conservation International. 23: 487–494.

Mellone, U., R. Limiñana, P. López-López and V. Urios. 2015b. Regional and age-dependent differences in the effect of wind on the migratory routes of Eleonora's falcon. Current Zoology. 61: 428–434.

Mellone, U., G. Lucia, E. Mallia and V. Urios. 2016. Individual variation in orientation promotes a 3000 km latitudinal change in wintering grounds in a long-distance migratory raptor. Ibis. 158: 887–893.

Mellone, U., P. López-López, R. Limiñana, G. Piasevoli and V. Urios. 2013b. The trans-equatorial loop migration system of Eleonora's falcon: differences in migration patterns between age classes, regions and seasons. Journal of Avian Biology. 44: 417–426.

Mellone, U., V. Urios, H. Rguibi-Idrissi, R. Limiñana, A. Benhoussa and P. López-López. 2012c. Ranging behaviour of Eleonora's Falcons *Falco eleonorae* during chick-rearing. Acta Ornitologica. 47: 195–198.

Mellone, U., J. De La Puente, P. López-López, R. Limiñana, A. Bermejo and V. Urios. 2013a. Migration routes and wintering areas of booted eagles *Aquila pennata* breeding in Spain. Bird Study. 60: 409–413.

Mellone, U., J. De la Puente, P. López-López, R. Limiñana, A. Bermejo and V. Urios. 2015a. Seasonal differences in migration patterns of a soaring bird in relation to environmental conditions: a multi-scale approach. Behavioral Ecology and Sociobiology. 69: 75–82.

Mellone, U., B. Yáñez, R. Limiñana, A.R. Muñoz, D. Pavón, J.M. Gonzalez, V. Urios and M. Ferrer. 2011b. Summering staging areas of non-breeding Short-toed Snake Eagles *Circaetus gallicus*. Bird Study. 58: 516–521.

Mellone, U., R.H.G. Klaassen, C. García-Ripollés, R. Limiñana, P. López-López, D. Pavón, R. Strandberg, V. Urios, M. Vardakis and T. Alerstam. 2012a. Interspecific comparison of the performance of soaring migrants in relation to morphology, meteorological conditions and migration strategies. PLoS ONE. 7: e39833.

Melnikov, L. and A. Mischenko. 2019. Lesser Spotted Eagle *Clanga pomarina* Brehm, 1831 (*Aves: Acciandpitridae*) on the Eastern Border of the Present Range in Russia. Acta zoologica bulgarica, Suppl. 14: 85–89.

Melnikov, Y.I. 2008. The distribution and numbers of spotted eagle on the territory of Predbaikalye. pp. 195–214. *In:* Research and Conservation of the Greater and the Lesser Spotted Eagle in Northern Eurasia. Ivanovskii Gosudarstvennyi Universitet, Ivanovo.

Mendelssohn, H. and Y. Leshem. 1983. The status and conservation of vultures in Israel. pp. 86–98. *In*: Vulture Biology and Management, Wilbur and Jackson eds, London.

Méndez, M., J.A. Godoy and J.A. Donázar. 2015. Genetic analysis of Egyptian Vulture (*Neophron percnopterus*) in the Balkans and Turkey. BSPB, Sofia, 27 pp.

Mengoni, C., I. Zuberogoitia, N. Mucci, G. Boano, T. Urban, E. Guzzo and M. Sarà. 2018. Genetic variability in Peregrine Falcon populations of the Western Palaearctic region. Ornis Hungarica. 26: 12–26.

Meyburg, B.-U. 1974. Sibling aggression and mortality among nestling eagles. Ibis. 116: 224–228.

Meyburg, B.-U. 1996. The lesser spotted eagle *Aquila pomarina*: Population status and present state of research. pp. 377–387. *In:* Meyburg, B.-U. and R.D. Chancellor [eds.]. Eagle Studies. Berlin, London and Paris: WWG on Birds of Prey.

Meyburg, B.-U. 2001. On Cainism in the Lesser Spotted Eagle *Aquila pomarina*. Acta ornithoecologica. 4: 269–278.

Meyburg, B.-U. 2002. On Cainism in the lesser spotted eagle (*Aquila pomarina*) and a possible explanation for the phenomenon in this and other eagle species. pp. 53–61. *In:* Yosef, R., M.L. Miller and D. Pepler [eds.]. Raptors in the New Millennium. Eilat: Intern. Birding & Research Center.

Meyburg, B.-U. 2005. Migration and Persecution of Raptors in southern Turkey. Ornithologische Mitteilungen. 57: 12–16.

Meyburg, B.-U. and Z. Pielowski. 1991. Cainism in the Greater Spotted Eagle *Aquila clanga*. Birds of Prey Bulletin 4: 143–148.

Meyburg, B.-U. and C. Meyburg 1988. Lesser Spotted Eagle as an alpine hunter in the Caucasus. Journal for Ornithology. 129: 102–103.

Meyburg, B.-U. and C. Meyburg. 1999. The study of raptor migration in the Old World using satellite telemetry. pp. 2992–3006. *In:* Adams, N.J. and R.H. Slotow [eds.]. Proc. 22nd Int. Ornithol. Congr. BirdLife South Africa, Johannesburg, Republic of South Africa.

Meyburg, B.-U. and C. Meyburg. 2005. Tracking the endangered Greater Spotted Eagle. Tracker News. 6: 4.

Meyburg, B.-U. and M.R. Fuller. 2007. Satellite tracking. pp. 242–248. *In:* Bird, D.M. and K.L. Bildstein [eds.]. Raptor Research and Management Techniques, Hancock House Publishers, Surrey, Canada.

Meyburg, B.-U. and C. Meyburg. 2009b. Annual cycle, timing and speed of migration of a pair of Lesser Spotted Eagles (*Aquila pomarina*)—a study by means of satellite telemetry. Population Ecology of Raptors and Owls. 6: 63–85.

Meyburg, B.-.U. and C. Meyburg. 2009a. GPS-Satelliten-Telemetrie bei einem adulten Schwarzmilan (*Milvus migrans*): Aufenthaltsraum während der Brutzeit, Zug und Überwinterung. Populationsökologie Greifvogel und Eulenarten. 6: 243–284.

Meyburg, B.-U. and C. Meyburg. 2009c. Satelliten-Telemetrie beim Schreiadler (*Aquila pomarina*). Labus, Naturschutz im Landkreis Mecklenburg-Strelitz, Sonderheft. 13: 16–44.

Meyburg, B.-U. and C. Meyburg. 2010. Satellite tracking of eastern imperial eagles *Aquila heliaca*. *In:* The 6th International Conference on Asian Raptors, Ulaanbaatar, Mongolia.

Meyburg, B.-U. and C. Meyburg. 2013. Telemetrie in der Greifvogelforschung. Greifvoegel und Falknerei 2013: 26–60. Verlag J. Neumann-Neudamm AG, Melsungen.

Meyburg, B.-U. and P. Boesman. 2013. Steppe Eagle (*Aquila nipalensis*). *In*: del Hoyo, J., A. Elliott, J. Sargatal and D. Christie [eds.]. Handbook of the Birds of the World Alive, Lynx Edicions, Barcelona.

Meyburg, B.-U. and C. Meyburg. In prep. The flight altitudes of a long-distance migrant during the course of the year - Lesser Spotted Eagle (*Clanga pomarina*) also move in over 4.000m height.

Meyburg, B.-U. and G.M. Kirwan. 2016a. Eastern imperial eagle (*Aquila heliaca*). *In:* del Hoyo, J., A. Elliott, J. Sargatal, D.A. Christie and E. de Juana [eds.]. Handbook of the Birds of the World Alive. Lynx Edicions, Barcelona, Spain.

Meyburg, B.-U., W. Scheller and C. Meyburg. 1993. Satellite tracking of a juvenile Lesser Spotted Eagle (*Aquila pomarina*) during autumn migration. Journal for Ornithology. 134: 173–179.

Meyburg. B.-U., W. Scheller and C. Meyburg. 1995. Migration and wintering of the Lesser Spotted Eagle (*Aquila pomarina*): a study by means of satellite telemetry. Journal for Ornithology. 136: 401–422.

Meyburg, B.-U., C. Meyburg and C. Pacteau. 1996. Migration automnale d'un circaète Jean-le-Blanc *Circaetus gallicus* suivi par satellite. Alauda. 64: 339–344.

Meyburg, B.-U., C. Meyburg and J.C. Barbraud. 1998a. Migration strategies of an adult Short-toed Eagle *Circaetus gallicus* tracked by satellite. Alauda. 66: 39–48.

Meyburg, B.-U., W. Scheller and C. Meyburg. 2000. Migration and wintering of the Lesser Spotted Eagle *Aquila pomarina*: A study by means of satellite telemetry. Global Environmental Research. 4: 183–193.

Meyburg, B.-U., J. Matthes and C. Meyburg. 2002. Satellite-tracked Lesser Spotted Eagle avoids crossing water at the Gulf of Suez. British Birds. 95: 372–376.

Meyburg, B.-U., P. Paillat and C. Meyburg. 2003. Migration routes of Steppe Eagles between Asia and Africa: a study by means of satellite telemetry. The Condor. 105: 219–227.

Meyburg, B.-U., C. Meyburg and F. Franck-Neumann. 2007a. Why do female Lesser Spotted Eagles (*Aquila pomarina*) visit strange nests remote from their own? Journal of Ornithology. 148: 157–166.

Meyburg, B.-U., J. Matthes and C. Meyburg. 2010. Wintering ecology - Lesser Spotted Eagle and Red-billed Quelea. Falke. 57: 236–243.

Meyburg, B.-U., C. Meyburg and P. Paillat. 2012. Steppe Eagle migration strategies revealed by satellite telemetry. British Birds. 105: 506–519.

Meyburg, B.-U., G.M. Kirwan and E.F.J. Garcia. 2016b. Greater spotted eagle (*Clanga clanga*). *In:* del Hoyo, J., A. Elliott, J. Sargatal, D.A. Christie. and E. de Juana [eds.]. Handbook of the Birds of the World Alive. LynxEdicions, Barcelona, Spain.

Meyburg, B-U., Y. Angelov and S. Azar. 2020. A corridor of soaring bird migration in Lebanon on the Eastern Mediterranean flyway. Sandgrouse. 42: 46–58.

Meyburg, B.-U., X. Eichaker, C. Meyburg and P. Paillat. 1995. Migrations of an adult Spotted Eagle tracked by satellite. British Birds. 88: 357–361.

Meyburg, B.-U., T. Mizera, G. Maciorowski and J. Kowalski. 1997. Schelladler *Aquila clanga* brütet in partiellem Jugendgefieder. Limicola. 11: 82–87.

Meyburg, B.-U., M. Gallardo, C. Meyburg and E. Dimitrova. 2004a. Migrations and sojourn in Africa of Egyptian vultures (*Neophron percnopterus*) tracked by satellite. Journal of Ornithology. 145: 273–280.

Meyburg, B.-U., C. Meyburg, J. Matthes and H. Matthes. 2007b. Spring migration, late arrival, temporary change of partner and breeding success in the Lesser Spotted Eagle *Aquila pomarina*. Vogelwelt. 128: 21–31.

Meyburg, B.-U., F. Ziesemer, H.D. Martens and C. Meyburg. 2010. On the biology of Honey Buzzards (*Pernis apivorus*). Results of satellite tracking. pp. 21–24. *In:* 7th Int. Symp. Population Ecology of Raptors and Owls.

Meyburg, B.-U., P.W. Howey, C. Meyburg and C.D. Fiuczynski. 2011. Two complete migration cycles of an adult Hobby tracked by satellite. British Birds. 104: 2–15.

Meyburg, B.-U., C. Meyburg, F. Ziesemer and H. Martens. 2013. Migration and wintering strategies of adult Honey Buzzards *Pernis apivorus* from Germany revealed by satellite telemetry, 150. *In* 9th Conf. Eur. Orn. Union.

Meyburg, B.-U., P. Howey, C. Meyburg and R. Pretorius. 2017b. Year-round satellite tracking of Amur Falcon (*Falco amurensis*) reveals the longest migration of any raptor species across the open sea. *In*: British Ornithologists' Union Annual Conference 'From avian tracking to population processes'. University of Warwick.

Meyburg. B.-U., P. Paillat, C. Meyburg and M. McGrady. 2017c. Migration of a Peregrine Falcon *Falco peregrinus calidus* from Saudi Arabia to Cape Town as revealed by satellite telemetry. Ostrich. 2017: 1–4.

Meyburg, B.-U., T. Mizera, C. Meyburg and M.J. Mcgrady. 2018. Collision between a migrating lesser spotted eagle (*Clanga pomarina*) and an aircraft as detailed by fine-scale GSM-GPS telemetry data. Slovak Raptor Journal. 12: 55–62.

Meyburg, B.-U., C. Meyburg, T. Mizera, G. Maciorowski and J. Kowalski. 1998b. Greater Spotted Eagle wintering in Zambia. Africa. Birds & Birding. 3: 62–68.

Meyburg, B.-U., C. Meyburg, T. Bělka, O. Šreibr and J. Vrana. 2004b. Migration, wintering and breeding of a Lesser Spotted Eagle (*Aquila pomarina*) from Slovakia tracked by satellite. Journal of Ornithology. 145: 1–7.

Meyburg, B.-U., C. Meyburg, T. Mizera, G. Maciorowski and J. Kowalski. 2005. Family break up, departure, and autumn migration in Europe of a family of Greater Spotted Eagles (*Aquila clanga*) as reported by satellite telemetry. Journal of Raptor Research. 39: 462.

Meyburg, B.-U., K. Graszynski, T. Langgemach, P. Sömmer and U. Bergmanis. 2008. Cainism, nestling management in Germany in 2004–2007 and satellite tracking of juveniles in the Lesser Spotted Eagle (*Aquila pomarina*). Slovak Raptor Journal. 2: 53–72.

Meyburg, B.-U., U. Bergmanis, T. Langgemach, K. Graszynski, A. Hinz, I. Börner, C. Meyburg and W.M.G. Vansteelant. 2017a. Orientation of native versus translocated juvenile lesser spotted eagles (*Clanga pomarina*) on the first autumn migration. Journal of Experimental Biology. 220: 2765.

Meyer, S.K., R. Spaar and B. Bruderer. 2000. To cross the sea or to follow the coast? Flight directions and behaviour of migrating raptors approaching the Mediterranean Sea in autumn. Behaviour. 137: 379–399.

Meyer, S.K., R. Spaar and B. Bruderer. 2003. Sea crossing behaviour of falcons and harriers at the southern Mediterranean coasts of Spain. Avian Science. 3: 153–162.

Mezzavilla, F. 2006. Primi dati sulla migrazione primaverile dei rapaci sopra la pianura trevigiana (TV). Infomigrans. 17: 10.

Michelat, D. 2005. Une nouvelle espèce nicheuse pour l'avifaune française: l'Aigle pomarin *Aquila pomarina*. Ornithos. 12: 310–320.

Michelat, D. 2019. Dans l'intimité de l'unique couple de pomarin de France. Rapaces de France. 21: 35–40.

Michev, T., L. Profirov, K. Nyagolov and M. Dimitrov. 2011. The autumn migration of soaring birds at Bourgas Bay, Bulgaria. British Birds. 104: 16–37.

Michev, T.M., L.A. Profirov, B.T. Michev, L.A. Hristov, A.L. Ignatov, E.H. Stoynov and N.H. Chipev. 2018. Long-term Changes in Autumn Migration of Selected Soaring Bird Species at Burgas Bay, Bulgaria. Acta Zoologica Bulgarica. 70: 57–68.

Mihoub, J.B., O. Gimenez, P. Pilard and F. Sarrazin. 2010. Challenging conservation of migratory species: Sahelian rainfalls drive first-year survival of the vulnerable lesser kestrel *Falco naumanni*. Biological Conservation. 143: 839–847.

Mikula, P., M. Hromada and P. Tryjanowski. 2013. Bats and swifts as food of the European kestrel (*Falco tinnunculus*) in a small town in Slowakia. Ornis Fennica. 90: 178–185.

Miller, R.A., A. Onrubia, B. Martín, G.S. Kaltenecker, J.D. Carlisle, M.J. Bechard and M. Ferrer. 2015. Local and regional weather patterns influencing post-breeding migration counts of soaring birds at the Strait of Gibraltar, Spain. Ibis. 158: 106–115.

Miller, T.A., R.P. Brooks, M.J. Lanzone, D. Brandes, J. Cooper, J.A. Tremblay, J.A.Y. Wilhelm, A. Duerr and T.E. Katzner. 2016. Limitations and mechanisms influencing the migratory performance of soaring birds. Ibis. 158: 116–134.

Millhouser, P. 2016. Following the migration of lesser kestrels from Mongolia. Available at: https://tinyurl.com/sspmhux (accessed November 2019).

Millon, A. and V. Bretagnolle. 2008. Predator population dynamics under a cyclic prey regime: Numerical responses, demographic parameters and growth rates. Oikos. 117: 1500–1510.

Milobog, Y.V. 2012. *Falconiformes* of the steppe zone of Ukraine: Species composition, territorial distribution, number dynamics and protection. PhD Thesis. Kryvvy Rih National University, Ukraine.

Mirski, P. 2008. Wiosenny przelot myszołowa (*Buteo buteo*) i innych ptaków szponiastych (Falconiformes) Doliną Środkowej Wisły. pp. 33–38. *In:* Chętnicki, W. [ed.]. Materiały XI Ogólnopolskiego Przeglądu Działalności Kół Naukowych Przyrodników, Białystok.

Mischenko, A.L. and V.N. Melnikov. 2019. Lesser Spotted Eagle *Clanga pomarina* Brehm, 1831 (Aves: Accipitridae) in Russia: recent data on numbers and distribution. Acta Zoologica Bulgarica, Suppl. 14: 71–75.

Mission Migration. 2018. Annual reports: Red Kite. https://www.migraction.net. (accessed 15/8/2018).

Mitrofanov, O.B. 2015. To the distribution of the SteppeEagle in the Altai reserve. pp. 337–338. *In:* Kovshar, A.F. [ed.]. Abstracts 14th International. Ornithol. Conf. of Northern Eurasia,vol. 1.

Mitropolsky, O.V., E.R. Fotteler and G.P. Tretyakov. 1987. Order Falconiformes. pp. 123–247. *In:* Matchanov, N.M., A.K. Sagitov and R.N. Meklenburtsev [eds.]. Birds of Uzbekistan, vol. 1. Fan Publisher, Tashkent, USSR (in Russian).

MME – Birdlife Hungary, 2020. www.satellitetracking.eu (accessed 23/3/2020).

MME – Sabuko, 2020. www.satellitetracking.eu (accessed 23/3/2020).

Molina, B. 2015. El milano real en España. III Censo Nacional. Población invernante y reproductora en 2014 y método de censo. SEO/BirdLife, Madrid.

Monsarrat, S., S. Benhamou, F. Sarrazin, C. Bessa-Gomes, W. Bouten and O. Duriez. 2013. How predictability of feeding patches affects home range and foraging habitat selection in avian social scavengers? PLoS ONE 8: e53077.

Monti, F., J.M. Dominici, R. Choquet, O. Duriez, G. Sammuri and A. Sforzi. 2014. The Osprey reintroduction in Central Italy: dispersal, survival and first breeding data. Bird Study. 61: 465–473.

Monti, F., F. Delfour, V. Arnal, S. Zenboudji, O. Duriez and C. Montgelard. 2018a. Genetic connectivity among osprey populations and consequences for conservation: philopatry versus dispersal as key factors. Conservation Genetics. 19: 839–851.

Monti, F., O. Duriez, J.M. Dominici, A. Sforzi, A. Robert, L. Fusani and D. Grémillet. 2018b. The price of success: integrative long-term study reveals ecotourism impacts on a flagship species at a UNESCO site. Animal Conservation. 21: 448–458.

Monti, F., O. Duriez, V. Arnal, J.M. Dominici, A. Sforzi, L. Fusani, D. Gremillet and C. Montgelard. 2015. Being cosmopolitan: evolutionary history and phylogeography of a specialized raptor, the osprey *Pandion haliaetus*. BMC Evolutionary Biology. 15: 255.

Monti, F., D. Grémillet, A. Sforzi, G. Sammuri, J.M. Dominici, R. Triay Bagur, A. Muñoz Navarro, L. Fusani and O. Duriez. 2018d. Migration and wintering strategies in vulnerable Mediterranean Osprey populations. Ibis. 160: 554–567.

Monti, F., D. Grémillet, A. Sforzi, J.M. Dominici, R. Triay Bagur, A. Muñoz Navarro, L. Fusani, R.H.G. Klaassen, T. Alerstam and O. Duriez. 2018c. Migration distance affects stopover use but not travel speed: contrasting patterns between long- and short-distance migrating ospreys. Journal of Avian Biology. 49: e01839.

Morabito, A., D. Ricciardi and M. Gustin. 2013. La migrazione post-riproduttiva nel Parco Nazionale d'Aspromonte (RC). Settembre 2013. Infomigrans. 32: 3.

Moreau, R.E. 1972. The Palaearctic-African Bird Migration Systems, Academic Press, London.

Moreno-Rueda, G. and M. Pizarro. 2007. Snake species richness and shrubland correlate with the short-toed eagle (*Circaetus gallicus*) distribution in south-eastern Spain. Annales Zoologici Fennici. 44: 314–320.

Morganti, M., R. Ambrosini and M. Sarà. 2019. Different trends of neighbouring populations of Lesser Kestrel: effects of climate and other environmental Conditions. Population Ecology. 61: 300–314.

Morozov, N.S. 2007. Changes in the timing of migration and winter records of the Common Buzzard (*Buteo buteo*) in the central part of European Russia: The effect of global warming? Zoologicheskiy Zhurnal. 86: 1336–1355.

Mougeot, F. and V. Bretagnolle. 2006. Breeding biology of the Red Kite *Milvus milvus* in Corsica. Ibis. 148: 436–448.

Mueller, T., R.B. O'Hara, S.J. Converse, R.P. Urbanek and W.F. Fagan. 2013. Social Learning of Migratory Performance. Science. 341: 999–1002.

Mullié, W.C. 2009. Birds, locusts, and grasshoppers. pp. 202–223. *In*: Zwarts, L., R. Bijlsma, J. van der Kamp and E. Wymega [eds.]. Living on the Edge: Wetlands and Birds in a Changing Sahel. KNNV Publishing, Zeist, The Netherlands.

Mullié, W.C. and Y. Guèye. 2010. Does bird predation enhance the impact of Green Muscle® (*Metarhizium acridum*) used for grasshopper control? Journal of Orthoptera Research. 19: 139–155.

Mundy, P.J., D. Butchart, J. Ledger and S. Piper. 1992. The Vultures of Africa, Academic Press, London.

Muñoz, A.R., D. Chamorro, A.G. Toxopeus, V. Venus, W. Bouten and A.K. Skidmore. 2016. One complete migration cycle of an adult Griffon Vulture: from southern Spain to Senegambia as revealed by high-resolution GPS tracking technology. *In:* Proceedings African Congress for Conservation Biology. El Jadida, Morocco.

Muriel, R., M. Ferrer, E. Casado and C.P. Calabuig. 2010. First successful breeding of reintroduced Ospreys *Pandion haliaetus* in mainland Spain. Ardeola. 57: 175–180.

Nachtigall, W. 2008. Der Rotmilan (*Milvus milvus*, L. 1758). *In:* Sachsen und Sudbrandenburg – Untersuchungen zu Verbreitung und Okologie. Dissertation, Univ. Halle-Wittenberg.

Naoroji, R. 2011. Birds of prey of the Indian subcontinent, Om Books International, New Delhi, India.

Nathan, R., O. Spiegel, S. Fortmann-Roe, R. Harel, M. Wikelski and W.M. Getz. 2012. Using tri-axial acceleration data to identify behavioral modes of free-ranging animals: general concepts and tools illustrated for griffon vultures. Journal of Experimental Biology. 215: 986–996.

Negro, J.J., M. De La Riva and J. Bustamante. 1991. Pattern of winter distribution and abundance of Lesser Kestrel (*Falco naumanni*) in Spain. Journal of Raptor Research. 25: 30–35.

Newton, I. 1986. The Sparrowhawk. T. & A.D. Poyser, London.

Newton, I. 1995. Relationship between breeding and wintering ranges in Palearctic-African migrants. Ibis. 137: 241–249.

Newton, I. 2008. The Migration Ecology of Birds. Academic Press, London, UK.

Newton, I. 2010a. Population ecology of raptors. T. & A.D. Poyser, London.

Newton, I. 2010b. Bird Migration, Collins, London.

Newton, I. 2016. Migration ecology of raptors. pp. 36–48. *In*: Forsman, D. [ed.]. Raptors of Europe, North Africa and the Middle East. Christopher Helm, London.

Newton, I., E. Meek and B. Little. 1978. Breeding ecology of the Merlin in Northumberland. British Birds. 71: 376–398.

Nielsen, B.P. and S. Christensen. 1970. Observations on the autumn migration of raptors in Lebanon. Ornis Scandinavica. 1: 65–73.

Nijman, V. 2004. Magnitude and timing of migrant raptors in central Java, Indonesia. Ardea. 92: 161–168.

Nikolov, S.C. 2015. High juvenile mortality during migration in a declining population of a long-distance migratory raptor. Ibis. 157: 545–557.

Nilsson, C., R.H. Klaassen and T. Alerstam. 2013. Differences in speed and duration of bird migration between spring and autumn. American Naturalist. 181: 837–845.

Nisbet, I.C.T. and T.C. Smout. 1957. Autumn observations on the Bosphorus and Dardanelles. Ibis. 99: 483–499.

Nisbet, I.C.T., P.R. Evans and P.P. Feen. 1961. Migration from Morocco into Southwest Spain in relation to weather. Ibis. 103: 349–372.

Noga, M., L. Vadel and R. Slobodník. 2017. Review and summary of red-footed falcon (*Falco vespertinus*) observations during migration periods in Slovakia. Slovak Raptor Journal. 11: 51–67.

Noskov, G.A. 2016. Buzzard *Buteo buteo*. pp. 280–283. *In:* Noskov, G.A., T.A. Rymkevich and A.R. Gaginskaya [eds.]. Migration of birds of North-West Russia. ANO LA 'Proffessional', Saint Petersburg.

Nourani, E. and N.M. Yamaguchi. 2017. The effects of atmospheric currents on the migratory behavior of soaring birds: a review. Ornithological Science. 16: 5–15.

Nourani, E., N.M. Yamaguchi and H. Higuchi. 2017. Climate change alters the optimal wind-dependent flight routes of an avian migrant. Proceedings of the Royal Society B. 284: 20170149.

Nourani, E., N.M. Yamaguchi, A. Manda and H. Higuchi. 2016. Wind conditions facilitate the seasonal water-crossing behaviour of Oriental Honey-buzzards *Pernis ptilorhynchus* over the East China Sea. Ibis. 158: 506–518.

Nourani, E., W.M.G. Vansteelant, P. Byholm and K. Safi. 2020. Dynamics of the energy seascape can explain intra-specific variations in sea-crossing behaviour of soaring birds. Biology Letters. 16: 20190797.

Nualsri, C., M. Siponen, H.J. Smit, D. Allen and R. DeCandido. 2010. Khao Dinsor - Thailand Raptor Migration Summary 2010. Technical Report.

Nygård, T. 1999. Long term trends in pollutant levels and shell thickness in eggs of merlin in Norway, in relation to its migration pattern and numbers. Ecotoxicology. 8: 23–31.

Ochapovsky, V.S. 1967. Materials on the fauna of birds of the Krasnodar Territory. Ph.D. Dissertation. University, Krasnodar, Russia (in Russian).

Ogada, D.L., F. Keesing and M.Z. Virani. 2011. Dropping dead: causes and consequences of vulture population declines worldwide. Annals of the New York Academy of Sciences: 1–15.

Ogada, D., P. Shaw, R.L. Beyers, R. Buij, C. Murn, J.M. Thiollay, C.M. Beale, R.M. Holdo, D. Pomeroy, N. Baker, S.C.Krüger, A. Botha, M.Z. Virani, A. Monadjem and A.R.E. Sinclair. 2015. Another continental vulture crisis: Africa's vultures collapsing toward extinction. Conservation Letters. 9: 89–97.

Olea, P.P. 2001. Post-fledging dispersal in the endangered Lesser Kestrel *Falco naumanni*. Bird Study. 48: 110–115.

Olea, P.P., R. Vera, A. de Frutos and H. Robles. 2004. Pre-migratory communal roosts of the Lesser Kestrel in the boreal summer. Journal of Raptor Research. 38: 278–282.

Ollé, A., F. Trabalon and M. Bertran. 2015. A review of occurrences of the Pallid Harrier *Circus macrourus* in the Western Mediterranean: a new migrant and wintering species. Catalan Journal of Ornithology. 31: 7–14.

Ollé, A., V. Estrada-Devesa and M. Gil-Velasco. 2016. Northern Peregrines *Falco peregrinus peregrinus* and *Falco peregrinus calidus*: two poorly-known taxa in Catalonia. Butlletí del CAC. 1: 11–25.

Onrubia, A. 2016. Spatial and temporal patterns of soaring birds migration through the Strait of Gibraltar. PhD Thesis, University of León.

Onrubia, A. and B. Martin. 2017. Migración por el Estrecho de Gibraltar. pp. 59–62. *In:* Urios, V.A., J. Bermejo, J. Vidal-Mateo and J. De la Puente [eds.]. Migración y ecología espacial de la población española de águila calzada. Monografía n. 2 del programa Migra. SEO/BirdLife. Madrid.

Onrubia, A., G. Muñoz, L. Barrios, A. De la Cruz and A.R. Muñoz. 2011b. Forest raptor migration over the Strait of Gibraltar, pp. 288–297. *In:* Zuberogoitia, I. and J.E. Martínez [eds.]. Ecology and Conservation of European Forest-Dwelling Raptors. Diputación Foral de Bizkaia.

Onrubia, A., A.-R. Muñoz, G.M. Arroyo, J. Ramírez, A. de la Cruz, L. Barrios, B.-U. Meyburg, C. Meyburg and T. Langgemach. 2011a. Autumn migration of Lesser Spotted Eagle *Aquila pomarina* in the Strait of Gibraltar: accidental or regular? Ardea. 99: 113–116.

Oppel, S., A.N. Powell and D.L. Dickson. 2009. Using an algorithmic model to reveal individually variable movement decisions in a wintering sea duck. Journal of Animal Ecology. 78: 524–531.

Oppel, S., P. Iankov, S. Mumun, G. Gerdzhikov, M. Iliev, S. Isfendiyaroglu, C. Yeniyurt and E. Tabur. 2014. Identification of the best sites around the gulf of Iskenderun, Turkey, for monitoring the autumn migration of Egyptian Vultures *Neophron percnopterus* and other diurnal raptors. Sandgrouse. 36: 240–249.

Oppel, S., V. Dobrev, V. Arkumarev, V. Saravia, A. Bounas, E. Kret, M. Velevski, S. Stoychev and S.C. Nikolov. 2015. High juvenile mortality during migration in a declining population of a long-distance migratory raptor. Ibis. 157: 545–557.

Oppel, S., V. Dobrev, V. Arkumarev, V. Saravia, A. Bounas, A. Manolopoulos, E. Kret, M. Velevski, G.S. Popgeorgiev and S.C. Nikolov. 2017. Landscape factors affecting territory occupancy and breeding success of Egyptian Vultures on the Balkan Peninsula. Journal of Ornithology. 158: 443–457.

Oppel, S., V. Dobrev, V. Arkumarev, V. Saravia, A. Bounas, E. Kret, M. Velevski, S. Stoychev, A. Orihuela-Torres, P. Perales, D. Rosado and J.M. Pérez-García. 2017. Feeding ecology of the Common Kestrel *Falco tinnunculus* in the south of Alicante (SE Spain). Revista Catalana d'Ornitologia. 33: 10–16.

Orta, J., D.A. Christie and E.F.J. Garcia. 2018. Red Kite (*Milvus milvus*). *In*: del Hoyo, J., A. Elliott, J. Sargatal, D.A. Christie and E. de Juana [eds.]. Handbook of the Birds of the World Alive, Lynx Edicions, Barcelona, Spain (retrieved from https://www.hbw.com/node/52977 on 13 July 2018).

Orta, J., J.S. Marks, G.M. Kirwan. 2019. Oriental Honey-buzzard (*Pernis ptilorhynchus*). *In:* del Hoyo, J., A. Elliott, J. Sargatal, D.A. Christie and E. de Juana [eds.]. Handbook of the Birds of the World Alive. Lynx Edicions, Barcelona, Spain (retrieved from https://www.hbw.com/node/52959 on 17 October 2019).

Orta, J., P.F.D. Boesman and J.S. Marks. 2020. Eurasian Kestrel (*Falco tinnunculus*), version 1.0. *In*: Billerman, S.M., B.K. Keeney, P.G. Rodewald and T.S. Schulenberg [eds.]. Birds of the World. Cornell Lab of Ornithology, Ithaca, NY, USA. https://doi.org/10.2173/bow.eurkes.01.

Orta, J., J.S. Marks, E.F.J. Garcia and G.M. Kirwan. 2020. Black Kite (*Milvus migrans*), version 1.0. *In:* del Hoyo, J., A. Elliott, J. Sargatal, D.A. Christie and E. de Juana [eds.]. Birds of the World, Cornell Lab of Ornithology, Ithaca, NY, USA, https://doi.org/10.2173/bow.blakit1.01.

Orta, J., G.M. Kirwan, P.F.D. Boesman, E.F.J. Garcia and J.S. Marks. 2020. Eurasian Griffon (*Gyps fulvus*), version 1.0. *In:* del Hoyo, J., A. Elliott, J. Sargatal, D.A. Christie and E. de Juana [eds.]. Birds of the World, Cornell Lab of Ornithology, Ithaca, NY, USA.

Ostapenko, M.M., G.P. Tretyakov, E. Shernazarov and E.N. Lanovenko. 1976. The birds migrations in south-western Kyzylkum. pp. 127–142. *In:* Gavrilov, E.I. [ed.]. Migration of Birds of Asia. Nauka Press, Alma-Ata, Kazakhstan (in Russian).

Österlöf, S. 1977. Migration, wintering areas and site tenacity of the European Osprey (*Pandion haliaetus haliaetus*). Ornis Scandinavica. 8: 61–78.

Østnes, J.E., R.T. Kroglund, O. Kleven and T. Nygĺrd. 2019. Migratory patterns of Ospreys (*Pandion haliaetus*) from central Norway. Ornis Fennica. 96: 101–111.

Özbahar, I. and M. Gül. 2018. Censuses and Observations from a Raptor Migration Bottleneck, Hatay Turkey: Phenology and Threats. Raptors Conservation 2018 Supplementum 1: 56.

P.R. Monte San Bartolo. 2015. XVIII campo sulla migrazione pre ripro duttiva dei rapaci nel Parco Naturale Regionale del Monte San Bartolo (PU) Primavera 2015. Infomigrans. 35: 7.

Paesani, G. and L. Vanni. 2006. La migrazione post-riproduttiva dei rapaci sull'Isola d'Elba (LI). Infomigrans. 18: 4.

Pajunen, M., M. Siponen, H.J. Smit, K. Sutasha, C. Nualsri, P. Round, D. Allen and R. DeCandido. 2012. Khao Dinsor—Thailand Raptor Migration Summary 2011. Technical Report.

Palasthy, J. and B.-U. Meyburg. 1973. On the food of the Lesser Spotted Eagle *Aquila pomarina* in Eastern Slovakia under atypic climatic conditions. Ornithologische Mitteilungen. 25: 61–72.

Palatitz, P., P. Fehérvári, S. Solt and B. Barov. 2009. European Species Action Plan for the Red-footed Falcon *Falco vespertinus*. BirdLife International/MME, Budapest, Hungary.

Palatitz, P., S. Solt and P. Fehérvári [eds.]. 2018. THE BLUE VESPER. Ecology and Conservation of the Red-footed Falcon. MME, Budapest, Hungary.

Palma, L. and P. Beja. 1994. Autumm migration of raptors through Sagres (SW Portugal). pp. 179–185. *In:* Meyburg, B.-U. and R.D. Chancellor [eds.]. Raptor Conservation Today: Proceedings of the 4th World Conference on Birds of Prey and Owls. WWWGBP, Berlin and Pica Press, London.

Palmer, R.S. [ed.]. 1988. Handbook of North American birds, vol. 4, Yale Univ. Press, New Haven.

Palomino, D. 2016. Milano negro—*Milvus migrans*. *In:* Salvador, A. and M.B. Morales [eds.]. Enciclopedia Virtual de los Vertebrados Españoles. Museo Nacional de Ciencias Naturales, Madrid. http://www.vertebradosibericos.org/.

Palomino, D. and B. Molina. 2012. Águila calzada *Aquila pennata*. pp. 188–189. *In*: Del Moral, J.C., B. Molina, A. Bermejo and D. Palomino [eds.]. Atlas de las aves en invierno en España 2007–2010. Ministerio de Agricultura, Alimentación y Medio Ambiente y SEO/BirdLife. Madrid.

Pandolfi, M. and F. Piccinetti. 2008. Osservazione della migrazione primaverile dei rapaci nelle Isole Tremiti e nel Promontorio del Gargano (Parco Nazionale del Gargano) 2008. Infomigrans. 21: 5.

Pandolfi, M. and L. Sonet. 2003. Migrazione di rapaci lungo la cosa adriatica (Parco Naturale San Bartolo, 1998–2001). Fenologia e comportamento delle specie del genere *Circus*. Avocetta. 27: 57–59.

Panuccio, M. 2005. Protection of migratory raptors in the Mediterranean. Sustainable Mediterranean. 35: 13–14.

Panuccio, M. 2011. Wind effects on visible raptor migration in Spring at the Strait of Messina, Southern Italy. Journal of Raptor Research. 45: 88–92.

Panuccio, M. and N. Agostini. 2006a. Spring passage of second-calendar-year Honey Buzzards at the Strait of Messina. British Birds. 99: 95–96.

Panuccio, M. and N. Agostini. 2006b. Comments on the roosting behaviour of Marsh Harriers during migration. British Birds. 99: 365–367.

Panuccio, M. and N. Agostini. 2006c. Spring migration of Pallid (*Circus macrourus*) and Montagu's harriers (*Circus pygargus*) in relation to sex and age classes at two watchsites of the central Mediterranean. Buteo. 15: 3–10.

Panuccio, M. and N. Agostini. 2010a. Comparison of the water crossing behaviour of Western Marsh Harriers (*Circus aeruginosus*) and European Honey Buzzards (*Pernis apivorus*) during autumn migration. Chinese Birds. 1: 30–35.

Panuccio, M. and N. Agostini. 2010b. Timing, age classes and water crossing behaviour of Black Kites (*Milvus migrans*) during spring migration across the Central Mediterranean. The Ring. 32: 55–61.

Panuccio, M. and N. Agostini. 2012. The status of second-calendar-year Honey-buzzards in Europe – continues (response to Corso 2012). British Birds. 105: 486.

Panuccio, M., N. Agostini and B. Massa. 2002. Crossing the Tyrrhenian Sea: spring migration of Marsh Harriers (*Circus aeruginosus*), sex classes and relation to wind conditions. Vogelwarte. 41: 271–275.

Panuccio, M., N. Agostini and B. Massa. 2004. Spring raptor migration at Ustica, southern Italy. British Birds. 97: 400–414.

Panuccio, M., N. Agostini and U. Mellone. 2005a. Autumn migration strategies of honey buzzards, black kites, marsh and montagu's harriers over land and over water in the Central Mediterranean. Avocetta. 29: 27–32.

Panuccio, M., M. Gustin and G. Bogliani. 2011. A comparison of two methods for monitoring migrating broad-winged raptors approaching a long water crossing. Avocetta. 35: 13–17.

Panuccio, M., N. Agostini and G. Premuda. 2012. Ecological barriers promote risk minimisation and social learning in migrating short-toed snake eagles. Ethology Ecology and Evolution. 24: 74–80.

Panuccio, M., N. Agostini and C. Barboutis. 2013b. Raptor migration in Greece: a review. Avocetta. 37: 1–7.

Panuccio, M., G. Chiatante and D. Tarini. 2013c. Two different migration strategies in response to an ecological barrier: Western Marsh Harriers and juvenile European Honey Buzzards crossing the central-eastern Mediterranean in autumn. Journal of Biological Research-Thessaloniki. 19: 10–18.

Panuccio, M., U. Mellone and L. Muner. 2013d. Differential wintering area selection in Eurasian Marsh Harrier (*Circus aeruginosus*): a ringing recoveries analysis. Bird Study. 60: 52–59.

Panuccio, M., G. Bogliani and G. Dell'Omo. 2018b. From breeding to wintering sites: flocking behavior allows fledging black kites to find safe flyway and stopover-sites. pp. 87. *In:* Abstracts Raptor Research Foundation Conference, South Africa.

Panuccio, M., B. Ghafouri and E. Nourani. 2018c. Is the slope between the Alborz mountains and Caspian Sea in northern Iran a bottleneck for migrating raptors? Journal of Raptor Research. 52: 530–533.

Panuccio, M., N. Agostini, L. Baghino and G. Bogliani. 2013a. Visible migration of Short-toed Snake Eagles: interplay of weather and topographical features. Journal of Raptor Research. 47: 60–68.

Panuccio, M., N. Agostini, U. Mellone and G. Bogliani. 2014. Circannual variation in movement patterns of the Black Kite (*Milvus migrans migrans*): a review. Ethology Ecology and Evolution. 26: 1–18.

Panuccio, M., B. D'Amicis, E. Canale and A. Roccella. 2005b. Sex and age ratios of Marsh Harriers *Circus aeruginosus* wintering in central-southern Italy. Avocetta. 29: 13–17.

Panuccio, M., A. Duchi, G. Lucia and N. Agostini. 2017. Species-specific behaviour of raptors migrating across the Turkish Straits in relation to weather and geography. Ardeola. 64: 305–324.

Panuccio, M., N. Agostini, G. Bogliani and G. Dell'Omo. 2018a. Migrating raptor counts: the need for sharing objectives and field protocols, and the benefits of using radar. Bird Study. 65: S77–S84.

Panuccio, M., G. Lucia, N. Agostini, D. Ottonello and G. Bogliani. 2015. Motion capacity, geography and ecological features explain the present distribution of a migratory top predator. Ecological Research. 30: 181–190.

Panuccio, M., C. Barboutis, G. Chiatante, A. Evangelidis and N. Agostini. 2016a. Pushed by increasing air temperature and tailwind speed: weather selectivity of raptors migrating across the Aegean Sea. Ornis Fennica. 93: 159–171.

Panuccio, M., B. Martin, M. Morganti, A. Onrubia and M. Ferrer. 2016b. Long-term changes in autumn migration dates at the Strait of Gibraltar reflect population trends of soaring birds. Ibis. 159: 55–65.

Panuccio, M., V. Stanzione, C. Catoni, M. Santini and G. Dell'Omo. 2016c. Radar tracking reveals influence of crosswinds and topography on migratory behavior of European honey buzzards. Journal of Ethology. 34: 73–77.

Panuccio, M., N. Agostini, L. Nelli, G. Andreou and S. Xirouchachis. 2019a. Factors shaping distribution and abundance of raptors wintering in two large Mediterranean islands. Community Ecology. 20: 93–103.

Panuccio, M., G. Dell'Omo, G. Bogliani, C. Catoni and N. Sapir. 2019b. Migrating birds avoid flying through fog and low clouds. International Journal of Biometeorology. 63: 231–239.

Panuccio, M., M. Gustin, G. Lucia, G. Bogliani and N. Agostini. 2019c. Flight altitude of migrating European Honey Buzzards facing the open sea. Ornithological Science. 18: 49–57.

Panuccio, M., N. Agostini, S. Wilson, G. Lucia, J. Ashton-Booth, G. Chiatante, U. Mellone and S. Todisco. 2006. Does the Honey-buzzard feed during migration? British Birds. 99: 367–368.

Panuccio, M., N. Agostini, G. Lucia, U. Mellone, S. Wilson, J. Ashton-Booth, G. Chiatante and S. Todisco. 2010. Local weather conditions affect migration strategies of adult Western Honey Buzzards (*Pernis apivorus*) through an isthmus area. Zoological Studies. 49: 651–656.

Parchami-Araghi, M., F. Eskandari and E. Gilasian. 2015. Avian wound myiasis caused by *Calliphora vicina* Robineau-Desvoidy (*Diptera: Calliphoridae*) in an immature migrating Eastern Imperial Eagle (*Aquila heliaca* Savigny) (*Aves: Accipitridae*) in southwestern Iran. Journal of Veterinary Science and Technology. 6: 212.

Parfenov, E.A. 2006. To the fauna of the rare species of Falconiformes of the Caucasian Mineral Waters Region. pp. 140–144. *In:* Ishchenko, V.M. [ed.]. Problems of the Development of Biology and Ecology in the North Caucasus. Proceedings of Conference, Stavropol State University, Stavropol, Russia (in Russian).

Pavlov, A.N. and A.V. Molodovsky. 1962. About birds of prey of Mangyshlak and their numbers. Zoological Journal. 41: 951–954 (in Russian).

Pavón, D., R. Limiñana, V. Urios, A. Izquierdo, B. Yáñez, M. Ferrer and A. de la Vega. 2010. Autumn migration of juvenile Short-toed Eagles *Circaetus gallicus* from southeastern Spain. Ardea. 98: 113–117.

Pearson, D., G. Backhurst and C. Jackson. 2014. The study and ringing of Palaearctic birds at Ngulia Lodge, Tsavo West National Park, Kenya, 1969–2012: an overview and update. Scopus Special Suppl. 4: 1–80.

Pearson, D.J. and P.C. Lack. 1992. Migration patterns and habitat use by passerine and near-passerine migrant birds in eastern Africa. Ibis. 134: 89–98.

Pelle, L., A. Siclari and A. Scuderi. 2014. La migrazione post riproduttiva nel Parco Nazionale d'Aspromonte (RC) agosto-ottobre 2014. Infomigrans. 34: 3.

Pennycuick, C.J. 2008. Modelling the flying bird, Academic Press, London, UK.

Pepler, D. and R. Matin. 2001. The lesser kestrel (*Falco naumanni*) in Africa: migration, conservation status and dangers: a review. pp. 318–336. *In:* Toledano, F.G. and M. Corroto Matesanz [eds.]. Proceedings of the IV Lesser Kestrel National Congress Madrid: Consejeria de Medio Ambiente y Grefa.

Pererva, V.I. 1986. Brief reports on the Steppe Eagle. pp. 126–127. *In*: Gvozdev, E.V. [ed.]. Rare Animals of Kazakhstan. Nauka. Alma-Ata, Kazakhstan (in Russian).

Pererva, V.I. and A.Yu. Blokhin. 1981. Assessment of the death of rare species of birds of prey on power lines. pp. 36–39. *In:* Flint, V.E. [ed.]. Biological aspects of protection of rare animals: Proceedings. Institute of Nature Protection and Reserves Moscow, Russia.

Pérez-García, J.M., U. Sellis and Ü. Väli. 2014. Winter ranging behaviour of a greater spotted eagle (*Aquila clanga*) in southeast Spain during four consecutive years. Slovak Raptor Journal. 8: 123–128.

Perlman, Y. 2015. Assessment of several globally threatened species in Israel for IUCN update 2015. BirdLife's Globally Threatened Bird Forums (https://globally-threatened-bird-forums.birdlife.org/2016/11/pallid-harrier-circus-macrourus-downlist-from-near-threatened-to-least-concern/).

Péron, G., C.H. Fleming, O. Duriez, J. Fluhr, C. Itty, S. Lambertucci, K. Safi, E.L.C. Shepard and J.M. Calabrese. 2017. The energy landscape predicts flight height and wind turbine collision hazard in three species of large soaring raptor. Journal of Applied Ecology. 54: 1895–906.

Peshev, H., E. Stoynov, D. Parvanov and A. Grozdanov. 2018. Seasonal and spatial dynamics of the population of the Griffon Vulture in Southwestern Bulgaria. Acta Zoologica Bulgarica. 12: 67–75.

Peshkov, B.I. 1976. Some features of the ecology of the Steppe Eagle in the eastern part of the range (Transbaikalia). pp. 100–103. *In*: Valyus, M.I. and R.S. Volskis [eds.]. Materials for the 2nd All-Union Conference "Species and its productivity in the area". Vilnius, Lithuania (in Russian).

Peterson, R.T. 1934. A field guide to the birds, Houghton Mifflin, Boston.

Petrov, V.Yu. 2006. Steppe Eagle - *Aquila nipalensis* (Temminck, 1828). pp. 82–84. *In:* Irisova, N.L. [ed.]. Red Book of Altai Territory (Animals). Barnaul, Russia (in Russian).

Petrovich, Z.O. and K.O. Redinov. 2008. Identification of sites of congregation of Red-footed falcons (*Falco vespertinus*) during autumn migration in the north-western Black Sea region. pp. 286–289. *In:* Gavrilyuk, M.N. [ed.]. New Research on birds of prey and owls. Materials of the III International scientific conference "Birds of prey of Ukraine". Kriviy Rig, Ukraine, Publisher (in Russian).

Petty, S.J., D.I.K. Anderson, M. Davidson, B. Little, T.N. Sherrat, C.J. Thomas and X. Lambin. 2003. The decline of Common Kestrels *Falco tinnunculus* in a forested area of northern England: the role of predation by Northern Goshawks *Accipiter gentilis*. Ibis. 145: 472–483.

Pfeiffer, T. and B.U. Meyburg. 2009. Migratory and wintering behaviour of the Red Kite *Milvus milvus* in Thuringia (Germany) as revealed by Satellite Telemetry. Vogelwarte. 47: 171–187.

Phipps, W.L., P. López-López, E.R. Buechley, S. Oppel, E. Álvarez, V. Arkumarev, R. Bekmansurov, O. Berger-Tal, A. Bermejo, A. Bounas, I.C. Alanís, J. de la Puente, V. Dobrev, O. Duriez, R. Efrat, G. Fréchet, J. García, M. Galán, C. García-Ripollés, A. Gil, J.J. Iglesias-Lebrija, J. Jambas, I.V., Karyakin, E. Kobierzycki, E. Kret, F. Loercher, A. Monteiro, J. Morant Etxebarria, S.C. Nikolov, J. Pereira, L. Peške, C. Ponchon, E. Realinho, V. Saravia, Ç.H. Şekercioğlu, T. Skartsi, J. Tavares, J. Teodósio, V. Urios and N. Vallverdú. 2019. Spatial and temporal variability in migration of a soaring raptor across three continents. Frontiers in Ecology and Evolution. 7: 1–14.

Pilard, P., G. Jarry and V. Lelong. 2009. Suivi et conservation du dortoir de Faucons crécerellettes et d'Elanions naucler de la région de Kaolack (Sénégal). pp. 159–161. *In:* Pilard, P. [ed.]. Actes du VII Congrès International sur le Faucon crécerellette 2009. LPO Service Edition, Almendralejo, Spain.

Pilard, P., J.M. Thiollay and G. Rondeau. 2004. Données sur l'hivernage du Faucon crécerellette *Falco naumanni* en Afrique de l'Ouest. Alauda. 72: 323–328.

Pilard, P., M. Bourgeois and D. Sylla. 2017. Localisation des quartiers d'hivernage et phènologie des migrations prènuptiale et postnuptiale chez la population francaise du faucon crecerellette *Falco naumanni* à l'aide de geolocateurs. Alauda. 85: 1–28.

Pilard, P., M. Bourgeois, B. Remy and N. Saulnier. 2016. French Lesser Kestrel Action Plan. Abstract book of the International Lesser Kestrel Expert Workshop in Plovdiv, Bulgaria, 4–8.10.2016 http://greenbalkans.org/birdsofprey/lesserkestrellife/en/International_Expert_Workshop_-c150.

Pilyuga, V.I. 1991. New data on the nesting of endangered birds of prey in the Odessa region and adjacent territories. pp. 139–164. *In:* Korzyukov, A.I. [ed.]. Rare Birds of the Region of the Northern Black Sea. Lybid Press, Kiev-Odessa, Ukraine (in Russian).

Pleske, T. 1888. Revision der turkestanischen Ornis nach Sammlungen des verstorbenen Konservators V. Russov. St.-Petersb. 58 p. (Mem. Acad. Sci. St.-Petersb. Ser. 7, t. 36, n. 3) (in Russian).

Podofillini, S., J.G. Cecere, M. Griggio, M. Corti, E.L. De Capua, M. Parolini, N. Saino, L. Serra and D. Rubolini. 2019. Benefits of extra food to reproduction depend on maternal condition. Oikos. 128: 943–959.

Poessel, S.A., E.A. Bragin, P.B. Sharpe, D.K. Garcelon, K. Bartoszuk and T.E. Katzner. 2018. Movements and landscape use of Eastern Imperial Eagles *Aquila heliaca* in Central Asia. Bird Study. 65: 208–218.

Pokrovsky, I., E. Dorothée, R.A. Ims, O. Kulikova, N. Lecomte and N.G. Yoccoz. 2014. Diet, nesting density, and breeding success of rough-legged buzzards (*Buteo lagopus*) on the Nenetsky Ridge, Arctic Russia. Polar Biology. 37: 447–457.

Polakowski, M., L. Jankowiak, Z. Kasprzykowski, G. Bela, A. Kosmicki, A. Janczyszyn, A. Niemczyk and D. Kilon. 2014. Autumn migratory movements of raptors along the southern Baltic coast. Ornis Fennica. 91: 39–47.

Polivanov, V.M., N.N. Polivanova and O.A. Vitovich. 1985. Visible migration of birds across Teberdinskii Zapovednik. pp. 19–33. *In:* Polivanov, V.M. [ed.]. The Birds of North-Western Caucasus. General Directorate of Hunting and Reserves under the Council of Ministers of the RSFSR, Moscow, Russia (in Russian).

Polivanov, V.M., O.A. Vitovich, V.A. Telpov and I.V. Tkachenko. 1999. Notes on the meetings of the Imperial Eagle in the Central Caucasus. pp. 88. *In:* Belik, V.P. [ed.]. Royal Eagle: Distribution, State of Populations and Prospects for the Protection of the Imperial Eagle (*Aquila heliaca*) in Russia. Series: Rare Bird Species, vol. 1. Russian Bird Conservation Union, Moscow, Russia (in Russian).

Poole, A.F. 1989. Ospreys: a Natural and Unnatural History, Cambridge University Press, Cambridge.

Poole, A.F. 2019. Ospreys: The Revival of a Global Raptor, Johns Hopkins Uuniversity Press, Baltimore.

Popov, V.V. 2000. The Red-footed falcon *Falco vespertinus* in the Baikal region. Russian Ornithological Journal, Express Edition. 123: 25–27. (in Russian).

Popov, V.V. and V.G. Maleev. 2008. Reduction of the number of some common bird species in the Upper Angara area, 216–230. *In:* Fauna and Ecology of Animals in Central Siberia and the Far East. Interuniversity scientific proceeding. Issue. 5. Publisher, Krasnoyarsk, Russia (in Russian).

Portenko, L.A. 1928. Sketch of the fauna of birds of Podolsk Governorate. Bulletin MOIP. Otd. Biol. 37: 92–204.

Porter, R. and A. Suleiman. 2012. The Egyptian Vulture *Neophron percnopterus* on Socotra, Yemen: population, ecology, conservation and ethno-ornithology. Sandgrouse. 34: 44–62.

Porter, R. and I. Willis. 1968. The autumn migration of soaring birds at the Bosphorus. Ibis. 110: 520–536.

Porter, R. and M. Beaman. 1985. A resume of raptor migration in Europe and the Middle East. Conservation Studies on Raptors. ICBP Technical Publication. pp. 237–242. *In:* Newton, I. and R.D. Chancellor [eds.]. Conservation Studies on Raptors, ICBP Technical Publication 5., Cambridge.

Porter, R. and S. Christensen. 1987. The autumn migration of raptors and other soaring birds in North Yemen. Sandgrouse. 9: 121–124.

Porter, R., I. Willis, S. Christensen and B.P. Nielsen. 1974. Flight identification of European raptors, T. and A.D. Poyser, London, U.K.

Poslavsky, A.N. 1986. The Imperial Eagle. pp. 134. *In:* Gvozdev, E.V. [ed.]. Rare Animals of Kazakhstan. Nauka Press, Alma-Ata, Kazakhstan (in Russian).

Potapov, E.R. 1997. What determines the population density and reproductive success of rough-legged buzzards (*Buteo lagopus*) in the Siberian tundra? Oikos. 78: 362–376.

Prakash, V. 1988. Greater spotted eagle (*Aquila clanga*) breeding in Keoladeo National Park, Bharatpur. Journal of Bombay Natural History Society. 85: 418.

Premuda G. 2017a. Versilia, Alpi Apuane – Campo rapaci primavera. Infomigrans. 39: 8–9.

Premuda G. 2019b. Versilia – Alpi Apuane. Campo Rapaci Autunno. Infomigrans. 44: 11–13

Premuda, G. 2017b. Versilia-Alpi Apuane campo rapaci 2017b. La ventennale autunnale. Infomigrans. 40: 10.

Premuda, G. 2018. Versilia – Alpi Apuane. Campo rapaci 2018: la ventennale autunnale. Infomigrans. 42: 11–13.

Premuda, G. 2019a. Versilia – Alpi Apuane 2019. Campo rapaci primavera. Infomigrans. 43: 7–9.

Premuda, G. and L. Baghino. 2004. La migrazione autunnale dell'Aquila minore, *Hieraaetus pennatus*, attraverso la Penisola Italiana. Rivista Italiana Ornitologia. 74: 103–115.

Premuda, G. and L. Baghino. 2012. Spring arched migration of black kite *Milvus migrans* over the Apuane Alps (Tuscany). Avocetta. 36: 103–106.

Premuda, G., U. Mellone and L. Cocchi. 2004. Osservazioni sulla modalità della migrazione primaverile dei rapaci a Capo d'Otranto. Avocetta. 28: 33–36.

Premuda, G., L. Cocchi and R. Raneri. 2007c. La migrazione primaverile dei rapaci a Cap Bon (Tunisia). Infomigrans. 19: 2.

Premuda, G., N. Agostini, L. Cocchi and R. Molajoli. 2007a. Primo campo di osservazione della migrazione primaverile dei rapaci in Sardegna. Aves Ichnusae. 8: 3–11.

Premuda, G., N. Agostini, L. Cocchi and R. Molajoli. 2007a. Primo campo di osservazione della migrazione primaverile dei rapaci in Sardegna. Aves Ichnusae. 8: 3–11.

Premuda, G., N. Agostini, L. Cocchi and R. Molajoli. 2007a. Primo campo di osservazione della migrazione primaverile dei rapaci in Sardegna. Aves Ichnusae. 8: 3–11.

Premuda, G., A. Belosi, F. Viviani and M. Franchini. 2015. Short-toed Eagle *Circaetus gallicus* population monitoring at the Apuane Alps migration watch-site (Tuscany). Avocetta. 39: 5–12.

Premuda, G., L. Baghino, M. Gustin and M. Borioni. 2010. Data on spring migration of immature Short-toed Eagles Circaetus gallicus through the Central Mediterranean route (Italy, Tunisia). Avocetta. 34: 65–68.

Premuda, G., M. Gustin, M. Pandolfi, L. Sonet and M. Cento. 2008. Spring raptor migration along the Adriatic coast (Italy): a comparative study over three sites. Avocetta. 32: 13–20.

Premuda, G., L. Baghino, T. Guillosson, M. Jardin, M. Tirado and V. Esteller. 2007b. A remarkable case of circuitous autumn migration of the Booted Eagle *Hieraaetus pennatus* through the western and central Mediterranean. Ardeola. 54: 349–357.

Prevost, Y.A. 1983. The moult of the Osprey. Ardea. 71: 199–209.

Prieta, J. 2015. Red-footed falcon (*Falco vespertinus*) in Extremadura. http://aves-extremadura.blogspot.com/2015/05/el-cernicalo-patirrojo-en-extremadura.html [accessed 15/9/2015].

Prins, H.H.T. and T. Namgail [eds.]. 2017. Bird Migration Across the Himalayas, Cambridge University Press.

Prommer, M., J. Bagyura, J. Chavko and M. Uhrin. 2012. Migratory movements of Central and Eastern European Saker Falcons (*Falco cherrug*) from juvenile dispersal to adulthood. Aquila. 119: 111–134.

Ptaki Silesiana. 2020. Monitoring lotów. http://ptaki-silesiana.pl/1/strona-glowna.html. Accessed in 31.03.2020.

Puan, C.L., C.A. Yeap, K.C. Lim, A.T. Lim, S.S. Khoo and N. Cheung. 2014. Northbound Migration Count of Raptors at Tanjung Tuan, Peninsular Malaysia: Magnitude, Timing, and Flight Behavior. Journal of Raptor Research. 48: 162–172.

Puzanov, I.I., V.I. Kozlov and G.P. Kiparisov. 1955. Animals of the Gorky Region (Vertebrates). 2nd ed. Gorky, Russia (in Russian).

Raab, R., I. Literak, C. Schutz, P. Spakovszky, J. Steindl and N. Schonemann. 2017. GPS-based telemetry studies of Central European Red Kites *Milvus milvus* – methodical difficulties and analytical opportunities based on the initial results. Ornithologische Mitteilungen. 7/8: 245–260.

Raine, A. 2007. La migrazione autunnale nell'isola di Malta: risultati del Campo Rapaci. Malta. Infomigrans. 20: 2–3.

Raine, A. 2008. La migrazione autunnale a Malta: risultati del Campo Rapaci 2008. Malta. Infomigrans. 22: 2–3.

Raine, A. and R. Vella. 2009. La migrazione post riproduttiva a Malta: risultati del Campo Rapaci 2009. Malta. Infomigrans. 24: 2–3.

Raine, A. and R. Vella. 2010. La migrazione primaverile a Malta: risultati del campo organizzato da BirdLife Malta (11–30 aprile 2010). Infomigrans. 25: 4–5.

Ramírez, J. and M. Panuccio. 2019. Flight feather moult in Western Marsh Harriers during autumn migration. Avian Research. 10: 7.

Ramirez, J., F. Zufiaur and M. Guerrero. 2019. Spring migration of Eurasian Griffon Vultures across the Strait of Gibraltar: number, timing and age composition. Ardeola. 66: 113–118.

Rayaleh, H., M. McGrady, E.O. Abdillahi and A.M. Darar. 2013. Spring raptor migration across the Bab el Mandeb Straits and fitting of GPS PTT to Egyptian vulture –Djibouti Side- February 28 to March 14, 2013. Djibouti Nature, Djibouti, Djibouti Nature.

Rebassa, M. 1995. La migraciò postnupcial de rapinyaires a l'illa de Cabreara: tretsgenerals. Anuari Ornitològic de les Balears 1995.

Rebassa, M. and L. Ventoso. 2015. La migració de rapinyaires per Mallorca i Cabrera. Es Busqueret. 40: 6–11.

Région Aquitaine. 2018. Programme en faveur du Balbuzard pêcheur en Aquitaine 2018-2021, Bilan d'activite 2018. Technical report, pp. 31.

Réglade M.A., A.P. Auvinen, M. Bouzin and J. Terraube. 2015. Les stationnements hivernaux du Busard pâle *Circus macrourus* en France (2006–2015). Ornithos. 22: 312–325.

Restani, M. 2000. Age-specific stopover behavior of migrant Bald Eagles. Wilson Bulletin. 112: 28–34.

Ricciardi, D., R. Garavaglia, D. Ardizzone, G. Chiofalo, J.P. Fyott, R. Vella, I. Adami, S. Cutini and A. Giordano. 2009. Stretto di Messina (versante siciliano): 26° anno di attività 1 aprile - 27 maggio 2009. Infomigrans 23: 2–3.

Ricciardi, D., R. Garavaglia, D. Ardizzone, G. Chiofalo, J.P. Fyott, R. Vella, I. Adami, S. Cutini, A. Giordano and F. Adragna. 2010. Stretto di Messina (versante siciliano): 27° anno di attività 1 aprile - 23 maggio 2010. Infomigrans. 25: 2–3.

Riedstra, B., C. Dijkstra and S. Daan. 1998. Daily energy expenditure of male and female marsh harrier nestlings. The Auk. 115: 653–641.

Ristow, D. 2004. On the insect diet of Eleonora's Falcon *Falco eleonorae* and its importance for coloniality. pp. 705–712. *In*: R.D. Chancellor and B.-U. Meyburg [eds.]. Raptors Worldwide.

Roberts, J.L. 1979. Observations of the migration of raptors and other large soaring birds in Bulgaria, 1975–1978. Ibis. 121: 301–312.

Roberts, T.J. 1991. The birds of Pakistan, Oxford University Press, Karachi, Pakistan.

Rodríguez, A., J.J. Negro, J. Bustamante, J. Fox and V. Afanasyev. 2009. Geolocators map the wintering grounds of threatened Lesser Kestrels in Africa. Diversity and Distributions. 15: 1010–1016.

Rodríguez, B., A. Rodríguez, M. Siverio and F. Siverio. 2013. Conservation implications of past and present nesting habitat selection of the endangered Osprey *Pandion haliaetus* population of the Canary Islands. Ibis. 155: 891–897.

Rodríguez, C. and J. Bustamante. 2003. The effect of weather on Lesser Kestrel breeding success: can climate change explain historical population declines? Journal of Animal Ecology. 72: 793–810.

Rodríguez, C., K. Johst and J. Bustamante. 2006. How do crop types influence breeding success in Lesser Kestrels through prey quality and availability? A modelling approach. Journal of Applied Ecology. 43: 587–597.

Rogacheva, E.V. 1988. Birds of Central Siberia. Science Publishing House, Moscow, Russia. (in Russian).

Rogacheva, E.V., E.E. Syroechkovsky, O.V. Bursky, A.A. Moroz and B.I. Sheftel. 1988. Birds of the central siberian biosphere reserve. Part 1. Non-passerines, 42–96. *In*: Rogacheva, E.V. [ed.]. Protection and Rational use of Fauna and Ecosystems of the North Yenisei, City, Russia. Publisher (in Russian).

Román, J.R., A. Onrubia, G. Muñoz, R. Benjumea, A. Román Muñoz, A. de la Cruz, M. González, D. Cuenca, L. Barrios, C. Torralvo, J. Elorriaga, F. Zufiaur, B. Pérez, M. León and V. Cortés. 2009. La migración del Cernícalo Primilla *Falco naumanni* a través del estrecho de Gibraltar. pp. 151–154. *In:* Pilard, P. [ed.]. Actas del VII Congreso Internacional sobre el Cernícalo Primilla. LPO Mission Rapaces. Imprimerie Lagarde, France.

Roselaar, C.S. and A. Mansour. 2009. Review of Rare Birds in Iran, 1860s–1960s. Podoces. 4: 1–27.

Rosén, M. and A. Hedenstrom. 2002. Soaring flight in the Eleonora's Falcon (*Falco eleonorae*). The Auk. 119: 835–40.

Rosén, M., A. Hedenström, A. Badami, F. Spina and S. Åkesson. 1999. Hunting flight behaviour of the Eleonora's falcon *Falco eleonorae*. Journal of Avian Biology. 30: 342–350.

Roth, T. 2008. Outward (autumn) bird migration at the Southeastern Peninsula and Cape Greco, Cyprus: the phenologies of regular migrants. Sandgrouse. 30: 77–89.

Roth, T. and A. Corso. 2007. Cyprus – the autumn 2005 raptor migration at the southeast peninsula (terminating at Cape Greco). Sandgrouse. 29: 79–90.

RRRCN. 2014. Migration study of eagles of southern Siberia in 2014. http://rrrcn.ru/en/migration/ (accessed 15 September 2019).

RRRCN. 2016. Migration study of eagles of southern Siberia in 2016. http://rrrcn.ru/en/migration/ (accessed 15 September 2019).

RRRCN. 2020. Russian Raptor Research and Conservation Network. Ringing Data Base. http://rrrcn.ru/en/ringing/bd (accessed 31 March 2020).

RSPB. 2008. https://www.rspb.org.uk/our-work/conservation/satellite-tracking-birds/tracking-ospreys/#.

Rudebeck, G. 1950. Studies on bird migration based on field studies in southern Sweden. Berlingska boktryckeriet, Lund.

Rudnick, J.A., T.E. Katzner, E.A. Bragin, O.E. Rhodes and J.A. Dewoody. 2005. Using naturally shed feathers for individual identification, genetic parentage analyses, and population monitoring in an endangered Eastern Imperial Eagle (*Aquila heliaca*) population from Kazakhstan. Molecular Ecology. 14: 2959–2967.

Rus, A.I., A.E. Duerr, T.A. Miller, J.R. Belthoff and T.E. Katzner. 2017. Counterintuitive roles of experience and weather on migratory performance. The Auk. 134: 485–497.

Russian Raptor Research and Conservation Network. 2017. Satellite and GSM tracking. http://rrrcn.ru/en/telemetry.

Rustamov, E.A. and A.K. Kurbanov. 2011. Steppe Eagle. pp. 248–249. *In:* Annabairamov, V. [ed.]. Red Book of Turkmenistan, vol. 2. Invertebrates and vertebrates. Ashgabat, Turkmenistan (in Russian).

Ruxton, G.D. and D.C. Houston. 2004. Obligate vertebrate scavengers must be large soaring fliers. Journal of Theoretical Biology. 228: 431–436.

Ryabstev, V. and T. Katzner. 2007. Severe declines of Eastern Imperial Eagle *Aquila heliaca* populations in the Baikal region, Russia: a modern and historical perspective. Bird Conservation International. 17: 197–209.

Ryabstev, V.V. 1999. The ecology of the Imperial Eagle in the Baikal region. pp. 122–128. *In:* Belik, V.P. [ed.]. Royal Eagle: Distribution, State of Populations and Prospects for the Protection of the Imperial Eagle (*Aquila heliaca*) in Russia. Series: Rare Bird Species, vol. 1. Russian Bird Conservation Union, Moscow, Russia (in Russian).

Ryabtsev, V.V. 1997. The greater spotted eagle *Aquila clanga* in the Baikal Region. Russian Ornithological Journal. 24: 3–8.

Ryabtsev, V.V. 2000. Eagles of Baikal. Talcy, Irkutsk.

Ryabtsev, V.V. 2016. Trends in raptor numbers of Front-Baikalia during 1930–2012. pp. 188–193. *In:* Birds of Prey in the Northern Eurasia: problem and adaptations in current environment, Sochi.

Safriel, U. 1968. Bird migration at Elat, Israel. Ibis. 110: 283–320.

Sakoulis, A. 2001. The status of the Imperial Eagle *Aquila heliaca* in Greece. Acrocephalus. 22: 105–108.

Salewski, V., H. Schmaljohann and M. Herremans. 2005. New bird records from Mauritania. Malimbus. 27: 22–39.

Salikhbaev, H.S., D.Y. Kashkarov and A. Sharipov. 1970. The Birds. pp. 42–99. *In:* Sultanov, G.S. [ed.]. Ecology of Vertebrate Animals of Nuratau Ridge. Fan Press, Tashkent, Uzbekistan (in Russian).

Salvan, J. 1967. Contribution à l'étude des oiseaux du Tchad. L'Oiseau et R.F.O. 38: 53–85.

Sammut, M. 2005. Marsh Harriers roosting in trees. British Birds. 98: 314–316.

Sammut, M. and E. Bonavia. 2004. Autumn raptor migration over Buskett, Malta. British Birds. 97: 318–322.

Sammut, M., N. Fenech and J.A. Pirotta. 2013. Autumn raptor migration over Malta. British Birds. 106: 217–223.

Sanchez-Zapata, J.A., J.A. Donazar, A. Delgado, M. Forero, O. Ceballos and F. Hiraldo. 2007. Desert locust outbreaks in the Sahel: resource competition, predation and ecological effects of pest control. Journal of Applied Ecology. 44: 323–329.

Sanchis-Serra, A., C. Real Margalef, J.V. Morales Pérez, I.M. Pérez Ripol, C. Tormo-Cuñat, Y. Carrión-Marco, G. Pérez Jordà, A. Ribera-Gómez, J. Bolufer-Marqués and V. Villaverde-Bonilla. 2014. Towards the identification of a new taphonomic agent: An analysis of bone accumulations obtained from modern Egyptian vulture (*Neophron percnopterus*) nests. Quaternary International. 330: 136–149.

Sándor, A. and B.P. Anthony. 2018. Mapping the conflict of raptor conservation and recreational shooting in the Batumi Bottleneck, Republic of Georgia. Journal of Threatened Taxa. 10: 11850–11862.

Sándor, A., J. Jansen and W.M.G. Vansteelant. 2017. Understanding hunter's habits and motivations for shooting raptors in the Batumi raptor-migration bottleneck, southwest Georgia. Sandgrouse. 39: 2–15.

Sangha, H.S. and S.S. Poonia. 2016. Recovery of a ringed juvenile Eastern Imperial Eagle *Aquila heliaca* at Sardarshahr, Thar Desert, India. Indian Birds. 11: 17–18.

Sangha, H.S., S.S. Dhumal and S. Ovalekar. 2014. The first breeding record of the Saker Falcon *Falco cherrug milvipes* for the Indian Subcontinent in Ladakh, Jammu & Kashmir. Indian Birds. 9: 146–148.

Santos, C.D., J.P. Silva, A.R. Muñoz, A. Onrubia and M. Wikelski. 2020. The gateway to Africa: what determines sea crossing performance of a migratory soaring bird at the strait of Gibraltar? Journal of Animal Ecology. 89: 1317–1328.

Santos, C.D., F. Hanssen, A.R. Muñoz, A. Onrubia, M. Wikelski, R. May and J.P. Silva. 2017. Match between soaring modes of black kites and the fine-scale distribution of updrafts. Scientific Reports. 7: 6421.

Sanz-Aguilar, A., F. De Pablo and J. Donázar. 2015a. Age-dependent survival of island vs. mainland populations of two avian scavengers: delving into migration costs. Oecologia. 179: 405–414.

Sanz-Aguilar, A., A. Cortés-Avizanda, D. Serrano, G. Blanco, O. Ceballos, J.M. Grande, and J.A. Donázar. 2017. Sex- and age-dependent patterns of survival and breeding success in a long-lived endangered avian scavenger. Scientific Reports. 7: 40204.

Sanz-Aguilar, A., J.A. Sánchez-Zapata, M. Carrete, J.R. Benitez, E. Ávila, R. Arenas and J.A. Donázar. 2015b. Action on multiple fronts, illegal poisoning and wind farm planning, is required to reverse the decline of the Egyptian vulture in southern Spain. Biological Conservation. 187: 10–18.

Saparmuradov, D. and A. Eminov. 1994. The abundance of some rare species of birds in the Middle Amu Darya during the period of seasonal migrations. pp. 43–45. *In:* Rare and Poorly Studied Birds of Uzbekistan and adjacent territories. Materials 1 Republican Ornithological Conf. Tashkent, Uzbekistan, December 6–7 (in Russian).

Sarà, M. 2003. The colonization of Sicily by the Black Kite (*Milvus migrans*). Journal of Raptor Research. 37: 167–172.

Sarà, M. 2010. Climate and land-use changes as determinants of Lesser Kestrel *Falco naumanni* abundance in Mediterranean cereal steppes (Sicily). Ardeola. 57: 3–22.

Sarà, M. and M. Di Vittorio. 2003. Factors influencing the distribution, abundance and nest-site selection of an endangered Egyptian vulture (*Neophron percnopterus*) population in Sicily. Animal Conservation. 6: 317–328.

Sarà, M. and S. Bondì. 2016. Satellite tracking on migration and wintering of Sicilian Lesser Kestrel. International Lesser Kestrel Expert Workshop in Plovdiv, Bulgaria, 4-8.10.2016. http://greenbalkans.org/birdsofprey/lesserkestrellife/en/International_Expert_Workshop_-c150.

Sarà, M., D. Campobello, L. Zanca and B. Massa. 2014. Food for flight: Pre-migratory dynamics of the Lesser Kestrel *Falco naumanni*. Bird Study. 61: 29–41.

Sarà, M., S. Bondì, A. Bermejo, M. Bourgeois, M., Bouzin, J. Bustamante, J. de la Puente, A. Evangelidis, A. Frassanito, E. Fulco, G. Giglio, G. Gradev, M. Griggio, L. López-Ricaurte, P. Kordopatis, S. Marin, J. Martínez, R. Mascara, U. Mellone, S. Pellegrino, P. Pilard, S. Podofillini, M. Romero, M. Gustin, N. Saulnier, L. Serra, A. Sfougaris, V. Urios, M. Visceglia, K. Vlachopoulos, L. Zanca, J.G. Cecere and D. Rubolini. 2019. Broad-front migration leads to strong migratory connectivity in the lesser kestrel (*Falco naumanni*). Journal of Biogeography. 46: 2663–2677.

Saravia, V., E. Kret and V. Dobrev and S.C. Nikolov. 2016. Assessment of mortality causes for the Egyptian Vulture (*Neophron percnopterus*) in Bulgaria and Greece (1997–2015). Hellenic Ornithological Society, Athens, Greece, 6 pp.

Saurola, P. 1977. Suomalaisten hiirihaukkojen muuttoreitit. Lintumies. 12: 44–53.

Saurola, P. 2005. Monitoring and conservation of finnish Ospreys, *Pandion haliaetus*, in 1971–2005. *In*: Status of raptor populations in eastern fennoscandia. Proceedings of the Workshop, Kostomuksha, Karelia, Russia, November: 8–10.

Saurola, P. 2007. Monitoring and conservation of finnish Ospreys, *Pandion haliaetus*, in 1971–2005. pp. 8–10. *In:* Koskimies, P. and N.V. Lapshin [eds.]. Status of raptor populations in eastern fennoscandia, Proceedings of the Workshop, Kostomuksha, Karelia, Russia, November.

Saurola, P. 2012. An overview of monitoring for raptors in Finland. Acrocephalus. 33: 203–215.

Saurola, P. 2014. Finnish satellite ospreys. LUOMUS Finnish Museum of Natural History. Available at http://www.luomus.fi/en/finnish-satellite-ospreys.

Saurola, P., J. Valkama and W. Velmala. 2013. The Finnish Bird Ringing Atlas, vol. I, Suomen Rengastusatlas I, Finnish Museum of Natural History, Helsinki.

Scalco, L., G. Biffoni, L. Bajno, N. Grasso and F. Sartirana. 2019. La migrazione post-riproduttiva in Valle Stura (CN): anno 2019. Infomigrans. 44: 18–21.

Schaub, T., R.H.G. Klaassen, A. Schlaich and B.J. Koks. 2018. Sleeper dikes and unexpected detours. First results on habitat use and migratory behaviour of Rough-legged Buzzards wintering in the Netherlands. 151st DO-G Annual Convention Heidelberg. DOI 10.13140/RG.2.2.26670.36169.

Scherler, P., B. Naef-Daenzer, A. Bach and M.U. Grüebler. 2015. Mechanisms of population regulation over time and space of Swiss Red Kites. pp. 47–48. *In:* De la Puente, J. and V. De la Torre [eds.]. Abstracts Book of the II International Symposium on the Red Kite. Binaced, Huesca, Spain. SEO-Monticola & FAB.

Schifferli, A. 1967. Vom Zug schweizerischer Schwarzer Milane *Milvus migrans* nach Ringfunden. Ornithologische Beobachter. 64: 34–51.

Schindler, S., K. Poirazidis, C. Ruiz, C. Scandolara, B. Cárcamo, C. Eastham and G. Catsadorakis. 2015. At the crossroads from Asia to Europe: spring migration of raptors and black storks in Dadia National Park (Greece). Journal of Natural History. 49: 285–300.

Schlaich, A.E. 2011. Stopover site ecology of Montagu´s Harrier (*Circus pygargus*) in East-Morocco. MSc Thesis, University of Oldenburg, Germany.

Schlaich, A.E. 2019. Migrants in double jeopardy: Ecology of Montagu's Harriers on breeding and wintering grounds. PhD Thesis, University of Groningen, the Netherlands.

Schlaich, A.E., R.H.G. Klaassen, W. Bouten, V. Bretagnolle, B.J. Koks, A. Villers and C. Both. 2016. How individual Montagu's Harriers cope with Moreau's Paradox during the Sahelian winter. Journal of Animal Ecology. 85: 1491–1501.

Schmaljohann, H. 2018. Proximate mechanisms affecting seasonal differences in migration speed of avian species. Scientific Reports. 8: 4106.

Schmid, H. 2000. Getrennte Wege: Der Herbstzug von juvenilen und adulten Wespenbussarden *Pernis apivorus*–eine Synthese. Ornithologische Beobachter. 97: 191–222.

Schmid, H., T. Steuri and B. Bruderer. 1986. Zugverhalten von Mäusebussard *Buteo buteo* und Sperber *Accipiter nisus* im Alpenraum. Ornithologische Beobachter. 83: 111–134.

Schmidt-Rothmund, D., R. Dennis and P. Saurola. 2014. The Osprey in the Western Palearctic: Breeding Population Size and Trends in the Early 21 st Century. Journal of Raptor Research. 48: 375–386.

Schneider-Jacoby, M. 2001. Lastovo a new bottleneck site for the migratory Honey Buzzards (*Pernis apivorus*). Acrocephalus. 22: 163–165.

Scholer, M.N., B. Martín, M. Ferrer, A. Onrubia, M.J. Bechard, G.S. Kaltenecker and J.D. Carlisle. 2016. Variable shifts in the autumn migration phenology of soaring birds in southern Spain. Ardea. 104: 83–93.

Schröpfer, L. 2008. Raroh velký. *Falco cherrug*. Sokol rároh. Saker Falcon. pp. 291–292. *In:* Cepák, J., P. Klvaňa, J. Škopek, L. Schröpfer, M. Jelínek, D. Hořák, J. Formánek and J. Zárybnický [eds.]. Atlas migrace ptáků České a Slovenské republiky. Czech and Slovak bird migration atlas. Aventinum, Praha.

Schweizer, M. and M. Mitropolskiy. 2008. The occurrence of Crested Honey Buzzard *Pernis ptilorhynchus* in Uzbekistan and Tajikistan and its status in Central Asia. Sandgrouse. 30: 161–163.

Semenov, N.M., A.V. Agafonov, D.S. Rezinko and A.A. Rozhkov. 1959. Breeding and abundance of the steppe eagle in the steppes of the southern Stalingrad and Northern Astrakhan regions. pp. 159–163. *In*: Formozov, A.N. [ed.]. Geography of the Population of Terrestrial Vertebrates and Methods of its Study. Moscow, Russia.

SEO/BirdLife. 2012a. Análisis preliminar del banco de datos de anillamiento de aves del Ministerio de Agricultura, Alimentación y Medio Ambiente, para la realización de un atlas de migración de aves de España. SEO/BirdLife-Fundación Biodiversidad. Madrid.

SEO/BirdLife 2012b. Atlas de las aves en invierno en España 2007–2010. Ministerio de Agricultura, Alimentación y Medio Ambiente-SEO/ BirdLife. Madrid.

Sergio, F. 2003. Relationship between laying dates of black kites *Milvus migrans* and spring temperatures in Italy: rapid response to climate change? Journal of Avian Biology. 34: 144–149.

Sergio, F. and G. Bogliani. 1999. Eurasian hobby density, nest area occupancy, diet, and productivity in relation to intensive agriculture. The Condor. 101: 806–817.

Sergio, F. and V. Penteriani. 2005. Public information and territory establishment in a loosely colonial raptor. Ecology. 86: 340–346.

Sergio, F., P. Pedrini and L. Marchesi. 2003. Adaptive selection of foraging and nesting habitat by black kites (*Milvus migrans*) and its implications for conservation: a multi-scale approach. Biological Conservation. 112: 351–362.

Sergio F., R.G. Bijlsma, G. Bogliani and I. Wyllie. 2001. *Falco subbuteo* Hobby. BWP Update. 3(3): 133–156.

Sergio, F., J. Blas, M. Forero, N. Fernández, J.A. Donázar and F. Hiraldo. 2005. Preservation of wide-ranging top predators by site-protection: Black and red kites in Doñana National Park. Biological Conservation. 125: 11–21.

Sergio, F., A. Tanferna, R. De Stephanis, L.L. Jiménez, J. Blas and F. Hiraldo. 2017. Migration by breeders and floaters of a long-lived raptor: implications for recruitment and territory quality. Animal Behaviour. 131: 59–72.

Sergio, F., G. Tavecchia, A. Tanferna, J. Blas, G. Blanco and F. Hiraldo. 2019. When and where mortality occurs throughout the annual cycle changes with age in a migratory bird: individual vs population implications. Scientific Reports. 9: 1–8.

Sergio, F., A. Tanferna, R. de Stephanis, L.L. Jiménez, J. Blas, G. Tavecchia, D. Preatonit and F. Hiraldo. 2014. Individual improvements and selective mortality shape lifelong migratory performance. Nature. 515: 410–413.

Serrano, D., J.L. Tella, M.G. Forero and J.A. Donázar. 2001. Factors affecting breeding dispersal in the facultatively colonial lesser kestrel: individual experience vs. conspecific cues. Journal of Animal Ecology. 70: 568–578.

Serrano, D., J.L. Tella, J.A. Donázar and M. Pomarol. 2003. Social and individual features affecting natal dispersal in the colonial lesser kestrel. Ecology. 84: 3044–3054.

Shamoun-Baranes J., F. Liechti and W.M.G. Vansteelant. 2017. Atmospheric conditions create freeways, detours and tailbacks for migrating birds. Journal of Comparative Physiology. 203: 509–529.

Shamoun-Baranes, J., Y. Leshem, Y. Yom-Tov and O. Liechti. 2003. Differential Use of Thermal Convection by Soaring Birds over Central Israel. The Condor. 105: 208–218.

Sharma, A.K., m. Saini, S.D. Singh, V. Prakash, A. Das, R.B. Dasan, S. Pandey, D. Bohara, T.H. Galligan, R.E., Green, D. Knopp and R.J. Cuthbert. 2014. Diclofenac is toxic to the Steppe Eagle *Aquila nipalensis*: widening the diversity of raptors threatened by NSAID misuse in South Asia. Bird Conservation Internationl. 24: 282–286.

Sharma, P. and K.S.G. Sundar. 2009. Counts of Steppe Eagles *Aquila nipalensis* at a carcass dump in Jorbeer, Rajasthan, India. Forktail. 25: 161–163.

Shevchenko, V.L. 1986. Steppe Eagle in the Northern Caspian region. pp. 117–123. *In:* Rare Animals of Kazakhstan. Materials for the 2nd edition of the Red Book of the Kazakh SSR Alma-Ata, Kazakhstan (in Russian).

Shevchenko, V.L., E.I. Gavrilov, V.A. Naglov, A.K. Fedosenko and O.M. Tatarinova. 1978. About the avifauna of the Volga-Urals interfluve (birds of prey and owls). pp. 99–117. *In:* Biology of Birds in Kazakhstan. Proceedings of the Institute of Zoology of Kazakhstan. V. 38. Publisher, Alma-Ata, Kazakhstan.(in Russian).

Shirihai, H. and D.A. Christie. 1992. Raptor migration at Eilat. British Birds. 85: 141–186.

Shirihai, H., E. Dovrat, D.A. Christie and A. Harris. 1996. The Birds of Israel, Academic Press, London.

Shirihai, H., R. Yosef, D. Alon, G.M. Kirwan and R. Spaar. 2000. Raptor migration in Israel and the Middle East, International Birding & Research Center, Eilat, Israel.

Shiu, H.-J., K.-i Tokita, E. Morishita, E. Hiraoka, Y. Wu, H. Nakamura and H. Higuchi. 2006. Route and site fidelity of two migratory raptors: Grey-faced Buzzards *Butastur indicus* and Honey-buzzards *Pernis apivorus*. Ornithological Science. 5: 151–156.

Shmeleva, G.P. and S.P. Haritonov. 2016. Migrations of diurnal raptors on materials of Russian Bird Ringing Centre. Ivanovo State University, Ivanovo.

Shurulinkov, P., G. Daskalova, N. Chakarov, K. Hristov, S. Dyulgerova, Y. Gocheva, S. Cheshmedzhiev, M. Madzharov and I. Dimchev. 2011. Characteristics of soaring birds' spring migration over inland SE Bulgaria. Acrocephalus. 32: 29–43.

Sierdsema, H. and E.E. van Loon. 2008. Filling the gaps: using count survey data to predict bird density distribution patterns and estimate population sizes. Revista Catalana d'Ornitologia. 24: 88–99.

Sikora, A. and P. Zieliński. 2013. Occurrence of Greater Spotted Eagle *Aquila clanga* in Pomerania compared with its occurrence in Poland. Ptaki Pomorza. 4: 121–133.

Simmons, K.E.L. 1951. Raptor migration in the Suez area: autumn 1949-spring 1950. Ibis. 93: 402–406.

Simmons, R.E. 1997. Steppe Eagle. pp. 180. *In:* Harrison, J.A., D.G. Allan, L.G. Underhill, M. Herremans, A.J. Tree, V. Parker and C.J. Brown [eds.]. The atlas of South African birds. Vol. 1: Non-passerines. BirdLife South Africa and Avian Demography Unit, Johannesburg, South Africa.

Simmons, R.E. 2000. Harriers of the world: Their behaviour and ecology, Oxford University Press, Oxford.

Simmons, R.E. 2005. Steppe Eagle *Aquila nipalensis*. pp. 528–529. *In:* Hockey, P.A.R., W.R.J. Dean and P.G. Ryan [eds.]. Roberts Birds of Southern Africa. 7th ed. Trustees of the John Voelcker Bird Book Fund, Cape Town, South Africa.

Siponen, M., K. Sutasha, A. Forsten, R. DeCandido, C. Nualsri, P. Round, A. Lindholm and W. Phatara-Atikom. 2013. Khao Dinsor—Thailand Raptor Migration Summary 2012. Technical Report.

Siverio, M., P. Lopez-Suarez, F. Siverio, B. Rodrıguez, N. Varo-Cruz, and L. F. Lopez-Jurado 2014. Density, nest site characteristics and breeding rates of the Osprey (*Pandion haliaetus*) in the southern limit of its range in the Western Palearctic (Boa Vista, Cape Verde Islands). African Journal of Ecology. 52: 50–58.

Siverio, M., F. Siverio, B. Rodríguez and J.C. Del Moral. 2018. El águila pescadora en España y Portugal: población invernante 2016–2017, reproductora en 2018 y método de censo. SEO/BirdLife. Madrid.

Smith, N.G., D.L. Goldstein and G.A. Bartholomew. 1986. Is long-distance migration possible for soaring hawks using only stored fat? The Auk. 103: 607–611.

Snow, D.W. 1968. Movements and mortality of British Kestrels *Falco tinnunculus*. Bird Study. 15: 65–83.

Snow, D.W. and C.M. Perrins. 1998. The Birds of the Western Palearctic. Concise edition, Oxford University Press, Oxford.

Sokolov, V., A. Sokolov and A. Dixon. 2018. Migratory movements of Peregrine falcons *Falco peregrinus*, breeding in the Yamal peninsula, Russia. Ornis Hungarica. 26: 222–231.

Sokolov, V., N. Lecomte, A. Sokolov, Md.L. Rahman and A. Dixon. 2014. Site fidelity and home range variation during the breeding season of peregrine falcons (*Falco peregrinus*) in Yamal, Russia. Polar Biology. 37: 1621–1631.

Somov, N.N. 1897. Ornithological fauna of Kharkiv province. Trudy Kharkivskogo obshchestva ispytatelej prirody, Supplement to vol. 26: 149–150.

Sonet, L., V. Amatiello, M. Pandolfi and M. Balsamo. 2009. La migrazione primaverile nel Parco Naturale del Monte San Bartolo (PS). Infomigrans. 23: 7.

Sonin, V.D. and S.I. Lipin. 1980. Seasonal aspects of ecology of the Imperial Eagle at the Baikal region. pp. 122–124. *In:* Kibalchich, O.A. [ed.]. Seasonal Rhythm of Rare and Endangered Species of Plants and Animals. Proceedings of Conference. Moscow Branch of Geographical Society of the USSR, Moscow, Russia (in Russian).

Sørensen, I.H., A.E. Schlaich, R.H.G. Klaassen, H. Heldbjerg and B.J. Koks. 2017. Rare case of an adult male Montagu's Harrier *Circus pygargus* over-summering in West Africa, as revealed by GPS tracking. Journal of Ornithology. 158: 753–760.

Sotnikov, V.N. 1999. Birds of the Kirov region and adjacent territories. V. 1. Non-passerines. Publisher, Kirov, Russia (in Russian).

Spaar, R. 1997. Flight strategies of migrating raptors; a comparative study of interspecific variation in flight characteristics. Ibis. 139: 523–535.

Spaar, R. and B. Bruderer. 1996. Soaring migration of Steppe Eagles *Aquila nipalensis* in southern Israel: flight behaviour under various wind and thermal conditions. Journal of Avian Biology. 27: 289–301.

Spaar, R. and B. Bruderer. 1997. Migration by flapping or soaring: flight strategies of Marsh, Montagu's and Pallid harriers in southern Israel. The Condor. 99: 458–469.

Spaar, R., H. Stark and F. Liechti. 1998. Migratory flight strategies of Levant Sparrowhawks: time or energy minimization. Animal Behaviour. 56 : 1158–1197.

Spangenberg, E.P. and G.A. Feigin. 1936. The birds of Low Syrdarya and adjacent areas. pp. 41–184. *In:* Makarov, V.N. [ed.]. Proceedings of Zoological Museum of Moscow University, vol. 3. Biomedgiz, Moscow, Russia (in Russian).

Spiegel, O., R. Harel, W.M. Getz and R. Nathan. 2013. Mixed strategies of griffon vultures (*Gyps fulvus*) response to food deprivation lead to a hump-shaped movement pattern. Movement Ecology. 1: 5.

Spiegel, O., R. Harel, A. Centeno-Cuadros, O. Hatzofe, W.M. Getz and R. Nathan. 2015. Moving beyond curve fitting: using complementary data to assess alternative explanations for long movements of three vulture species. American Naturalist. 185: E44–E54.

Spina, F. and S. Volponi, S. 2008. Atlante della Migrazione degli Uccelli in Italia, vol. 1: non-Passeriformi. ISPRA – MATTM, Roma.

Stach, R., S. Jakobsson, C. Kullberg and T. Fransson. 2012. Geolocators reveal three consecutive wintering areas in the thrush nightingale. Animal Migration. 1: 1–7.

Stanciu, CR., R. Zaharia, G.B. Chisamera, I. Cobzaru, V.D. Gavril and D. Murariu. 2017. Migration Strategies of Common Buzzard (*Buteo Buteo* Linnaeus, 1758) in Dobruja. Travaux du Muséum National d'Histoire Naturelle "Grigore Antipa". 60: 1–9.

Stanciu, C.-R., R. Zaharia, G.-B. Chişamera, I. Cobzaru, V.-D. Gavril, R.-V. Popescu-Mirceni and D. Murariu. 2017. Aspects regarding raptors migration over the Black Sea. Travaux du Muséum National d'Histoire Naturelle "Grigore Antipa". 60: 1–20.

Stara, K., L. Sidiropoulos and R. Tsiakiris. 2016. Bound eagles, evil vultures and cuckoo horses. Preserving the bio-cultural diversity of carrion eating birds. Human Ecology. 44: 751–764.

Stark, H. and F. Liechti. 1993. Do Levant Sparrowhawks *Accipiter brevipes* also migrate at night? Ibis. 135: 233–236.

Stawarczyk, T., T. Cofta, Z. Kajzer, J. Lontkowski and A. Sikora. 2017. Rare birds of Poland. Studio B&W, Sosnowiec.

Steinfatt, O. 1932. Der Bosporus als Landbrücke für den Vogelzug zwischen Europa und Kleinasien dargestellt auf Grund von Beobachtungen im Herbst 1931. Journal für Ornithologie. 80: 354–383.

Sternalski, A. and V. Bretagnolle. 2010. Experimental evidence of specialised phenotypic roles in a mobbing raptor. Behavioral Ecology and Sociobiology. 64: 1351–1361.

Sternalski, A., F. Mougeot and V. Bretagnolle. 2012. Adaptive significance of permanent female mimicry in a bird of prey. Biology Letters. 8: 167–170.

Storino, P., S. Santagati, A. Siclari, G. Martino, E. Muscianese, M. Policastrese, A. Scuderi, S. Urso and S. Tralongo. 2017. Studio della migrazione post-riproduttiva di uccelli rapaci e cicogne nel parco nazionale dell'Aspromonte – Stagione 2017. Infomigrans. 40: 5–7.

Stoychev S.A., D.A. Demerdzhiev, D.D. Dobrev, S.D. Spasov, B.-U. Meyburg and G.S. Popgeorgiev. 2018. Dispersal and mortality of eastern imperial eagles from Bulgaria. pp. 135–136. *In*: Abstracts of the II International Scientific and Practical Conference "Eagles of Palearctic: Study and Conservation". Raptors Conservation Suppl. 1. Sibecocenter, Novosibirsk, Russia.

Stoychev, S., D. Demerdzhiev, S. Spasov, B.-U. Meyburg and D. Dobrev. 2014. Survival rate and mortality of juvenile and immature eastern imperial eagles (*Aquila heliaca*) from Bulgaria studied by satellite telemetry. Slovak Raptor Journal. 8: 53–60.

Strahm, W. and D. Landenbergue. 2013. La réintroduction du Balbuzard pêcheur *Pandion haliaetus* en Suisse romande. Nos Oiseaux. 60: 123–142.

Strandberg, R. 2013. Ageing, sexing and subspecific identification of Osprey, and two WP records of American Osprey. Dutch Birding. 35: 69–87.

Strandberg, R. and T. Alerstam. 2007. The strategy of fly–and–forage migration, illustrated for the osprey (*Pandion haliaetus*). Behavioral Ecology and Sociobiology. 61: 1865–1875.

Strandberg, R., R.H.G. Klaassen and K. Thorup. 2009d. Spatio-temporal distribution of migrating raptors: a comparison of ringing and satellite tracking. Journal of Avian Biology. 40: 500–510.

Strandberg, R., T. Alerstam, M. Hake and N. Kjellen. 2008a. Short-distance migration of the Common Buzzard *Buteo buteo* recorded by satellite tracking. Ibis. 151: 200–206.

Strandberg, R., T. Alerstam, M. Hake and N. Kjellén. 2009a. Short-distance migration of the Common Buzzard *Buteo buteo* recorded by satellite tracking. Ibis. 151: 200–206. Strandberg, R., R.H.G. Klaassen, P. Olofsson and T. Alerstam. 2009c. Daily travel schedules of adult Eurasian Hobbies *Falco subbuteo* – variability in flight hours and migration speed along the route. Ardea. 97: 287–295.

Strandberg, R., R.H.G. Klaassen, M. Hake and T. Alerstam. 2009e. How hazardous is the Sahara Desert crossing for migratory birds? Indications from satellite tracking of raptors. Biology Letters. 6: 297–300.

Strandberg, R., M. Hake, R.H.G. Klaassen and T. Alerstam. 2012. Movements of Immature European Honey Buzzards *Pernis apivorus* in Tropical Africa. Ardea. 100: 157–162.

Strandberg, R., R.H.G. Klaassen, M. Hake, P. Olofsson and T. Alerstam. 2009b Converging migration routes of Eurasian hobbies *Falco subbuteo* crossing the African equatorial rain forest. Proceedings of the Royal Society B. 276: 727–733.

Strandberg, R., R.H.G. Klaassen, M. Hake, P. Olofsson, K. Thorup and T. Alerstam. 2008b. Complex timing of Marsh Harrier *Circus aeruginosus* migration due to pre- and post-migratory movements. Ardea. 96: 159–171.

Strautman, F.I. 1963. Birds of the western regions of the Ukraine, vol. 1, Publishing House of Lviv University, Lviv, Russia (in Russian).

Stresemann, E. 1954. Zur Frage der Wanderungen des Eleonorenfalken. Vogelwarte. 17: 182–183.

Strick, J., P. Vercammen, J. Judas and O. Combreau. 2011. Satellite tracking of a rehabilitated Greater Spotted Eagle *Aquila clanga*. Zoology in the Middle East. 54: 103–106.

STRIX. 2012. Developing and testing the methodology for assessing and mapping the sensitivity of migratory birds to wind energy development, BirdLife International, Cambridge.

Subedi, T. and R. DeCandido. 2013. Raptor migration summary for autumn 2012, central Nepal. Nepal Raptor Conservation Program, Himalayan Nature and Nepalese Ornithological Union. Kathmandu, Nepal.

Subedi, T. and R. DeCandido. 2014. Raptor migration summary for autumn 2013, central Nepal. Nepal Raptor Conservation Program, Himalayan Nature and Nepalese Ornithological Union. Kathmandu, Nepal.

Subedi, T.R. and S. Gurung. 2018. Raptor Migration Summary, Thoolakharka Nepal – Autumn 2018. Nepal Raptor Conservation Program, Himalayan Nature and Nepalese Ornithological Union. Kathmandu, Nepal.

Subedi, T.R., S. Gurung and H.S. Baral. 2018. Migration count of Steppe Eagle at Thoolakharka watch site, Nepal. Raptors Conservation. Supplement. 1: 94.

Subedi, T.R., R. DeCandido, H.S. Baral, S. Gurung, S. Gurung, C.L. Puan and S.A.M. Sah. 2017. Population structure and annual migration pattern of Steppe Eagles at Thoolakharka watch site, Nepal, 2012–2014. Journal of Raptor Research. 51: 165–171.

Sugasawa, S. and H. Higuchi. 2019. Seasonal contrasts in individual consistency of oriental honey buzzards' migration. Biology Letters. 15: 20190131.

Sultana J. and C. Gauci. 1982. A new guide to the birds of Malta, The Ornithological Society, Malta.

Sunyer, C. and J. Viñuela. 1990. Migración e invernada del Esmerejón en España. Ardeola. 37: 279–290.

Survillo, A.V., N.B. Sandzhiev, S. Ts. Ulyumdzhiev, G.M. Cherevichenko, V.D. Kolesnikov and M.O. Mastyukov. 1977. On the abundance and ecology of the Steppe Eagle in the central regions of Kalmykia. pp. 247–248. *In:* Neshcheret, V.A. [ed.]. VII All Union Ornithol. Conf. Abstracts of Reports and Messages. Part 2. Naukova Dumka. Kiev, Ukraine (in Russian).

Sutherland, W.J. 1998. Evidence for flexibility and constraint in migration systems. Journal of Avian Biology. 29: 441–446.

Sutherland, W.J. and D.J. Brooks. 1981. The autumn migration of raptors, storks, pelicans, and spoonbills at the Belen pass, southern Turkey. Sandgrouse. 2: 1–21.

Swinhoe, R. 1870. Zoological notes of a journey from Canton to Peking and Kalgan. Proceedings Zoological Society of London. 38: 427–451.

Sylven, M. 1978. Interspecific relations between sympatrically wintering Common Buzzards *Buteo buteo* and Rough-legged Buzzard *Buteo lagopus*. Ornis Scandinavica. 9: 109–115.

Symes, C.T. and S. Woodborne. 2010. Migratory connectivity and conservation of the Amur Falcon *Falco amurensis*: a stable isotope perspective. Bird Conservation International. 20: 134–148.

Tal, T. and D. Jansen. 2019. Spring report 2019. Batumi Raptor Count, Georgia. URL: https://www.batumiraptorcount.org/spring-report-2019 (last accessed 30 March 2020).

Tanferna, A., L. López-Jiménez, J. Blas, F. Hiraldo and F. Sergio. 2012. Different location sampling frequencies by satellite tags yield different estimates of migration performance: pooling data require a common protocol. PLoS ONE. 7: e49659431.

Tanferna, A., L. López-Jiménez, J. Blas, F. Hiraldo and F. Sergio. 2013. Habitat selection by Black kite breeders and floaters: implications for conservation management of raptor floaters. Biological Conservation. 160: 1–9.

Tavares-Nunes, G., L. Sander-Hoffmann, B. Macena, G.A. Bencke and L. Bugoni. 2015. A black kite *Milvus migrans* on the Saint Peter and Saint Paul Archipielago, Brazil. Revista Brasileira de Ornitologia. 23: 31–35.

Tella, J.L. and M.G. Forero. 2000. Farmland habitat selection of wintering Lesser Kestrels in a Spanish pseudosteppe: implications for conservation strategies. Biodiversity and Conservation. 9: 433–441.

Tella, J.L., J.J. Negro, J.A. Donázar and F. Hiraldo. 1996. Costs and benefits of urban nesting in the lesser kestrel (*Falco naumanni*). pp. 53–60. *In*: Bird, D.M., D. Varland and J.J. Negro [eds.]. Raptors in Human Landscapes. Academic Press, London.

Terrasse, M. 2006. Évolution des déplacements du Vautour fauve en France et en Europe. Ornithos. 13: 273–299.

Terrasse, J.F., F. Sarrazin, J.P. Choisy, C. Clemente, S. Henriquet, P. Lecuyer, J.L. Pinna and C. Tessier. 2004. A success story: the reintroduction of Griffon *Gyps fulvus* and Black *Aegypius monachus* Vultures in France. pp. 127–145. *In:* Chancellor, R.D. and B.-U. Meyburg [eds.]. Raptors worldwide. World Working Group on Birds of Prey and Owls and MME/BirdLife Hungary. Berlin and Budapest.

Terrasse, M. and M. Thauront. 1988. The vultures of Mali. Vulture News. 20: 4–7.

Terraube, J. and B. Arroyo. 2011. Factors influencing diet variation in a generalist predator across its range distribution. Biodiversity and Conservation. 20: 2111–2131.

Terraube, J., B. Arroyo, A. Bragin, E. Bragin and F. Mougeot. 2012b. Ecological factors influencing the breeding distribution and success of a nomadic, specialist predator. Biodiversity and Conservation. 21: 1835–1852.

Terraube, J., F. Mougeot, T. Cornulier, A. Verma, A. Gavrilov and B. Arroyo. 2012a. Broad wintering range and intercontinental migratory divide within a core population of the near-threatened pallid harrier. Diversity and Distributions. 18: 401–409.

Thake, M.A. 1983. Marsh Harrier migration through Malta in autumn in relation to weather. Il Merill. 22: 1–6.

Thévenot, M., R. Vernon and P. Bergier. 2003. The Birds of Morocco. An Annotated Check-list. British Ornithologists' Union, Tring, UK.

Thibault, J.C. and V. Bretagnolle. 2001. Monitoring, research and conservation of Osprey *Pandion haliaetus* on Corsica, Mediterranean, France. Vogelwelt. 122: 173–178.

Thibault, J.C., R. Triay, P. Beabrun, D. Boukhalfa, J.M. Dominici and A. Torre. 1996. Osprey (*Pandion haliaetus*) in the Mediterranean: characteristics of a resident population with a patchy distribution. *In*: Muntaner, J. and J. Mayol [eds.]. Biologia y Conservacion de las Rapaces Mediterraneas. Monografias, n°4. SEO. Madrid.

Thiollay, J.M. 1975. Migration des printemps au Cap Bon (Tunisie). Nos Oiseaux. 33: 109–121.

Thiollay, J.M. 1977. Importances des populations de rapaces migrateurs en Mediterranee Occidentale. Alauda. 45: 115–121.

Thiollay, J.M. 2006a. Severe decline of large birds in the Northern Sahel of West Africa: a long-term assessment. Bird Conservation International. 16: 353–365.

Thiollay, J.M. 2006b. The decline of raptors in West Africa: long-term assessment and the role of protected areas. Ibis. 148: 240–254.

Thiollay, J.M. 2007. Raptor population decline in West Africa. Ostrich. 78: 405–413.

Thomson, A.L. 1931. On "abmigration" among the ducks: an anomaly shown by the results of bird-marking. *In* Proc. Int. Ornithol. Congr. 7: 382–388.

Thorup, K., R.A. Holland, A.P. Tøttrup and M. Wikelski. 2010. Understanding the Migratory Orientation Program in Birds: Extending Laboratory Studies to Study Free-flying Migrants in a Natural Setting. Integrative and Comparative Biology. 50: 315–322.

Thorup, K., T. Alerstam, M. Hake and N. Kjellén. 2003. Bird orientation: compensation for wind drift in migrating raptors is age dependent. Proceedings of the Royal Society B. 270: S8–S11.

Tilba, P.A. 2010. The migration of the Red-footed falcon in the Northwest Caucasus. Strepet. 5: 94–98 (in Russian).

Toffoli, R. and C. Bellone. 1996. Osservazioni sulla migrazione autunnale dei rapaci diurni sulle Alpi Marittime. Avocetta. 20: 7–11.

Tomé, R., H. Costa and D. Leitão. 1998. A migração outonal de aves planadoras na região de Sagres—resultados da campanha de 1994. SPEA, Lisbon.

Tøttrup, A.P., R.H.G. Klassen, M.W. Kristensen, R. Strandberg, Y. Vardanis, Å. Lindström, C. Rahbek, T. Alerstam and K. Thorup. 2012b. Drought in Africa Caused Delayed Arrival of European Songbirds. Science. 338: 1307.

Tøttrup, A.P., R.H.G. Klassen, R. Strandberg, K. Thorup, M.W. Kristensen, P.S. Jorgensen, J. Fox, V. Afanasyev, C. Rahbek and T. Alerstam. 2012a. The annual cycle of a trans-equatorial Eurasian-African passerine migrant: different spatio-temporal strategies for autumn and spring migration. Proceedings of the Royal Society B. 279: 1008–1016.

Trabalon, F. and R. García. 2015. A long-term study of differential postnuptial migration timing between age classes in the Honey Buzzard *Pernis apivorus* in the NE Iberian Peninsula. Revista Catalana d' Ornitologia. 31: 18–22.

Treep, J., G. Bohrer, J. Shamoun-Baranes, O. Duriez, R. Prata de Moraes Frasson and W. Bouten. 2016. Using high-resolution GPS tracking data of bird flight for meteorological observations. Bulletin American Meteorological Society. 97: 951–961.

Triay, R. and M. Siverio. 2008. El aguila pescadora en España. Poblacion en 2008 y metodo de censo. SEO/Birdlife. Madrid.

Trierweiler, C. and B.J. Koks. 2009. Montagu's Harrier *Circus pygargus*. pp. 312–327. *In*: Zwarts, L., R. Bijlsma, J. van der Kamp and E. Wymega [eds.]. Living on the Edge: Wetlands and Birds in a Changing Sahel. KNNV Publishing, Zeist, The Netherlands.

Trierweiler, C., B.J. Koks, R.H. Drent, K.M. Exo, J. Komdeur, C. Dijkstra and F. Bairlein. 2007. Satellite tracking of two Montagu's Harriers (*Circus pygargus*): Dual pathways during autumn migration. Journal of Ornithology. 148: 513–516.

Trierweiler, C., R.H.G. Klaassen, R.H. Drent, K. Exo, J. Komdeur, F. Bairlein and B.J. Koks. 2014. Migratory connectivity and population-specific migration routes in a long-distance migratory bird. Proceedings of the Royal Society B. 281: 20132897.

Trierweiler, C., W.C. Mullié, R.H. Drent, K.M. Exo, J. Komdeur, F. Bairlein, A. Harouna, M. de Bakker and B.J. Koks. 2013. A Palaearctic migratory raptor species tracks shifting prey availability within its wintering range in the Sahel. Journal of Animal Ecology. 82: 107–120.

Tsvelykh, A.N. and V.E. Panjushkin. 2002. Zimovki chernogo korshuna (*Milvus migrans*) in Ukraine. Vestnik Zoologii. 36: 81–83.

Ueta, M. and H. Higuchi. 2002. Difference in migration pattern between adult and immature birds using satellites. The Auk. 119: 832–835.

Ueta, M. and V.V. Ryabstev. 2001. Migration routes of four juvenile Imperial Eagles *Aquila heliaca* from the Baikal region of eastern Russia. Bird Conservation International. 11: 93–99.

Ulfstrand, S. 1977. Plumage and size variation in Swedish common buzzards *Buteo buteo* L. (Aves, Accipitriformes). Zoologica Scripta. 6: 69–75.

Underhill-Day, J. 2002. Marsh Harrier *Circus aeruginosus*. pp. 225–226. *In:* Wernham, C.V., M.P. Toms, J.H. Marchant, J.A. Clark, G.M. Siriwardena and S.R. Baillie [eds.]. The Migration Atlas: Movements of the Birds of Britain and Ireland, T. and A.D. Poyser, London, U.K.

Üner, Ö., K.A. Boyla, E. Bacak, E. Birel, İ. Çelikoba, C. Dalyan, E. Tabur and Ü. Yardim. 2010. Spring migration of soaring birds over the Bosphorus, Turkey, in 2006. Sandgrouse. 32: 20–33.

Urcun, J.P. and J. Bried. 1998a. The autumn migration of Red Kite *Milvus milvus* through the Pyrenees. pp. 641–654. *In:* Chancellor, R.D., B.U. Meyburg and J.J. Ferrero [eds.]. Holarctic Birds of Prey. Proceedings of an International Conference. ADENEX & WWGBP, Merida & Berlin.

Urcun, J.P. and J. Bried. 1998b. The autumn migration of raptors through the Pyrenees. pp. 655–680. *In*: Chancellor, R.D., B.-U. Meyburg and J.J. Ferrero [eds.]. Holartic Birds of Prey. Proceedings of an International Conference. ADENEX & WWGBP, Merida & Berlin.

Urios, V., A. Bermejo, J. Vidal-Mateo and J. De la Puente [eds.]. 2017. Migración y ecología espacial de la población española de águila calzada. Monografía n. 2 del programa Migra. SEO/BirdLife. Madrid.

Uysal, I. and M. Tosunoğlu. 2018. Migration of raptor birds across the Gallipoli peninsula/Dardanalles, Turkey. American Journal of Innovative Research and Applied Sciences. 6: 288–293.

Väli, Ü. 2010. Successful breeding of a ten-year-old hybrid spotted eagle *Aquila clanga* × *A. pomarina* retaining immature plumage characters. Ardea. 98: 235–241.

Väli, Ü. 2015a. Monitoring of spotted eagles in Estonia in 1994–2014: Stability of the lesser spotted eagle (*Aquila pomarina*) and decline of the greater spotted eagle (*A. clanga*). Slovak Raptor Journal. 9: 55–64.

Väli, Ü. 2015b. Timing of breeding and migration of the Common Buzzard (*Buteo buteo*) in Estonia. Hirundo. 2: 29–42.

Väli, Ü. 2017. Home range size and breeding dispersal of a common buzzard (*Buteo buteo*). Slovak Raptor Journal. 11: 111–116.

Väli, Ü. and A. Lõhmus. 2000. The Greater Spotted Eagle and its conservation in Estonia. Hirundo Supplementum. 3: 1–50.

Väli, Ü. and A. Lõhmus. 2002. Parental care, nestling growth and diet in a Spotted Eagle *Aquila clanga* nest. Bird Study. 49: 93–95.

Väli, Ü. and A. Lohmus. 2004. Nestling characteristics and identification of the lesser spotted eagle *Aquila pomarina*, greater spotted eagle *A. clanga*, and their hybrids. Journal of Ornithology. 145: 256–263.

Väli, Ü. and O. Vainu. 2015. Short-distance migration of Estonian Common Buzzards *Buteo buteo*. Ringing & Migration. 30: 81–83.

Väli, Ü and U. Sellis. 2016. Migration patterns of the Osprey *Pandion haliaetus* on the Eastern European-East African flyway. Ostrich. 87: 23–28

Väli, Ü., U. Sellis and O. Vainu. 2015. Where do our buzzards fly? Eesti Loodus. 66: 703–709 (in Estonian).

Väli, Ü., P. Mirski, U. Sellis, M. Dagys and G. Maciorowski. 2018. Genetic determination of migration strategies in large soaring birds: evidence from hybrid eagles. Proceedings of the Royal Society B. 285: 20180855.

Väli, Ü., V. Dombrovski, M. Dzmitranok, G. Maciorowski and B.-U. Meyburg. 2019. High genetic diversity and low differentiation retained in the European fragmented and declining Greater Spotted Eagle (*Clanga clanga*) population. Scientific Reports. 9: 3064.

Väli, Ü., E. Nellis, E. Lelov, I. Tammekänd, A. Tuule and E. Tuule. 2014. Distribution, abundance and habitat for wintering birds of prey in Estonia. Hirundo. 27: 14–35.

Väli, Ü., V. Dombrovski, R. Treinys, U. Bergmanis, S.J. Daroczi, M. Dravecky, V. Ivanovski, J. Lontkowski, G. Maciorowski, B.U. Meyburg, T. Mizera, R. Zeitz and H. Ellegren. 2010. Widespread hybridization between the Greater Spotted Eagle *Aquila clanga* and the Lesser Spotted Eagle *Aquila pomarina* (Aves: Accipitriformes) in Europe. Biological Journal of the Linnean Society. 100: 725–736.

Vallotton, L. 2011. Possible first record of a Greater Spotted Eagle *Aquila clanga* in Botswana. Gabar. 22: 28–30.

Van den Bossche, W. and L. Lens. 1994. Soaring bird migration at the Bosphorus (Turkey): the need for a multi-station survey. Le Gerfaut. 84: 51–62.

van Maanen, E., I. Goradze, A. Gavashelishvili and R. Goradze. 2001. Trapping and hunting of migratory raptors in western Georgia. Bird Conservation International. 11: 77–92.

Van Zyl, A. 1994. A comparison of the diet of the Common Kestrel *Falco tinnunculus* in South Africa and Europe. Bird Study. 41: 127–130.

Vanni, L. and G. Paesani. 2007. La migrazione post-riproduttiva sull'Isola d'Elba (LI). Parco nazionale Arcipelago toscano. Infomigrans. 20: 6.

Vansteelant, W.M.G., J. Kekkonen and P. Byholm. 2017a. Wind conditions and geography shape the first outbound migration of juvenile honey buzzards and their distribution across sub-Saharan Africa. Proceedings of the Royal Society B. 284: 20170387.

Vansteelant, W.M.G., B. Verhelst, J. Shamoun-Baranes, W. Bouten, E.E. van Loon and K.L. Bildstein. 2014. Effect of wind, thermal convection, and variation in flight strategies on the daily rhythm and flight paths of migrating raptors at Georgia's Black Sea coast. Journal of Field Ornithology. 85: 40–55.

Vansteelant, W.M.G., J.Z. Shamoun-Baranes, J. van Diermen, W. van Manen and W. Bouten. 2017b. Soaring across continents: decision-making of a soaring migrant under changing atmospheric conditions along an entire flyway. Journal of Avian Biology. 48: 887–896.

Vansteelant, W.M.G., J. Shamoun-Baranes, W. van Manen, J. van Diermen and W. Bouten. 2017c. Seasonal detours by soaring migrants shaped by wind regimes along the East Atlantic Flyway. Journal of Animal Ecology. 86: 179–191.

Vansteelant, W.M.G., R. Klaassen, R. Strandberg, K. Janssens, F. T'Jollyn, W. Bouten, B.J. Koks and A. Anselin. 2020a. Western Marsh Harriers *Circus aeruginosus* from nearby breeding areas migrate along comparable loops but on contrasting schedules in the West African-Eurasian flyway. Journal of Ornithology. 161: 953–965.

Vansteelant, W.M.G., W. Bouten, R.H.G. Klaassen, B.J. Koks, A.E. Schlaich, J. van Diermen, E.E. van Loon and J.Z. Shamoun Baranes. 2015. Regional and seasonal flight speeds of soaring migrants and the role of weather at hourly and daily scales. Journal of Avian Biology. 46: 25–39.

Vansteelant, W.M.G, J. Wehrmann, D. Engelen, J. Jansen, B. Verhelst, R. Benjumea, S. Cavaillès, T. Kaasiku, B. Hoekstra and F. de Boer. 2020b. Accounting for differential migration strategies between age groups to monitor raptor population dynamics in the eastern Black Sea flyway. Ibis. 162: 356–372.

Vardanis, Y., R.H.G. Klaassen, R. Strandberg and T. Alerstam. 2011. Individuality in bird migration: routes and timing. Biology Letters. 7: 502– 505.

Vardanis, Y., J.A. Nilsson, R.H. Klaassen, R. Strandberg and T. Alerstam. 2016. Consistency in long-distance bird migration: contrasting patterns in time and space for two raptors. Animal Behaviour. 113: 177–187.

Varshavsky, S.N. 1957. Seasonal movements and migration of birds, predominantly raptors, in the Northern Priaralye. pp. 69–84. *In*: Proceedings of II Baltic Ornithological Conference, Moscow. USSR Academy of Sciences, Moscow, Russia (in Russian).

Velevski, M., S.C. Nikolov, Ben Hallmann, V. Dobrev, L. Sidiropulos, V. Saravia, R. Tsiakiris, V. Arkumarev, A. Galanaki, T.H. Kominos, K. Stara, E. Kret, B. Grubač, E. Lisičanec, T. Kastritis, D. Vavylis, M. Topi, B. Hoxha and S. Oppel. 2015. Population decline and range contraction of the Egyptian Vulture *Neophron percnopterus* on the Balkan Peninsula. Bird Conservation International. 25: 440–450.

Verhelst, B., J. Jansen and W. Vansteelant. 2011. South West Georgia: an important bottleneck for raptor migration during autumn. Ardea. 99: 137–146.

Verma, A. 2005. Winter ecology of harriers roosting at Akola, Maharashtra, India. Zoos' Print Journal. 20: 1943–1947.

Vetrov, V.V. and Y.V. Milobog. 2008. Distribution of the imperial eagle (*Aquila heliaca* Sav.) in the steppe zone of Ukraine. pp. 51–55. *In:* Gavrilyuk, M.N. and G.V. Fesenko [eds.]. Recent Research on Falcons and Owls: Proceedings of the Third International Scientific Conference [Birds of Prey of Ukraine]. Krivoy Rog, Kiev, Ukraine (in Russian).

Vidal-Mateo, J. 2019. Movement ecology of migratory raptors. PhD Thesis. University of Alicante, Alicante.

Vidal-Mateo, J., J. De la Puente, U. Mellone, A. Bermejo and V. Urios. 2015. Migration routes and wintering areas of wintering Red Kites in Spain. pp. 25. *In:* De la Puente, J. and V. De la Torre [eds.]. Abstracts Book of the II International Symposium on the Red Kite. Binaced, Huesca, Spain. SEO-Monticola & FAB.

Vidal-Mateo, J., U. Mellone, P. López-López, J. De la Puente, C. García-Ripollés, A. Bermejo and V. Urios. 2016. Wind effects on the migration routes of trans-Saharan soaring raptors: geographical, seasonal and interspecific variation. Current Zoology. 62: 89–97.

Village, A. 1990. The Kestrel, T. & AD Poyser, London.

Vincze, O., C.I. Vágási, P.L. Pap, C. Palmer and A.P. Møller. 2019. Wing morphology, flight type and migration distance predict accumulated fuel load in birds. Journal of Experimental Biology. 222: jeb183517.

Viñuela, J. and J.P. Veiga. 1992. Importance of rabbits in the diet and reproductive success of black kites in southwestern Spain. Ornis Scandinava. 23: 132–138.

Virani, M.Z., C. Kendall, P. Njoroge and S. Thomsett. 2011. Major declines in the abundance of vultures and other scavenging raptors in and around the Masai Mara ecosystem, Kenya. Biological Conservation. 144: 746–752.

Viter, S.G. 2013. On the status of stay and the biology of eagles in the Kharkov District, Ukraine. Raptors Conservation. 27: 102–121 (in Russian).

Viter, S.G. 2018. Characteristic of Birds of Prey Migration in the Kharkov Region, Ukraine. Raptors Conservation. 36: 73–107.

Vittery, A. 1983. Movements of Palaearctic raptors in the Ethiopian Rift Valley. Scopus. 7: 1–9.

Vlachopoulos, K., P. Kordopatis, A. Evangelidis, C. Angelidis, S. Polymeros, J. Fric and A. Sfougaris. 2016. First insights of small and large scale movements of the Lesser Kestrel (*Falco naumanni*) population in Thessaly plain, Greece, using different telemetry techniques. International Lesser Kestrel Expert Workshop, Plovdiv, Bulgaria, 4–8.10.2016. http://greenbalkans.org/birdsofprey/lesserkestrellife/en/International_Expert_Workshop_-c150.

Walls, S.S. and R.E. Kenward. 1995. Movements of radio-tagged Common Buzzards *Buteo buteo* in their first year. Ibis. 137: 177–182.

Walls, S.S. an R.E. Kenward 1998. Movements of radio-tagged Buzzards *Buteo buteo* in early life. Ibis. 140: 561–568.

Walls, S. and R. Kenward. 2020. The Common Buzzard, T. & A.D. Poyser, London. Walter, H. 1979. Eleonora's falcon. Adaptations to prey and habitat in a social raptor. London: The University of Chicago Press.

Wahl, R. and C. Barbraud 2014. The cemography of a newly established Osprey *Pandion haliaetus* population in France. Ibis. 156: 84–96.

Warshawskii, S.N. 1957. Seasonal movements and migration of birds, predominantly raptors, in the Northern Priaralye. pp. 69–84. *In:* Kumary, E.V. [ed.]. Proceedings of II Baltic Ornithological Conference. Publishing House of the USSR Academy of Sciences, Moscow, Russia (in Russian).

Wassink, A. and G.J. Oreel. 2007. The Birds of Kazakhstan. De Cocksdrop, Texel.

Watson, D. 1977. The Hen Harrier.Poyser, Berkhamsted.

Webster, M.S., P.P. Marra, S.M. Haig, S. Bensch and R.T. Holmes. 2002. Links between worlds: unravelling migratory connectivity. Trends in Ecology & Evolution. 17: 76–83.

Wehrmann, J., F. de Boer, R. Benjumea, S. Cavaillès, D. Engelen, J. Jansen, B. Verhelst and W.M.G. Vansteelant. 2019. Batumi Raptor Count: autumn raptor migration count data from the Batumi bottleneck, Republic of Georgia. Zookeys. 836: 135–157.

Weidensaul, S. 2018. A Galaxy of Falcons: Witnessing the Amur Falcon's Massive Migration Flocks. https://www.allaboutbirds.org/news/a-galaxy-of-falcons-witnessing-the-amur-falcons-massive-migration-flocks/[accessed 8/7/2020].

Weiss, N. 2018. How fast does the Steppe Eagle population decline?: survey results from Eilat, Israel. Raptors Conservation Suppl. 1: 92–93.

Welch, G. and H. Welch. 1988. The autumn migration of raptors and other soaring birds across the Bab-el-Mandeb straits. Sandgrouse. 10: 26–50.

Welch, G. and H. Welch. 1989. Autumn Migration across the Bab-el-Mandeb Straits. pp. 123–126. *In:* Meyburg, B.-U. and R.D. Chancellor [eds.]. Raptors in the Modern World. WWGBP: Berlin, London and Paris.

Welch, G. and H. Welch. 1991a. Spring raptor observations from Djibouti. Ornithological Society of the Middle East Bulletin. 26: 25–27.

Welch, G. and H. Welch. 1991b. The autumn migration of the Steppe Eagle *Aquila nipalensis.* Sandgrouse. 13: 24–33.

Welch, G. and H. Welch. 1992. Preliminary observations of raptor migration along the Al Hada escarpment near Taif, Kingdom of Saudi Arabia. Bulletin of the Ornithological Society of the Middle East. 28: 5–10.

Welch, G. and H. Welch. 1998. Raptor migration Bab al Mandab, Yemen. Spring 1998. Phoenix. 15: 11–12.

Welch, G. and H. Welch. 1999. Raptor migration Bab el Mandeb, Yemen - spring 1998. Phoenix. 5: 11–12.

Wells, D.R. 1999. The Birds of the Thai-Malay Peninsula, vol. 1. Non-Passerines. Academic Press, London, UK.

Wernham, C., M. Toms, J. Marchant, .. Clark, G. Siriwardena and S. Baillie [eds.]. 2002. The migration Atlas: movements of the birds of Britain and Ireland, T. & A.D. Poyser, London.

Weatherhead, P.J. and M.R.L. Forbes. 1994. Natal philopatry in passerine birds: genetic or ecological influences? Behavioral Ecology. 5: 426–433.

White, C.M. 1994. Family falconidae. pp. 216–275. *In*: Del Hoyo, J., A. Elliot and J. Sargatal [eds.]. Handbook of Birds of the World: New World vultures to Guinea fowl. Barcelona, Linx Edition.

Wilk, T., R. Bobrek and A. Pępkowska-Król. 2018. Jesienna migracja ptaków szponiastych *Accipitriformes* i sokołowych *Falconiformes* w polskiej części Karpat. Ornis Polonica. 59: 1–16.

Williamson, K. The migration of the Icelandic Merlin. British Birds. 47: 434–441.

Wimpfheimer, D., B. Bruun, S. Baha el Din and M. Jennings. 1983. The migration of birds of prey in the northern Red sea area. Report of the 1982 Suez Study. Holy Land Conservation Fund, New York.

Wink, M. and D. Ristow. 2000. Biology and molecular genetics of Eleonora's Falcon *Falco eleonorae*, a colonial raptor of Mediterranean islands (24). pp. 653–658. *In:* Chancellor, R.D. and B.-U. Meyburg [eds.]. Raptors at Risk. Surrey, BC., Canada: Hancock House and World Working Group on Birds of Prey.

Wink, M., H. Sauer-Gürth and D. Pepler. 2004. Phylogeographic relationships of the Lesser Kestrel (*Falco naumanni*) in breeding and wintering quarters inferred from nucleotide sequences of the mitochondrial cytochrome b gene. pp. 505–510. *In*: Chancelor, R.D. and B.-U. Meyburg [eds.]. Raptors Worldwide, Berlin: The World Working Group on Birds of Prey and Owl.

Witkowski, J. 1989. Breeding biology and ecology of the marsh harrier *Circus aeruginosus* in the Barcycz valley, Poland. Acta Ornithologica. 25: 223–320.

Wright J., S. Cavaillès and W.M.G. Vansteelant. 2019. Photographic evidence of a 2nd calendar-year female European Honey Buzzard *Pernis apivorus* on autumn migration in the Western Palearctic. Sandgrouse. 41: 101–104.

Xirouchakis, S.M. and G. Andreou. 2009. Foraging behaviour and flight characteristics of Eurasian griffons *Gyps fulvus* in the island of Crete, Greece. Wildlife Biology. 15: 37–52.

Yamaguchi, N., Y. Arisawa, Y. Shimada and H. Higuchi. 2012. Real-time weather analysis reveals the adaptability of direct sea-crossing by raptors. Journal of Ethology. 30: 1–10.

Yamaguchi, N.M., E. Hiraoka, N. Fijikata and H. Higuchi. 2017. Migration routes of satellite-tracked Rough-legged Buzzards from Japan: the relationship between movement patterns and snow cover. Ornithological Science. 16: 33–41.

Yamaguchi, N., K.I. Tokita, A. Uematsu, K. Kuno, M. Saeki, E. Hiraoka, K. Uchida, M. Hotta, F. Nakayama, M. Takahashi, H. Nakamura and H. Higuchi. 2008. The large-scale detoured migration route and the shifting pattern of migration in Oriental honey-buzzards breeding in Japan. Journal of Zoology. 276: 54–62.

Yáñez B., A.-R. Muñoz, K.L. Bildstein, I. Newton, A.G. Toxopeus and M. Ferrer. 2014. Individual variation in the over-summering areas of immature Short-toed Snake Eagles *Circaetus gallicus*. Acta Ornitholologica. 49: 137–141.

Yasko, A.O. 2017. Ornithological observations in the city of Aktau. Russian Ornithological Journal, Express Edition. 26: 382–394. (in Russian).

Yom-Tov, Y. 1984. On the difference between the spring and autumn migrations in Eilat, southern Israel. Ringing & Migration 5: 141–144.

Yosef, R. 1995. Spring 1994 raptor migration at Eilat, Israel. Journal of Raptor Research. 29: 127–134.

Yosef, R. 1996. Raptors feeding on migration at Eilat, Israel: opportunistic behavior or migratory strategy? Journal of Raptor Research. 30: 242–245.

Yosef, R. and D. Alon. 1997. Do immature Palearctic Egyptian Vultures *Neophron percnopterus* remain in Africa during the northern summer? Vogelwelt. 118: 285–290.

Yosef, R. and L. Fornasari. 2004. Simultaneous decline in Steppe Eagle (*Aquila nipalensis*) populations and Levant Sparrowhawk (*Accipiter brevipes*) reproductive success: coincidence or a Chernobyl legacy? Ostrich. 75: 20–24.

Yosef, R., P. Tryjanowski and K.L. Bildstein. 2002. Spring migration of adult and immature buzzards (*Buteo buteo*) through Elat, Israel: timing and body size. Journal of Raptor Research. 36: 115–120.

Yosef, R., G. Verdoorn, A. Helbig and I. Seibold. 2000. A new subespecies of the booted eagle from southern Africa, inferred from biometrics and mitochondrial DNA. pp. 43–49. *In:* Chancellor, R.D. and B.-U. Meyburg [eds.]. Raptors at Risk: Proceedings of the V World Conference on Birds of Prey and Owls, Midrand, Johannesburg, South Africa, 4–11 August 1998. WWGBP & Hancock House Publishers, Berlin & Surrey.

Yousef, R., L. Fornasari, P. Tryjanowski, M.J. Bechard, S. Kaltenecker and K. Bildstein. 2003. Differential spring migration of adult and juvenile Levant Sparrowhawks (*Accipiter brevipes*) through Eilat, Israel. Journal of Raptor Research. 37: 31–36.

Zakharov, V.D. 2006. Birds of the Southern Urals (species composition, distribution, abundance). Miass, city, Ukraine (in Russian).

Zalles, J. and K.L. Bildstein [eds.]. 2000. Raptor watch: a global directory of raptor migration sites, BirdLife Conservation Series No. 9.

Zannetos, S.P., Y. Zevgolis and T. Akriotis. 2018. First record of Crested (or Crested-type) Honey Buzzard *Pernis ptilorhynchus* for Greece. Bulletin of the British Ornithologists' Club. 138: 386–389.

Zarudny, N.A. 1916. The Birds of Aral Sea. Proceedings of the Turkestan branch of Russian Geographical Society, V. 12: 229 pp. (in Russian).

Zavyalov, E.V., G.V. Shlyakhtin, V.V. Piskunov, L.A. Lebedeva, V.G. Tabachishin, A.L. Podolskii, E.I. Sarantseva, A.A. Bauynov, N.N. Yakushev and I.B. Kochetova. 1999. Birds of prey in the Saratov Region: Golden Eagle. Berkut. 8: 21–45 (in Russian).

Zduniak, P., R. Yosef, T.H. Sparks, H. Smit and P. Tryjanowski. 2010. Rapid advances in the timing of the spring passage migration through Israel of the Steppe Eagle *Aquila nipalensis*. Climate Research. 42: 217–222.

Zefania, S. 2001. Observation of Sooty and Eleonora's falcons in Madagascar. pp. 151–159. *In*: Wings over Africa. International Seminar on Bird Migration: Research, Conservation, Education and Flight Safety. April–May 2001. Tel Aviv, Israel: Society for Protection of Nature.

Zijlstra, M. 1987. Bruine Kiekendief *Circus aeruginosus* in Flevoland in de winter. Limosa, 60: 57–62.

Zijlstra, M., S. Daan and J. Bruinenberg-Rinsma. 1992. Seasonal variationin the sex ratio of marsh harrier *Circus aeruginosus* broods. Functional Ecology. 6: 553–559.

Zimmerman, D.A., D.J. Pearson and D.A. Turner. 2020. Birds of Kenya and Northern Tanzania, Christopher Helm, London.

Zorrilla, I., R. Martinez, M.A. Taggart and N. Richards. 2015. Suspected flunixin poisoning of a wild Eurasian Griffon Vulture from Spain. Conservation Biology. 29: 587–592.

Zu-Aretz, S. and Y. Leshem. 1983. The sea, a trap for gliding birds. Torgos. 5: 16–17.

Zub, K., E. Pugacewicz, B. Jędrzejewska and W. Jędrzejewski. 2010. Factors affecting habitat selection by breeding Lesser Spotted Eagles *Aquila pomarina* in Northeastern Poland. Acta Ornithologica. 45: 105–114.

Zubarovsky, V.M. 1977. The fauna of Ukraine, vol. 5. *In:* Voinstvenskii, M.A. [ed.]. The Birds. Issue 2 – Birds of Prey. Naukova Dumka, Kiev, Ukraine (in Ukrainian).

Zuberogoitia, I., J. Zabala, J.E. Martínez, J.A. González-Oreja and P. López-López. 2014. Effective conservation measures to mitigate the impact of human disturbances on the endangered Egyptian vulture. Animal Conservation. 17: 410–418.

Zuberogoitia, I., J.A. Gonzalez-Oreja, J.E. Martinez, J. Zabala, I. Gomez and P. López-López. 2013. Foraging movements of Eurasian griffon vultures: implications for supplementary feeding management. European Journal of Wildlife Research. 59: 421–429.

Zuberogoitia, I., J.A. Martínez, J. Zabala, J.E. Martínez, I. Castillo, A. Azkona and S. Hidalgo. 2005. Sexing, ageing and moult of Buzzards *Buteo buteo* in a southern European area. Ringing & Migration. 22: 153–158.

Zwarts, L., R.G. Bijlsma, J. van der Kamp and E. Wymenga. 2009. Living on the edge: wetlands and birds in a changing Sahel. Zeist: KNNV Publishing.

Index

D

E

F

G

H

I

K

L

T

U

V

W

Z

Color Plates

Fig. 1. Michele Panuccio during fieldwork in Panarea (Aeolian islands, Italy), spring 2006 (© Patrizia Bonelli).

Fig. 2. Adult Egyptian vultures roosting beside the Strait of Gibraltar (Spain), during post-breeding migration (© John Wright).

Fig. 3. Adult European honey buzzard crossing the Strait of Messina (Italy) during spring migration (© Michele Panuccio).

Fig. 4. Adult male European honey buzzard exhausted during the crossing of the Sahara desert in Mauritania (© Nicolas Van Ingen).

Fig. 5. Griffon vulture landing after crossing the Strait of Gibraltar (Spain) under strong sidewinds during spring (© Javi Elorriaga).

Fig. 6. Griffon vultures concentrating around wind farms before crossing the Strait of Gibraltar (Spain) during post-breeding migration (© Paco Azufiaur).

Fig. 7. Short-toed snake eagle arriving in Europe after crossing the Strait of Gibraltar (Spain) during pre-breeding migration (© Michele Panuccio).

Fig. 8. Booted eagle crossing the Strait of Gibraltar (Spain) during pre-breeding migration (© Ugo Mellone).

Fig. 9. Juvenile steppe eagle in Oman (© Markus Varesvuo).

Fig. 10. Juvenile Eurasian sparrowhawk during post-breeding migration, showing full crop, Finland (© Markus Varesvuo).

Fig. 11. Adult male Montagu's harrier tagged with GPS data-logger, the Netherlands (© Joris Van Alphen).

Fig. 12. Adult female Montagus' harrier during boreal winter in Sahel (Senegal; © Joris Van Alphen).

Fig. 13. Adult black kite during post-breeding migration at the Strait of Gibraltar (Spain; © Yeray Seminario).

Fig. 14. Black kites waiting for the drop of strong *Levante* winds during post-breeding migration at the Strait of Gibraltar (Spain; © Alejandro Onrubia).

Fig. 15. Flock of steppe buzzards (with two black kites) during post-breeding migration in Israel (© Markus Varesvuo).

Fig. 16. Adult female lesser kestrel tagged with 5g Argos transmitter, just before release, Alicante (Spain; © Ugo Mellone).

For Product Safety Concerns and Information please contact our EU representative GPSR@taylorandfrancis.com Taylor & Francis Verlag GmbH, Kaufingerstraße 24, 80331 München, Germany

Batch number: 10732790

Printed by Printforce, the Netherlands